The Anthropocene

This book is devoted to the Anthropocene, the period of unprecedented human impacts on Earth's environmental systems, and illustrates how geographers envision the concept of the Anthropocene.

This edited volume illustrates that geographers have a diverse perspective on what the Anthropocene is and represents. The chapters also show that geographers do not feel it necessary to identify only one starting point for the temporal onset of the Anthropocene. Several starting points are suggested, and some authors support the concept of a time-transgressive Anthropocene. Chapters in this book are organized into six sections, but many of them transcend easy categorization and could have fit into two or even three different sections. Geographers embrace the concept of the Anthropocene while defining it and studying it in a variety of ways that clearly show the breadth and diversity of the discipline.

This book will be of great value to scholars, researchers, and students interested in geography, environmental humanities, environmental studies, and anthropology.

The chapters in this book were originally published as a special issue of the journal *Annals of the American Association of Geographers*.

David R. Butler is Texas State University System Regents' Professor Emeritus, and University Distinguished Professor Emeritus in the Department of Geography at Texas State University, USA. His research interests include geomorphology in the Anthropocene, zoogeomorphology, dendrogeomorphology, and mountain environments and environmental change, especially in the Rocky Mountains.

The Anthropocene

Edited by
David R. Butler

Routledge
Taylor & Francis Group

LONDON AND NEW YORK

First published 2022
by Routledge
2 Park Square, Milton Park, Abingdon, Oxon, OX14 4RN

and by Routledge
605 Third Avenue, New York, NY 10158

Routledge is an imprint of the Taylor & Francis Group, an informa business

British Library Cataloguing-in-Publication Data
A catalogue record for this book is available from the British Library

ISBN13: 978-1-032-07668-3 (hbk)
ISBN13: 978-1-032-07669-0 (pbk)
ISBN13: 978-1-003-20821-1 (ebk)

DOI: 10.4324/9781003208211

Typeset in Goudy Oldstyle Std
by codeMantra

Publisher's Note
The publisher accepts responsibility for any inconsistencies that may have arisen during the conversion of this book from journal articles to book chapters, namely the inclusion of journal terminology.

Disclaimer
Every effort has been made to contact copyright holders for their permission to reprint material in this book. The publishers would be grateful to hear from any copyright holder who is not here acknowledged and will undertake to rectify any errors or omissions in future editions of this book.

Contents

PART 6
The Anthropocene and Geographic Education

Citation Information

The chapters in this book were originally published in the journal, *Annals of the American Association of Geographers*, volume 111, issue 3 (2021). When citing this material, please use the original page numbering for each article, as follows:

For any permission-related enquiries please visit:
http://www.tandfonline.com/page/help/permissions

Notes on Contributors

Paul C. Adams is Professor of Geography at the University of Texas at Austin, USA. His research occupies the intersection of media studies, communication theory, and human geography. He considers how perceptions, representations, actions, and infrastructures are intertwined through mediated communications. His books include *Geographies of Media and Communication* (2009).

Deghyo Bae is Professor of Civil and Environmental Engineering and President of Sejong University, Seoul, South Korea. His research areas include atmosphere and surface runoff interactions, flood forecasting, satellite hydrology, and GIS-based water resource engineering.

Péter Bagoly-Simó is Full Professor and the Chair of Geography Education in the Department of Geography at Humboldt-Universitat zu Berlin, Germany. His research interests include educational media and textbooks, geographical knowledge, and education for sustainable development.

Sydney N. Bailey is Graduate Student in the Department of Geography at the University of Missouri, Columbia, USA. Her research focuses on climate–vegetation interactions in the U.S. Rocky Mountains.

Timothy P. Beach is Guggenheim, Dumbarton Oaks, and AAAS Fellow. He holds a Centennial Chair in US Mexican Relations and is a Professor and directs the Soils and Geoarchaeology Labs at UT Austin, USA. His research ranges from soil profiles to watersheds, and from the Pleistocene to the present, especially in the Maya and Mediterranean worlds.

Jacob Bendix is Professor in the Department of Geography and the Environment at Syracuse University, USA. His research interests include plant biogeography and fluvial geomorphology, with particular interest in disturbance ecology and the interactions between ecological and geomorphological processes and patterns.

Mia M. Bennett is Assistant Professor in the Department of Geography and School of Modern Languages & Cultures (China Studies Program) at The University of Hong Kong, Hong Kong. Her research critiques the politics of development in regions commonly thought of as frontiers, namely, the Arctic.

Nicolas T. Bergmann recently earned his doctorate from Montana State University, Bozeman, USA. His research interests lie at the intersections of political ecology, historical geography, and environmental history.

Christine Biermann is Assistant Professor in the Department of Geography and Environmental Studies at the University of Colorado, Colorado Springs, USA. Her research addresses the ecosocial dynamics of forests and fisheries.

Robert M. Briwa is Assistant Professor of Geography within the Department of History at Angelo State University, San Angelo, USA. His research interests include historical and cultural geographies of the U.S. West and literary geographies.

David R. Butler is Texas State University System Regents' Professor Emeritus, and University Distinguished Professor Emeritus in the Department of Geography at Texas State University, USA. His research interests include geomorphology in the Anthropocene, zoogeomorphology, dendrogeomorphology, and mountain environments and environmental change, especially in the Rocky Mountains.

Steven J. Cardinal is Graduate Student in the Department of Geography at the University of Missouri, Columbia, USA. His research interests are repeat photography and subalpine forests in the U.S. Rocky Mountains.

Heejun Chang is Professor in the Department of Geography at Portland State University, USA. His research focuses on the combined impacts of climate change and urban development on water resource resilience, including water supply, quality, demand, floods, and water-related ecosystem services.

Marisa Cigliano is Graduate from the Environmental Program at the University of Vermont, Burlington, USA. Her research interests involve social geography, environmental justice, and feminist and decolonial political ecology.

Katherine R. Clifford is Researcher for the Western Water Assessment at the University of Colorado, Boulder, USA. Her research examines questions of environmental knowledge, regulation, and decision-making in the context of environmental change.

Susan L. Cutter is Carolina Distinguished Professor and Director of the Hazards & Vulnerability Research Institute in the Department of Geography at the University of South Carolina, Columbia, USA. Her research interests include vulnerability and resilience science, spatial and social inequalities, and hazards geospatial analytics.

Valeria Dattilo is Teaching Assistant of Semiotics and Philosophy of Language at the Department of Humanities at the University of Calabria, Arcavacata di Rende (CS), Italy. She holds a PhD in philosophy of communication and entertainment, theory, and history of languages from the University of Calabria.

Meredith J. DeBoom is Assistant Professor in the Department of Geography at the University of South Carolina, Columbia, USA. She is a political geographer whose research focuses on debates over development, distribution, extraction, and (geo)politics in southern Africa. Her current work examines how Africans are engaging with geopolitical and environmental changes to pursue their own political goals.

Francesco De Pascale is Teaching Assistant of Geography at the Department of Culture and Society at the University of Palermo, Italy. He holds a PhD in human geography from the University of Calabria after defending "the first thesis that deals specifically with geoethics", according to the IAPG.

Sisimac Duchicela is Doctoral Candidate in the Department of Geography and the Environment at the University of Texas at Austin, USA. Her research has focused on high mountain ecosystems, specifically of Andean forests, paramos, and punas. In her doctoral research, she is developing evaluation methods for ecological restoration practices used in Ecuador and Peru, with the goal of increasing human adaptability to global environmental change.

Nicholas P. Dunning is Professor in the Department of Geography & GIS at the University of Cincinnati, USA. His research interests include geoarchaeology and cultural ecology with a special focus in the Neotropics, especially the Maya Lowlands and West Indies.

Grant P. Elliott is Associate Professor of Geography at the University of Missouri, Columbia, USA. He is a biogeographer interested in studying ecological responses to climate change.

Sunyong Eom is Project Researcher in the Center for Spatial Information Science at the University of Tokyo, Chiba, Japan. His research interest covers spatial information science, land use planning, and transportation modeling for sustainable and resilient urban structure.

Madeleine Fairbairn is Assistant Professor in the Environmental Studies Department at the University of California Santa Cruz, USA. Her research interests include the financialization of farmland ownership and the emerging agri-food tech sector.

Albert E. Fulton II is Visiting Lecturer in the Department of Geography at the University of North Alabama, Florence, USA. His research interests include Quaternary paleoecology, pollen analysis, historical ecology, phytogeography, and prehistoric human–environment interactions.

Ben A. Gerlofs is Assistant Professor in the Department of Geography at the University of Hong Kong, Hong Kong SAR. His research interests include developing transnational perspectives on the aesthetics and politics of neighborhood change and exploring the political and social contexts and consequences of catastrophic urban environmental events.

Jaclyn Guz is PhD Candidate in the Graduate School of Geography at Clark University, Worcester, USA. Her research interests include forest dynamics, climate change, and ecosystem management.

John Harrington Jr. is Independent Scholar and Professor Emeritus at Kansas State University, Manhattan, USA. His research interests include climatology, human dimensions of global change, GIScience, applied geography, geographic thought, and geography education.

Jeffrey Hoelle is Associate Professor in the Department of Anthropology at the University of California, Santa Barbara, USA. His research focuses on human–environment interactions in Amazonia, with a focus on cattle raising, deforestation, and ideologies of cultivation.

Chang-yu Hong is Research Fellow at the Jeju Research Institute, Korea. His area of concentration is sustainability, socio-hydrology, and environmental planning, with a focus on issues of stakeholder governance and conflict resolution.

Erik Isberg is Doctoral Student in the Division of History of Science, Technology and Environment at KTH Royal Institute of Technology, Stockholm, Sweden. His research interests include the history of paleoclimatology, the history of planetary-scale environmental knowledge, and the scientific and intellectual roots of the Anthropocene.

Mark Jackson is Senior Lecturer in the School of Geographical Sciences at the University of Bristol, UK. His research interests include decolonial and postcolonial thought and practice; critical theory, with an emphasis on critique, posthumanism, materiality, modernity, and ethics; political ecology; and critical urbanisms.

L. Allan James is Distinguished Professor Emeritus in the Department of Geography at the University of South Carolina, Columbia, USA. His research interests include fluvial responses to episodic sedimentation, interactions between anthropogenic sediment and flood risks, and applications of geospatial science in geomorphology.

Nicholas C. Kawa is Assistant Professor in the Department of Anthropology at the Ohio State University, Columbus, USA. His research centers on human relationships to plants, soils, and bodily waste in Amazonia and the U.S. Midwest.

Lisa C. Kelley is Assistant Professor in the Department of Geography and Environmental Sciences at the University of Colorado, Denver, USA. Her research combines political ecology, critical agrarian studies, and remote sensing to address diverse questions related to agrarian and forest change.

Anna Klimaszewski-Patterson is Assistant Professor in the Department of Geography at California State University, Sacramento, USA. Her research interests include reconstructing past environments and visualizing spatial processes through models and augmented or virtual reality.

Dominik Kulakowski is Professor of Geography in the Graduate School of Geography at Clark University, Worcester, USA. His research interests are centered on forest ecology and management, and include the causes and consequences of forest disturbances, interactions and feedbacks among ecological processes, effects of climate on tree demography, and effective ecosystem management in the face of global environmental change.

Thomas Barclay Larsen is Instructor of Geography at the University of Northern Iowa, Cedar Falls, USA. His research interests include human–environment relations, the Anthropocene, geographic thought, geography education, and the unity of knowledge.

Rebecca Lave is Professor in the Department of Geography at Indiana University, Bloomington, USA. Her research combines political economy, science and technology studies, and fluvial geomorphology to study the ecosocial dynamics of rivers and streams.

Carolina Levis is Postdoctoral Fellow in the Graduate Program in Ecology at the Federal University of Santa Catarina, Brazil. Her research interests include domesticated forests, Amerindian forest legacies, and Amazonian ecology.

Sheryl Luzzadder-Beach is Centennial Professor and the Director of the Water Quality and Environmental Hydrology Lab in Geography and the Environment at the University of Texas at Austin, USA. She is a past president of the American Association of Geographers. Her research interests include hydrology, geoarchaeology, and geostatistics, pursued in landscapes from Mesoamerica to the Mediterranean.

Samuel A. Markolf is Assistant Research Professor in the Department of Civil, Environmental, and Sustainable Engineering at Arizona State University, Tempe, USA. His research broadly focuses on urban and infrastructure systems, with an emphasis on analyzing the extent to which interconnected SETS can enhance (or hinder) resilience.

Scott Mensing is Gibson Professor of Geography and Foundation Professor in the Department of Geography at University of Nevada, Reno, USA. His research interests include reconstruction of past environments and the impact preindustrial societies have had in shaping ecology.

Harlan Morehouse is Senior Lecturer in the Department of Geography at the University of Vermont, Burlington, USA. His research interests lie at the intersection of political ecology and environmental philosophy, with a current focus on attunement and soundscape ecology.

Christopher T. Morgan is Associate Professor in the Department of Anthropology at the University of Nevada, Reno, USA. His research interests include the archaeology of hunter-gatherer adaptations to climate change and the evolution of human socioeconomic systems.

Lindsay Naylor (she/her) is Associate Professor in the Department of Geography at the University of Delaware, Newark, USA, and is the co-facilitator of the Embodiment Lab. She is the author of the award winning book *Fair Trade Rebels: Coffee Production and Struggles for Autonomy in Chiapas*.

Emily Reisman is Assistant Professor in the Department of Environment and Sustainability at the University at Buffalo, USA. Her research concerns the political ecology of agricultural knowledge, most recently in almond production and emerging agri-food technologies.

Daniel D. Richter is Professor of Soils and Ecology in the Nicholas School of the Environment at Duke University, Durham, USA. His research interests include how humanity is transforming Earth's soils, the biogeochemistry of Earth's critical zone, and particularly in North America's Southern Piedmont.

Wonsuh Song is part-time Lecturer in the School of Education at Waseda University, Tokyo, Japan. She has been researching weathering of rocks by microorganisms, but recently expanded the scope of her research to environmental effects caused by water, such as flood, drought, and wildfires.

Sverker Sörlin is Professor of Environmental History in the Division of History of Science, Technology and Environment, and its KTH Environmental Humanities Laboratory at KTH Royal Institute of Technology, Stockholm, Sweden. His research interests include the historical and geographical science politics of climate change with a focus on the cryosphere, and the history of global environmental governance.

J. Anthony Stallins is Professor in the Department of Geography at the University of Kentucky, Lexington, USA. His research focuses on the nature of boundaries among organisms, the intersection of science and social theory, and the middle ground of geographic thought.

William R. Travis is Associate Professor in the Department of Geography at the University of Colorado, Boulder, USA. His research focuses on human interaction with hazards and risks, especially in a changing global environment.

Michael A. Urban is Associate Professor in the Department of Geography at the University of Missouri, Columbia, USA. Research interests include human drivers of environmental change, related concepts such as the Anthropocene, and exploring how public policy responds to flood control, agricultural development, and climate change.

Dana Veron (she/her) is Associate Professor in the Department of Geography & Spatial Sciences and the School of Marine Science and Policy at the University of Delaware, Newark, USA. Her research focuses on the physical interaction between the lower atmosphere and the land–sea–ice surfaces, including coastal winds, cloud–radiation processes, and polar energy balance.

Antoinette M. G. A WinklerPrins is Deputy Division Director of the Division of Behavioral and Cognitive Sciences in the Directorate for Social, Behavioral, and Economic Sciences at the U.S. National Science Foundation. She is also Adjunct Professor of Environmental Sciences and Policy at the Johns Hopkins University, Baltimore, USA. Her research interests include the formation of cultural landscapes, anthrosols, urban agriculture, and smallholder livelihoods.

Catherine H. Yansa is Associate Professor in the Department of Geography, Environment, and Spatial Sciences at Michigan State University, East Lansing, USA. Her research interests include paleoecology, pollen and plant macrofossil analyses, plant colonization and succession after deglaciation, and prehistoric human–environment interactions.

Kenneth R. Young is Professor in the Department of Geography and the Environment of the University of Texas at Austin, USA. He investigates the ecological and social dimensions of biodiversity conservation, aiming to produce policy-relevant research.

David J. Yu is Assistant Professor with joint appointments in the Lyles School of Civil Engineering and the Department of Political Science at Purdue University, West Lafayette, USA. His research focuses on the resilience and governance of coupled systems (e.g., engineered–social, socio hydrological, or social–ecological) in which some natural or humanmade commons are shared by many and managed through collective choice.

Kathryn Yusoff is Professor of Inhuman Geography in the School of Geography at Queen Mary University of London, UK. She is the author of *A Billion Black Anthropocenes or None, Minneapolis* (2018), a special issue on "Geosocial Formations and the Anthropocene" (with Nigel Clark) in *Theory Culture and Society*, and *Geologic Life: Inhuman Intimacies and the Geophysics of Race* (forthcoming).

Introduction: The Anthropocene

David R. Butler

This special issue of the *Annals of the American Association of Geographers* is devoted to the Anthropocene, the period of unprecedented human impacts on Earth's environmental systems. The articles contained in this special issue illustrate that geographers have a diverse perspective on what the Anthropocene is and represents. The articles also show that geographers do not feel it necessary to identify only one starting point for the temporal onset of the Anthropocene. Several starting points are suggested, and some authors support the concept of a time-transgressive Anthropocene. Articles in this issue are organized into six sections, but many of them transcend easy categorization and could easily have fit into two or even three different sections. Geographers embrace the concept of the Anthropocene while defining it and studying it in a variety of ways that clearly show the breadth and diversity of the discipline. *Key Words: Anthropocene, environmental change, environmental degradation, geographic education, human impact, natural hazards, physical geography.*

The term *Anthropocene* originated in the year 2000 (Crutzen and Stoermer 2000), denoting the concept that humans have become the primary environmental influence on Earth systems. Although the concept of humans as the primary environmental influence on Earth systems has been around since at least the mid-1800s (Marsh 1864), this viewpoint slowly increased throughout the twentieth century (e.g., Sherlock 1922; Thomas 1956; Nir 1983) and has accelerated dramatically in the twenty-first century. New journals have appeared with a primary focus on the Anthropocene, including *Anthropocene* (first appearing in 2013) and *The Anthropocene Review* (which began publication in 2014). All of this begs the questions of what, precisely and exactly, is the Anthropocene, and when did it begin?

Geologists are attempting to define the Anthropocene as a new geological epoch, as "a potential formal unit of the Geological Time Scale, whether by a physical reference point, that is, a Global Boundary Stratigraphic Section and Point (GSSP, or 'golden spike'), as is typical for chronostratigraphic boundaries in the Phanerozoic, or by a numerical age, a Global Standard Stratigraphic Age (GSSA)" (Zalasiewicz and Waters 2017). As of this writing, the geologists' working group on the Anthropocene has not yet come to a conclusion or agreement as to the starting date (Zalasiewicz et al. 2017). Candidates for the starting date include the onset of agriculture in the early Holocene, the changes in the Earth's atmosphere associated with the period of European expansion and colonization, the onset of the Industrial Revolution, and the onset of the Nuclear Age (Figure 1) and the so-called Great Acceleration of widespread environmental change and deposition of nuclear fallout from atmospheric testing of nuclear weapons. Other environmental changes including human impacts on animal communities (Butler 2018; Zerboni and Nicoll 2019) and the extent of surface and subsurface mining activities and degradation (e.g., Sherlock 1922) have also been examined in the context of potentially identifying an Anthropocene starting point.

This special issue of the *Annals of the American Association of Geographers* makes it clear that geographers do not feel bound by the formal constraints of identifying a starting date or even a definition for the Anthropocene that geologists must employ in identifying a stratigraphic boundary. A call for papers on the Anthropocene for this special issue produced a wide array of contributions representing the broad diversity that is our discipline. This issue contains twenty-nine articles chosen from a greater number of submitted abstracts for potential papers. The articles include conceptual considerations and definitions of the Anthropocene (seven articles), historical perspectives on the concept prior to its definition in 2000 (three articles), physical geography (both classic and critical; six articles), examinations

Figure 1. The dawn of the so-called Nuclear Age, one proposed starting date for the Anthropocene, occurred on 16 July 1945, at the Trinity Site in southern New Mexico when the world's first atomic bomb was detonated on the northern end of what is today the White Sands Missile Range. The Trinity bomb was detonated about 30 m above the ground atop a steel tower. The site is memorialized by the stark 4-m-high stone monolith shown here. Photo by author.

of natural hazards and disasters (three articles), the environment and environmental degradation (eight articles), and geographic education (two articles). Some articles could have been categorized in more than one of these artificial groupings, and each one should be read bearing in mind that these categories have no "hard and fast" boundaries.

Definitions and Conceptual Considerations

The first seven articles make very clear how broad and unconstrained geographers are in their consideration of the concept and definition of the Anthropocene. Stallins (this issue) uses topology to illustrate that the Anthropocene should be viewed as a moving window of human and natural processes rather than a fixed point in time. De Pascale and Dattilo (this issue) use geoethics and Peircean semiotics to develop a triangle heuristic as a metaphor for the Anthropocene. Hoelle and Kawa (this issue)

identify two periods of the Anthropocene. Their "old Anthropocene" begins with human food production, whereas their "new Anthropocene" coincides with the start of the Industrial Revolution. Yusoff (this issue) seeks to examine the Anthropocene through lenses of race and geopower, arguing that geopower is the product of historical geologies of race.

Adams (this issue) argues that geographers must study the power of words to intervene more effectively in the Anthropocene. He examines stakeholders in the portion of Texas underlain by the Ogallala aquifer in the context of language and its relationship with the concept of the Anthropocene. Reisman and Fairbairn (this issue) use agri-food systems to examine and foster an understanding and a challenging of the concepts of the Anthropocene. The final article in this section, by Jackson (this issue), uses examples from Afro-Caribbean and Indigenous geographies and scholarships to better understand and to perhaps refuse the concept of the rationalizations emanating from Anthropocene horizons.

Historical Perspectives on the Anthropocene

Three articles on the historical perspectives and precursors to the formal definition of the Anthropocene illustrate the history of how geographers have examined the concept of human impact on the environment long before the term *Anthropocene* came into being. Bendix and Urban (this issue) discuss George Perkins Marsh's (1864) classic book *Man and Nature* and illustrate its historical influence and importance in shaping how many (especially physical) geographers conceptualize the Anthropocene and its modern interconnections among geomorphic, biotic, and human elements of the environment. Sörlin and Isberg (this issue) provide a distinctly European perspective on how studies in glaciology, palynology, and biogeography set the groundwork for integrating understandings of time and the human–earth relationship within the context of the Anthropocene. Larsen and Harrington (this issue) conclude this section with an examination of the premise of the Anthropocene through studying early modern as well as contemporary geographic thought.

Physical Geography and the Anthropocene

The articles in physical geography illustrate how numerous starting points for the Anthropocene can be posited, based on a variety of geomorphic, stratigraphic, palynological, and vegetative evidence. The role of critical physical geography (CPG) is also discussed.

James et al. (this issue) use three specific examples of long-term anthropogeomorphic change recorded in bottomland alluviums to illustrate three distinctly different potential starting dates for the Anthropocene based on alluvial stratigraphy and related landforms. Elliott et al. (this issue) examine upper treeline in the southern Rocky Mountains and show how hotter, drier conditions (especially on south-facing slopes) caused by Anthropocene-related climate change are enveloping and affecting upper treeline along topoclimatic gradients. Fulton and Yansa (this issue) use paleoecological and archaeological records from the lower Great Lakes region of northeastern North America to identify three periods of developing and progressively intensifying anthropogenic influence, each of which could be considered a starting point for a Paleoanthropocene. A pre-Columbian Anthropocene in California is described by Klimaszewski-Patterson et al. (this issue), who argue for a flexible, anthropologically and ecologically informed conceptualization of the Anthropocene that is time-transgressive rather than based on a stratigraphically distinct starting date. Luzzadder-Beach et al. (this issue) present detailed light detection and ranging and multiproxy ecological and pedological verification for an early Anthropocene in Mesoamerica, again illustrating how geographers do not feel constrained by the need for a worldwide, distinct, stratigraphic starting point that geologists require to embrace the concept of the Anthropocene.

The final article in this section, by Biermann et al. (this issue), takes a bibliometric analysis approach from a CPG perspective to examine the nature of academic articles on the Anthropocene published between 2002 and 2019. They posit that all biophysical questions are also social and showcase the utility of a CPG approach for analysis of the Anthropocene.

Natural Hazards, Disasters, and the Anthropocene

The articles on natural hazards and disasters continue the trend of being unconstrained by a specific starting date or even a definition of the Anthropocene. Natural hazards of a diverse group ranging from floods to earthquakes are examined through the Anthropocene lens, and the social and economic costs and inequalities come to the fore. Cutter (this issue) describes the changing nature of hazard and disaster risk from local to global scales in the Anthropocene and shows that those risks have accelerated since the "golden spike" of C^{14} in 1964. Gerlofs (this issue) examines the Mexico City earthquakes of 1985 and 2017 as they relate to the politics of geology and political economy in the Anthropocene. The final article in this section, by Chang et al. (this issue), presents case studies of urban flood risk management in the Anthropocene from three cities, in the United States, South Korea, and Japan. They illustrate the utility of resilience and social learning perspectives for analyzing and interpreting the various cities' strategies in coping with flooding over time.

The Environment and Environmental Degradation

Climate change, nature conservation, soil and vegetation impacts, and the degradation of natural landscapes are examined in this section, with at times depressing clarity. The section begins with an article by WinklerPrins and Levis (this issue). They examine Amazonia as an Anthropogenic space, via studying three forms of landscape transformation there—anthrosols, cultural or domesticated forests, and anthropogenic earthworks. Their goal is to understand the past in Amazonia and to guide its conservation. Guz and Kulakowski (this issue) describe and study the ecological resilience of forests in the Anthropocene and compare and contrast management of protected versus intensely used forests through the lens of resilience theory. The article by Young and Duchicela (this issue) considers the state of biodiversity conservation in the changing world of the Anthropocene. The authors suggest that it is appropriate to rethink implications for sustainability and human–nature relationships in general and for biodiversity in particular as a result of human-induced Anthropocene stress(es).

Bergmann and Briwa (this issue) investigate the role visual imagery plays in shaping geographical imaginations of the Anthropocene, through the use of a unique set of posters of U.S. National Parks. (Rothstein n.d.). These posters evoke the aesthetic of the toxic sublime, which the authors suggest is appropriate for the new wilderness landscapes of the Anthropocene. DeBoom (this issue) develops a theoretical framework—climate necropolitics—for developing integrated analyses of the distribution of both environmental and social violence in the Anthropocene. She draws on fieldwork using multiple methods and illustrates the applied value of climate necropolitics through a case study of the Chinese Communist Party's ecological civilization.

The next two articles in this section examine the effects of the Anthropocene on Arctic landscapes. Morehouse and Cigliano (this issue) examine glacial recession in the Anthropocene and describe how the loss of glacial ice needs to be considered not only in quantifiable terms of amount of ice lost but in terms of the effects of ice loss from the human perspective. They discuss this perspective through considering everyday encounters, generational experiences, and stories generated at the interface of ice and culture.

Bennett (this issue) continues the examination of Arctic ice and discusses how humanity needs to refocus its gaze and redefine its perceptions of beauty to accommodate the rapidly developing ice-free landscapes of the Anthropocene. This section concludes with an article by Clifford and Travis (this issue), who examine the politics of data management for emerging environmental conditions in the Anthropocene. They call for a need to construct new perspectives as to what constitutes a "normal" or "abnormal" environmental event.

The Anthropocene and Geographic Education

The final two articles examine the role of geographic education and how it represents the Anthropocene in curricula and determine how students can find their "fit" in the Anthropocene. Bagoly-Simo (this issue) illustrates how the geography curricula in lower secondary education in fifty countries represent the Anthropocene and discusses commonalities and differences among the numerous curricula. Finally, Naylor and Veron (this issue) consider the question of how best to teach the environmental changes of the Anthropocene to assist students in determining where they fit in the Anthropocene. They suggest that teaching from both a climate science and from a social or cultural perspective assists in this process and aids an understanding of the place of geographic education in universities in the Anthropocene.

Concluding Remarks

Regardless, then, of which specific definition or start time is chosen, geographers seem to agree that the Anthropocene is upon us. The strength of the discipline is shown by its ability to escape the constricting bonds of required definitions and time frames for the Anthropocene. The Anthropocene can be studied as a starting point or as an ongoing, time-transgressive phenomenon. The Anthropocene is a concept that geography and geographers clearly embrace, and the articles of this special issue hopefully illustrate this well.

Acknowledgments

Putting together a special issue of a journal is a somewhat daunting proposition in the best of times.

Doing so in the midst of a global pandemic is truly challenging. Yet, the authors of the articles contained herein persevered despite lockdowns, transitions to virtual classrooms, illnesses among families and selves, home schooling, and other unprecedented extended parenting needs. Reviewers also heroically stepped forward when asked, despite the increased burdens on their time and mental and physical health. I am extremely grateful to all of the authors and reviewers who have made this special issue possible.

In addition, special recognition should be given to Jennifer Cassidento, Managing Editor of the *Annals*, who worked with me to keep the special issue on track and whose gentle reminders were always given with a smiling attitude. Thanks also to Dr. Stephen Hanna, the AAG Cartographic Editor, who was called on heavily during the global pandemic, and Lea Cutler, Production Editor at Taylor & Francis.

I also gratefully acknowledge the efforts of Camilla Montoya, Public Affairs Specialist at the White Sands Missile Range (WSMR) in New Mexico, for arranging permission for my visit to the Trinity Site prior to the semiannual "open house" at the site in October 2017. Thanks are also due to Drew Hamilton, Public Affairs Specialist, and Robert Carver, Chief of Public Affairs, both from WSMR, for accompanying me to the Trinity Site and providing background and commentary that was very useful.

References

Butler, D. R. 2018. Zoogeomorphology in the Anthropocene. *Geomorphology* 303:146–54. doi: 10.1016/j.geomorph.2017.12.003.

Crutzen, P. J., and E. F. Stoermer. 2000. The Anthropocene. *Global Change Newsletter* 41:17–18.

Marsh, G. P. 1864. *Man and nature; or, physical geography as modified by human action.* New York: Scribner.

Nir, D. 1983. *Man, a geomorphological agent.* Dordrecht, The Netherlands: D. Reidel.

Rothstein, H. n.d. *National Parks 2050.* Accessed January 15, 2021. https://www.hrothstein.com/national-parks-2050

Sherlock, R. L. 1922. *Man as a geological agent—An account of his action on inanimate nature.* London: HF and G Witherby.

Thomas, W. L., Jr., ed. 1956. *Man's role in changing the face of the earth.* Chicago: The University of Chicago Press.

Zalasiewicz, J., and C. N. Waters. 2017. Arguments for a formal global boundary stratotype section and point for the anthropocene. In *Encyclopedia of the anthropocene*, ed. M. Goldstein. Oxford, UK: Elsevier Science & Technology.

Zalasiewicz, J., C. N. Waters, C. P. Summerhayes, A. P. Wolfe, A. D. Barnosky, A. Cearreta, P. Crutzen, E. Ellis, I. J. Fairchild, A. Gałuszka, et al. 2017. The Working Group on the Anthropocene: Summary of evidence and interim recommendations. *Anthropocene* 19:55–60. doi: 10.1016/j.ancene.2017.09.001.

Zerboni, A., and K. Nicoll. 2019. Enhanced zoogeomorphological processes in North Africa in the human-impacted landscapes of the Anthropocene. *Geomorphology* 331:22–35. doi: 10.1016/j.geomorph.2018.10.011.

Part 1

Definitions and Conceptual Considerations

The Anthropocene: The One, the Many, and the Topological

J. Anthony Stallins (iD)

Given the many discourses about markers for the Anthropocene, those peripheral to one's academic niche might elicit indifference or even dismissal. Conversely, a shallow pluralism can take root in which any Anthropocene demarcation matters equally as others. I propose a more diplomatic coexistence of ideas regarding the Anthropocene boundary issue. In this perspective, the choice of when to delineate the Anthropocene's start and how to signify its presence is analogous to a modifiable areal unit problem. Boundaries can be drawn from a range of anthropogenic phenomena. Geographic subdisciplines have acquired distinctive ways of sublimating socioecological patterns and processes into a timestamp. Less attention, however, is given to how their respective temporal modes and ensuing markers of anthropogenic change overlap and relate to one another. I show how topology, as invoked in the biophysical sciences and social theory, integrates these temporalities of the Anthropocene. The Anthropocene can be framed as a cusp catastrophe, a folded surface in which different modes of change emerge from and coexist with each other. These trajectories of change, the gradual, the threshold driven, and those exhibiting hysteresis, encapsulate the interdependencies among past, present, and future invoked across different delineations of the Anthropocene. The Anthropocene might be less a fixed point in time as it is a moving window where human and natural processes are folded into one another. An Anthropocene represented as a folded surface rather than a timeline incorporates the importance of unpredictably productive responses to the present Anthropocene moment. *Key Words: Anthropocene, diplomacy, hysteresis, pragmatism, topology.*

From its inception, the concept of the Anthropocene has been a debate about boundaries. Among geographers, these boundaries have often corresponded with subdisciplinary affinities. Critical geographers target the rise of colonialism and global capitalism. Biophysical geographers identify the uptick in the extent of agriculture 10,000 years ago or the peak signature of radioactive fallout from nuclear bomb testing. These and other markers, however, do more than signify academic self. Intent is also implicit to their designation. Any Anthropocene boundary prioritizes a particular view of the past that steers anticipation of a future and the kinds of actions we take in the present (Anderson 2010). Consequently, the choice of an Anthropocene marker can be made to support the perception of the long-term influence of humans. Anthropocene boundary work can make appeals to our ecomodernist hopes for a flourishing of new ways to live on Earth (Ellis 2015). Anthropocene boundaries might even be rejected because they mask underlying social processes and no clear date identifies when humans became geophysical agents (Bauer and Ellis 2018).

In this manner, the concept of Anthropocene allows people to reinforce and perpetuate preferred views about the implications of human interaction with the Earth. As Castree (2017) noted, "What counts as epochal change is a matter of perspective, since it emerges from judgements about when quantitative change morphs from qualitative transformation" (289). The inevitable dichotomies that result, like the good versus a bad Anthropocene, drive conversation toward confusion as individuals argue preferred versions of an Anthropocene concept and its markers. Philosophical and political perspectives become entangled with scientific measures of human impacts and proposed geological stratigraphic units for the start of the Anthropocene (Autin 2016). As the number of demarcations and interpretations of the Anthropocene have grown, perhaps so, too, has the temptation to advocate for one's favored temporal representation of the Anthropocene and the discourse surrounding it.

Following Anderson (2019), I take the position that Anthropocene markers are representations in relation. They are entangled and coevolving rather than isolated. Any individual demarcation of the

Anthropocene is lived with in the midst of other events and processes. Accordingly, the pertinent question might not be when the Anthropocene began but how to summarize the relationality of its many demarcations (Castree et al. 2014). Such a pluralistic approach aims for a diplomatic coexistence of boundaries, a negotiation among parties often distrustful of each other (Castree 2015a, 2015b). For example, designating a stratigraphic marker for the Anthropocene has been productive in ways beyond earth science. Initial proposals to reduce the Anthropocene down to a geologic unit fueled wide-ranging debate on the causes and correlates of the Anthropocene. Similarly, knowledge of how racial and class inequities signal the Anthropocene can be useful in ways that do not obviate all of the practices of global change science (Castree 2015a, 2015b). Yet Anthropocene demarcations are often presented as single agential cuts (Barad 2007) along partisan lines of academic identity. As an alternative, Anthropocene boundaries should be conceived as productive insofar as their relational character is foregrounded as difference but not necessarily contradiction. Following Conway (2019), rather than "dissolving all antagonisms, such that a sea of mutuality might then rise," this perspective on the Anthropocene diplomacy seeks "to dispel unnecessary antagonisms so that necessary ones might come to the fore. It explores the folds that are possible so as to enable the cuts that are necessary" (22).

From this point of view, the Anthropocene can be conceived as a temporal analogue of the modifiable areal unit problem (MAUP), a central concept of geography. Yet this is not simply the observation that the Anthropocene began at different times in different places. Instead, having contrasting, even conflicting demarcations of time (and space) to infer the Anthropocene can be viewed as constructive. In this sense, the MAUP and its counterpart in time, the modifiable temporal unit problem, are not just fallacies of interpretation, the usage most geographers associate with this concept. As I have argued (Stallins 2012), these modifiable unit issues encapsulate the topological character of environmental problem solving performed by organisms. Biological life operates through a constant engagement with the modifiable unit challenge of making predictable sense (cuts) out of the shifting boundaries that define its environment. Accordingly, for humans,

any single Anthropocene boundary is a local negotiation that reduces environmental pattern and process down to a point or interval on a timeline. Yet it is only through collective negotiation among many different demarcations of the Anthropocene that this boundary work can become more fully productive.

As an example, one proposed marker for the Anthropocene is the Orbis spike of the early 1600s (Lewis and Maslin 2015). This boundary sublimates the economic processes of European colonization into a timestamp of abruptly lowered carbon dioxide concentrations due to forest regrowth after the genocidal depopulation of the Americas. As a biophysical marker, the Orbis spike also signals a start to the colonialist, global-scale transformations of people and landscapes that continues today. That diffuse political and economic processes become less visible in the stratigraphic designations of the Anthropocene underscores the challenge of finding representations of the Anthropocene that accommodate what seem to be subdisciplinary irreconcilabilities. Yet it is not that we do not understand how these environmental signals and economic processes relate (Saldanha 2019). The challenge is linking the geometries of process and form encompassing colonial power and carbon dioxide levels in Antarctic ice cores. Hence, the Anthropocene is also a task of visualization, of cartographic imagination to bring together objects and processes that defy traditional mapping.

In this article, I show how a topological approach inspired by this reformulation of MAUP lessens some of the partisanship of defending any one Anthropocene boundary from among many. Topology provides a means to represent how different temporal modes of environmental and social change jointly contribute to a more continuous form for the Anthropocene. By allowing many perspectives to exist in relation, topology avoids reifying a single Anthropocene boundary as preeminent; that is, as guilty of committing the fallacy version of MAUP. This hews to Castree's (2015a) advice about how the Anthropocene moment should avoid narratives that "risk perpetuating an emaciated conception of reality wherein Earth systems and social systems are seen as knowable and manageable if the 'right' ensemble of expertise is achieved" (1). This topological view emphasizes a diplomatic coexistence or "presence" (Kaika 2018) among the different temporalities of the Anthropocene rather than any final delineation of origin or single best marker.

Topology in the Biophysical Sciences and Social Theory

Formally, topology is a branch of mathematics that studies shapes. Topologists treat shapes as spaces whose coordinates are not necessarily contained within a Cartesian coordinate system. Instead, they are intrinsic to the surface itself. Topologists focus on what aspects of a shape remain constant, such as its dimensionality or number of edges, when the surface is deformed. In topological data analysis, high-dimensional visualizations are created from large data sets of many interacting variables. Analysis of the shape of these data provides insight into the relationships among variables. The shape of this data cloud is abstract, but the surfaces and distances within it convey insights about relationships among real-world processes.

Topology has a long history in the biophysical sciences through the use of ball and cup diagrams and fitness surfaces (Inkpen and Petley 2001). These topological ways of seeing can be mathematically formal as well as descriptive and conceptual (Prager and Reiners 2009). One of the more well-known topological forms invoked in the biophysical and social sciences is the cusp catastrophe (Zeeman 1976; Graf 1979; Thorn and Welford 1994). This shape formalizes how a few variables can interact and create a surface where distances between points on it represent transitions among different states. These transitions can range from gradual to sudden depending on the location on the surface. Studies on lakes, coral reefs, oceans, forests, and arid lands have shown that smooth gradual change can be interrupted by sudden drastic switches to a contrasting state (Scheffer et al. 2001). Anthropocene scholars have invoked these abrupt transitions to suggest that the Earth might irreversibly tip and lock into a degraded state once planetary boundaries are exceeded (Barnosky et al. 2012). It is now recognized, however, that the potential for these large jumps in state is more varied in space and time (Dakos et al. 2015). Gradual and less sudden threshold dynamics can coexist with them.

For social theorists, topology is a way of thinking about relationality, space, and movement without mathematical constraints. Topology in human geography provides a way to map out how people and things change and how they relate without quantification (Martin and Secor 2014). Social theorists have used topology metaphorically to account for how presences and absences no longer correspond to measures of physical proximity. For example, power can extend itself in ways that are nonterritorial in the sense that its reach is present in quieter but more pervasive forms irrespective of traditional measures of proximity (Allen 2016). The social relations of home are topological in that they are a collection of attachments that consist of people, places, ideas, and things that are both near and far (Kallio 2016). In human geography, topology has come to be a shorthand for the contextual, relational constitution of the world that defies physical proximity and spaces defined by absolute distances. Social theorists invoke topology as metaphor and rhetorical construct in accounts of the Anthropocene. Their Plantationocene and the Capitalocene encompass the destructive structural logics of resource depletion and petrochemical dependency embedded in the world system of capitalism (Davis et al. 2019). The Chthulucene is a foil to the Anthropocene, a multispecies unfolding and "tentacularity" connecting disparate realms of life in potentially collaborative and creative webs of kinship (Haraway 2015).

Topology provides a means to integrate the pluralistic and often competing delineations of Anthropocene. The diversity of qualitative and quantitative interpretations of the Anthropocene and the markers for them defy conventional space and time boundary making. Topological approaches, however, can meld temporal perspectives on the Anthropocene in ways that timelines and absolute measures of space and time cannot.

Shapes and Surfaces of the Anthropocene

As a conceptual rather than formal mathematical topology, the Anthropocene can be represented as a cusp catastrophe demarcated by three axes (Figure 1). One axis represents the variable of time. A second axis is ecological malleability. This variable conveys the degree to which ecological systems can become entrained by humans. High ecological malleability denotes a socioecological system in which a subset of nonhuman organisms and processes are readily shaped by humans. Low ecological malleability implies ecological systems

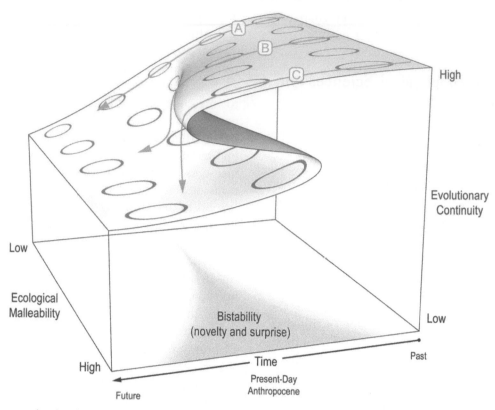

Figure 1. Cusp catastrophe for the Anthropocene with three temporal trajectories: (A) gradual, (B) threshold, and (C) hysteretic. Changes in the shape of the distortion ellipses along each trajectory represent how past, present, and future inform one other. Consistent ellipse shape along a trajectory indicates predictable correlations among past, present, and future. Distortions indicate less predictability of the future based on the past and present. The gray shaded area represents where novelty and surprise arise from bistability and tipping.

with properties that resist human incorporation or domestication, an unruliness of nature. The third axis represents evolutionary continuity. This response variable is a measure of the depth of time comprising evolutionary development in the absence of human impacts. Lower positions on this axis represent a greater divergence from past nonhuman evolutionary context. Higher positions represent systems more temporally continuous with nonhuman environmental change and ecological processes. These three axes demarcate regions on the surface of this fold that parallel a trend of increasing human impacts through time, from the past, to a present-day Anthropocene, and a future.

The movement from an unmodified, evolutionarily conserved past to the ecologies of the humanized present traces multiple paths on this surface. Although many trajectories are possible along the contours of this surface, the three predominant ones shown in Figure 1 range from the gradual, to threshold-driven, and a path that exhibits irreversibility, or hysteresis, associated with sudden changes in state.

These three trajectories also possess contrasts in how the past is correlated with present and how they anticipate the future. The changing shapes of the distortion ellipses projected on this surface represent the past, present, and future correlate and inform one another. For example, for a ball moving along a gradual trajectory (A), change is relatively predictable. The unchanged shape of the distortion ellipses conveys a consistent predictability in how the past informs the present and the future through time. With threshold trajectories (B), the past has much less capacity to inform the present and the future, as represented in the shrinking of the distortion ellipse where the surface bends inward in the center. With hysteresis (C), the distortion ellipses stretch more in one direction, indicating how the past can inform the present only within increasingly narrow bounds. Trajectories here become more blind to the future. These three predominant trajectories and their contrasts in how past, present, and future relate to one another give form to the many Anthropocenes that coexist as one.

Gradual Anthropocenes

Although the Anthropocene is often described as sudden, others have argued for a longer run-up to the present. For this trajectory, its low ecological malleability results in greater evolutionary continuity and a stronger correlation of the past with the present and future in the accumulation of human impacts. Early agricultural societies by 3,000 years ago had set in motion the large-scale anthropogenic modification of soil and biota (Jenny et al. 2019; Stephens et al. 2019). Human behavior has been a long-term ecological driver of plant and animal evolution for at least 50,000 years (Sullivan, Bird, and Perry 2017). These gradual trajectories reflect how some environmental and evolutionary constraints have not vanished in the Anthropocene. Aspects of our environment, nonhuman organisms, and human behavior and biology can exhibit a resistance to anthropogenic influences. For example, our mammalian qualities have not suddenly disappeared because we focus more on sociocultural identities and have secured the title of planetary engineer (Laist 2015). In these ways, a gradualist framing of the Anthropocene reinserts nature into the Anthropocene moment. Even the governmental and economic systems operative today remain correlated over a length of time with the past. Their futures are path dependent. As many critical geographers recognize, new envisionings of society do not easily escape the ruts left by old economic orders just because their radically transformative potentials are recognized. The colonial past remains entrenched in different economic and political forms. Defining the Anthropocene as a point on a timeline obscures many biophysical as well as social features of the past that are continuous and predictably correlated with the present and near future (Figures 2A–2B).

Threshold Anthropocenes

Many of the variables of the post–World War II Great Acceleration map as threshold trajectories. For this trajectory, malleability increases and allows greater deviation from historical precedents as axis position shifts toward less evolutionary continuity. These more industrialized Anthropocenes have sharper deviations from long-term trends, as exemplified by rapid increases in the number of humans on Earth, in rates of fossil fuel consumption, and in the pace of environmental change. Consequently,

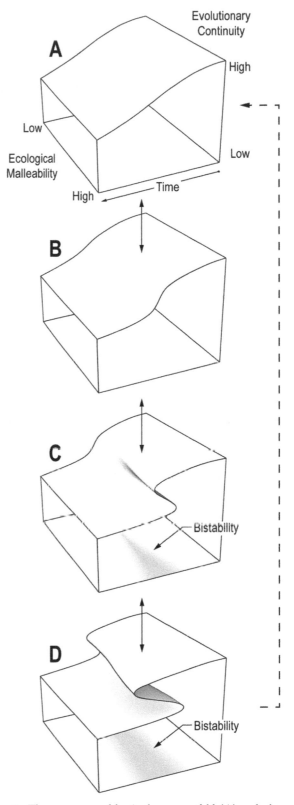

Figure 2. The emergence of the Anthropocene fold: (A) gradual transition surface; (B) initiation of threshold change where malleability is high; (C) hysteresis develops as system is managed for predictability, thereby increasing potential for sudden change in state; (D) with greater folding, more bistability and uncertainty but also greater potential for novelty and surprise. Sequence is not intended to be deterministically developmental.

rainfall, temperature, and river flows no longer have a stable mean around which they predictably fluctuate. This loss of stationarity, an inability to predict the future based on the past, is often taken to be the defining feature of the Anthropocene. The Great Acceleration, however, forms out of the velocity of a gradual past. Its threshold trajectories and loss of stationarity have a dependence on the degree to which the details of past can be resolved. This dependence arises because the accumulation of knowledge and technology that makes this acceleration possible also helps resolve the details of environmental histories extending back millennia. In other words, the more we know about the past, the more we know about the threshold slope leading to the future and what constitutes stationarity.

This dependency of thresholds on the resolution of the gradualist past is not confined to biophysical characterizations of the Anthropocene. With this rapid accumulation of knowledge has come a broadening in awareness of the inequities of the Anthropocene, of who has gained and who has borne the costs of this acceleration not only in the present but also in the past. To imagine a more equitable decolonized political future analogously depends on interpretations of history and awareness of the global present to serve as anticipatory guides. As seen on the Anthropocene surface (Figures 2A–2C), threshold trajectories are continuous with gradualist trajectories of change. Threshold and gradual trajectories inform one another whether they highlight biophysical or social and political interpretations of the Anthropocene.

Hysteretic Anthropocenes

Axis positions for this trajectory signal the capacity of human systems to utilize ecological malleability to the extent that abrupt shifts to novel systems can occur. With hysteresis, threshold change is delayed. Tipping points are eventually reached, resulting in a sudden jump to a historically novel state. These transitions can be irreversible. The tipping metaphor has gained prominence in many fields. Human economic systems can undergo these kinds of critical transitions (Battiston et al. 2016). Climate also has a propensity to tip irreversibly (Steffen et al. 2018). The potential for tipping has been overstated, however, particularly for ecological systems. Tipping might

be more variable in space and time than early studies suggested, largely because the experimental designs that informed them downweighted the role of spatial heterogeneity (Kefi et al. 2013). As reflected in Figures 2C–2D, tipping and hysteresis coexist with and emerge from gradual and threshold trajectories of change.

The relationships between past, present, and future are more uncertain on the hysteretic region of the Anthropocene surface. This uncertainty is a trade-off, though, for the generation of novelty. The hysteresis fold demarcates a region where multiple states manifest within the same general conditions. In this bistable region between tipping points, seemingly oppositional states can coexist (Figure 3). The good Anthropocene of the technological optimists and ecomodernists, as achieved through scientific mitigation of human impacts, coexists with the protectionist and cautionary outlook for a bad Anthropocene. Robbins (2020) framed this as a coexistence between the forward-looking ecomodernist's more-is-less world and a skeptic's look backward to a less-is-more world. Similarly, bistability allows for conservation to coexist as preservationist, neoliberal, or decolonialist (Collard, Dempsey, and Sundberg 2015). In this bistable region, the Capitalocene coexists with the Chthulucene. Through the property of bistability, hysteresis can foster a mosaic of contrasting, even seemingly contradictory social and biophysical states despite their proximity on the Anthropocene fold.

Due to its propensity to tip and to allow different states to coexist, the Anthropocene fold produces not only problems but also solutions and new ways for humans to encounter and modify nature. The "Anthropo-scene," according to Lorimer (2017), is unique in that it makes possible novel forms of knowledge and sets the stage for new arrangements for knowledge production to emerge. It is where humans construct their sociocultural niche through constant experimental action and reaction (Ellis 2015). It is on this folded region of the surface that the past coexists with the emergence of new biophysical and sociopolitical entities and unexpected events in ways that generate the ongoing reconfiguration of the world.

This bistable region should not be viewed as the only source of solutions, however. Gradual trajectories lessen the generation of novelty, but they are more likely to provide deeper evolutionary solutions,

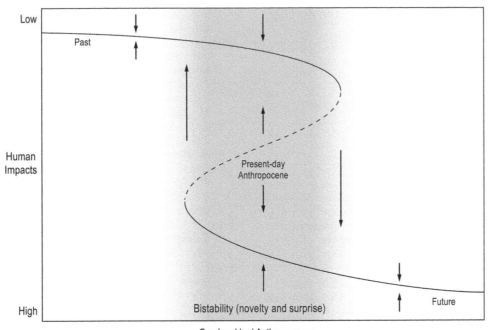

Figure 3. Bistability, novelty, and the tensions of the Anthropocene. Arrows indicate the direction in which the system tends to develop in response to feedbacks. The dashed line indicates a dynamically unfavored region as the feedbacks tend to direct organization to one of two states. Time is along the horizontal axis.

ones that have been recognized by indigenous cultures and conserved through time. In sum, a wide range of solutions, traditional as well as novel, can be present at the same time across this surface (Figure 2C). Too much bistability (Figure 2D) or too little (Figure 2A) would not offer as wide a range of solutions for humans to manage their influence on Earth.

Coda: Topology and Pragmatism

The present-day Anthropocene consists of multiple trajectories of change. It is relational with characteristics of paradox, pluralism, and perspectivism (de la Cadena and Blaser 2018; Wells 2018; Fagan 2019). Instead of a timeline emanating from the past that crosses some threshold, the Anthropocene is an involution, a topological folding over of human and natural processes. This Anthropocene is less a fixed point in time than it is a moving window where the fluxes of nature coexist with the cultures of nature. It is an evolving, propagating boundary where human sociocultural processes shape and are shaped by ecological theory and practice.

This topological interpretation of the Anthropocene is built on more than the truism that boundaries are impermanent and imprecise. Instead, it conveys how the conflicting boundary interpretations that animate the modifiable temporal and areal unit problems are not necessarily fallacies to avoid. They leverage productive differences to negotiate new yet temporary meanings from among antagonisms. As an attempt at diplomacy, topology negotiates among the oneness and manyness of debates over Anthropocene boundaries and signifiers. As for the hazard of reifying any one definition of the Anthropocene and a marker for it, care should also be taken as to how the pluralism of the Anthropocene is construed. Philosophical pragmatists would hold that this pluralism should not be a simplistic celebration of the many, nor the grounds for hardening one's favored belief and discourse. Instead, this pluralism provides "first and foremost a pragmatics, an experimental, exploratory and unpredictably productive response to our present moment" (Savransky 2019, 5).

Acknowledgment

In addition to my anonymous reviewers, my thanks go to William Reihard (https://williamreinhard.

artstation.com/) for providing his skills in the development of the three-dimensional graphics.

ORCID

J. Anthony Stallins http://orcid.org/0000-0002-3911-828X

References

Allen, J. 2016. *Topologies of power: Beyond territory and networks*. London and New York: Routledge.

Anderson, B. 2010. Preemption, precaution, preparedness: Anticipatory action and future geographies. *Progress in Human Geography* 34 (6):777–98. doi: 10.1177/0309132510362600.

Anderson, B. 2019. Cultural geography II: The force of representations. *Progress in Human Geography* 43 (6):1120–32. doi: 10.1177/0309132518761431.

Autin, W. J. 2016. Multiple dichotomies of the Anthropocene. *The Anthropocene Review* 3 (3): 218–30. doi: 10.1177/2053019616646133.

Barad, K. M. 2007. *Meeting the universe halfway: Quantum physics and the entanglement of matter and meaning*. Durham, NC: Duke University Press.

Barnosky, A. D., E. A. Hadly, J. Bascompte, E. L. Berlow, J. H. Brown, M. Fortelius, W. M. Getz, J. Harte, A. Hastings, P. A. Marquet, et al. 2012. Approaching a state shift in Earth's biosphere. *Nature* 486 (7401): 52–58. doi: 10.1038/nature11018.

Battiston, S., J. D. Farmer, A. Flache, D. Garlaschelli, A. G. Haldane, H. Heesterbeek, C. Hommes, C. Jaeger, R. May, and M. Scheffer. 2016. Complexity theory and financial regulation. *Science* 351 (6275):818–19. doi: 10.1126/science.aad0299.

Bauer, A. M., and E. C. Ellis. 2018. The Anthropocene divide obscuring understanding of social-environmental change. *Current Anthropology* 59 (2):209–27. doi: 10.1086/697198.

Castree, N. 2015a. Geography and global change science: Relationships necessary, absent, and possible. *Geographical Research* 53 (1):1–15. doi: 10.1111/1745-5871.12100.

Castree, N. 2015b. Unfree radicals: Geoscientists, the Anthropocene, and left politics. *Antipode* 49 (Suppl. 1):52–74. doi: 10.1111/anti.12187.

Castree, N. 2017. Anthropocene: Social science misconstrued. *Nature* 541 (7637):289. doi: 10.1038/541289c.

Castree, N., W. M. Adams, J. Barry, D. Brockington, B. Buscher, E. Corbera, D. Demeritt, R. Duffy, U. Felt, K. Neves, et al. 2014. Changing the intellectual climate. *Nature Climate Change* 4 (9):763–68. doi: 10.1038/nclimate2339.

Collard, R. C., J. Dempsey, and J. Sundberg. 2015. A manifesto for abundant futures. *Annals of the Association of American Geographers* 105 (2):322–30. doi: 10.1080/00045608.2014.973007.

Conway, P. R. 2019. The folds of coexistence: Towards a diplomatic political ontology, between difference and contradiction. *Theory Culture & Society* 37 (3):23–47. doi: 10.1177/0263276419885004.

Dakos, V., S. R. Carpenter, E. H. van Nes, and M. Scheffer. 2015. Resilience indicators: Prospects and limitations for early warnings of regime shifts. *Philosophical Transactions of the Royal Society B-Biological Sciences* 370 (1659):10. doi: 10.1098/rstb.2013.0263.

Davis, J., A. A. Moulton, L. Van Sant, and B. Williams. 2019. Anthropocene, Capitalocene, Plantationocene? A manifesto for ecological justice in an age of global crises. *Geography Compass* 13 (5):e12438–15. doi: 10.1111/gec3.12438.

de la Cadena, M., and M. Blaser. 2018. *A world of many worlds*. Durham, NC: Duke University Press.

Ellis, E. C. 2015. Ecology in an anthropogenic biosphere. *Ecological Monographs* 85 (3):287–331. doi: 10.1890/14-2274.1.

Fagan, M. 2019. On the dangers of an Anthropocene epoch: Geological time, political time and post-human politics. *Political Geography* 70:55–63. doi: 10.1016/j.polgeo.2019.01.008.

Graf, W. L. 1979. Catastrophe theory as a model for change in fluvial systems. In *Adjustments of the fluvial system*, ed. D. D. Rhodes and E. J. Williams, 13–32. London: Allen and Unwin.

Haraway, D. 2015. Anthropocene, Capitalocene, Plantationocene, Chthulucene: Making kin. *Environmental Humanities* 6 (1):159–65. doi: 10.1215/22011919-3615934.

Inkpen, R., and D. Petley. 2001. Fitness spaces and their potential for visualizing change in the physical landscape. *Area* 33 (3):242–51. doi: 10.1111/1475-4762.00028.

Jenny, J.-P., S. Koirala, I. Gregory-Eaves, P. Francus, C. Niemann, B. Ahrens, V. Brovkin, A. Baud, A. E. K. Ojala, A. Normandeau, et al. 2019. Human and climate global-scale imprint on sediment transfer during the Holocene. *Proceedings of the National Academy of Sciences* 116 (46):22972–76. doi: 10.1073/pnas.1908179116.

Kaika, M. 2018. Between the frog and the eagle: Claiming a "scholarship of presence" for the Anthropocene. *European Planning Studies* 26 (9): 1714–27. doi: 10.1080/09654313.2018.1484893.

Kallio, K. P. 2016. Living together in the topological home. *Space and Culture* 19 (4):373–89. doi: 10.1177/1206331216631290.

Kefi, S., V. Dakos, M. Scheffer, E. H. Van Nes, and M. Rietkerk. 2013. Early warning signals also precede non-catastrophic transitions. *Oikos* 122 (5):641–48. doi: 10.1111/j.1600-0706.2012.20838.x.

Laist, R. 2015. Why I identify as mammal. *New York Times*, October 24. Accessed October 24, 2019. https://opinionator.blogs.nytimes.com/2015/10/24/why-i-identify-as-mammal/.

Lewis, S. L., and M. A. Maslin. 2015. Defining the Anthropocene. *Nature* 519 (7542):171–80. doi: 10.1038/nature14258.

Lorimer, J. 2017. The Anthropo-scene: A guide for the perplexed. *Social Studies of Science* 47 (1):117–42. doi: 10.1177/0306312716671039.

Martin, L., and A. J. Secor. 2014. Towards a post-mathematical topology. *Progress in Human Geography* 38 (3):420–38. doi: 10.1177/0309132513508209.

Prager, S. D., and W. A. Reiners. 2009. Historical and emerging practices in ecological topology. *Ecological Complexity* 6 (2):160–71. doi: 10.1016/j.ecocom.2008.11.001.

Robbins, P. 2020. Is less more … or is more less? Scaling the political ecologies of the future. *Political Geography* 76:102018. doi: 10.1016/j.polgeo.2019.04.010.

Saldanha, A. 2019. A date with destiny: Racial capitalism and the beginnings of the Anthropocene. *Environment and Planning D-Society & Space* 38 (1):12–34. doi: 10.1177/0263775819871964.

Savransky, M. 2019. The pluralistic problematic: William James and the pragmatics of the pluriverse. *Theory, Culture & Society*. Advance online publication. doi: 10.1177/0263276419848030.

Scheffer, M., S. Carpenter, J. A. Foley, C. Folke, and B. Walker. 2001. Catastrophic shifts in ecosystems. *Nature* 413 (6856):591–96. doi: 10.1038/35098000.

Stallins, J. A. 2012. Scale, causality, and the new organism–environment interaction. *Geoforum* 43 (3):427–41. doi: 10.1016/j.geoforum.2011.10.011.

Steffen, W., J. Rockstrom, K. Richardson, T. M. Lenton, C. Folke, D. Liverman, C. P. Summerhayes, A. D. Barnosky, S. E. Cornell, M. Crucifix, et al. 2018. Trajectories of the Earth system in the Anthropocene. *Proceedings of the National Academy of Sciences* 115 (33):8252–59. doi: 10.1073/pnas.1810141115.

Stephens, L., D. Fuller, N. Boivin, T. Rick, N. Gauthier, A. Kay, B. Marwick, C. Geralda Armstrong, C. Michael Barton, T. Denham, et al. 2019. Archaeological assessment reveals Earth's early transformation through land use. *Science* 365 (6456):897–902. doi: 10.1126/science.aax1192.

Sullivan, A. P., D. W. Bird, and G. H. Perry. 2017. Human behaviour as a long-term ecological driver of non-human evolution. *Nature Ecology & Evolution* 1 (3):1–11. doi: 10.1038/s41559-016-0065.

Thorn, C. E., and M. R. Welford. 1994. The equilibrium concept in geomorphology. *Annals of the Association of American Geographers* 84 (4):666–96. doi: 10.1111/j.1467-8306.1994.tb01882.x.

Wells, J. 2018. Mind the gap: Bridging the two cultures with complex thought. *Ecological Complexity* 35:81–97. doi: 10.1016/j.ecocom.2017.11.001.

Zeeman, E. C. 1976. Catastrophe theory. *Scientific American* 234 (4):65–83. doi: 10.1038/scientificamerican0476-65.

The Geoethical Semiosis of the Anthropocene: The Peircean Triad for a Reconceptualization of the Relationship between Human Beings and Environment

Francesco De Pascale🆔 and Valeria Dattilo

This article seeks to reconcile, as well as operationalize, two different methodological approaches on the basis of some important basic affinities: geoethics and Peircean semiotics. For this purpose, Peirce's triangle is conceived as a "translator mechanism" to parse the human–planet relationship that cannot be dealt with through actions in pairs but must be considered as a triadic relationship in which geoethics comes into play to develop a new relationship between human beings and environment. Following this, the triangle heuristic will employ the vertices Geoethics–Illness of the Earth–Society as a metaphor of the Anthropocene era through the lens of Peircean semiotics. This triangle method will help investigate some research questions: (1) Is planet illness an icon, index, or a symbol of the negative impact of the society? (2) When do we encounter environmental phenomena constituting images of planet illness? (3) What is the salient perspective from which to study the phenomenon of the Anthropocene? In discussing these issues, the authors call into play the concept of noosphere and propose a new ethical framework guiding human behavior toward the environment. *Key Words: Anthropocene, geoethics, Hippocratic triangle, noosphere, Peirce's semiotics.*

This article proposes a nuanced, semiotic conception of the relationship between human beings and the environment. On the one hand, the analysis challenges a dualistic approach that considers nature as something autonomous or independent of human action (Vogel 2015, 2016). On the other, the article extends and nuances the argument that in the Anthropocene era, there are no longer examples of a completely separate and independent nature from humans (Chakrabarty 2012; Hamilton, Bonneuil, and Gemenne 2015; Latour 2015; Hamilton 2017). Rather, the article combines the perspectives offered by geoethics and semiotics to examine the processes of the Anthropocene era and offer an intermediate position that considers environment itself as a semiotic product. In the first move, the article draws on the heuristic of Peirce's triangle, a fundamental theory of semiotics. In the second move, the analysis turns to the geoethical paradigm as a possible framework for parsing the processes of the Anthropocene era and for instituting a planetary Hippocratic oath. Together, these analytical frameworks will guide the survey to answer the following research questions:

(1) Is planet illness an icon, index, or a symbol of the negative impact of the society? (2) When do we encounter environmental phenomena constituting images of planet illness? (3) What is the salient perspective from which to study the phenomenon of the Anthropocene?

The Anthropocene: An Introduction to the Concept in the Human and Social Sciences

The concept of the Anthropocene, currently controversial in geology due to issues concerning temporal and stratigraphic limits, has, in the human and social sciences, been a driving force for the development of new research paradigms that run parallel to the declensions of the posthuman, political ecology, and the environmental humanities, as well as a medium to strengthen the link between environmental research and sociopolitical commitment (Farrier 2019; Yusoff 2019). In addition, the term's reference to *anthropos*—that is, to human without distinction—has given rise to a new wave of reflections, theoretical constructions and deconstruction of the relationship between nature and culture, human and nonhuman, genders, cultures, and, more broadly, the alleged essence or authenticity of something like "human beings" (Bogues 2006; McKittrick 2015; Baranzoni, Lucci, and Vignola 2016). The Anthropocene, although not without its limitations, still remains a productive interdisciplinary tool, creating a meeting place between socioanthropological, historical–political, philosophical, and ethical perspectives on questions of environmental change, agency, and ethics. The term also captures and calls forth a cross-disciplinary approach—one that allows us to formulate a solid framework of ethical, social, and cultural values—useful to the geoscience community as well as humanities scholars. Responding to the complexity of the Anthropocene, as a critical term and a potential geological epoch, both research communities extend their pedagogical and research inquiry, their more traditional disciplinary foci, and integrate questions explored earlier by ethicists and philosophers with an analysis of issues that affect society as a whole. Within this wider framework, geoethics (Němec 2005; Wyss and Peppoloni 2014; Peppoloni, Bobrowsky, and Di Capua 2015; Nikitina 2016; Gundersen 2017; Peppoloni et al. 2017) is concerned with the ethical implications of the most relevant phenomena of our time, from climate change to environmental pollution and from the exploitation of resources to disasters caused by hazards and vulnerability (De Pascale and Dattilo 2019). Specifically, geoethics brings the social, human, and natural sciences together in different ways to confront current ecological crises from various ethical, cultural, philosophical, political, social, and biological perspectives (Oppermann and Iovino 2017). Putting the concept of noosphere in the Anthropocene era in dialogue with Peircean semiotics, as the article goes on to demonstrate, allows us to analyze the signs that represent the processes related to these crises and the wider relationship between ethics and semiotics in times of environmental crises.

Conceptual and Methodological Framework

In its broadest sense, semiotics studies signs both in the field of communication (signs intended as tools for communicating, for exchanging information) and in the field of knowledge (signs understood as mediators of reality that surrounds us; Eco 1975). In the humanities, semiotics, unlike linguistics, also studies nonlinguistic signs. This article, building on the distinction made by Eco (1975), focuses on general semiotics that takes into account the sign in the cognitive field, relating it to the notion of the Anthropocene. To this end, the article draws on Peirce's (1931–1935) notion of the sign. Peircean semiotic theory should not be understood simply as a theory of interpersonal communication, a limited and limiting view of his theory. Rather, Peirce applied semiotic categories typical of human beings to the whole universe. According to Peirce, the sign is a category that embraces the whole universe (Peirce 1931–1935). To be perceived as "real," extralinguistic objects must be seen as signs, and the concepts of objects must be considered semiotically. This leads to the assertion that environment and, in general, the universe are also semiotic products.

Within this much more inclusive definition (Peirce 1931–1935), the sign is not limited to the linguistic-communicative sphere only but must be considered as a cognitive tool. There are three elements that make up the sign. Translated into graphic terms, the structure that comes out of Pierce's vision of the sign is a triangle that requires the presence of three elements: *representamen*, or sign *tout court*, object, and interpretant, at the vertices of the triangle (Figure 1). The definition of

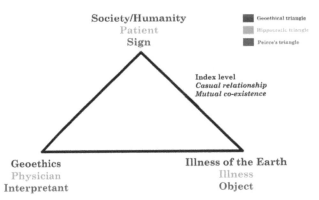

Figure 1. All of the elements that characterize the triangles of Peirce, Hippocrates, and geoethics. *Sign* or *representamen*, *object*, and *interpretant* are the elements of Peirce's triangle (red writing). *Physician*, *illness*, and *patient* are the factors affecting the Hippocratic triangle in medicine (green writing). *Geoethics*, *Illness of the Earth*, and *Society/Humanity* are the elements of the geoethical triangle proposed by the authors (blue writing). The causal relationship and the mutual coexistence between Society and Illness of the Earth correspond, in Peirce's triangle, to a causal relationship between sign and object (index level). Sources: Peirce (1931–1935, 1958), Matteucci et al. (2012), De Pascale and Dattilo (2019).

sign implies the beginning of the process of semiosis. Semiosis is defined as "an action, or influence, which is, or involves, a cooperation of three subjects, such as a sign, its object, and its interpretant, this tri-relative influence not being in any way resolvable into actions between pairs" (Peirce 1931–1935, 5.484). A sign can stand in the place of something else, in someone's eyes, only because this relationship is mediated by an interpretant, which is another sign that translates and explains the previous one. This process, which lasts indefinitely, is called "unlimited semiosis" (Eco 1975). It is an infinite series of interpretations that can have as its final interpretant what Peirce calls *habit*, or the point of arrival of unlimited semiosis. The starting point of semiosis (i.e., of the process of creation, of formation of meaning) is the external reality: a dynamic object. The sign (or *representamen*) refers (stands for) to a dynamic object. The immediate object does not correspond with the dynamic object (which is, instead, the object in itself, the real one). This happens because a sign (a *representamen*) always represents something from a certain point of view. The immediate object (content) is the way in which the dynamic object (real object) is focused (Proni 1990; Pisanty and Pellerey 2004). The interpretant is not the interpreter. The interpreter is the one who grasps the link between sign and object; the interpretant is a second sign that highlights in what sense it can be said that a certain sign refers to a given object.

For example, the word *table*, which is the sign or *representamen*, refers to the part that is meant; that is, the content (immediate object) that is "a piece of furniture consisting of a horizontal plane supported mostly by four legs". The dynamic object is the table intended as a real or extralinguistic object. The interpreter is the one (the individual) who grasps the relationship between the word *table* and the table intended as a real object. The interpretant is the idea of the individual that characterizes and motivates this relationship. This understanding depends on the idea in the interpreter's mind. To put it simply, a chef might interpret the word or sign table in a different way than a pupil at school.

Precisely because Peirce does not reduce semiotics to a theory of communicative acts, Peirce's triad can also be productively used to analyze environmental phenomena that have a human recipient, as, for example, in the case of climatic symptoms (Eco 1975) or the manifestations of anthropogenic climate change in the Anthropocene. What are these manifestations? Many researchers argue that the Anthropocene represents a new division of geological time, claiming that human activity, through its use of fossil fuels, has warmed the planet, raised sea levels, degraded the ozone layer, and acidified the oceans (Crutzen and Stoermer 2000; Crutzen 2002; Zalasiewicz et al. 2008; Bonneuil and Fressoz 2013; Monastersky 2015). This means that, to take just one of these examples, the degraded ozone layer signifies that something has happened. It then becomes necessary to understand what this is a sign of.

The relationship between human beings and the environment cannot be reduced to action between two elements but must be considered as a triadic relationship, one in which geoethics would come into play as *habit* (final logical interpretant) and as an attempt to develop a new relationship between human beings and the environment. Geoethics can be defined as the investigation and reflection on a human's operational behavior toward the geosphere (Peppoloni, Bobrowsky, and Di Capua 2015). As a field, geoethics draws from the work of Leopold, who, already in the 1940s, proclaimed the need to develop a new relationship between humans and the environment. Leopold identified the concept of "conservation" as an ethical criterion indicating "a state of harmony between human and the Earth" (Leopold 1949, 202). For Leopold, the premise of ethics is to forge an awareness that the individual is a member of a community of interdependent parts. Hence, for him, the role of *Homo sapiens* must change from conqueror of the Earth to a mere member and citizen of his community.

Equally, geoethics appeals to the noosphere, a concept proposed by Le Roy and de Chardin, then amplified by Vernadsky (1987), and seen as a third stage of development. Stage one is the physical geosphere of inanimate matter, stage two is the biosphere of animate or biological matter, and stage three is the noosphere, as proposed by Vernadsky. The term denotes a "commonsense" biosphere, meaning that the functioning, dynamics, and evolution of the Earth must largely be determined by reasonable human activity, an activity that focuses its efforts on creation rather than destruction (Vernadsky 1987). At the core of the concept of noosphere, however, is not the universality of the human species but the universality of common human faculties, such as intellect, language, and intelligence, as well as the ability to appeal to responsibility. It is precisely to these capacities and to the sense of responsibility that geoethics appeals in addressing the problems related to adaptation and mitigation of the effects of climate change. This appeal implies the need for reviewing our ways of "building our future" and taking care of the place where we dwell (De Pascale and Dattilo 2019; De Pascale and Roger 2020). Consequently, by combining Peircean semiotics and geoethical concerns, this article aims to propose a semiotic theory of the Anthropocene that opens up to a much wider considerations of the range of signs and possible explanations for attitudes toward and perceptions of the environment.

Geoethics and Semiotics for a Reconceptualization of the Human–Environment Relationship

The first step toward such a joint geoethical and semiotic approach to the Anthropocene and the questions of responsibility and agency that it calls forth involves a reconceptualization of the human–environment relationship. Inspired by the work of Matteucci et al. (2012), we propose a tripartite division of signs in a Peircean triangle with the respective vertices: Geoethics, Illness of the Earth, and Society/Humanity (Figure 1). In contrast to the work of these scholars, who attribute ethical obligations solely to the field of geology (Figure 1), however, we contend that geoethical responsibility needs to be represented by a synthesis between geology, geography, and philosophy. By first extending this ethical obligation to the field of geoethics and combining it with a semiotic approach, in the second move we can consider the meaning and function of this geoethical triangle and the analogies with the Hippocratic triangle.

According to Peirce, mental images correspond to the iconic level, where the relationship between sign and object is based on similarity (Farinelli 2004). When, instead, something is said to be the index of something else, it means that something is related to something else from a causal point of view. The third level of interpretation coincides, for Peirce, with the symbolic one. The symbol is linked to the

object by virtue of a shared convention of a recognized mental association. To explain this better, we can give some examples. A photograph of a table is not a table but an icon, because it resembles the table. Written numbers (e.g., 5) are symbols, because there is nothing inherent in that figure that represents five of anything; it is just a convention that we all follow. Natural signs are part of the index category and are the footprints in the sand, the smoke that rises from the fire. These signs communicate by cause–effect connection (i.e., the lightning is the index of a possible storm).

This definition of the sign helps us understand how a certain number of environmental phenomena and processes, including climate change, constitute a historical sign that stands for planet illness and the impacts that global societies have on the planet. Peirce's statements, if analyzed from an ecological perspective, lead to a confirmation that such a relationship cannot be an icon or symbol of the impacts that such a society has on the Earth. Therefore, disasters are an immediate effect, an index's sign, of a society's negative impact (e.g., increasing environmental pollution, building abuses, overexploitation of mineral resources, etc.; De Pascale and Dattilo 2019). In this context, the Anthropocene geological era placed greater emphasis on the incidence and direct responsibility of human (social, political, and economic) factors in catastrophic events and in the changes in the dynamics of the Earth system. Thus, a semiotic link of indical type is established, based on the consequences and coexistences between the signs Society and Illness of the Earth.

Geoethics and semiotics, seen in an interdisciplinary perspective, envision a radical and strategic ecology, capable of mirroring another conceivable world. Geoethics, as an emergent discipline, can help us reflect on planetary illness, a discipline that can contribute to the construction of an ethical and social knowledge, strengthening the idea that planetary health is a common heritage to be shared. Furthermore, this field of inquiry can foster a culturally sensitive renewal and way of relating to the planet, through a growing sensitivity to the defense of human life and health and wealth of all Earth systems. In particular, geoethics and semiotics can provide useful indications to develop new behavioral responses, such as the ethical and ecological reorganization of the economy, politics, and society to address the problems, processes, and threats manifested by the Anthropocene.

From the Hippocratic Triangle to the Geoethical Triangle

Building on these respective analytical contributions of geoethics and semiotics to the thinking of justice and responsibility in the Anthropocene, our proposed triad considers the existence of a relationship and the mutual implications between Geoethics, Society, and the Illness of the Earth and possesses an analogy with the Hippocratic oath triangle. The Hippocratic triangle represents the vision of an important figure in the history of medicine, Hippocrates of Kos (c. 460 BC–c. 370 BC), according to which the doctor is at the service of the art and science of medicine and, conversely, the practice of the patient helping the doctor fight his or her illness. In fact, art is made up of three terms: doctor, patient, and illness (Hippocrates 1657). Although this has been the basis of the medical profession, in recent times this relationship has become more complex due to the increased acknowledgment of the role of the community and, more generally, society within this physician–illness–patient triangle (Matteucci et al. 2012). From a semiotic point of view, the doctor looks at the patient (a person who is ill) and sees the symptoms of an illness. In other words, the symptoms are the signs that indicate the illness. The doctor understands the illness via these signs and can then work directly on the illness.

By analogy, the triangle of Hippocrates, physician–illness–patient (Figure 1) corresponds to the relationships between Geoethics, Illness of the Earth, and the Society (Figure 1), where the Illness of the Earth is represented by the extreme natural events that are actually or potentially harmful to humans and nonhumans and by the equally harmful effects of the impacts of humans on our planet. Geoethics is the lens that we can use to read the signs, just as the doctor uses her or his knowledge of medicine to read the signs of the symptoms in the patient. Therefore, the intervention of the geologist or geographer and also of the philosopher on issues concerning Illness of the Earth can be seen as parallel to the role of the doctor toward the patient and, more generally, toward the health of the population.

The relationship between Society and the Illness of the Earth, for its part, visible in the proposed triangle (Figure 1), would correspond, in Peirce's triangle (Figure 1), to a causal relationship between sign and object. The consequences in terms of human life or material damage to the society are an indicator of Illness of the Earth. The latter, though, is also an indication of the negative impact of the society. So, in response to the first research question (Is planet illness an icon, index, or a symbol of the negative impact of the society?), planet illness can be seen as an index of the negative impact of society. Specifically, Illness of the Earth is no longer simply the immediate object within Peirce's triangle but a dynamic object that is knowable through the unlimited semiosis and the continuous change of subject and habit, intended as regular dispositions to act (De Pascale and Dattilo 2019). To be precise, this is a relationship of mutual implications (and not of univocal consequentiality) between humanity or society and nature or planet. Therefore, the effect affects the cause and, in turn, becomes the cause itself. Consequently, we can answer the second research question: When do we encounter environmental phenomena constituting images of planet illness? For example, human CO_2 emissions (fossil fuels, represented in the geoethical triangle by Society because they are an inherent product of the society) are a cause of global warming (Illness of the Earth), but global warming itself has obvious repercussions and effects on human existence (e.g., with the occurrence of disasters related to hurricanes, tropical storms, floods, heat waves, and droughts).

Moreover, this analysis of images of planet illness allows us also to consider the relationship between risks, vulnerability, and disasters, with a number of scholars arguing that society is capable of transforming the effects of an extreme event into disaster (e.g., Alexander 1993, 2013; Gaillard 2007; Weichselgartner and Kelman 2015). At first glance, the use of the word *disaster* would be symptomatic of the link of univocal consequentiality that is identified between natural hazards and damage to the anthropic component. Natural hazards can be defined as "those elements in the physical environment, harmful to human beings and caused by forces extraneous to them" (Burton and Kate 1964, 413; Butler 1988, 212). Hurricanes, landslides, avalanches, droughts, and floods are hazards that take the name and the representation of disasters whenever they have a strong impact on humans and their activities, capable of causing a radical change in the previous setup. The term *disaster* would then be connected to human activity, because it is linked to the consequences in terms of human lives or material damage suffered by the communities following the occurrence of calamitous events (De Pascale and Dattilo 2019). For example, an earthquake in a desert or a hurricane in the open sea, without the presence of human beings, cannot be considered a disaster. In this first case, society is positioned as a passive actor that undergoes the effects of natural hazards (planet illness).

Disasters, however, could also be seen as manifestations of Society/Humanity within this triangle, with this repositioning placing emphasis on the processual and constructed, and oft sustained, nature of vulnerability. Disasters are caused by vulnerability and result from an interaction between a hazard (e.g., an earthquake, tsunami, hurricane, landslide, or flood) and a vulnerable community. According to Kelman et al. (2016), "Vulnerability refers to the propensity to be harmed, in this case by a hazard, and to be unable to deal with that harm alongside the social processes creating and maintaining that propensity" (129). Vulnerability is "a manifestation of uneven connectedness and a direct expression of a wider, not always immediately visible and potentially catastrophic, susceptibility to environmental hazards and phenomena" (Mika 2019, 12–13). In sum, interpreting disasters as part of the Society/Humanity element of the triangle shifts attention to society and human actions and behaviors that, in an active way, are responsible for creating social vulnerability and turning an extreme event into disaster.

Therefore, in the geoethical–semiotic approach on which the aforementioned triangle is based, a relationship of mutual co-implication and the responsibility of anthropic factors in calamitous events clearly emerges. Their effects, then, have repercussions on humanity itself, establishing thus a link of consequentiality and mutual implication between Society and Illness of the Earth. The interpretant in the proposed triangle is the field of geoethics, which focuses important responsibilities on society, from which, in turn, the ethical importance of actions is derived, similar to the way the medical mind of the doctor focuses on the patient to read the signs that lead to an understanding of her or his illness. Also, the importance of terminology and the urgency of the scholarly work of geoethics are made clear with the scholar's obligation to propose and denounce wrong actions and behaviors. Hence, the parallelism between the Hippocratic obligations of doctors toward patients and that of geologists, geographers, and philosophers toward the Earth and society.

Conclusions

Environment is not intended as a formless material, as an aesthetic and ethical object of contemplation, but as part of us, as part of humanity. Unlike the Pleistocene, Olocene, and all previous geological epochs, the Anthropocene is characterized primarily by the impact of humans on the dynamics of the planet. To such an extent, it is marked by the role of human species, as a decisive factor in the balance between the sustainability of the planet and climate change. The purpose of this article was to analyze the relationship between humanity and the planet through the lens of geoethics and semiotics of Peircean tradition: two different approaches with conceptual affinities. To this end, one of the key concepts of semiotics, Peirce's triangle, has been engaged as a "translator mechanism" to create, along with the concept of the noosphere, an ethics of responsibility and care for the Earth. This relationship, in our opinion, cannot be resolved in binary actions but through concepts such as the Peircean triadic paradigm. In addressing the third and final research question— What is the salient perspective from which to study the phenomenon of the Anthropocene?—our answer is from the presence of humanity on the Earth; in other words, from the vantage point of human nature, at the intersections where humans and environment meet and find confluence. It is only by starting from a definition of human nature that is no longer understood as a dichotomic relationship but as two global entities involved in a symbiotic relationship of reciprocal implications that it is possible, as we have argued, to infer positive action and an ethics of care. It would, in fact, be a curious setback for those who deal with the so-called hard sciences not to take into account the contribution of human and social sciences in the interpretation and study of the threats and processes of the Anthropocene (Holm and Travis 2017). This does not mean considering all environmental phenomena as anthropogenic events. It would mean acknowledging, however, in Peirce's (1931–1935) words, that "every scientific explanation of a natural phenomenon is a hypothesis that there is something in nature to which the human reason is analogous; and that it really is so all the successes of science in its applications to human convenience are witnesses" (1.316).

Finally, the processes of the present era can be addressed by a new geoethical semiosis of the Anthropocene, in which the semiotic triad is represented by Humanity as the sign, Illness of the Earth as the object in question, and Geoethics in the role of interpretant. Peirce's triangle can be employed here as a means to help understanding, as a heuristic device (a simple tool to help us understand complex situations). Geoethics, in dialogue with semiotics, can work toward the necessary dialogue between human and natural sciences and their respective attempts to respond to the times of the Anthropocene. In effect, geoethics also faces the responsibility and the challenge of transferring effective information and adequate education to deal with global environmental challenges. In so doing, geoethics can equally bring together those working in hard and social sciences in concert with humanities scholars, all of whom have an ethical obligation to be at the forefront of the battle against global warming and other manifestations of negative anthropogenic impact on the Earth.

Acknowledgments

The authors thank Dr. Kasia Mika, Dr. Ian Robinson, and Dr. Charles Travis for the language review. The research article is the result of work shared by both authors. Specifically, Valeria Dattilo is the author of the first two sections and Francesco De Pascale is the author of the next three sections. Conclusions are attributable to both authors.

ORCID

Francesco De Pascale ⓘD http://orcid.org/0000-0002-3845-3422

References

Alexander, D. E. 1993. *Natural disasters*. London: UCL Press.

Alexander, D. E. 2013. Resilience and disaster risk reduction: An etymological journey. *Natural Hazards and Earth System Sciences* 13 (11):2707–16. doi:10.5194/nhess-13-2707-2013.

Baranzoni, S., A. Lucci, and P. Vignola. 2016. L'Antropocene: Fine, medium o sintomo dell'uomo? *Lo Sguardo—Rivista di Filosofia* 22 (3):5–9.

Bogues, A., ed. 2006. *After man, towards the human: Critical essays on Sylvia Wynter*. London: Global Distributor.

Bonneuil, C., and J. Fressoz. 2013. *L'événement Anthropocène. La Terre, l'histoire et nous*. Paris: Éditions du Seuil.

Burton, I., and R. W. Kate. 1964. The perception of natural hazards in resource management. *Natural Resources Journal* 3:412–41.

Butler, D. R. 1988. Teaching natural hazards: The use of snow avalanches in demonstrating and addressing geographic topics and principles. *Journal of Geography* 87 (6):212–25. doi: 10.1080/00221348808979790.

Chakrabarty, D. 2012. Postcolonial studies and the challenge of climate change. *New Literary History* 43 (1):1–18. doi:10.1353/nlh.2012.0007.

Crutzen, P. J. 2002. Geology of mankind. *Nature* 415 (6867):23. doi: 10.1038/415023a.

Crutzen, P. J., and E. F. Stoermer. 2000. The Anthropocene. *IGBP Newsletter* 41:17–18.

De Pascale, F., and V. Dattilo. 2019. La sémiosis de l'Anthropocène: Pour une réinterprétation de la relation entre l'homme et la nature par le biais de la géoéthique. *Rivista Geografica Italiana* 126 (2):23–40. doi:10.3280/RGI2019-002002.

De Pascale, F., and J. C. Roger. 2020. Coronavirus: An Anthropocene's hybrid? The need for a geoethic perspective for the future of the Earth. *AIMS Geosciences* 6 (1):131–34. doi:10.3934/geosci.2020008.

Eco, U. 1975. *Trattato di Semiotica Generale*. Milano, Italy: Bompiani.

Farinelli, F. 2004. *Geografia: Un'introduzione ai modelli del mondo*. Torino, Italy: Einaudi.

Farrier, D. 2019. *Anthropocene poetics: Deep time, sacrifice zones, and extinction*. Minneapolis: University of Minnesota Press.

Gaillard, J. C. 2007. Resilience of traditional societies in facing natural hazards. *Disaster Prevention and Management: An International Journal* 16 (4):522–44. doi: 10.1108/09653560710817011.

Gundersen, L. C., ed. 2017. *Scientific integrity and ethics in the geosciences*. Hoboken, New Jersey: Wiley, American Geophysical Union.

Hamilton, C. 2017. *Defiant Earth: The fate of humans in the Anthropocene*. Cambridge, MA: Polity.

Hamilton, C., C. Bonneuil, and F. Gemenne, eds. 2015. *The Anthropocene and the global environmental crisis: Rethinking modernity in a new epoch*. London and New York: Routledge.

Hippocrates. 1657. *Opere*. Genevae: Typis & sumptibus Samuelis Chouët.

Holm, P., and C. Travis. 2017. The new human condition and climate change: Humanities and social science perceptions of threat. *Global and Planetary Change* 156:112–14. doi:10.1016/j.gloplacha.2017.08.013.

Kelman, I., J. C. Gaillard, J. Lewis, and J. Mercer. 2016. Learning from the history of disaster vulnerability and resilience research and practice for climate change. *Natural Hazards* 82 (Suppl. 1):129–43. doi: 10.1007/s11069-016-2294-0.

Latour, B. 2015. *Face à Gaïa: Huit conférences sur le nouveau régime climatique. Les Empêcheurs de penser en rond*. Paris: La Découverte.

Leopold, A. 1949. *A Sand County almanac*. New York: Oxford University Press.

Matteucci, R., G. Gosso, S. Peppoloni, S. Piacente, and J. Wasowski. 2012. A Hippocratic oath for geologists? *Annals of Geophysics* 55 (3):365–69. doi: 10.4401/ag-5650.

McKittrick, K. 2015. *Sylvia Wynter: On being human as praxis*. Durham, NC: Duke University Press.

Mika, K. 2019. *Disasters, vulnerability, and narratives: Writing Haiti's futures*. London and New York: Routledge.

Monastersky, R. 2015. First atomic blast proposed as start of Anthropocene. *Nature*. Accessed October 20, 2019. http://www.nature.com/news/first-atomic-blast-proposed-as-start-of-anthropocene-1.16739.

Němec, V. 2005. *Developing geoethics as a new discipline*. Accessed October 15, 2019. http://www.bgs.ac.uk/agid/Downloads/VN05Geoethics.pdf.

Nikitina, N. K. 2016. *Geoethics: Theory, principles, problems*. 2nd ed. Moscow, Russia: Geoinformmark.

Oppermann, S., and S. Iovino. 2017. The environmental humanities and the challenges of the Anthropocene. In *Environmental humanities: Voices from the Anthropocene*, ed. S. Oppermann and S. Iovino, 1–21. London: Rowman & Littlefield.

Peirce, C. S. 1931–1935. *Collected papers of Charles Sanders Peirce*. Vols. 1–6. Cambridge, MA: Belknap Press of Harvard University Press.

Peirce, C. S. 1958. *Collected papers of Charles Sanders Peirce*. Vols. 7–8. Cambridge, MA: Belknap Press of Harvard University Press.

Peppoloni, S., P. Bobrowsky, and G. Di Capua. 2015. Geoethics: A challenge for research integrity in geosciences. In *Integrity in the global research arena*, ed. N. Steneck, M. Anderson, S. Kleinert, and T. Mayer, 287–94. Singapore: World Scientific.

Peppoloni, S., G. Di Capua, P. Bobrowsky, and V. Cronin, eds. 2017. Geoethics at the heart of all geoscience. *Annals of Geophysics* 60.

Pisanty, V., and R. Pellerey. 2004. *Semiotica e interpretazione*. Milan, Italy: Bompiani.

Proni, G. 1990. *Introduzione a Peirce*. Milan, Italy: Bompiani.

Vernadsky, V. I. 1987. *Filosofskie Mysli Naturalista* [Philosophical ideas of a naturalist]. Moscow: Nauka.

Vogel, S. 2015. *Thinking like a mall: Environmental philosophy after the end of nature*. Cambridge, MA: The MIT Press.

Vogel, S. 2016. "Nature" and the (Built) Environment. In *The Oxford Hanbook of Environmental Political Theory*, eds. T. Gabrielson, C. Hall, J.M. Meyer, D. Schlosberg, Oxford, Oxford University Press, 149–59.

Weichselgartner, J., and I. Kelman. 2015. Geographies of resilience: Challenges and opportunities of a descriptive concept. *Progress in Human Geography* 39 (3):249–67. doi:10.1177/0309132513518834.

Wyss, M., and S. Peppoloni, eds. 2014. *Geoethics, ethical challenges and case studies in Earth sciences*. Amsterdam: Elsevier.

Yusoff, K. 2019. *A billion black Anthropocenes or none*. Minneapolis: University of Minnesota Press.

Zalasiewicz, J., M. Williams, A. Smith, T. L. Barry, A. L. Coe, P. R. Bown, P. Brenchley, D. Cantrill, A. Gale, P. Gibbard, et al. 2008. Are we now living in the Anthropocene? *GSA Today* 18 (2):4–8. doi:10.1130/GSAT01802A.1.

Placing the *Anthropos* in Anthropocene

Jeffrey Hoelle and Nicholas C. Kawa

In this article, we review the place of "the human" in influential approaches to the Anthropocene to expose the diverse conceptualizations of humanity and human futures. First, we synthesize current research on humans as landscape modifiers across space and time, making a key distinction between the "old Anthropocene" (beginning with human food production) and the "new Anthropocene" (coinciding with the start of the Industrial Revolution). Second, we engage critical perspectives on the structuring effects of capitalist and colonialist systems—now periodized as the Capitalocene and Plantationocene, respectively—that have driven environmental degradation and human inequality over the past half-millennium. In the third section, we introduce alternative perspectives from anthropological and ethnographic research that confront the socioecological disruptions of capitalism and colonialism, drawing on indigenous Amazonian perspectives that have a more capacious understanding of the human—including species other than *Homo sapiens*. Finally, to conclude, we extend our analysis to a broader suite of visions for building socially and environmentally just futures captured in the framework of the pluriverse, which stands in strong contrast with the techno-modernist aspirations for the next stage in which humans become separated from Earth, in space. In recognizing these varied understandings of humanity, we hope to call attention to the diverse possibilities for human futures beyond the Anthropocene. *Key Words: Anthropocene, Capitalocene, human–environment interactions, Plantationocene, pluriverse.*

The Anthropocene has proven to be a useful—albeit controversial—concept for recognizing cumulative human impacts on the Earth's systems and for generating robust discussion regarding pending environmental collapse. The ongoing debates about the origins and causes of the Anthropocene also have important implications for how we address this crisis. Given the centrality of *anthropos*—the human—to the issue and the solution, we offer a focus on humans as agents of environmental change. By looking across time and space, we aim to question and scrutinize human–environment relations and associated structures, ideologies, events, and technologies that have contributed either directly or indirectly to the recognition of the Anthropocene. Through cross-cultural and cross-disciplinary examination, we also draw attention to the different ways in which humans have conceived of human–environment relationships, including the concept of nature and the place of nonhuman actors in our socioecological relations. Where and how we place the *anthropos* in the Anthropocene has implications for more than just scholarly debates or our understanding of human–environment relations over time. It also has potential consequences for how we

collectively imagine the human place on the planet, who gets counted under that umbrella of humanity, and how that vison should dictate the future of our socioecological relations on this planet and beyond.

This article examines the human through a variety of social scientific, humanistic, and interdisciplinary frameworks to offer a deeper understanding of just what is meant when we speak of the "age of humans," or the Anthropocene. As anthropologists, we center humans in the study of human–environment relations and environmental impacts across time and space and follow other critical scholars in our examination how humans are conceptualized in scholarly and political debates about developmental and environmental futures. The article has four main sections, each focused on human–environment relations and the systems in which they are embedded at distinct points in human history that contribute to or emerge from the Anthropocene. First, we draw on diverse disciplinary literatures to assess human management and landscape modification over time, identifying an "old Anthropocene" popularized in archaeology that stands in contrast to a "new Anthropocene" that is associated with the onset of the Industrial Revolution and modern technological expansion. We then delve into critical engagement with the Anthropocene from scholars who approach this new geological epoch through the structuring effects of global capitalism and colonialism over the past 500 years—now referred to as the Capitalocene and Plantationocene, respectively. In the third section, we begin to rethink the *anthropos* in the Anthropocene by drawing attention to how Amazonian indigenous perspectives fundamentally question humanity as a condition unique to *Homo sapiens*. Then, to conclude, we consider alternatives to capitalist and ecomodernist futures of ever-expanding economic growth and technological "progress" that eventually extend human life beyond the bounds of Earth. Recognizing that the human is many things across time and that more just and equitable futures are not only possible but necessary, we close with an examination of how decolonizing practices in the present offer a vision for a future world "in which many worlds fit" (Kothari et al. 2019, xxvii; see also Marcos 2002). In other words, by exposing the diverse conceptualizations of humanity, we highlight the diverse possibilities for human futures beyond the Anthropocene.

The Old and New Anthropocene

The concept of the Anthropocene is rooted in the simple notion that humans have fundamentally altered the planet. Since the concept was first introduced by Crutzen and Stoermer (2000), wide-ranging debate has opened regarding the origins of the Anthropocene and the precise activities and behaviors responsible for this planetary transformation. To simplify these scholarly debates, we contend that there are two different visions of this geological epoch's origins presented by scholars working in such diverse fields as geology, geography, history, and archaeology, among many others. In the most basic terms, there is an old Anthropocene and there is a new Anthropocene. The old Anthropocene is linked to the earliest forms of human landscape modification—from the manipulation of fire and early food production strategies to the development of agriculture (Glikson 2013; Stephens et al. 2019). The new Anthropocene, on the other hand, is squarely placed in the modern industrial era. Several of the advocates of the new Anthropocene see its origin at the dawn of industrialization (e.g., Steffen,

Crutzen, and McNeill 2007; Ellis et al. 2010), but others, like the Anthropocene Working Group (AWG), pin it to nuclear bomb testing in the 1950s (Zalasiewicz et al. 2015; Carrington 2016). Despite the differences between them (and variation within them), both the old and new Anthropocene share a recognition of humans as landscape managers and modifiers par excellence. What makes these two visions of the Anthropocene distinct are the forms of human impact on the planet deemed to be significant, as well as the different types of arguments and evidence they employ for delineating this new geological epoch.

When the Anthropocene began to gain steam as a concept at the beginning of the twenty-first century, it was first linked to the rise of European industrialization. Crutzen and Stoermer (2000) saw that changes to Earth's climate beginning with the accelerated release of greenhouse gases followed the onset of the industrial period around 1850. Soon after the term appeared in print, other scholars began to question this origin story of the (European) industrial Anthropocene. Notably, environmental scientist Ruddiman (2003) responded with the "early-anthropogenic hypothesis" that argued human alteration of the planet could be witnessed thousands of years ago with the start of crop and livestock domestication and the beginnings of agriculture—sometimes known as the Neolithic revolution (see also Ruddiman 2013). Ruddiman's model drew on both archaeological and climatological data to make its case. Specifically, he argued that atmospheric CO_2 began an anomalous increase 8,000 years ago coinciding with forest clearance in Eurasia resulting from early agriculture. A similar trend for atmospheric methane (CH_4) could be found around 5,000 years ago, which Ruddiman linked to the expansion of rice irrigation in Asia. In sum, Ruddiman contended that the origins of agriculture and the origins of the Anthropocene were one and the same.

Researchers in archaeology, in particular, began to support this view of an older Anthropocene, offering additional forms of evidence. Not only did forest clearing resulting from agriculture have impacts on the climate thousands of years ago but the growth of sedentary societies and the concentrated deposition of organic wastes ("middening") led to the formation of anthropogenic soils, which could serve as the "golden spikes" or markers of the Anthropocene (Certini and Scalenghe 2011). Even places like Amazonia, where domestication and agriculture took on different forms, offer evidence to support the early Anthropocene model, including the presence of anthropogenic forests, anthropogenic mounds, raised agricultural fields, and football field–sized geoglyphs (Schaan 2010; Levis et al. 2012; Watling et al. 2017). A recent synthesis of research by more than 250 archaeologists also supports this old Anthropocene model (Stephens et al. 2019). Although there is considerable variation in the time frame in which different world regions were altered by human food production, this study asserts that by 3,000 years ago, most of the planet was already transformed by hunter-gatherers, farmers, and pastoralists.

Despite significant support for the old Anthropocene model, it has led to deeper consideration of the scale and extent of human modification of the environment across time as well as the temporal variability of human impacts. Models of the new Anthropocene have identified geological signatures linked to anthropogenic activity that can be found in much greater ubiquity, albeit in thinner slices of time. These include everything from the remains of the modern broiler chicken (Bennett et al. 2018) to industrially produced microplastics that now blanket Earth and even can be found in deep ocean trenches (Zalasiewicz et al. 2016; see also Williams et al. 2016). The AWG, however, has argued that the sharpest of these signals comes from artificial radionuclides that spread globally via nuclear bomb testing in the early 1950s (see Zalasiewicz et al. 2015). It is this date—1950 A.D. specifically—that the AWG will submit to the International Commission on Stratigraphy in a formal proposal by 2021. In the meantime, there is still much to debate regarding the underlying behaviors, systems, and even ideologies driving current planetary changes.

The Capitalocene and Plantationocene

The questions about the origins of the Anthropocene are important for how we understand the problem as well as theorize and enact potential solutions. Although it is certain that the effects of the Anthropocene are evident in a number of indicators associated with the Great Acceleration in the mid-twentieth century (Steffen et al. 2011), the concept does little to explain what led us to this point. Critical social scientists in the fields of anthropology, geography, and sociology, in particular, draw attention to all that is obscured and erased by a term that

designates a generalized humanity as universally responsible for the present ecological crisis (Malm and Hornborg 2014; Hornborg 2017). Two of the most influential reconceptualizations of the Anthropocene—the Capitalocene and the Plantationocene—argue for attention to the enjoined systems that began to shape the world around 500 years ago: global capitalism and colonialism. Although these alternative proposals have different foci and arguments, they share an interest in how humans are differentially situated within broader political–economic systems undergirded by power-laden hierarchies that drive human social inequality and environmental destruction

Moore (2017), one of the principal proponents of the Capitalocene concept, critiqued the attribution of environmental change to the "human enterprise" as a "mighty, largely homogeneous, acting unit" (596). According to Moore (2017), early capitalism created patterns of power, capital, and nature that laid the groundwork for the commonly understood origins of the Anthropocene. Malm (2016), in *Fossil Capital: The Rise of Steam Power and the Roots of Global Warming*, similarly argued against the myth that humans are predestined to degrade the environment, because such a narrative ignores the structuring effects of capitalism.

The development of capitalism is intimately linked with colonialism, and a specific suite of socioecological relationships, beginning roughly in the mid-fifteenth century. In Haraway et al. (2016), another increasingly influential term was coined: the Plantationocene. The Plantationocene emphasizes the plantation as a central analytic for understanding the rationalized production system that requires the simultaneous exploitation of nature and human labor (Mintz 1986; McKittrick 2011; Li 2018; Paredes 2020).

The global environmental implications of colonialism and the emerging global world system can also be seen in the geological record. According to the Orbis Spike, global declines in atmospheric carbon dioxide between 1570 and 1620 A.D. were the result of massive Native American population declines following the spread of disease and violence under European colonialism (Lewis and Maslin 2015). For this reason, H. Davis and Todd (2017) argued that the starting date of the Anthropocene should coincide with the colonization of the Americas due to the reverberating effects of dispossession and genocide of Native peoples as well as the endurance of colonial ecocidal regimes.

As noted earlier, scholars who have proposed the ideas of the Capitalocene and Plantationocene assert that it is not necessarily all humans who are responsible for widespread ecological degradation on the planet but rather the capitalist system and global elites invested in the most rapacious forms of natural resource extraction and agro-industrial production. The Capitalocene and the Plantationocene emphasize how the diffuse, contextual relations between capitalism and assemblages of humans and nonhumans work together to produce environmental destruction. Both perspectives seek to bridge the divide and confront the ideologies that place *Homo sapiens* in a privileged position over other organisms, seeking to reinsert humanity back into the web of life (Moore 2017) and "make kin" across species lines (Haraway 2015).

The field of black geographies has productively examined how the Anthropocene and its derivatives like the Plantationocene have overlooked the legacies of racial politics and white supremacy inherent in plantation ecologies (J. Davis et al. 2019), articulating with research on the racialized geographies that persist and are obscured by the Anthropocene's generalized discourses about humanity (Pulido 2018; Whyte 2018; Yusoff 2018; Resnick forthcoming). Perspectives from ecofeminism, environmental racism, and environmental justice argue that reconnection requires confronting the linkages between capitalism, racism, and sexism that result in environmental degradation and disproportionately affect the poor, women, and people of color (Pellow 2007; Shiva 2016). Crucial to achieving socioenvironmental transformations is critical analysis of the factors directly affecting vulnerable populations but also the structures that perpetuate the parallel forms of domination against nature and humans, such as with colonial drives to "domesticate" nature and racialized subjects (Hage 2017) and the role of scientific rationalism in the exploitation of women and nature (Merchant 1980). As such, socially and environmentally just solutions are only achieved by diversifying and decolonizing anthropocentric hierarchies between human and nonhuman while also demanding recognition that environmental struggles are inextricably linked with social struggles for equality.

Rethinking "the Human" in the Age of Humans

As greenhouse gases accumulate in the Earth's atmosphere, industrial plastics become ubiquitous in the oceans, and our discards begin to form distinctive stratigraphic layers, we are coming to grips with the fact that we cannot so easily separate ourselves from our environs, much less exert full control over them. The Anthropocene thus presents a fundamental paradox—with the increased recognition of humanity's capacity to alter the environment, the separation between the human and nonhuman has grown increasingly fuzzy, and it is unclear who or what is really in control. This is why Lorimer (2012) argued that the Anthropocene essentially represents the nail in the coffin for the modern dichotomy between nature and culture. The question now is this: How might we—particularly in so-called Western industrial societies—think differently about our relations with the world around us? Perhaps even more important, how might this help us rethink the basic condition of humanity, or *anthropos*?

As the Anthropocene has gained greater recognition across scholarly disciplines, social scientists have grappled with the fact that "the human" is being drawn to the center of the perceived environmental crisis. In the process, it has actually prompted new methodological experiments and forms of theorization that attempt to "decenter" the human and draw other beings and entities into social scientific and humanistic analyses. Multispecies ethnography is one example of this shift that has been taken up by scholars in the environmental social sciences and humanities, which seeks to contextualize human lives within wider networks of relations with different organisms and nonhuman (or "more-than-human") others (Kirksey and Helmreich 2010; Haraway 2013; Van Dooren 2014; Tsing 2015).

In an even more radical rethinking of "the human" at the center of the Anthropocene, Viveiros de Castro (2017) called attention to the limited scope of European-American anthropocentrism. His concern was that it associates humanity with just one species alone: *Homo sapiens*. Although scholarly Western Enlightenment thinking promulgates this view in a manner that often goes unquestioned, Viveiros de Castro reminds us that many people understand that *Homo sapiens* is not the only species that is human. In the day-to-day lives of peoples across the world, particularly indigenous peoples, it is apparent that humanity is a shared quality, not an exclusionary one. As Viveiros de Castro (2017) remarked in his treatise *Cannibal Metaphysics*, "When everything is human, the human becomes a wholly other thing" (63).

Many anthropologists have noted that Amazonian indigenous peoples acknowledge diverse beings in the world as persons with subjective agencies, "each endowed with the same generic type of soul [or], same set of cognitive and volitional capacities" that allow them to see themselves as human (Viveiros de Castro 2004, 6; see also Vilaça 2005; Fausto 2008). Although humans might perceive other living forms as animals or plants or spirits, the framework of Amerindian perspectivism suggests that perception is borne out of bodily difference and positionality in intersubjective relations.

What, then, might an expansion beyond Western Enlightenment ideas about the human—or even simply personhood—do for Anthropocenic politics? What futures can be conjured when, as Wynter (2003) showed us, we challenge the overrepresentation of Western bourgeois "man" in scholarly thinking about human existence and human freedoms? It seems evident in this time of ecological crisis that rather than cling to a European–North American *Homo sapiens*–centered view of the world, new perspectives are very much needed (Krenak 2020). In *Human, All Too Human*, Nietzsche (1910) wrote, "Most people are far too much occupied with themselves to be malicious" (88). Our fear is that he was very much wrong. There is a subtle maliciousness found in the disregard for others and their place on the planet, and it is embedded in our very limited notion of who counts as human, whose lives matter, and whose lives are treated as dispensable.

Two Visions of Human Futures: The Pluriverse and the Space Age

By questioning the human, we are not arguing for one specific or monolithic vision of humanity. Instead, we insist that *anthropos* and its diverse forms should not be assumed or taken for granted. By considering different dimensions of humanity and the different possibilities of its constitution—including the notion that humanity is not synonymous with *Homo sapiens*—this can invite deeper engagement with two very distinct visions of human futures: an earthly pluriverse, or an escape to space.

In recent years, critical analyses have been combined with scholar-activist proposals for alternative socioecological futures, such as in the 2015 special issue of this journal (Braun 2015). Although there are many allied terms and approaches, here we discuss these under the umbrella of the "pluriverse" (Escobar 2018; Kothari et al. 2019). A number of social scientists have adopted this notion, but it is perhaps best encapsulated by the Zapatistas, who have argued that they are working toward "a world in which many worlds fit" (Marcos 2002, 80; see also de la Cadena and Blaser 2018).

In *Pluriverse: A Post-Development Dictionary*, Kothari et al. (2019) elaborated on this idea, describing the pluriverse as "a broad trans-cultural compilation of concepts, worldviews and practices from around the world, challenging the modernist ontology of universalism in favor of a multiplicity of possible worlds" (xvii). The pluriverse, then, includes experimental alternatives in the present, such as agro-ecology (Toledo 2019) and degrowth (Demaria and Latouche 2019; see also Kallis and March 2015). Attention to structural change in the present can contribute to "Civilizational Transitions" away from the dominant Western "capitalist hetero patriarchal modernity" toward a more socioecologically just world in the pluriverse (Escobar 2019, 121).

On the other extreme is a world of monocultures, resource extraction, and capital accumulation in the hands of a very limited swath of humanity that is not only a single species but a very limited portion of that one. Social scientists who have proposed the ideas of the Capitalocene and Plantationocene often overlap with perspectives associated with the pluriverse, in their critical assessment of the philosophical and structural underpinnings of dominant "reformist" approaches to addressing the climate crisis, such as ecomodernism, in which "knowledge and technology, applied with wisdom, might allow for a good, or even great, Anthropocene" (Asafu-Adjaye et al. 2015, 6). Such techno-centric approaches include geoengineering as a response to climate change (see Keith 2000) and transhumanism, in which humans achieve "the singularity"—the merging of "biological existence to technology" (Kurzweil 2005, 9). What ecomodernism and other reformist philosophies avoid is addressing the fundamental exploitation of the capitalist system and a lack of scrutiny of linear visions of development centered on the assumed universal benefits of technological solutions and economic growth (Sachs 1992; Escobar 2011). In other words, the future of the Anthropocene as currently conceived is one in which *anthropos* is treated as synonymous with *Homo sapiens* but in practice is a world that largely upholds the system of capitalist hetero-patriarchal modernity.

In *The Future of Humanity*, Kaku (2018) explained why planetary conquest is the next logical step for humans. According to Kaku (2018), it is the fate of *Homo sapiens* to "become like the gods" and "shape the universe in our image" (14). For the author, colonizing and terraforming other planets is an extension of inherent human "restlessness" harnessed through scientific inquiry and technological innovation. Kaku argued for this future in space by pitting humans against a hostile nature that must be escaped before it is too late, making little mention of the anthropogenic climate change that narrows the range of environmental futures in the Anthropocene. In the end, terraforming Mars seems like the next logical step for humanity, and certainly less audacious than finding a way to live together on Earth. Perhaps if we could redirect the "restlessness" of humans that Kaku projected to space and apply it to this world, a better future might be possible, as voices from the pluriverse argue. By drawing together different disciplinary perspectives, as well as the voices of activists and populations whose voices have historically been neglected or appropriated, alternatives to colonialism and capitalism might flourish.

Conclusion

How we place the *anthropos* in the Anthropocene matters for how we understand human nature and how we envision a collective future on the planet. If we see the Anthropocene as nothing more than an extension of humanity's innate tendency to modify and transform its surroundings, then such a view would seem to support continued and perhaps even more radical technological intervention into the Earth's systems and beyond, from geoengineering schemes to space colonization. As we already know, many techno-optimists are actively advocating for this vision of humanity and its future, from Elon Musk's SpaceX program to unprecedented schemes that involve spraying sulfate into the stratosphere to reflect solar radiation.

Rather than lobby for another extreme makeover of the Earth's systems or otherwise seek to escape from the planet entirely, we should think more ambitiously about how we address the Anthropocene both socially and politically. Of course, this would first require us to spend time thinking more deeply about our collective history on this planet and confront how we have come to this point of crisis. Indigenous scholars like Whyte (2018) have shown that the unfolding apocalypse associated with climate change and global environmental change more broadly is only seen as new to European settler colonial society. For indigenous peoples of the Americas and those whose lives were swept up in the horrors of the transatlantic slave trade, it has been ongoing for the last 500 years.

Some environmentalists today argue that due to the urgency and imminence of the climate crisis, it must command all of our attention if we hope to avoid planetary catastrophe. The question is whether we can truly address such a crisis without meaningfully addressing the forms of social and environmental inequality that brought us here in the first place. If we continue to see the world as divided in half, between nature and culture, then it is only logical that we will see violence against humans and nonhumans as separate problems. If we begin to see the world as one, then we might find that these problems are in fact one and the same.

Acknowledgment

We thank the anonymous reviewers and the editor of this special issue for their help in refining the article and its arguments.

References

Asafu-Adjaye, J., L. Blomquist, S. Brand, B. W. Brook, R. DeFries, E. Ellis, C. Foreman, D. Keith, M. Lewis, M. Lynas, et al. 2015. *An ecomodernist manifesto.* Accessed June 6, 2020. http://www.ecomodernism.org/

Bennett, C. E., R. Thomas, M. Williams, J. Zalasiewicz, M. Edgeworth, H. Miller, B. Coles, A. Foster, E. J. Burton, and U. Marume 2018. The broiler chicken as a signal of a human reconfigured biosphere. *Royal Society Open Science* 5 (12):180325. doi: 10.1098/rsos.180325.

Braun, B. 2015. Futures: Imagining socioecological transformation—An introduction. *Annals of the Association of American Geographers* 105 (2):239–43.

Carrington, D. 2016. The Anthropocene epoch: Scientists declare dawn of human-influenced age. *The Guardian*, August 29.

Certini, G., and R. Scalenghe. 2011. Anthropogenic soils are the golden spikes of the Anthropocene. *The Holocene* 21 (8):1269–74.

Crutzen, P. J., and E. F. Stoermer. 2000. The Anthropocene. *Global Change Newsletter* 41:12–13.

Davis, H., and Z. Todd. 2017. On the importance of a date, or, decolonizing the Anthropocene. *ACME: An International Journal for Critical Geographies* 16 (4):761–80.

Davis, J., A. A. Moulton, L. Van Sant, and B. Williams. 2019. Anthropocene, Capitalocene, … Plantationocene? A manifesto for ecological justice in an age of global crises. *Geography Compass* 13 (5):e12438. doi: 10.1111/gec3.12438.

de la Cadena, M., and M. Blaser, eds. 2018. *A world of many worlds*. Durham, NC: Duke University Press.

Demaria, F., and S. Latouche. 2019. Degrowth. In *Pluriverse: A post-development dictionary*, ed. A. Kothari, A. Salleh, A. Escobar, F. Demaria, and A. Acosta, 148–51. New Delhi: Tulika Books and Authorsupfront.

Ellis, E. C., K. Klein Goldewijk, S. Siebert, D. Lightman, and N. Ramankutty. 2010. Anthropogenic transformation of the biomes, 1700 to 2000. *Global Ecology and Biogeography* 19 (5):589–606.

Escobar, A. 2011. *Encountering development: The making and unmaking of the third world*. Princeton, NJ: Princeton University Press.

Escobar, A. 2018. *Designs for the pluriverse: Radical interdependence, autonomy, and the making of worlds*. Durham, NC: Duke University Press.

Escobar, A. 2019. Civilizational transitions. In *Pluriverse: A post-development dictionary*, ed. A. Kothari, A. Salleh, A. Escobar, F. Demaria, and A. Acosta, 121–23. New Delhi, India: Tulika Books and Authorsupfront.

Fausto, C. 2008. Donos demais: Maestria e domínio na Amazônia. *Mana* 14 (2):329–66.

Glikson, A. 2013. Fire and human evolution: The deep-time blueprints of the Anthropocene. *Anthropocene* 3:89–92.

Hage, G. 2017. *Is racism an environmental threat?* Cambridge, UK: Polity Press.

Haraway, D. J. 2013. *When species meet*. Minneapolis: University of Minnesota Press.

Haraway, D. J. 2015. Anthropocene, Capitalocene, Plantationocene, Chthulucene: Making kin. *Environmental Humanities* 6 (1):159–65.

Haraway, D. J., N. Ishikawa, S. F. Gilbert, K. Olwig, A. L. Tsing, and N. Bubandt. 2016. Anthropologists are talking—About the Anthropocene. *Ethnos* 81 (3):535–64.

Hornborg, A. 2017. Dithering while the planet burns: Anthropologists' approaches to the Anthropocene. *Reviews in Anthropology* 46 (2–3):61–77.

Kaku, M. 2018. *The future of humanity: Terraforming Mars, interstellar travel, immortality, and our destiny beyond Earth*. London: Allen Lane.

Kallis, G., and H. March. 2015. Imaginaries of hope: The utopianism of degrowth. *Annals of the Association of American Geographers* 105 (2):360–68.

Keith, D. W. 2000. Geoengineering the climate: History and prospect. *Annual Review of Energy and the Environment* 25 (1):245–84.

Kirksey, S. E., and S. Helmreich. 2010. The emergence of multispecies ethnography. *Cultural Anthropology* 25 (4):545–76.

Kothari, A., A. Salleh, A. Escobar, F. Demaria, and A. Acosta, eds. 2019. *Pluriverse: A post-development dictionary*. New Delhi: Tulika Books and Authorsupfront.

Krenak, A. 2020. *Ideas to postpone the end of the world*. Toronto: House of Anansi Press.

Kurzweil, R. 2005. *The singularity is near: When humans transcend biology*. New York: Penguin.

Levis, C., P. F. de Souza, J. Schietti, T. Emilio, J. L. Purri da Veiga Pinto, C. R. Clement, and F. R. C. Costa. 2012. Historical human footprint on modern tree species composition in the Purus-Madeira interfluve, central Amazonia. *PLoS ONE* 7 (11):e48559. doi: 10.1371/journal.pone.0048559.

Lewis, S. L., and M. A. Maslin. 2015. Defining the Anthropocene. *Nature* 519 (7542):171–80. doi: 10.1038/nature14258.

Li, T. M. 2018. After the land grab: Infrastructural violence and the "mafia system" in Indonesia's oil palm plantation zones. *Geoforum* 96:328–37. doi: 10.1016/j.geoforum.2017.10.012.

Lorimer, J. 2012. Multinatural geographies for the Anthropocene. *Progress in Human Geography* 36 (5):593–612.

Malm, A. 2016. *Fossil capital: The rise of steam power and the roots of global warming*. New York: Verso.

Malm, A., and A. Hornborg. 2014. The geology of mankind? A critique of the Anthropocene narrative. *The Anthropocene Review* 1 (1):62–69.

Marcos, S. 2002. *Our word is our weapon: Selected writings*. New York: Seven Stories Press.

McKittrick, K. 2011. On plantations, prisons, and a black sense of place. *Social & Cultural Geography* 12 (8):947–63. doi: 10.1080/14649365.2011.624280.

Merchant, C. 1980. *The death of nature: Women, ecology, and scientific revolution*. New York: Harper One.

Mintz, S. W. 1986. *Sweetness and power: The place of sugar in modern history*. New York: Penguin.

Moore, J. W. 2017. The Capitalocene, part I: On the nature and origins of our ecological crisis. *The Journal of Peasant Studies* 44 (3):594–630.

Nietzsche, F. 1910. *Human, all too human: A book for free spirits*. London: T. N. Foulis.

Paredes, A. 2020. Chemical cocktails defy pathogens and regulatory paradigms. In *Feral atlas: The more-than-human Anthropocene*, ed. A. L. Tsing, J. Deger, A. Keleman Saxena, and F. Zhou. Palo Alto, CA: Stanford University Press.

Pellow, D. N. 2007. *Resisting global toxics: Transnational movements for environmental justice*. Cambridge, MA: MIT Press.

Pulido, L. 2018. Racism and the Anthropocene. In *Future remains: A cabinet of curiosities for the Anthropocene*, ed. G. Mitman, M. Armiero, and R. Emmett, 116–28. Chicago: University of Chicago Press.

Resnick, E. Forthcoming. The limits of resilience: Managing waste in the racialized Anthropocene. *American Anthropologist*.

Ruddiman, W. F. 2003. The anthropogenic greenhouse era began thousands of years ago. *Climatic Change* 61 (3):261–93.

Ruddiman, W. F. 2013. The Anthropocene. *Annual Review of Earth and Planetary Sciences* 41 (1):45–68.

Sachs, W., ed. 1992. *The development dictionary: A guide to knowledge as power*. London: Zed Books.

Schaan, D. 2010. Long-term human induced impacts on Marajó Island landscapes, Amazon estuary. *Diversity* 2 (2):182–206.

Shiva, V. 2016. *Staying alive: Women, ecology, and development.* North Atlantic Books.

Steffen, W., P. J. Crutzen, and J. R. McNeill. 2007. The Anthropocene: Are humans now overwhelming the great forces of nature? *Ambio* 36 (8):614–22. doi: 10.1579/0044-7447(2007)36[614:TAAHNO]2.0.CO;2.

Steffen, W., J. Grinevald, P. Crutzen, and J. McNeill. 2011. The Anthropocene: Conceptual and historical perspectives. *Philosophical Transactions: Series A, Mathematical, Physical, and Engineering Sciences* 369 (1938):842–67. doi: 10.1098/rsta.2010.0327.

Stephens, L., D. Fuller, N. Boivin, T. Rick, N. Gauthier, A. Kay, B. Marwick, C. Geralda Armstrong, C. M. Barton, T. Denham, et al. 2019. Archaeological assessment reveals Earth's early transformation through land use. *Science* 365 (6456):897–902. doi: 10.1126/science.aax1192.

Toledo, V. M. 2019. Agro-ecology. In *Pluriverse: A post-development dictionary*, ed. A. Kothari, A. Salleh, A. Escobar, F. Demaria, and A. Acosta, 85–88. New Delhi: Tulika Books and Authorsupfront.

Tsing, A. L. 2015. *The mushroom at the end of the world: On the possibility of life in capitalist ruins.* Princeton, NJ: Princeton University Press.

Van Dooren, T. 2014. *Flight ways: Life and loss at the edge of extinction.* New York: Columbia University Press.

Vilaça, A. 2005. Chronically unstable bodies: Reflections on Amazonian corporalities. *Journal of the Royal Anthropological Institute* 11 (3):445–64.

Viveiros de Castro, E. 2004. Perspectival anthropology and the method of controlled equivocation. *Tipití: Journal of the Society for the Anthropology of Lowland South America* 2 (1):3–22.

Viveiros de Castro, E. 2017. *Cannibal metaphysics.* Minneapolis: University of Minnesota Press.

Watling, J., J. Iriarte, F. E. Mayle, D. Schaan, L. C. R. Pessenda, N. J. Loader, F. A. Street-Perrott, R. E. Dickau, A. Damasceno, and A. Ranzi. 2017. Impact of pre-Columbian "geoglyph" builders on Amazonian forests. *Proceedings of the National Academy of Sciences of the United States of America* 114 (8):1868–73. doi: 10.1073/pnas.1614359114.

Whyte, K. P. 2018. Indigenous science (fiction) for the Anthropocene: Ancestral dystopias and fantasies of climate change crises. *Environment and Planning E: Nature and Space* 1 (1–2):224–42. 10.1177/2514848618777621.

Williams, M., J. Zalasiewicz, C. N. Waters, M. Edgeworth, C. Bennett, A. D. Barnosky, E. C. Ellis, M. A. Ellis, A. Cearreta, P. K. Haff, et al. 2016. The Anthropocene: A conspicuous stratigraphical signal of anthropogenic changes in production and consumption across the biosphere. *Earth's Future* 4 (3):34–53.

Wynter, S. 2003. Unsettling the coloniality of being/power/truth/freedom: Towards the human, after man, its overrepresentation—An argument. *CR: The New Centennial Review* 3 (3):257–337.

Yusoff, K. 2018. *A billion black Anthropocenes or none.* Minneapolis: University of Minnesota Press.

Zalasiewicz, J., C. N. Waters, J. A. Ivar do Sul, P. L. Corcoran, A. D. Barnosky, A. Cearreta, M. Edgeworth, A. Gałuszka, C. Jeandel, R. Leinfelder, et al. 2016. The geological cycle of plastics and their use as a stratigraphic indicator of the Anthropocene. *Anthropocene* 13:4–17.

Zalasiewicz, J., C. N. Waters, M. Williams, A. D. Barnosky, A. Cearreta, P. Crutzen, E. Ellis, M. A. Ellis, I. J. Fairchild, J. Grinevald, et al. 2015. When did the Anthropocene begin? A mid-twentieth century boundary level is stratigraphically optimal. *Quaternary International* 383:196–203.

The Inhumanities

Kathryn Yusoff

This article proposes the inhumanities as an analytic to address the material confluences of race and environment in the epistemic construction of the humanities and social sciences. As the Anthropocene represents an explicit formation of political geology, from its inception as a means to frame a crisis of environmental conditions to the characterization of future trajectories of extinction, I argue that centering race is a way to reconceptualize and challenge the disciplinary approaches of the humanities, humanism, and the Anthropocene (e.g., the environmental humanities and geohumanities[1]). Foregrounding the conjoined historic geographies of racialization and ecological transformation through the discipline of geology, within the context of colonial and settler colonial extractivism, sets the conditions for thinking materially about decolonization as a geologic process. I make three interconnected points about the Anthropocene and inhumanities. First, the Anthropocene names a new field of geologically informed power relations that focus attention on the geographies of the inhuman, geologic forces, and the politics of nonlife. Second, the framing of the inhumanities forces a reckoning with the humanist liberal subject that orders the humanities: an invisible and indivisible white subject position that curates racialized geographies of environmental concern, impact, and futurity. Third, the inhumanities makes visible the historic double life of the inhuman as both matter and as a subjective racial category of colonial geographies and its extractive afterlives. In conclusion, I consider the emergence of geopower as a political technology of racial capitalism and governance of the present. Geopower, I argue, is the product of historical geologies of race that subtend a particular form of life marked by extractivism enacted on racialized geosocial strata. *Key Words: Anthropocene, environmental humanities, geology, inhuman, nonlife, race.*

What was this something, I asked myself, that needed as its own condition of existence the systemic impoverishment of the darker peoples of the world? The no less systemic inferiorization of the black and of other non-white peoples of the earth?

—Wynter (2000, 200)

As environmental concern and the designation of the Anthropocene foster new interdisciplinary configurations in the environmental and geohumanities, underpinned by targeted funding and programs, the human is often taken for granted as an accomplice in the designation of a field of concern, as the planet is taken as a presumed arena of action. Although climate change and mass transformation of the planetary geochemical systems do indeed call for a new understanding of the commons beyond geopolitical configurations of nation states, such modes of existence need scrutiny for how they mobilize geophysical changes of state to make new ontological claims on behalf of the elemental, volumetric, or subterranean, while erasing the historical forms of material differentiation that constitute the livability and modes of extinction that result in the "systemic inferiorization of black and of other non-white peoples of the earth" (Wynter 2000, 200). The human as a metaphysical–empirical concept is materially constituted through the scientific racism of paleontology in the seventeenth through nineteenth centuries and the geographies of colonialism that formed its praxis, a praxis that divided the human into its subcategories of human, subhuman, and inhuman, to extract and control the surface and subsurface of the earth. Racialized populations thus became a means to justify and govern this theft of persons and land. In the rush to secure white settler futurity (see Smith and Vasudevan 2017; Erickson 2020) in the context of environmental and climate change, the emergent field of Anthropocene studies and the environmental/geohumanities often assumes rather than problematizing the human that secures the concept of humanities at both a philosophical and epistemic level, disregarding the historical colonial geographies that materially delivered humanism, its structures of thought (the human and its "others" and in parallel the discourse of nature and its "others"), its organization of value (philosophy of natural philosophy, ethics, and racial capitalism), and its scientific and industrial institutions (that were funded, and thereby grounded, in the financial spoils of imperial enterprise).[2] As data from the UCL Legacies of British Slave-owner-ship[3] project has made abundantly clear, the financial payout to slave owners during abolition underpinned the Industrial Revolution in Britain and facilitated the development of its modes of capitalist production,[4] carbon economies, and scientific and educational institutions. Slavery thereby realized both the material transformation of the United Kingdom and its colonies and secured the establishment of its knowledge production and dissemination within global geographies.

Although pressure has been put on the figure of the human, its historicity, racialized, gendered, and sexualized forms (Wynter 2003; Walcott 2014; Jackson 2020), it is not just a hermeneutics of the human that is needed (e.g., from colonial man to Anthropocene man, master-subject "Man" to matter-subject "Anthropos") but, alongside, an unearthing of the very ground that materially constitutes the figure of the human through an examination of what might be termed the geologies of race. Geologies of race is a way to understand the conjoined material praxis of colonial terra- and subject-forming, its geosocial relation (Clark and Yusoff 2017), and its legacies in the present. The material incorporation of the European subject (and its settler colonial kin) in terms of value, accumulation, and subjective forms was defined against what was classified as fossil nature (indigeneity) and fossil energy (the enslaved) to transform the ecological and energetic organization of the world as a global geography. As the Anthropocene empirically describes a new field of geologically informed power relations that focuses attention on the geographies of the inhuman, geologic forces, and the politics of nonlife, it also represents an explicit formation of political geology that is racialized from its onset in the geologies of colonialism since 1492 (Lewis and Maslin 2015; Yusoff 2018c). My argument is not just about the recognition of geographies of colonialization as a spatial apparatus but as a set of interlocking affectual architectures and geophysical relations that constitute an antiblack and colonial earth in the present tense.

The Inhuman and the Anthropocene

In Césaire's (2000 [1972]) *Discourse on Colonialism*, he suggested that "at the very time when it most often mouths the word, the West has never been further from being able to live a true humanism—a humanism made to the measure of the world" (73). As the human is being measured in the world as a planetary geologic agent, Césaire's question remains pertinent to how the measure of this altering world is made, the organization of its disciplinary formations and institutions, and the natural philosophy through which an encounter with the future is staged. In another searing critique of humanism, Fanon (1961) tied the unrealized figure of a true humanism to the earth, as a wretched counterpoint, whereby the inhuman residues of the colonial project abide as discarded matter and the imposition of that subjective category on the discarded. In the context of securing natural resources and the wealth of settler societies through the colonial geoengineering of the planet, the inhumanity of the colonial subject forged a "Black and blackened" (to use Sharpe's [2018, xvi] term) subject category in the inhuman, a designation that materially functioned as the racialized understrata to the white surfaces of capital accumulation. Those blackened colonial afterlives in "modernity's project of unfreedom" (Walcott 2014, 94) are still very much present in the political geologies of climate change vulnerabilities, the wasting effects of racial capitalism, and neo-extractivist economies (Pulido 2016; Vergès 2017, 2019; Sealey-Huggins 2018; Gilmore 2002).

The narrative arc of humanism, Scott (2000) suggested in conversation with Wynter,

> is often told as a kind of European coming-of-age story. On this account, humanism marks a certain stage in Europe's consciousness of itself—that stage at which it leaves behind it the cramped intolerances of the damp and enclosed Middle Ages and enters, finally, into the rational spaciousness and secular luminosity of the Modern. As such, it forms a central, even defining, chapter in Europe's liberal autobiography. But that coming-of-age story has another aspect or dimension that is often relegated to a footnote, namely the connection between humanism and dehumanization. (119)

The Anthropocene discourse follows the same coming-of-age humanist script, searching for a material origin story that would explain the newly identified trajectory of the Anthropos; as a geologic configuration that grapples with the excess and waste of modernity's rationalities (see Hird 2012) and the material consequences of its imposition of reason in the dividuation of the earth. As Scott suggested, the birth of humanism is also the birth of Europe's colonial project, so that "humanism and colonialism inhabit the same cognitive-political Universe inasmuch as Europe's discovery of its Self is simultaneous with its discovery of its Others" (Scott 2000, 120). The Fanonian paradox of the invention of blackness has had much critical attention, yet the substance of othering, its geophysical properties rather than metaphysical dimensions, might also be thought to bear on the mobilization of forces of energy and minerality in the world, as well as the ongoing questions of environmental racism and injustice in material exposure to environmental events and toxic body burdens (Bullard 1990; Williams 2017, 2018; Woods 2017). That is to say that humanism has a material gravity and an anti–black and brown ground (Moten 2003; Roane 2018).

It is well noted that the Anthropocene as a concept is disrupting the binaries of nature and culture, human temporality and deep time, human history and prehistory, bio and geo, and rearranging temporal and spatial scales of analysis, but it also has unproblematically reinstated a pre- and postracial subject (Gunaratnam and Clark 2014). I want to suggest that the inhumanities is a means to reconceptualize and challenge the existing disciplinary approaches of the humanities that acknowledges rather than erases the collaborative junctures between metaphysical designations of the human and the geophysical praxis that bring subjective forms materially into being. This geologic praxis refers both to the extraction of natural and psychic resources for the maintenance of white heteropatriarchy, and how materiality regulates racist structures of extraction and subjugation. In the Anthropocene, the human often comes into view, organized as it is around the telos of a *coming catastrophe*. Although the human is a ritual object of self-flagellation for Western thought (through the Enlightenment and then its critiques of postcolonialism, poststructuralism, and posthumanism), it nonetheless remains a dominant figure that marshals the horizons of meaning with an irrepressible recurrence. Because this is where meaning has been made, as postcolonial thinkers such as Fanon, Glissant, Césaire, and Wynter understood, it is also where meaning gets unmade or remade, so it is both the target and site of possible emancipation with regard to racist structures of material engagement. Similarly, the inhuman is a site of traffic in subjective and material forms of life, so it is also a site of possibility for decolonizing the entanglement of race and geographic processes. As the category of the inhuman is positioned as ahistorical, or that which is placed outside of the time of racial capitalisms to propel its notion of time and space, so it holds the potential for reimagining the dimensions of time and space.

Alongside Fanon's (1961) decolonizing tract, *Wretched of the Earth*, the subterranean theorizing of the black radical tradition (e.g., C. L. R. James, Hortense Spillers, Saidiya Hartman, Fred Moten, Tina Campt, Tiffany Lethabo King) has shown that the ground (and grounding) of humanism is antiblack. The onto-epistemological organization of an antiblack earth matters for antiracist struggles beyond questions of territory and environmental justice, because it is an epistemic inscription that conditions structures of thought and their institutional manifestations. Race provided the imposition of material poverty with a moral justification, reallocating the burden of destitution onto a racial determinant rather than a systematic organization of material relations. Black and Indigenous studies have been theorizing a nonnormative materiality for years, precisely because the black and Indigenous subject is historically irreducible to the normative (humanist) subject. Furthermore, this antiblack and brownness is materially and theoretically made concurrently with the inscription of the earth through the grammars of geology within colonial praxis and its afterlives (Yusoff 2018c). The natality of this emergence can be said to constitute an ontological inscription of black, Indigenous, and earth as the ground for the "natural" right of whiteness to consolidate the planet for the generation of value (also see Bonds and Inwood 2016; Baldwin and Erickson 2020). The historic and geographical differences across which the questions of geology and race travel are specific and varied, and by no means exclusive, but across the plateaus of colonialism, race provided the political redemption for the white supremacy of matter. The discourse of "improvement" brought raced subjects and indigenous land theft into the economic terraforming project of natural resources.

This is to say that knowledge of the earth is also interned in a perspectivism and praxis of the supremacy of whiteness, which marks the capricious geography of the human; what Wynter (2000) referred to as "racially dominant white elite stratum" (126) that propagated the development of natural resources as a "secular telos of materiality redemption" (Wynter 1996, 300). Alongside, in different ways, scholars in Indigenous studies have argued and practiced for different ontological arrangements of the earth against settler colonial modes of extraction (Simpson 2017; Whyte 2018).

In the epochal imaginary of the Anthropocene there is a reconstitution of the subject as planetary, understood not through the conquest of space (as in the Enlightenment) but through the mobilization of an empire of geologic forces, what might be designated as the capitalization on modes of geopower (Grosz, Yusoff, and Clark 2017). Disrupting the homogenizing spatial identity of colonialism as primarily being about the traversing and settlement of territory, as if it were a relatively nonresistant earth for social and political relations to play out on, the mobility of geologic forces suggests anything but a nonresponsive ground. Thus, geographical concern with the horizontality of space and more recently with its verticality, subterranean spaces, and volumetrics does not yet attend to the change of geophysics, where *state* is a temporal geophysical condition rather than exclusively a spatial demarcation (i.e., the settler colonial state as a modality of ongoing colonial relations to materiality, time, and space). The geosocial formations of the Anthropocene are about the *tense* of geographic spaces in time rather than their expansiveness in space. If the ground on which the humanist subject is presumed to stand is shaken, it is not just providing a disruption to the spatialities that locate the preferred human as distinct from the violence of the wretched earth, but it is exposing that subject as a hermeneutic shell that neglects its material constitution (as one that only holds its elevated position because of the accumulated energy from systems of settler colonial violence and antiblackness). It is time, then, to imagine another subject capable of apprehending the differentiated and differentiating geoforces it is historically embedded within. That is, to ask can "we" stay in that state—in the *tense of geophysics*—without the fantasy of a materially autonomous subject that does not need the earth and the racialized forms of its extractive economies? As current material conditions are disrupting the ontology that imagines a subject that is not constituted by the earth, this normative subjectivity was not available nor desired for the many black and brown populations who are most closely involved in the actual fabrication of the world, in the mine, on the plantation, offshore, or underground. In the ongoing brutal apparatus of extraction, the contradictions of ethically hailing a humanist subject that is the praxis of the extinction of many worlds suggests that the psychic life of geology has yet to be fully examined in the Anthropocene.

Geologic forces constitute socialized worlds, yet they are often bracketed out because they introduce intricate and often contradictory spatial relations into the mix that problematize long-held spatial propositions about social relations and the organization of power and its anthropogenic origins. Geologic sovereignty requires an analysis of the relations of geopower, as they are differentiated through subjects and regulated by racial capitalism, which is in turn historically constituted by its colonial afterlives (see Pulido 2016, 2017; Pulido and Lara 2018; Tuana 2019). The Anthropocene seems to present a new geosocial formation of the human condition, but it has a much longer historical geography established in the wake of geology, made through the material classification of the inhuman, and its extractive grammars since 1492, where race became a modality that enabled material extraction and slavery instigated the collapse of the inhuman into the body politic of blackness. As geologic classification of earth minerals made matter as value, it also captured enslaved subjects in the brutal calculative enclosure of the inhuman as chattel. Extending work on the division of the world according to the technologies of race, race might equally be viewed as an ontological division of matter that established historically situated conditions of proximity to violent material forces without accumulation of value. Thus, technologies of race are equally geologies of race, whereby metaphysical designations have geophysical effects, establishing antiblack and brown gravities as the affective architecture of extraction.

The Inhumanities

The inhumanities forces a reckoning with the materialities of the humanist liberal subject that orders the humanities; an invisible and indivisible (white) liberal subject position that curates racialized geographies of environmental concern, practice, and futurity, from the material conditions of science policy to metaphoric assertions of environmental poetics. Although there has been attention to the crossings between the nonhuman and race (see

Jackson 2013; Luciano and Chen 2015; Muñoz 2015) and the dehumanizing effects of racial sciences, the relation between the inhuman and race—the inhumanities—is undertheorized.

The inhumanities refers to the classificatory systems and natural philosophies of thought that are desubjectifying (targeting particular populations into things, property, or properties of energy) and structurally subjugating (to black and brown life in the afterlives of slavery and imperialism). The dual effect of this subjection is to underground agency, in a body of matter rather than rights, under slavery and settler colonialism. Blackness was seen as a source of energy that was cultivated as a signifier and a carceral category to produce value through extraction in which race was implicit in the material reproduction of whiteness and settler futurity. At the heart of that racial formation was the extractive impulse, which builds capitalism into a globally functioning system that is world-altering. Race was the codification of the unequal racial distribution of geopowers, its accumulation and the placement of certain lives in material and psychic proximity to the inhuman, and therefore less conspicuous in the juridical and ethical recognition of a political subject.

As an analytic, the inhumanities puts the kinship between the extraction of bodies and the extraction of earth at the core of its concerns about the possibilities and potentialities of lives, conjoining genocide and ecocide as tenants in the colonial project (and its kin, settler colonialism and resource colonialism, or what Gomez-Barris [2019] called the "colonial Anthropocene"). This move from a biocentric paradigm to a geocentric one includes rethinking how the social and political configurations of subjectivity have been thought and experienced in relation to the earth. Extraction drives a particular formation of what and who gets constituted as a natural resource and where these activities of extraction take place in relation to the accumulation of value. Natural resources under colonialism and its afterlives in racial capitalism equal extraction plus capture, resulting in forms of epistemic and material enclosure for the production of value. We can see a parallel framing of personhood under slavery, whereby extraction (from Africa and precolonial relations) plus capture (through the technology of race) results in forms of material (spatial extraction and containment) and subjective (slave/native) enclosures. Developing a broader understanding of colonialism in relation to the integration between its knowledge and extraction networks, the emergence of geopower can be seen as a political and racial technology of governance of the present and as a set of forces that subtend the potentialities and exhaustion of life.

The political importance of reconceptualizing the inhuman in its material, epistemic, and conceptual forms is part of understanding the material transformation of the planet (Clark 2010) and how geologic grammars do geopolitical and geophysical work (Yusoff 2018c). This is also to understand how and where matter relations organize and arrange particular enduring forms of oppression, as extraction economies traverse subjective and material regimes of value. As McKittrick (2006) argued in *Demonic Grounds*, what would happen if we put black geographies at the center of our analysis? What kind of geography would that produce, and how would it shift the questions we ask? The mine and the plantation are the geosocial formations of the New World; what is important to notice is that they are both social institutions of extraction (the mine and the factory) and bio-geo engines that rearrange earth and ecologies. In other words, because of their repeatability across the Americas and empire, the material praxis of the mine and plantation completely rearrange the geochemical flows of the world, setting into place changed ecological and social relations that are racialized. The geophysics of the mine and plantation created the metaphysic conditions of the burden of being categorized as black. These geosocial arrangements are a *geophysics of power*.

What are often considered as spatial divisions in the conceptualization of political geography, and in particular the formation of planetary politics or planetary scale, are actually a question of material ontological division or "matter fix" that designates the location of agency on the side of biocentric life (which cleaves to a particular politics of life). It is the spatial arrangements of the divisions of materiality as agency (as active subject vs. fungible matter, which material enacts the master–slave relation) that organize an understanding of the arrangement of power as race (see Yusoff 2018a). This partiality of a preferred form of life characterizes, as Povinelli (2016) argued, the provinciality of Foucault's project in its conceptualization of a Western European genealogy. Sylvia Wynter, W.E.B. DuBois, and Achille

Mbembe all showed how that genealogy of man was underscored by the racial division of life and nonlife. Although Arendt (1944) argued that race names the connections between white settler states and European fascisms, she was unable to see colonization as anything other than a mirror to European thought and its practices in a "boomerang"[5] to European forms of racialization (see Gines 2014; Owens 2017). In Western philosophy, race names the point of quarterization between the inside and outside of (colonial) life. This schism between inhuman matter and subjects cleaves apart the biological and the geological using the signifying practices of race as a discourse located within particular bodies. This enables global geography to be claimed as universal (the planet), an exclusive domain that does not have to admit those that are not represented by the preferred figure of biologism (the humanistic subject; Silva 2007). Race is organized around biocentric codes, which are in turn underpinned by geologic grammars that stratify human origins as they do the origins of the planet, so there is no turning to the planet to do away with race, no planetary commons that is not at the same time in need of decolonization. The imperative to introduce a concept of geopolitics (see Mignolo 2011) that goes beyond a biologism divorced from the earth and its interrelation to forms of life, therefore, must be a compelling project for any antiracist practice.

Politics of Nonlife

The politics of nonlife have come into view through the lens of biopolitics: first, as a form of "thanatopolitics" in Foucault's thought, whereby governance is established over and through the "species body"; then as bare life in Agamben's (1998) designation of a productive relation to life that recognized the constitutive exclusions of states of exception as founding the "City of Men"; in "necropolitics" in Mbembe's formation of black life ("To exercise sovereignty is to exercise control over mortality and to define life as the deployment of manifestations of power"; Mbembe 2003, 12); and in the governance of the life–nonlife caesura in Povinelli's (2016, 5) account of "geonotopolitics" in the late liberal governance of settler colonial societies. These approaches variously engage Foucault's distinction of biopower forged in the transition from sovereign power to state power that constituted the new political subject, a subject constituted by the population and vicissitudes of biological life from the molecular to the pathologized, which assumes a normative biocentric human at its center. This biocentric human (Wynter 2003; Povinelli 2016) is complicated and constituted by race.

Although Agamben disagrees with a Foucauldian model in its succession and relation (Agamben arguing that sovereign power originates with the biopolitical body and that bare life constitutes political life), both establish a dialectical relation between the human as properly constituted as a political subject and its aberration. In its simplest iteration, there are forms of life on one side and nonlife on the other; nonlife that is constituted through death, and more recently in Mbembe and Povinelli's writing through forms of social death, exhaustion, and extinguishment, wherein nonlife emerges as a zone of governance. The gravitational pull that centers these divisions between life and nonlife is the human subject as it is conceived through a Western normative frame (as Mbembe and Povinelli aptly demonstrated in relation to black and indigenous life). This explains why Agamben's "bare life" can only take place in the concentration camps rather than on the plantation or in the hold of the slave ship: because the aberration can only be visited on a human body that is already coded human and whose recognition of such a status has slipped or is in abeyance, wherein biopolitical regimes act as an injunction or resolution of the potentiality of inclusion rather than its impossibility.

As Wynter has deftly shown, that human was constituted through its geographical outsides and racialized others from the onset, between what she called Man 1 (the rationalized political subject of the state, who escapes its prior ordering between heaven and earth in the chain of being to emerge as a political animal in the context of the state) and Man 2 (the secular successor to the political being, a biocentric being who is brought into view through evolutionary theory and eugenic codes of differentiation). Wynter argued that in the shift from the feudal to the bourgeois there is the lack of a claim to a nobility of blood, so the sociogenic mutation becomes a bioevolutionary claim (after Darwin) to a select eugenic line of descent (through genealogical account), what Wynter (2000) called the "governing sociogenetic principle … the master code of symbolic life and death … that is constitutive of the multiple and varying *genres* of the human in terms of which we can *experience ourselves as human*" (182–83, italics in original). Wynter's description of the modalities of "scientific humanism" outlines an epistemology of dispossession and conquest through the sciences and its imposition of monohumanism that are nonetheless defined by a particular version of overrepresented Western European man.

Wynter's critique assaults the presumptive "Referent-we" that necessitates the formation of the other who does not belong to that master class of species to organize the logos of its belonging. As McKittrick commented, "Particular (presently biocentric) macro-origin stories are overrepresented as the singular narrative through which the stakes of human freedom are articulated and marked. Our contemporary moment thus demands a normalized origin narrative of survival-through-ever-increasing-processes-of-consumption-and-accummulation" (Wynter and McKittrick 2015, 11). The human relies on a dialectic relation to the inhuman; that is, the racialized subject. Another way to imagine this is to historically situate the human in colonial geographies of extraction that produced what Fanon (2008 [1952]) called a "zone of nonbeing" (2), whereby the violent positioning of certain lives in relation to the biocentric norm of the human that frames the humanities is not a modifier of the human but the very action of its erasure. As McKittrick (2015) commented:

> These governing codes produced racialized/non-Europeans/nonwhite/New World/Indigenous/African peoples as first, fallen untrue Christians (in the fifteenth and sixteenth century) and, later, as biologically defective and damned (in the nineteenth century). I want to highlight Wynter's assertion of the ways in which our present conception of the human—and what it means to *be* human—delineates how colonial encounters, and thus the emancipatory breaches and thus reinventions of humanness. This is to say that at the nexus of theological punishment, colonial brutality, and imperial greed, underpinning the new sciences that recast how we perceive our physiology and our sociocultural systems—physics, astronomy, cartography, biology, and so forth—are the fallen and defective who put immense pressure on European ways of knowing the world. (143)

Alongside McKittrick's identification of the "emancipatory breaches" as a spacing in the human, Fanon's conceptualization reminds us of the spatial and subjective conflation of colonial understanding of vast populations of people and areas of earth designated as matter awaiting extraction. Yet, if the biopolitical frame is understood as always already constituted through an inhuman ground (or racialized subjects), even as it is prized apart by liberatory inventions that refute that designation, the inhuman is what secures the nonlife differentiation to produce the human as such (Foucault 2009, 2010). The inhuman, however, can never ever be incorporated into the political subject or being as such because of the material differentiation between figure (the political subject of juridical rights) and ground (the matter of resource to enable the becoming of the political subject, philosophically and materially). Thus, the politics and practices of making nonlife within a discourse of biopolitics, its sites of concern in terms of bodies (the alienated human subject; the subject without rights) and places (camps, zones of exception) remain exclusive within the domain of the human as it is designated by its partial humanism. According to Wynter (2000), this "partial humanism" organizes "*history* for Man, therefore, narrated and existentially lived as if it were the *history-for* the human itself" (198). Man thus represents himself as if he were the full scale and sense of the human and reproduces those epistemic modes of life as life (excluding the differential of the inhuman on which that life is built).

Adjacent to this theorizing of a praxis of the human sits a set of experiences and discourses articulated by black theory (black feminisms and Afropessimisms) that name black natality and its genealogical afterlives in the void rather than in the caesura between life and nonlife (i.e., born into slavery as property rather than person; see Hartman 1997; Moten 2003; Spillers 2003; Wilderson 2003; Weheliye 2014; Warren 2018). Black radical thought has highlighted the difference between populations targeted for erasure and those already erased; what Philip (2017) refers to as Bla_k, as the blank–black dynamic of erasure that marks black life from Middle Passage geographies to the present space of death-worlds characterized by carceral enclosures and terrorizing architectures (Mbembe 2003; Wilderson 2003).

The antiblack ground that secures the human's proposition is expropriating through the material differentiation of the life–nonlife boundary that splits the agentic and mute through a *matter fix*, which is also at the same time designated as a subjective boundary that enables further material subjugation.

This racialized grounding presses the violence of inhumane conditions into populations designated as structurally inhuman (through the exposure to the violence of environmental formations, wastes, and the "quasi-events" of their inundation and epigenetic instantiation). Another way to say this is that the partial biopolitical body also presupposes "white spatial formations" (Kwate and Threadcraft 2017, 535) that organize black space in a similar mode of devaluation and negative accumulation. For example, Kwate and Threadcraft (2017, 535) showed how the necropolitical rather than biopolitical functioning of black space conditions relations to medical care in black space, pathogenic environments, and environments that produce "excess death," policing, and the expansionist operations of the carceral state (see also Wright 2011).

The Anthropocene and its designation of an operative sphere of geopower signals the possibility of a potential breach in this division between nonlife governed as nature ("natural fruits" of slavery; the "natural" history of indigenous and "other" peoples that we find in the colonial museum alongside flora and fauna) or life governed as social (through the sociogenetic coding of overrepresented European man). To extend Wynter's formulation, the Anthropocene might see the introduction of *Man 3*, a geologically informed subject whose life forms are coded through the inhuman, whose constituting powers are realized through the mobilization and governance of geologic materials (minerals, geochemicals, water, air, fossil fuels, carbon, nitrogen, phosphorous, etc.) and the accrual of geopowers as a form of securing settler futurity (Trump's "freedom molecules").

The Double Life of the Inhuman

The inhumanities is a field in which the historic double life of the inhuman as both matter and as a racial category of liberal humanism are made visible as coconstituting discourses of modernity and the extractive geosocial relations that form the Anthropocene. Inhumanity is a state or quality of being inhuman or inhumane, or an inhuman or inhumane act. In short, this is a savage form of subjectivity that is organized as a category and thus subjugation for phenotypically designated populations through the operation of race[6] within the historical geographies of colonialism. As new forms of racialized beings were articulated through sixteenth- through nineteenth-century paleontology in

the context of colonialism, geology was also articulating new origins of the earth, as well as forming the material praxis of their rearrangement (through mining, ecological rearrangements and extractions, and forms of geologic displacements such as plantations, dams, fertilizers, crops, and introduction of "alien" animals). In Western philosophy, Adorno (1978 [1951], as discussed in the epigraph) introduced the idea of the inhuman society in *Minima Moralia: Reflections from Damaged Life*, citing that "life does not live" in the society of industrial war, whereby intimate acts of life should be thought of in relation to the catastrophes of the twentieth century. Although Adorno, along with many contemporary critiques, saw the descent into the inhumanity as located in his immediate historical context in World War II, the fall already anticipates its redemption, "to project negatively an image of utopia."

Whether catastrophe or utopia, the organization of worldly ethics around a Eurocentric subject continues to erase any responsibility for the prior catastrophes that were waged on behalf of and materially constituted the European ethical subject since 1492 and continue in ongoing settler colonialism and neoextraction. Although catastrophe is located in the realm of the imaginary in as much as the earth does not have catastrophes, it just has changed conditions, catastrophic displacements have nonetheless constituted the material and psychological lives of colonial subjects and the subjected (which is not to say that within those scenes of subjection, heroic endurances and poetics were not imaginatively formed that materially and physically refuted and resisted those colonial enclosures). Only by unsettling the normative frames of the human as an episteme of universality (that thereby continues to claim an expansionist [colonial] geography and refuses to acknowledge its history of inhumane acts through which such a figure was constituted) can the inhuman be encountered in the full sense of its existence, both as a historical geography of subjugation and as earth.

The inhuman is not simply an alienated form of the human, as it is most often encountered as a nonhuman referent that stands in for the animal form as a reminder of the animal root or pathologizing slur, but understood in its historic materially and situated occurrence, it is a form of differentiation in both matter and time, whereby the divisions and spacing of the genomic or organic principle of life are set in contrast to the inorganic. New modes of biologism attempted to declassify man as an animal and reclassify him as a cultural-ethno being through operative ontologies that produced hierarchies and ordering systems. The denotation of a nature–culture split that is organized through the institutionalization of reason (since Kant) was used to separate man from organic forms of life as both an overcoming of internal passions that constituted human experience and of external obstacles to an expansionist geography across territory and forms of life. Historically, this normative sphere of humanism was racist and specifically anti-black, and without challenging that history, it remains so, every time the universal or human is invoked. Some of the greatest challenges, of course, came from anticolonial thinkers struggling to make sense of their painful histories in their fullest terms, such as Fanon (1959, 1961), Césaire, Glissant, C. L. R. James and Wynter. As Wynter (2000) commented, "The degradation of concrete humans, that was/is the price of empire, of the kind of humanism that underlies it" (154).

For Wynter (2000), "what is called the West, rather than Latin-Christian Europe, *begins* with the founding of post-1492 Caribbean" (152). Wynter challenged the geographical imaginary that the Americas and Caribbean are somehow an epistemological *outside* to Western knowledge, in which people wrote

> not realizing that the condition of *their* being what they are today, and the condition of we being what we are today are totally interlinked. That you can't separate the strands of that very same historical process which has, by and large, enriched their lives and, at the same time, largely impoverished the lives of the majority of our [Caribbean] people. (Wynter 2000, 152)

There is a before to colonialism (and its spatiotemporalities), but there is no *going back before* colonialism, so in this sense, the need to battle humanism is a reckoning with the historic situatedness of Europe's imperial geographies, even as those "other" places were rendered as mere mirrors for Europe's soul. In conversation with Wynter, Scott referred to this as her admiration for "embattled humanism" (Scott 2000, 153).

> You know that you cannot turn your back on that which the West has brought in since the fifteenth century. It's transformed the world, and central to that has been humanism. But it's also that humanism against which Fanon writes [in *The Wretched of the Earth*] when he says, they talk about man and yet murder him everywhere on street corners. Okay. So it

is that embattled [humanism], one which challenges itself at the same time that you're using it to think with. (Wynter 2000, 154)

After Wynter, we might ask: What is the context that produces this geologic subject and brings it into being as a unified subject position? What geography allows this conception and materially sustains it as a geophysical state of stratified relations? This is another way of asking after the ground, or, how the concept of the inhuman grounding is a concrete material reality for black and brown subjects through the epistemic praxis of geographic and geologic institutions. Rather than taking white natural philosophers of reason as the site of the transformation of the human, if we begin in the colonial slave mine and plantation, the ground that sustains this production is the geologic of the inhuman (as nature, matter, and race). Without the geologic of the inhuman (as object and subject), the entire mechanism that produced the relation between Europe and the New World and the humanist subject falls away. Thus, the inhuman, as well as a source of subjection, is also a site of possibility, because that category of regression was always being transgressed, and other relations were being instigated that spoke to the possibility of other relations to the earth. For example, Wynter (2000, 165) discussed how the slave plot, where the enslaved grew food and cultivated other relations of temporality and belonging, existed as a threat because it spoke to other imaginations of geographies (see also McKittrick 2016). Everywhere that the inhumane is imposed, it is resisted by a humanness that highlights the dark contours of the humanist subject in its partiality. It is against this partiality that Césaire (2000 [1972]) imagined a fullness made to the "measure of the world" (73). Thus, the anticolonial critique is not simply a critique of the inadequacies of the human or a better humanism but a counterimaginary that opens up a fullness in the register of the world (Wynter 1984).

The inhuman names the paradigm of extraction that dominates the form of existence in the Anthropocene, but it is a limited one, as far as existing on a planet with one another in any mode of justice or natural jurisprudence. If the Anthropocene heralds a recognition of an ontological form of subjectivity hitherto excluded—a geologically informed and forming subject—then how might it be properly situated in steps to decolonize the humanist subject that informs its partial perspectivism? What becomes of this relation to the inhuman is the corrosive potential of the Anthropocene and its proper critical mode of address. Rather than the environmental and geohumanities (and its offspring, livable futures), "we" might better be organized around the institutes of the inhumanities. The inhumanities registers a commitment to dismantling the humanist subject and the white supremacy that characterized its geographic project of the differentiation of subjects and earth. Reshaping the institution of being is also a call to reimagine subjectivity and relation (Glissant 1990) in the context of the ground that sustains it (or the earth). Thus, the ontology of the humanist ground is disrupted by the Anthropocene thesis as (1) a stable ground on which social relations play out and (2) a neutral ground that is not constituted by subjugating relations. It is important to target the legitimating frame of the human as it becomes materially manifest in the environmental or geohumanities because of how it is being organized around the problem of the future of the world. The human as a biologically and culturally operative ontology denotes a preferred phenotypical subject and an institutionalized mode of life (namely, racial capitalism; see Saldanha 2020). If we take on the spatializing function of signification as a distancing between terms, we can see this subjectification as a means for geographic dispossession. Considering the inhumanities raises three important questions for Anthropocene research:

1. The reconstructing of the dual root of race and materiality in the geographies of environmental change and the doubling of race as metaphysical and geophysical claim in historical geographies of colonialism. The rigid racial hierarchies of humanism were cooked in the crucible of colonialism and functioned simultaneously as an expression of epidermal codes and geographies of dispossession. The grammar of geology in colonial projects—the inhuman—established the stability of the object of property for extraction, as subject and matter.

2. The locating of inherently racist structures and histories of materialism that go beyond the identification of territorial theft and material acquisition and reside in epistemes of geologic classification and valuation. To dismantle the grammar of geology first requires an understanding of how it functions as a mode of social and environmental ascription in an epistemic mode, alongside the planetary practice of extraction.

3. The primacy of a material (environmental) rather than ideological structuring of race—which in turn changes the spaces and structures of thought that are

marshaled for antiracist action. As such, race is organized around biocentric codes that are historically underpinned by geologic foundations about the story of life and earth, operationalized to effect ongoing regimes of geographic dispossession. There is no turning to the planet to do away with race, no planetary commons that is not at the same time in need of decolonization (Gabrys 2018). The inhuman is a starting point from which to rethink material redistribution of geopower and its racialized carceral modes. The inhumanities is a counterconceptualization of the environmental and geohumanities that foregrounds the role of the politics of nonlife and the figure of the inhuman as the political figure of an earthbound commons, that undoes an extractive account of matter because it must always ask if it is the global epistemic production of the privileging of an extractive account of matter that racializes and depletes subjective-environment relations.

A politics of nonlife that is not predicated on necropolitics or on a biopolitical mode of address is a way to think about the emergence of subjective modes and the earth; a way that is not already conditioned by the subjugation of biopolitical exclusion and the promise of inclusion held beyond the abysmal ground of dispossession and the material life of racial inequality. As there is no humanist subject without the intramural question of the materiality of race, there is no environmental or geohumanities without the question of racial justice. The inhuman is an epistemological site for the undoing of geographies of colonial materialities in the present tense across both subjective and earth relations.

Notes

1. Making claims about and against the environmental humanities obviously raises the question of what exactly is this entity, and a more interesting question, why has it emerged now, alongside the Anthropocene and in deference to the decisive politics of climate change? When new neologisms come along there is a tendency to gather everything under them and say this is what we have been doing all along. The Anthropocene is a case in point. It is climate change, extinction, and global environmental change all wrapped up into one. The rebirth of the humanist subject, or what I call *Anthropogenesis*, conjoins the environmental humanities and the Anthropocene. These disciplinary organizations are not particularly interesting in a genealogical sense but rather in terms of how they challenge thinking or become an exercise in "point and erase" in a political field of inquiry; that is, what is depoliticized and what is drawn attention to and how this focus both points

to concerns and erases painful histories. We should remember that the very same modern subject was birthed in the Lisbon Earthquake (1755) at that other seismic rearrangement of disciplines, wherein Kant fashioned his natural philosophy. Although Benjamin called Kant the first geographer because he first introduced lecture courses on geography as a natural science, replacing God and grounding rational thought in the earth, he had a moment of hesitancy of human supremacy before the void (Clark 2010). Kant could not quite let go of the vision of redemption through ascendancy. Within that ascendancy of reason, the preferred status of the subject of the Enlightenment was bound to a discourse of racialized hierarchies. Kant's legacy (see *History and Physiography of the Most Remarkable Cases of the Earthquake Which Towards the End of the Year 1755 Shook a Great Part of the Earth* (Kant 1994 [1756]) set in place a new judgment, predicated on the universal ... but; that is, the imagination of a universal subject, universal just not ... black. This adjudication is the ongoing negative dialectics of white supremacy that proliferates environments from nature scripts of colonial museums to freedom in the black outdoors. What perhaps is distinctive about environmental humanities, as opposed to cultures of climate change, is that science and policy are not at the center of the frame, thereby recentering the arts and humanities. Broadly, environmental humanities is an interdisciplinary field, most represented, like the Anthropocene, in Europe, the United Kingdom, and North America, with the exception of the African Environmental Humanities Network (Agbonifo 2014). In black studies, the term *ecology* is used instead of environment, perhaps to foreground the imbrication of social–environmental relations, but no terminology is neutral. It is useful to recall DeLoughrey's (2019, 70) work on the origins of ecology as centered around the Odum brothers' fieldwork in the Pacific, as part of the U.S. program that included human subject trials on the Marshallese Islanders. Environmental humanities includes a range of programs, fixed term research clusters, funded centers, and pedagogical programs and several journals. Most statements of purpose agree that the environmental humanities are a diverse and emergent field that is interdisciplinary and converges around the human and environment, in the context of environmental issues and concern. (For a comprehensive list, see Emergence of the EH commissioned by MISTRA [Sweden] in Nye et al. [2013]).

2. There are many emergent strands in black and Indigenous scholarship that address the "Black Outdoors" (series at Duke University Press, edited by Carter and Cervenak); "Black Ecologies" at Black Perspectives (https://www.aaihs.org/introducing-the-black-ecologies-series/; see also Roane and Hosbey, https://crdh.rrchnm.org/essays/v02-05-mapping-black-ecologies/); and work critical to the whiteness of the environmental and geohumanties such as DeLoughrey, Neimanis, and Rose, for example.

3. See https://www.ucl.ac.uk/lbs/.
4. For those who would rebrand the Anthropocene the Capitalocene, Wynter's rebuke to the a priori conditions of production is well made:

> It is not primarily the mode of production—capitalism—that controls us, although it controls us at the overtly empirical level through the institution of the free market system, and the everyday practices of its economic system. But you see, for those to function, the processes of their functioning must be *discursively* instituted, regulated *and* at the same time normalized, legitimated. So what I am going to suggest is that what institutes, regulates, normalizes and legitimates, what then controls us, is instead the *economic* conception of the human—Man—that is produced by the disciplinary discourses of our now planetary system of academia as the first purely secular and operational public identity in human history. ... In order to be unified in *economic* terms we have to first produce an *economic* conception of being human. ... This is why, however much abundance we can produce, we cannot solve the problem of poverty and hunger. Since the goal of our mode of production is *not* to produce for human beings in general, it's to provide the material conditions of existence for the production and reproduction of our present conception of being human: to secure the well-being, therefore, of those of us, the global middle classes, who have managed to attain to its ethno-class criterion. (Wynter 2000, 160)

That is to say, there is no fossil capitalism without the engine of race.

5. Arendt argued that nonlife in the camps comes back from the colonial encounter, as a consequence, rather than a modality that is exported there through an expansionist geographic logic that secures its freedom through practices of unfreedoming others.
6. Wynter (2000) argued that the

> bio-climatically phenotypically differentiated Color Line, one drawn in W.E.B. DuBois's terms "the lighter and darker races" of humankind, and at its most extreme between White and Black. This is, as a line made both conceptually and institutionally unbreachable, with this thereby giving rise to an issue, which as Aimé Césaire of the Francophone Caribbean island of Martinique pointed out in his letter of resignation from the French Communist Party, in 1956, was one whose historically instituted singularity, that to which we gave the name of *race*, could not be made into a subset *of any other issue*, but had instead to be theoretically identified and fought on its own terms. (3)

References

Adorno, T. 1978 [1951]. *Minima moralia*, trans. E. F. N. Jephcott. London: Verso.

Agamben, G. 1998. *Homo sacer: Sovereign power and bare life*, trans. D. Heller-Roazen. Stanford, CA: Stanford University Press.

Agamben, G. 2005. *State of exception*, trans. K. Attell. Chicago: The University of Chicago Press.

Agbonifo, J. 2014. African Network of Environmental Humanities: First bold steps! *African Historical Review* 46 (2):151–52. doi: 10.1080/17532523.2014.943995.

Arendt, H. 1944. Race-thinking before racism. *The Review of Politics* 6 (1):36–73. doi: 10.1017/S0034670500002783.

Baldwin, A., and B. Erickson. 2020. Introduction: Whiteness, coloniality, and the Anthropocene. *Environment and Planning D: Society and Space* 38 (1):3–11. doi: 10.1177/0263775820904485.

Bonds, A., and J. Inwood. 2016. Beyond white privilege: Geographies of white supremacy and settler colonialism. *Progress in Human Geography* 40 (6):715–33. doi: 10.1177/0309132515613166.

Bullard, R. 1990. *Dumping in Dixie: Race, class, and environmental quality*. Boulder, CO: Westview.

Césaire, A. 2000 [1972]. *Discourse on colonialism*. New York: Monthly Review Press.

Clark, N. 2010. *Inhuman nature*. London: Sage.

Clark, N., and K. Yusoff. 2017. Geosocial formations and the Anthropocene. *Theory, Culture & Society* 34 (2–3):3 23. doi: 10.1177/0263276416688946.

DeLoughrey, E. 2019. *Allegories of the Anthropocene*. Durham, NC: Duke University Press.

Erickson, B. 2020. Anthropocene futures: Linking colonialism and environmentalism in an age of crisis. *Environment and Planning D: Society and Space* 38 (1):111–28. doi: 10.1177/0263775818806514.

Fanon, F. 1959. *L'An Cinq, de la Révolution Algérienne* [A dying colonialism]. Paris: François Maspero.

Fanon, F. 1961. *Les Damnés de la Terre* [The wretched of the earth]. Paris: François Maspero.

Fanon, F. 2008 [1952]. *Black skin, white masks*. London: Pluto Press.

Foucault, M. 2009. *Security, territory, population*. New York: Picador.

Foucault, M. 2010. *The birth of biopolitics*. New York: Picador.

Gabrys, J. 2018. Becoming planetary. *E-Flux*. Accessed July 20, 2020. https://www.eflux.com/architecture/accumulation/217051/becoming-planetary/.

Gilmore, R. 2002. Fatal couplings of power and difference: Notes on racism and geography. *The Professional Geographer* 54 (1):15–24. doi: 10.1111/0033-0124.00310.

Gines, K. 2014. *Hannah Arendt and the Negro question*. Bloomington: Indiana University Press.

Glissant, E. 1990. *Poetics of relation*. Minneapolis: University of Minnesota Press.

Gomez-Barris, M. 2019. Book review essay–The colonial anthropocene: Damage, remapping, and resurgent resources. https://antipodeonline.org/2019/03/19/the-colonial-anthropocene/

Grosz, E., K. Yusoff, and N. Clark. 2017. An interview with Elizabeth Grosz: Geopower, inhumanism and the biopolitical. *Theory, Culture & Society* 34 (2–3):129–46. doi: 10.1177/0263276417689899.

Gunaratnam, Y., and N. Clark. 2014. Pre-race post-race: Climate change and planetary humanism. *Dark Matter* 9:1. Accessed July 20, 2020. http://www.darkmatter101.org/site/2012/07/02/pre-race-post-race-climate-change-and-planetary-humanism/.

Hartman, S. 1997. *Scenes of subjection: Terror, slavery, and self-making in nineteenth-century America*. Oxford, UK: Oxford University Press.

Hird, M. J. 2012. Knowing waste: Towards an inhuman epistemology. *Social Epistemology* 26 (3–4):453–69. doi: 10.1080/02691728.2012.727195.

Jackson, Z. I. 2013. Animal: New directions in the theorization of race and posthumanism. *Feminist Studies* 39 (3):669–85.

Jackson, Z. I. 2020. *Becoming human: Matter and meaning in an antiblack world*. New York: New York University Press.

Kant, I. 1994 [1756]. *History and physiography of the most remarkable cases of the earthquake which towards the end of the year 1755 shook a great part of the Earth in four neglected essays*. Hong Kong: Philopsychy Press.

Kwate, N., and S. Threadcraft. 2017. Dying fast and dying slow in black space: Stop and frisk's public health threat and a comprehensive necropolitics. *Du Bois Review: Social Science Research on Race* 14 (2):535–56. doi: 10.1017/S1742058X17000169.

Lewis, S., and M. Maslin. 2015. Defining the Anthropocene. *Nature* 519 (7542):171–80. doi: 10.1038/nature14258.

Luciano, D., and M. Chen. 2015. Introduction: Has the queer ever been human? *GLQ: A Journal of Lesbian and Gay Studies* 21 (2–3):183–207. doi: 10.1215/10642684-2843215.

Mbembe, A. 2003. Necropolitics. *Public Culture* 15 (1):11–40. doi: 10.1215/08992363-15-1-11.

McKittrick, K. 2006. *Demonic grounds: Black women and the cartographies of struggle*. Minneapolis: University of Minnesota Press.

McKittrick, K., ed. 2015. *Sylvia Wynter: On being human as praxis*. Durham, NC: Duke University Press.

McKittrick, K. 2016. Rebellion/invention/groove. *Small Axe* 20 (149):79–91. doi: 10.1215/07990537-3481558.

Mignolo, W. 2011. Geopolitics of sensing and knowing: On (de)coloniality, border thinking and epistemic disobedience. *Postcolonial Studies* 14 (3):273–83. doi: 10.1080/13688790.2011.613105.

Moten, F. 2003. *In the break: The aesthetics of the black radical tradition*. Minneapolis: University of Minnesota Press.

Muñoz, J. 2015. Theorizing queer inhumanisms: The sense of brownness. *GLQ: A Journal of Lesbian and Gay Studies* 21 (2–3):209–13.

Nye, D., L. Rugg, J. R. Fleming, and R. Emmett. 2013. *The emergence of the environmental humanities*. Sweden: MISTA. https://www.mistra.org/wp-content/uploads/2018/01/Mistra_Environmental_Humanities_May2013.pdf

Owens, P. 2017. Racism in the theory canon: Hannah Arendt and "the one great crime in which America was never involved." *Millennium: Journal of International Studies* 45 (3):403–24. doi: 10.1177/0305829817695880.

Philip, M. N. 2017. *Bla_k: Essays and interviews.* Toronto: Book*hug Press.

Povinelli, E. 2016. *Geontologies: A requiem to late liberalism.* Durham, NC: Duke University Press.

Pulido, L. 2016. Flint, environmental racism, and racial capitalism. *Capitalism Nature Socialism* 27 (3):1–16. doi: 10.1080/10455752.2016.1213013.

Pulido, L. 2017. Geographies of race and ethnicity II: Environmental racism, racial capitalism and state-sanctioned violence. *Progress in Human Geography* 41 (4):524–33. doi: 10.1177/0309132516646495.

Pulido, L., and J. Lara. 2018. Reimagining the "justice" in environment justice: Radical ecologies, decolonial thought, and the black radical tradition. *Environment and Planning E: Nature and Space* 1 (1–2):76–98. doi: 10.1177/2514848618770363.

Roane, J. T. 2018. Plotting the black commons. *Souls* 20 (3):239–66. doi: 10.1080/10999949.2018.1532757.

Saldanha, A. 2020. A date with destiny: Racial capitalism and the beginnings of the Anthropocene. *Environment and Planning D: Society and Space* 38 (1):12–34. doi: 10.1177/0263775819871964.

Scott, D. 2000. The re-enchantment of humanism: An interview with Sylvia Wynter. *Small Axe* 8:119–207.

Sealey-Huggins, L. 2018. "The climate crisis is a racist crisis": Structural racism, inequality and climate change. In *The fire now: Anti-racist scholarship in times of explicit racial violence*, ed. A. Johnson, R. Salisbury, and B. Kamunge, 99–113. London: Zed.

Sharpe, C. 2018. Foreword: The heat and the burdens of the day. In *The fire now: Anti-racist scholarship in times of explicit racial violence*, ed. A. Johnson, R. Salisbury, and B. Kamunge, xv–0. London: Zed.

Silva, D. 2007. *Towards a global idea of race.* Minneapolis: University of Minnesota Press.

Simpson, L. B. 2017. *As we have always done: Indigenous freedom through radical resistance.* Minneapolis: University of Minnesota Press.

Smith, S., and P. Vasudevan. 2017. Race, biopolitics, and the future: Introduction to the special section. *Environment and Planning D: Society and Space* 35 (2):210–21. doi: 10.1177/0263775817699494.

Spillers, H. 2003. *Black, white, and in color: Essays on American literature and culture.* Chicago: University of Chicago Press.

Tuana, N. 2019. Climate apartheid: The forgetting of race in the Anthropocene. *Critical Philosophy of Race* 7 (1):1–31. doi: 10.5325/critphilrace.7.1.0001.

Vergès, F. 2017. Racial capitalism. In *Futures of black radicalism*, ed. G. Johnson and A. Lubin. A, 72–82. London: Verso.

Vergès, F. 2019. Capitalocene, waste, race, and gender. *E-Flux* 100 (May). Accessed July 20, 2020. https://www.e-flux.com/journal/100/269165/capitalocene-waste-race-and-gender/.

Walcott, R. 2014. The problem of the human: Black ontologies and "the coloniality of our being." In *Postcoloniality–decoloniality–black critique*, ed. S. Broek

and C. Junker, 93–108. Frankfurt, Germany: Campus Verlag.

Warren, C. 2018. *Ontological terror.* Durham, NC: Duke University Press.

Weheliye, A. 2014. *Habeas viscus: Racializing assemblages, biopolitics, and black feminist theories of the human.* Durham, NC: Duke University Press.

Whyte, K. P. 2018. Indigenous science (fiction) for the Anthropocene: Ancestral dystopias and fantasies of climate change crises. *Environment and Planning E: Nature and Space* 1 (1–2):224–42. doi: 10.1177/2514848618777621.

Wilderson, F., III. 2003. The prison slave as hegemony's (silent) scandal. *Social Justice* 3 (2):18–27.

Williams, B. 2017. Articulating agrarian racism: Statistics and plantationist empirics. *Southeastern Geographer* 57 (1):12–29. doi: 10.1353/sgo.2017.0003.

Williams, B. 2018. "That we may live": Pesticides, plantations, and environmental racism. *Environment and Planning E: Nature and Space* 1 (1–2):243–67. doi: 10.1177/2514848618778085.

Woods, C. 2017. *Development interrupted.* London: Verso.

Wright, M. W. 2011. Necropolitics, narcopolitics, and femicide: Gendered violence on the Mexico–U.S. border. *Signs* 36 (3):707–31. doi: 10.1086/657496.

Wynter, S. 1984. The ceremony must be found: After humanism. *Boundary 2* 12 (3):19–70. doi: 10.2307/302808.

Wynter, S. 1996. Is "development" a purely empirical concept or also teleological? A perspective from "We the underdeveloped." In *Prospects for recovery and sustainable development in Africa*, ed. A. Y. Yansane, 299–316. Westport, CT: Greenwood.

Wynter, S. 2000. Human being as noun? Or being human as praxis? Towards the autopoetic turn/overturn: A manifesto. Accessed July 20, 2020. https://www.scribd.com/document/329082323/Human-Being-as-Noun-Or-Being-Human-as-Praxis-Towards-the-Autopoetic-Turn-Overturn-A-Manifesto#from_embed.

Wynter, S. 2003. Unsettling the coloniality of being/power/truth/freedom: Towards the human, after man, its overrepresentation—An argument. *CR: The New Centennial Review* 3 (3):257–337. doi: 10.1353/ncr.2004.0015.

Wynter, S., and K. McKittrick. 2015. Unparalleled catastrophe for our species? Or, to give humanness a different future: Conversations. In *Sylvia Wynter: On being human as praxis*, ed. K. McKittrick, 9–89. Durham, NC: Duke University Press.

Yusoff, K. 2018a. The Anthropocene and geographies of geopower. In *Handbook on the geographies of power*, ed. J. Agnew and M. Coleman, 203–16. London: Edward Elgar.

Yusoff, K. 2018b. *A billion black Anthropocenes or none.* Minneapolis: University of Minnesota Press.

Yusoff, K. 2018c. Politics of the Anthropocene: Formation of the commons as a geologic process. *Antipode* 50 (1):255–76. doi: 10.1111/anti.12334.

Language and Groundwater: Symbolic Gradients of the Anthropocene

Paul C. Adams iD

This article argues that geographers must study the power of words as integral parts of human–environment relationships, with particular attention to local meanings, to intervene more effectively in the Anthropocene. Words are important tools by which people come to understand environmental changes and develop plans to facilitate mitigation and adaptation or, alternatively, to postpone these responses. This project considers the portion of Texas underlain by the Ogallala aquifer as a system of communication, exploring stakeholder articulations through in-depth interviews. The semiotic concepts of gradients, grading, degradation, and grace are employed to facilitate consideration of how verbal articulations intersect with resource use, conservation, anthropogenic environmental change, and action within a highly conservative political context. *Key Words: Anthropocene, environmental semiotics, groundwater, language.*

We are tool-using animals, belatedly realizing that our tools affect the planet on which we live. This recognition entails nothing less than a protracted crisis, with political, economic, philosophical, and religious components (Callison 2014). Among the tools caught up in this crisis are words, semiotic devices by which people understand environmental changes and develop plans to facilitate mitigation and adaptation or, alternatively, to postpone these responses (Hulme 2008; Lakoff 2010; Boykoff 2019). Geographers must therefore study the power of words as integral parts of human–environment relationships, at scales from the global to the local, to intervene more effectively in the Anthropocene. This article demonstrates a semiotic approach to the Anthropocene, taking words about groundwater as semiotic tools that both help and hinder sustainability.

Consider how groundwater is put into words. An informational booklet by the U.S. Geological Survey (USGS) contains the following description: "On a regional scale, the configuration of the water table commonly is a subdued replica of the land-surface topography" (USGS 1999, 6). These words evoke gently rounded uplands of groundwater hidden beneath visible hills and mountains. Later in the booklet, the impact of a well on an unconfined aquifer is explained in more technical language: "dewatering of the formerly saturated space between grains or in cracks or solution holes takes place. This dewatering results in significant volumes of water being released from storage per unit volume of earth material in the cone of depression" (USGS 1999, 14). Here we find specialized terms, "dewatering" and "cone of depression," with the latter giving three-dimensional form to the anthropogenic change indicated by the former. As Tuan (1991) noted in regard to the Mississippi River, a name can in effect "be said to have created the [hydrological] system by making the entire river, and not just the parts visible to observers on the ground, accessible to consciousness" (688–89). There are also legal terms like "rule of capture," the principle whereby "a landowner legally owns as much groundwater as he [sic] can pump, subject to prohibitions against waste and malice, without liability to any neighbors whose wells go dry as a result" (Welles 2013, 486). Another important verbal tool is "desired future conditions" (DFC), which is a "quantitative description … of the desired condition of the groundwater resources in a management area at one or more specified future times" (Texas Water Code 2019, Sec. 36.001). This exemplifies a name with "the creative power to call something into being … to impart a certain character to things" (Tuan 1991, 688).

What will be explored here is how such verbal tools imply *figure–ground relations*, which in turn indicate what is taken for granted and what is worthy of notice. The article examines the semiotics of groundwater, but the same approach could be employed with any other aspect of the environment. Of particular interest here are environmental *gradients*—significant differences (across space) or changes (through time). By examining how language embodies gradients we can better understand semiotic tools working on and in the Anthropocene.

The article begins with an introduction to the study site, followed by a discussion of theory and methodology. The body of the article interprets the meanings of production and consumption, conservation and waste as key indications of semiotic processes in the Anthropocene.

Study Site

The study site includes the portion of Texas underlain by the Ogallala aquifer: 36,500 square miles in the northernmost part of the state (Figure 1). Much of the water in this formation was deposited thousands of years ago, and replenishment of the Texas portion is less than a quarter inch (6 mm) per year (Reedy et al. 2008). It is being drawn down more than a foot (30 cm) per year in significant portions of twenty Texas counties (George, Mace, and Petrossian 2011; McGuire 2017). The water table has fallen by as much as 300 feet in some areas and, at the current rate of drawdown, the region's irrigation-based economy will collapse or undergo massive transformation by the end of this century.

The region is famous as an emblem of environmental crisis. Along with portions of adjacent states, this part of Texas experienced severe drought and catastrophic agricultural failure in the 1930s Dust Bowl. Today, much of the same region is covered by circles of corn, wheat, sorghum, and cotton a half-mile or mile in diameter. A local narrative holds that as material technology diffused into the region (center-pivot irrigation systems with affordable wells and downhole pumps, better plows, and cultivators, hybrid seeds, chemical fertilizers, herbicides, and insecticides), technological change transformed this place from Dust Bowl to cropland. It is equally valid, however, to assert that after the Dust Bowl people learned to read the environment differently and communicate its potentials and constraints in more productive (although not necessarily more sustainable) ways. These readings shaped the diffusion of various agricultural technologies, relations between people, and patterns on the landscape.

Such environmental readings employ *gradients*—ranges of difference with both physical and symbolic attributes, knitting together time and space. One such gradient is total annual precipitation, which ranges from twenty-three inches (585 mm) to eighteen inches (460 mm) along an east–west transect of the study site. A more complex set of spatial gradients exists below the earth's surface, where an undulating layer of sand, gravel, silt, and clay holds the Ogallala aquifer. This formation varies in thickness, depth, and composition, creating different levels of groundwater access for farmers from neighboring counties, and even from neighboring properties. There are also temporal gradients in water, like the annual oscillation between rain and snow, dry winters and somewhat less dry summers, or the annual cycle in which cones of depression (drawdown cones) are deepened and broadened by pumping during the growing season and then partially recover each winter. This, in turn, leads to another temporal gradient, the overall drawdown of groundwater throughout the region.

Gradients are translated into language in many ways. According to one verbal formulation, those who extract a natural resource from a finite supply can be called *consumers* engaged in *consumption*; their actions result in eventual *depletion* of the resource. According to an alternative formulation, the same actors are *producers* engaged in *production*; the result of their actions is *development* of the resource. Farmers, ranchers, and water conservation administrators throughout Texas favor the second set of terms when talking about groundwater. This word choice can be pursued to see how semiotic processes shape resource use. Beyond this, we can explore how stakeholders charged with conserving water manage to articulate that conservation goal despite the narrow definition of waste they uphold. Doing so reveals gradients of several sorts: articulated gradients of language, experienced gradients in time, and constraining and enabling gradients in social and physical environments.

Method and Theory

Extended interviews were conducted in 2018 and 2019 with thirty-four stakeholders in ten counties scattered across the study site. Subjects included nineteen farmers, two ranchers, seven officials in groundwater conservation districts, four agricultural extension agents, and the director of a metropolitan water utility. The total duration of the interviews was thirty-five hours. Participants answered questions about groundwater, including its value, usefulness, and management in the High Plains landscape, as well as questions about weather, climate, and sense of place. Interviews were interpreted using a

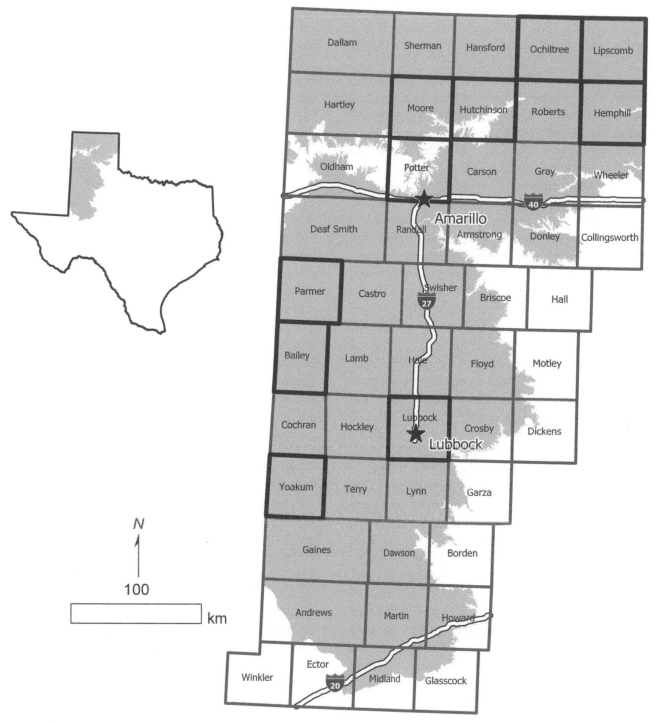

Figure 1. The Texas portion of the Ogallala aquifer (shaded) with highlighted boundaries around the ten counties where interviews were conducted. *Source:* Base map redrawn by Danielle A. Ruffe, after George, Mace, and Petrossian (2011, 51).

methodology drawing on communication geography, place attachment research, and environmental semiotics (Adams 2016; Kockelman 2016a, 2016b, 2016c; Smith 2018).

This methodology assumes that verbal constructs structure human perception and sense of place, but place has a reciprocal power over verbal meaning (Tuan 1991; Evernden 1992; Cronon 1996). Words can highlight or obscure environmental risks (Whorf 1941). Words also ascribe value to things, marking them as resources, whether human or nonhuman (Kockelman 2016a). Environmentally relevant words include general-purpose terms like "nature" (Evernden 1992) and "wilderness" (Cronon 1996), as well as terms of specific interest here: "water table," "drawdown," "right of capture," and "desired future conditions." Therefore, in an agricultural landscape, new material technologies (e.g., hybrid seeds, irrigation systems, Global Positioning System–guided combines, herbicides) do not simply alter the landscape; changes in technologies are linked both to human agency and the material environment through semiotic constructs.

Based on the writings of Ferdinand de Saussure, Charles Sanders Peirce, and others who followed in their footsteps, semiotics is the analysis of meaningful associations and distinctions. Semiotic equivalences and differences give structure to language and other aspects of culture, including human relations with natural phenomena. The Saussurian approach is most accessible and will be presented first by way of a brief introduction. We can think of a linguistic sign as composed of a signifier and a signified, each of which treats certain things as equivalent and certain things as different: The visual shapes of the letters in *water* are treated as equivalent to the sounds of the spoken word *water*, despite the manifest differences in these signifiers. Signifiers point to a signified that, in the case of water, is the chemical substance H_2O that is found on earth in solid, liquid, and gaseous states and is understood as the same thing (same signified) even when bound up in living organisms or below the earth surface. Signs have syntagmatic and paradigmatic relations to other signs. *Syntagmatic* relations are the grammar of signs, what goes with what, in what order or combination, whereas *paradigmatic* relations govern which signifiers can be substituted and how that affects meaning (Saussure [1916] 1983). The focus of the article can now be clarified as a study of paradigmatic relations between production and consumption, conservation and waste in a particular place.

Semiotic analysis problematizes the elusive, shifting, and fuzzy lines between linguistic constructs as they are built into verbal expressions and mapped onto phenomena in the world. Such analysis provides a glimpse of the constructedness of relations, not only between signs (as potentially substitutable things with constructed differences) but also between distinctions operating simultaneously at levels including signification, perception, and action. To integrate semiotics with action we must turn from

Saussure to Peirce (see Peirce and Hoopes 1991). For example, if one goes to fill a water bottle, then reads a sign "nonpotable water" and walks away without filling the bottle, the change between intended and final actions reveals the fulfillment of the sign's function, which Peirce calls the sign's *interpretant.* Both Tuan's interest in "language and the making of place" (Tuan 1991) and Cronon's concern about "the trouble with wilderness" (Cronon 1996) stem from an awareness that signs are not mere labels but also imply interpretants. That is to say that semiotic processes are geographical because they are place-making processes.

Signs are also place specific. If one drives on Interstate 10 from California to Florida, one starts on a "freeway" and ends on a "highway"; the signifier for a limited-access, multilane road changes as one moves from place to place. Similarly, departing from California, the plural of "you" is "you," but somewhere along the way the plural of "you" becomes "y'all." Thus, the sign "you" has a narrower signified in Alabama than in California. In this article, the signifiers "production" and "consumption" are assumed to map different signifieds depending on place and situation.

Insofar as place is generally understood by geographers in terms of location, locale, and sense of place (Agnew and Duncan 1989), signs do not merely vary with regard to location; they also help constitute locale and sense of place. Particular uses of words engage social and psychological processes of inclusion and exclusion, self-identity, and subjectivity (Adams 2017). Saying "y'all" in the South identifies one as an insider; it signifies who and what the speaker is. Likewise, describing well water extraction as "production" in Texas signifies that the speaker adopts local terminology relating to groundwater and is of the linguistic community. Insofar as place attachment has a communal dimension involving the "*expressive* (or symbolic) meaning of places to which people are attached" (Smith 2018, 6), such place-specific semiotic peculiarities are central to place attachment.

Place-based studies of semiotic processes are therefore needed to clarify placemaking, place attachment, and human–environment relations. In that interest, we will move between legal terminology of the Texas Water Code and excerpts from interviews with local stakeholders, particularly the directors of groundwater conservation districts (GCDs). Their attempts to grapple with environmental change will be interpreted in terms of gradients, grading, degradation, and grace. Gradients build on Peircean analysis and are "the way relative degrees (or quantities) of relevant dimensions (or qualities) vary over space, in time, or across individuals" (Kockelman 2016b, 406). People make active use of gradients, changing them according to perceived needs and interests, a type of action we can call *grading*. Grading involves individual human actions, as when a well owner creates a drawdown cone. It can also involve collective actions, as when hundreds of irrigation farmers, industrial users, and municipalities across a region all contribute to the drawdown of an aquifer.

By understanding actions relative to gradients as figure–ground relations, new light is shed on human–environment relations. The "figure is that entity whose degree (along some dimension) is being graded; and the ground is that entity whose degree (along the same dimension) is being used to grade" (Kockelman 2016b, 392). Grounds of semiotic comparison include things that range through space, as well as things that change through time. Semiotically speaking, the ground is what is taken for granted:

> For example, when I say, "the rains were heavy," you don't just need to know that I am talking about rains (as opposed to cellphones, stars, or trains); you also need to know what counts as a heavy rain around here, for people like us, engaged in an activity like this, given recent events and future plans as much as past experiences. (Kockelman 2016b, 397)

Environmental communication therefore draws on, and perpetuates, shared understandings of what is typical or normal, simultaneously indexing what is changing or unexpected, such as the disappearance of a useful or beneficial gradient. *Degradation* describes such a negative consequence of grading, and the attempt to preserve a valued gradient can be called *grace* (Kockelman 2016b, 2016c).

Insofar as "decline management is a primary goal of water-resource management" (Emel and Roberts 1995, 672) the community-organized resource regimes administered by GCDs in Texas are manifestations of this sort of grace. As White (1961) maintained, collective decisions about water can potentially be improved by expanding the range of choice from which policies are chosen. A semiotic approach to geography in the Anthropocene suggests new choices within a particular "socially and historically structured context" (Wescoat 1987, 51), facilitating more resilient articulations of each place's hydrosocial choices (Perramond 2016) relative to its manifestations of gradient-maintaining grace.

In this epistemological context, the Anthropocene can be understood as a period in which people increasingly encounter degradation of useful gradients, grading gets out of control, grace is in chronically short supply, environmental degradation becomes more widespread, and stakeholders search for verbal and visual language that will help them to intervene (Moser and Boykoff 2013). We now move to research findings from West Texas and the Panhandle, with attention to water consumption and production, waste and conservation.

Words and Water

Consumption or Production?

As signs, consumption and production are closely tied to grading and degradation. Generally, consumption depletes, degrades, or uses up something useful, whereas production creates, increases, or mobilizes something useful; the former is a shift toward absence, whereas the latter is a shift toward plenitude. This semiotic relationship varies geographically and historically, however. The Texas State Water Code (henceforth the Water Code) avoids the terms *consumption*, *extraction*, and *depletion* when referring to human use of groundwater (Texas Water Code 2019). In chapter 36 of the Water Code, there are seventy separate references to water involving words related to *produce*, including "production from water wells," "producing of wells," "water produced," "a well that produces the majority of its water," "groundwater that an aquifer is capable of producing," and so on. In striking contrast, the term *consumption* appears only once in the 39,541-word document. Behind the legalese (a product of the time in which the Water Code was written as well as its subsequent revisions) lurks a cornucopian model of the world in which people only create or augment hydrological resources, never depleting, degrading, or exhausting those resources. In Texas, an artesian well and a pumped well both "produce" water, reflecting a disenchanted, economistic worldview diametrically opposed to earlier understandings of the hydrologic cycle, where water moving through the environment was read as a sign of divine providence, supernatural power, and sacred perfection (Tuan 1968).

All of the Texas water district administrators who were interviewed employed the term *production* to refer to water extracted from the Ogallala aquifer, calling farmers and ranchers with working wells *producers*. Although these administrators were clearly dedicated to the goal of groundwater conservation, their production-oriented language is an unrecognized obstacle to reaching their goals and objectives; it positions water use on a temporal gradient—a slope from less to more, from lack to potential—which fails to reflect drawdown. When Becky,[1] the manager of a GCD east of Amarillo, described challenges facing water conservation districts she said: "[T]hey have people who can produce a lot, and they have people who can't produce very much." When Patricia, the manager of a GCD west of Lubbock, spoke about limits on water use she explained: "You still can produce the water, but you're gonna have to use maybe more than one well to get that production so that that smaller capacity pump is in the hole." When Jacob, the general manager of a large, centrally located GCD, pointed to a model of the aquifer, he said: "[A] well here versus a well where there's larger gravel, those two wells are going to have a different *production capability*." Through such spontaneous verbal articulations, the terms *produce*, *producer*, *and production* are drawn from the water code and transformed into practice. The term *produce* indeed serves as a general word to describe water sourced from a well, even if it flows on its own, as from an artesian well. The term is applied whether one is obtaining water from a well that replenishes or from a well that does not. This broad semiotic mapping implies that wells generally participate in a gradient (or gradients) tending toward abundance, potency, and value. Officially sanctioned words are missing if one wants to talk about groundwater degradation and exhaustion in Texas.

Beyond the normative question of how we should speak and write to better manage scarce resources lies a broader semiotic question: How can we articulate reality to better reflect increasing scarcity in the Anthropocene? Water districts in the study area have been innovative communicators. They have developed physical and digital models of the aquifer, technical reports, maps and manuals (Emel and Roberts 1995), lessons for local schools, and even trailers outfitted with interactive displays of hydrological processes. The weight of linguistic habit, however, continues to obstruct their communications about groundwater.

Conservation, Waste, and the Law

We turn now to another word with an interesting career in the Panhandle and West Texas. In the study region, responses to the term *conservation* range from neutral to positive, despite the region's conservative politics. This is due in part to the fact that when Texas added the "Conservation Amendment" to the State Constitution in 1917, conservation had a distinctly different meaning than it does today. It included the capture of surface water and the drilling of wells (Green 1973; Mace 2016). Conservation evolved after the Dust Bowl, when relatively erodible land was taken out of production by the federal Soil Bank program, which was renamed the Conservation Reserve Program (CRP) in the 1980s. This program sends over $74 million in federal funds to this part of Texas each year to support fallowing some 2 million acres (U.S. Department of Agriculture 2019), linking conservation not only to soil preservation but also to household financial security. Meanwhile, the state has enabled and encouraged the creation of GCDs at the local level. Approximately 100 of these GCDs are now recognized by the state, each guided by a locally elected board of directors for the purpose of managing groundwater. It is not surprising, therefore, that Panhandle conservatives support conservation. Conservation has been performed and articulated here in terms of resource capture, federal subsidies (CRP payments), and local governance (GCDs), all animated by a reigning logic of efficient resource capitalization (Trigilio 2016; Opie, Miller, and Archer 2019).

The CRP and GCDs can slow groundwater depletion. The latter often enforce setbacks from property lines when drilling wells, limit water extraction to a certain number of gallons per minute or acre-feet per year, and set the minimum distance allowed between adjacent wells. GCD planning tools also include desired future conditions; for example, 50/50 (50 percent of groundwater left after fifty years): a temporal gradient (drawdown) in the form of a policy objective linked to spatial gradients (varying groundwater availability across the GCD) and determined through public debate. Unfortunately, in many cases such conservation efforts are sufficiently lenient to accommodate the current rates of depletion.

One of the main functions of a GCD is nonetheless to prevent the waste of groundwater. One might therefore expect conservation and waste to be articulated as opposing philosophies. Oddly, conservation and waste are not coded semiotically as opposites in the study site. "Waste" is defined in the state's Water Code as "the flowing or producing of wells from a groundwater reservoir if the water produced is not used for a beneficial purpose" or "willfully or negligently causing, suffering, or allowing groundwater to escape into any river, creek, natural watercourse, depression, lake, reservoir, drain, sewer, street, highway, road, or road ditch, or onto any land other than that of the owner of the well" (Texas Water Code § 36.001). Under Texas water law, then, waste does not mean consuming water too quickly. In Jacob's words, "In Texas the legislature has stated that allowing water to escape your property, that constitutes waste. OK, so you need to keep it on your property." Well water crossing a property line in a ditch then sinking into the ground is waste, but well water moving across the same property line in a bottle for sale as drinking water, or in a tanker truck for use in fracking,[2] is not considered waste. One can also allow well water to flow in an existing waterway, but this requires a "bed and banks permit." State law in effect condones two related forms of "capture"—territorial capture and capitalist capture—as the opposite of "waste," although neither necessarily involves using less water. The determining factor is whether groundwater is being used for some "beneficial" purpose on the user's property or elsewhere or, alternatively, if the water is flowing without regard to human objectives.

The closely related "rule of capture" dates to 1904 (*Houston & Texas Central Railway Co. v. East* 98 Tex. 146, 81 S.W. 279 [1904]) and depends on the common law principle that every landowner in the state has a right to take, for use or sale, all of the water that he or she can capture; the state "recognizes that a landowner owns the groundwater below the surface of the landowner's land as real property" (Texas Water Code 2019 § 36.002). This official recognition implies that a subterranean flow of water from Property A to Property B is not recognized as seizure of Owner A's property by Owner B even though a well on Property B, operated by Owner B, might be causing or accelerating that flow. This legal territorialization of water (Perramond 2016) has the odd effect that the water one owns is constantly changing, because the Ogallala aquifer flows at a rate of about a foot (30 cm) per day, which adds up to 122 yards (109 m) per year, and local flows across (under) property lines can be much faster in response to drawdown cones (Quinn and Woodward 2015). This territorial definition of water rights creates profound contradictions, causing water to appear as a "badly behaved substance" (Emel, Roberts, and Sauri 1992, 38). Those whose job it is to manage water manifest these contradictions between territorialized property and material property as part of their place-based subjectification as environmental actors (Emel, Roberts, and Sauri 1992).

Playing by the Rules. Speaking with Troy, the general manager at a GCD that has implemented unusually comprehensive water use regulations, I asked whether he received any resistance from landowners. He replied:

> Oh yes. All the time! Let's, let's be straight about this: The water under your land is coming from somebody else's land. And somebody else owned it at one time. The rule of capture allows you to continue to pump and not really have to worry about the guys around you, except for the groundwater conservation district. So if you're telling me that you should be able to just pump whatever, and the hell with everybody else around you, that doesn't … that is not groundwater management. And yeah, I've heard that before!

This answer expressed personal commitment to groundwater management but left open the question of how Troy managed to defend the need for regulation. After further prompting, he explained:

> We hold everyone to the same account. If you go look at our rules in that book you won't see any difference in public water supply water rules compared to irrigators or industrial users. We treat everyone the same. And the reason is that we do want to have something left in fifty years or forty years.

Equal treatment is one way of dealing with differences among stakeholders, although owners of more land can extract more water, so equality does not necessarily mean equity. Troy's next comment invoked gradients in a different way, pointing out that some of the oldest landowners in the area were following the rules, so others should be able to adapt at least as well. He then followed up with this:

> Something else about this area I really like: We're real conservative. … I've used that as one of the things to say [to people who argue against regulation]: Look, everybody else out there is playing by the rules and seems it's not bothering them, so what's your problem?

Here, interestingly, the region's extreme conservatism (on the far right of political gradients) is taken as a sign that resource users desire equal application of rules. Although the association between conservatism and commitment to equality is debatable, Troy's comments indicate the discursive opportunity to link conservation to a conservative sense of place. Not only is sense of place "deeply politicized as people defend a sense of place rooted in one narrative and dismiss countervailing narratives as distortions and delusions" (Adams 2017, 5081), but narratives employ signs, and signs are interpreted in place-specific ways. Where regulation is rejected, conservation can be presented in other terms, such as protecting private property or preserving fairness.

Who You're Gonna Sit at Church With. As previously explained, in Texas waste is not necessarily the opposite of conservation. GCD administrators frame their role primarily around the preservation of peace and order, which can be thought of in Kockelman's terms as a kind of grace, in this case an effort to preserve valued social (as opposed to geophysical) gradients. Patricia, the general manager of a one-county conservation district, said:

> A lot of people think water districts are out preventing you from getting to your private property rights, and that's such a misconception because we're actually protecting you from the people who are producing next to you. … By our spacing regulations, [your neighbor's wells] aren't interfering with what's going on under your land. And so if you choose not to irrigate your property for so many years, um—of course, with gravity and the way the aquifer flows there is some [loss of water to one's neighbors], with the rule of capture with Texas—but, for the most part the way we're spacing out [wells] so that that cone of depression doesn't go underneath your property, you're protected from that [loss of water to one's neighbors].

Patricia further articulated a perspective on water that reflects the GCD's role in terms of community and morality:

> I know my producers and they know me. They know our office and that one-on-one communication. They know they can call me if they have a question. They know our board because it's who you're gonna sit at church with on Sunday morning and have those real conversations if they have an issue.

Conservation is articulated in these place-based terms as caring, neighboring, and leading a moral life. This place-based discourse engages the local value placed on community order, thus securing cooperation and buy-in from local stakeholders.

Conclusion

What people say does not reveal its full meaning until we drill down into the underlying semiotic gradients. In West Texas and the Panhandle, tensions between production and consumption, conservation and waste point to stakeholders' locally coded understandings of resource management. These words are ways of interpreting evolving human– environment relationships. They reflect gradients of groundwater in space, grading and degradation of hydrological resources through time, and the grace of achieving conservation goals through local commitment to the ideal of a peaceful, fair, stable, and moral community.

Like an aquifer, the currents of a linguistic underworld can be charted and its flows can be followed. Semiotic analysis helps to map the human–environment relations in a place. It reveals how the powerful text of a law circulates through environmental agents like conservation administrators and local water users, crossing boundaries and defying capture. The questions implied by this approach are not just about the human power to shape the environment but also about meaningful differences, and differences in meaning, and local forms of grace flowing below the surface, slowing degradation and hastening acceptance of place-based understandings of conservation. If the Anthropocene is a time when people's role in shaping the environment has come to the fore, then we must be aware that people are themselves shaped by an environment of language that channels their thoughts and actions.

ORCID

Paul C. Adams http://orcid.org/0000-0002-9303-0027

Notes

1. Names of interview respondents have been changed, as per University of Texas Institutional Review Board Exempt Protocol Number 2018-05-0099.
2. *Fracking* is a common term for hydraulic fracturing, a technique in which water and various "proppants" are injected into the oil-bearing formation under high pressure to facilitate the extraction of oil and gas.

References

Adams, P. C. 2016. Placing the Anthropocene: A day in the life of an enviro-organism. *Transactions of the Institute of British Geographers* 41 (1):54–65. doi: 10. 1111/tran.12103.

Adams, P. C. 2017. Place. In *The international encyclopedia of geography: People, the earth, environment, and technology*, ed. D. Richardson, vol. X, 5073–85. New York: Wiley-Blackwell and the Association of American Geographers.

Agnew, J. A., and J. S. Duncan, eds. 1989. *The power of place: Bringing together geographical and sociological imaginations*. London and New York: Routledge.

Boykoff, M. 2019. *Creative (climate) communications: Productive pathways for science, policy and society*. Cambridge, UK: Cambridge University Press.

Callison, C. 2014. *How climate change comes to matter: The communal life of facts*. Durham, NC: Duke University Press.

Cronon, W. 1996. The trouble with wilderness. *Environmental History* 1 (1):47–25. doi: 10.2307/3985059.

Emel, J., and R. Roberts. 1995. Institutional form and its effect on environmental change: The case of groundwater in the southern High Plains. *Annals of the Association of American Geographers* 85 (4):664–83. doi: 10.1111/j.1467-8306.1995.tb01819.x.

Emel, J., R. Roberts, and D. Sauri. 1992. Ideology, property, and groundwater resources: An exploration of relations. *Political Geography* 11 (1):37–54. doi: 10. 1016/0962-6298(92)90018-O.

Evernden, N. 1992. *The social creation of nature*. Baltimore: The Johns Hopkins University Press.

George, P. G., R. E. Mace, and R. Petrossian. 2011. Aquifers of Texas. Report 380, Texas Water Development Board, Austin, TX.

Green, D. E. 1973. *Land of the underground rain: Irrigation on the Texas High Plains, 1910–1970*. Austin: University of Texas Press. doi: 10.1086/ahr/79.4.1280-a.

Hulme, M. 2008. Geographical work at the boundaries of climate change. *Transactions of the Institute of British Geographers* 33 (1):5–11. www.jstor.org/stable/30131204. doi: 10.1111/j.1475-5661.2007.00289.x.

Kockelman, P. 2016a. *The chicken and the quetzal: Incommensurate ontologies and portable values in Guatemala's cloud forest*. Durham, NC: Duke University Press.

Kockelman, P. 2016b. Grading, gradients, degradation, grace: Part 1: Intensity and causality. *HAU: Journal of Ethnographic Theory* 6 (2):389–423. doi: 10.14318/hau6.2.022.

Kockelman, P. 2016c. Grading, gradients, degradation, grace: Part 2: Phenomenology, materiality, and cosmology. *HAU: Journal of Ethnographic Theory* 6 (3):337–65. doi: 10.14318/hau6.3.022.

Lakoff, G. 2010. Why it matters how we frame the environment. *Environmental Communication* 4 (1):70–81. doi: 10.1080/17524030903529749.

Mace, R. 2016. So secret, occult, and concealed: An overview of groundwater management in Texas. Paper presented at Conference Law of the Rio Grande, Water Law Institute, CLE International, Santa Fe, NM, April 14–15.

McGuire, V. L. 2017. Water-level and recoverable water in storage changes, High Plains Aquifer, predevelopment to 2015 and 2013–15. Scientific Investigations Report 2017-5040, U.S. Geological Survey, Reston, VA.

Moser, S. C., and M. T. Boykoff, eds. 2013. *Successful adaptation to climate change: Linking science and policy in a rapidly changing world*. London and New York: Routledge.

Opie, J., C. Miller, and K. L. Archer. 2019. *Ogallala: Water for a dry land*. 3rd ed. Lincoln: University of Nebraska Press.

Peirce, C. S., and J. Hoopes. 1991. *Peirce on signs: Writings on semiotic*. Chapel Hill: University of North Carolina Press.

Perramond, E. P. 2016. Adjudicating hydrosocial territory in New Mexico. *Water International* 41 (1):173–88. doi: 10.1080/02508060.2016.1108442.

Quinn, J. A., and S. L. Woodward. 2015. *Earth's landscape: An encyclopedia of the world's geographic features*, vol. 1. Santa Barbara, CA: ABC-CLIO.

Reedy, R. C., S. Davidson, A. Crowell, J. Gates, O. Akasheh, and B. R. Scanlon. 2008. Groundwater recharge in central High Plains of Texas: Roberts and Hemphill Counties. Report to TWDB, Austin, TX.

Saussure, F. de. [1916] 1983. *Course in general linguistics*. London: Duckworth.

Smith, J. S., ed. 2018. *Explorations in place attachment*. London and New York: Routledge.

Texas Water Code. 2019. Water code. Chapter 36. Groundwater conservation districts. Accessed May 12, 2020. https://statutes.capitol.texas.gov/Docs/WA/htm/WA.36.htm#36.113.

Trigilio, M. L. 2016. *Written on water: A modern tale of a dry west*. Mill Valley, CA: Green Planet Films.

Tuan, Y.-F. 1968. *The hydrologic cycle and the wisdom of God: A theme in geoteleology*. Toronto, ON: University of Toronto Press.

Tuan, Y.-F. 1991. Language and the making of place: A narrative-descriptive approach. *Annals of the Association of American Geographers* 81 (4):684–96. https://www.jstor.org/stable/2563430. doi: 10.1111/j.1467-8306.1991.tb01715.x.

U.S. Department of Agriculture, Farm Service Agency. 2019. Conservation Reserve Program statistics. Accessed May 12, 2020. https://www.fsa.usda.gov/programs-and-services/conservation-programs/reports-and-statistics/conservation-reserve-program-statistics/index.

U.S. Geological Survey. 1999. Sustainability of groundwater resources. U.S. Geological Survey Circular 1186, Reston, VA.

Welles, H. 2013. Toward a management doctrine for Texas groundwater. *Ecology Law Quarterly* 40 (2):483–515. https://www.jstor.org/stable/24113736

Wescoat, J. L., Jr. 1987. The "practical range of choice" in water resources geography. *Progress in Human Geography* 11 (1):41–59. doi: 10.1177/030913258701100103.

White, G. F. 1961. The choice of use in resource management. *Natural Resources Journal* 1 (1):23–40. https://digitalrepository.unm.edu/nrj/vol1/iss1/2.

Whorf, B. L. 1941. The relation of habitual thought and behavior to language. In *Language, culture, and personality: Essays in memory of Edward Sapir*, ed. L. Spier, A. I. Hallowell, and S. S. Newman, 75–93. Menasha, WI: Sapir Memorial Publication Fund.

Agri-Food Systems and the Anthropocene

Emily Reisman (iD) and Madeleine Fairbairn (iD)

Understanding the Anthropocene—as both a set of physiological phenomena and as an existential crisis of modernity—requires interrogating Earth-changing transformations in food and agriculture. Agri-food systems are not only at the core of alarming environmental trends; they also offer opportunities to directly engage important challenges to the Anthropocene concept. Many human geographers and other social scientists have raised critiques of the Anthropocene designation as glossing over social inequities, codifying a separation of humans from their environments, and naturalizing current transformations as complete and irreversible. In this article we interrogate the intersection between agri-food studies and critical Anthropocene scholarship, arguing that agri-food systems serve as a through line to competing Anthropocene origin stories, a source of theoretical insight for the complexity of human–environment relations, and a site of agency for engaging alternative futures. First, we examine four of the most commonly proposed starting points for the Anthropocene epoch, arguing that a focus on food and agriculture at each historical moment reveals the limits, frictions, and social unevenness of anthropogenic change. Second, we highlight theoretical tools from critical agrarian studies that help build a more complex understanding of agriculture and the Anthropocene, emphasizing the active role of agroecosystems and the centrality of structural inequalities to agroenvironmental change. Finally, we examine how food- and agriculture-related social movements are working to forge more livable futures by accounting for precisely the matters that characterizations of the Anthropocene as an epoch of global human dominance frequently overlook: socioecological unity and political economic difference. *Key Words: agriculture, Anthropocene, critical agrarian studies, food, political ecology.*

Agriculture is everywhere. It produces not only food but also fiber (to make cloth, rope, and paper), fuel (to power buildings, vehicles, and machinery), "fun" (in the form of coffee, tea, spices, and intoxicants), and pharmaceuticals. It is perhaps not surprising, therefore, that the environmental impacts associated with agriculture are frequently referenced as evidence that we have entered a new geologic time period—a "human dominated geological epoch" known as the Anthropocene (Lewis and Maslin 2015, 171). Rapid increases in water use, fertilizer contamination, deforestation, greenhouse gas emissions, and, of course, human population are among the most prominent metrics leveraged in support of the Anthropocene designation. Alarming statistics, often displayed as a series of exponential growth charts (Steffen, Crutzen, and McNeill 2007), portray agriculture's expanding environmental toll. Agriculture, it appears, supplies ample evidence that the "age of humans" (Lynas 2011) has arrived.

The Anthropocene designation is far from settled fact, however, sparking intense debate within the social sciences, including human geography (Castree 2014; Biermann et al. 2016; Veland and Lynch 2016; Preiser, Pereira, and Biggs 2017). Attributing global change to a universalized human "Anthropos," many critical scholars argue, risks ignoring the fact that some human groups have contributed far more to globally problematic transformations than others. It also tends to elide the role of structural inequalities along the lines of race, gender, class, geography, and more in producing these changes (Malm and Hornborg 2014), as well as the long-standing eco-social crises experienced by indigenous communities (H. Davis and Todd 2017). Indeed, the category of human itself has historically anchored social hierarchies (Weheliye 2014; Yusoff 2018) and thus might be ill equipped to define a more desirable future. Singling out a species as irreversibly dominant might inhibit action by naturalizing and depoliticizing ecological crises (Swyngedouw and Ernstson 2018), normalizing narratives of control as progress (Simpson 2020), institutionalizing human mastery (Crist 2016), and reifying a false division between humans and the biophysical world of which we are a part (Clark

2011). Social scientists, particularly those influenced by science and technology studies, also observe that writings on the Anthropocene tend to gloss over the diverse ways in which other organisms and objects shape and are coconstituted with human action (Braun 2008). By now, these critiques should be impossible to ignore, yet existing scholarship on agri-food systems and the Anthropocene often misses the opportunity to address them.

Although a rich and growing body of research has drawn attention to the central role of agriculture in global environmental change—from soils to water to atmospheric carbon—it has not always entered into conversation with critiques of the Anthropocene concept. Perhaps the most prominent scholarly effort to address agri-food systems in this new presumed epoch, for instance, the EAT–*Lancet* Commission report on Food in the Anthropocene, accepts the Anthropocene designation at face value and proposes a "universal healthy reference diet" for minimizing the impact of a period in which "humanity" has become the "dominating driver of change" (Willett et al. 2019, 450). Although recognizing the centrality of agri-food systems to forging a better future, this work pays little attention to the most controversial aspects of the Anthropocene concept: the universalization of human beings at the species level despite highly unequal contributions to global change and the role of structural inequalities in exacerbating environmental harm (Malm and Hornborg 2014). At the same time, some agri-food scholars deploy the term Anthropocene to add urgency to long-standing efforts, such as agrobiodiversity conservation (Zimmerer et al. 2019), alternative food networks (Beacham 2018), ethical business management (Blok 2018), and food justice (Jones 2019), or to reflect the eco-anxieties of their research subjects (Sexton 2018), without substantively addressing critiques of the Anthropocene concept. This is surprising considering the vibrant tradition of agrarian change scholarship that wrestles with precisely the themes that critics argue are ignored by the Anthropocene designation: the role of uneven power relations in producing environmental change; the flaws in modernist narratives of human mastery; and the problematic divide between human beings and the world of which we are a part.

We argue that critical agri-food scholarship provides a powerful vantage point from which to engage with the Anthropocene idea—one that precludes historical oversimplifications, enriches theoretical insights, and keeps alternative futures squarely in view. Rather than wholeheartedly accepting or rejecting the term Anthropocene, we embrace its controversial nature as a source of insight. The Anthropocene concept is contentious because it calls for a single, all-encompassing global story that risks erasure of alternatives. Focusing on agri-food systems as a central theme for understanding Anthropocene debates helps to hold several narratives simultaneously while keeping central concerns within the field of view. Examining the histories and theories of agrarian change alongside Anthropocene discourse, we argue, illustrates the limits, friction, and unevenness of human influence while underscoring the inseparability of people with their environments. Moreover, if the Anthropocene concept is to mobilize action for reversing problematic trends, it must address agriculture in a way that does not treat it only as a set of *impacts* to be avoided but rather as a site of political economic *processes* to be accounted for and reimagined. In this article we open up such possibilities by offering an integration of agri-food studies and critical Anthropocene scholarship, arguing that agri-food systems can serve as a critically engaged through line to competing Anthropocene origin stories, a source of theoretical insight for the complexity of human–environment relations, and a site of agency for forging alternative futures.

Agri-Food Systems as Historical Through Line

The meaning attributed to the Anthropocene rests heavily on the date chosen to mark its beginning. No matter which historical moment is chosen to mark the start of the Anthropocene, agriculture and food serve to complicate the story of global transformations by "mankind" (Crutzen 2002). We deploy the term *through line* to describe a "common or consistent element or theme shared by items in a series or by parts of a whole" (*Merriam-Webster* 2020), underscoring the continuity between seemingly disparate features. We begin by arguing that a focus on the through line of agri-food system transformations reveals the role of social complexity in each of four prominent origin stories for the Anthropocene: the dawn of agriculture, the emergence of capitalism, the Industrial Revolution, and the Great Acceleration of the mid-twentieth century. Although these are not the only temporal markers under consideration for the start of the Anthropocene, they have received the most scholarly attention and to some degree fall along disciplinary lines, with plant domestication largely favored by archaeologists (B. D. Smith and Zeder 2013), early capitalism by political economists (Altvater et al. 2016), the Industrial Revolution by atmospheric chemists (Steffen, Crutzen, and McNeill 2007), and the Great Acceleration by geologists (Zalasiewicz et al. 2015). Regardless of their relative merits for marking a new epoch, at each historical moment a closer look at agri-food systems highlights the necessity of acknowledging the limits, frictions, and unevenness of human influence on Earth systems.

The rise of agriculture, or Neolithic Revolution, is one proposed marker for the onset of the Anthropocene that, upon closer inspection, reveals the oversights that come of treating history as linear and of treating humans as an undifferentiated group affecting the globe. As archaeologists have shown, the transition to farming has been far from complete or irreversible. There is no clear mark indicating "before" and "after" the transition to agriculture, as groups shifted food provisioning strategies based on their circumstances, often with agriculture as a last resort (Head 2014). Labeling societies as either foragers or agriculturalists, even for a single moment in time, is also falsely dualistic, because many groups have lived from both wild and cultivated foods for thousands of years (B. D. Smith 2001). Where agriculture did arise, it is often associated with forced labor and the use of violence by elites to retain populations attempting to flee the disease and periodic famine brought on by farming's dense settlements and instability (Cohen 2009; Scott 2017), a clear indication that treating all humans as equally responsible for agriculture's impacts neglects deep inequalities. Importantly, foraging continues today (Jordan 2014), and its association with lack of "civilization" has proven deeply damaging, showing how the idea of a linear and complete conversion to agriculture is not only inaccurate but ignores real harm caused by such assumptions. For example, Native Americans in what is now called California "tended the wild," managing land for food production without plant or animal domestication, until forced to farm by settler colonists who deemed their nonagricultural lifeways inferior, expendable, and threatening (Anderson 2013). In considering the Anthropocene, a closer look at agriculture's origins indicates that the transition to farming is not unidirectional, comprehensive, or universally attributable to human groups, and thus neither are its biogeochemical effects.

A second influential origin story for the Anthropocene is the development of capitalism and its attendant practices of colonial conquest. A focus on agri-food systems during this period highlights the need to recognize that not all humans "dominate" the Earth equally and that capitalist transformations remain patchy. Without attempting to summarize extensive theorizations on the topic (Haraway 2015; Altvater et al. 2016; Moore 2017, 2018), we focus here on how the agrarian change inherent to capitalism's origins demonstrates that human-induced environmental changes are not only uneven but fundamentally predicated on inequality. The origin of capitalism is marked by the late sixteenth-century English enclosures in which elites appropriated peasant land to graze sheep and engage in the highly profitable wool trade, while peasant farmers became landless and thus obliged to engage in wage labor (Perelman 2000). Surplus capital and social dislocations intensified European colonial expansion in pursuit of raw materials and new markets, widely disseminating a plantation model of agriculture premised on racialized enslaved labor (Burnard 2019) while devastating indigenous foodways (Whyte 2018). In short, this was not a shift in the relationship between all humans and all natural resources. Instead, the change was driven by relatively small groups of elites and it relied on social hierarchies that exploited some for the gain of others. Importantly, capitalism's agrarian transformations were met with significant resistance, including within the plantation itself (J. Davis et al. 2019), and its dominance remains patchy to this day (Tsing 2015) as nonmarket exchange and economic diversity persist (Gibson-Graham 2006). The role of the agri-food system in capitalism and colonial expansion is undoubtedly world-changing, yet it also underscores how capitalism's effects on the Earth were rooted in social hierarchy and are neither uniformly distributed nor complete.

A third contender, and one of the earliest proposals for an Anthropocene "golden spike,"[1] is the Industrial Revolution, when greenhouse gas emissions began to rise. Here again, agrarian dynamics highlight that not all humans participated equally and that social dislocations were necessary conditions for industrialization's environmental changes. Although the Industrial Revolution is often considered an urban phenomenon, the proliferation of factories was, in fact, tightly linked with transformations in the countryside. Population migration into the cities created a "metabolic rift" whereby nutrients transported from the farm as produce no longer cycled back to replenish the soil's fertility (Foster 1999). The resulting soil exhaustion sparked a fertilizer industry that now impairs water quality around the world. The most iconic early industrial machine, the loom, rapidly transformed wool and cotton into global commodities, accelerating demand for raw materials, which in turn prompted expansion of colonial control in India and slavery in the U.S. South (Bailey 1994; Beckert 2014). Additionally, the Industrial Revolution's legacy of pollution, atmospheric and otherwise, would not have been possible without cheap agricultural products from the colonies subsidizing factory workers' low wages (Mintz 1986). Like both agriculture and capitalism, industrialization is a patchy transformation rife with resistance and not practiced by all peoples. Much as Luddites destroyed looms, farm laborers actively protested the rural poverty enabled by grain threshing machinery (Griffin 2015). The Industrial Revolution's fossil fuel fingerprint is undoubtedly impactful on a global scale, but its entanglement with rural transformations in Europe and beyond underscores the friction and disparities inherent to its ecological transformations.

Finally, in May 2019, the Anthropocene Working Group of the International Commission on Stratigraphy voted to establish the mid-twentieth century as the temporal marker for the Anthropocene. This periodization has been justified both for its pragmatism—atomic radiation serves as a clear and measurable indicator worldwide, although industrial chicken bones (C. E. Bennett et al. 2018) or plastics (Zalasiewicz et al. 2016), among other indicators, might be equally ubiquitous markers of this moment— and for its association with a set of changes known as the Great Acceleration: a steep rise in greenhouse gas levels, surface temperature, ocean acidification, coastal nitrogen, human population, real gross domestic product, and additional indicators of global change (Steffen, Crutzen, and McNeill 2007). Although diagrams of planetary transformation during this period have begun to acknowledge the geographic unevenness of these indicators by distinguishing between the environmental impacts of industrialized nations, BRIC (Brazil, Russia, India, China), and nonindustrialized countries (see, e.g., Steffen et al. 2015), an understanding of agrarian change during this period reveals that accelerated changes midcentury were rooted in inequalities not only between countries but within them. The environmental degradation associated with the industrialization of agriculture goes hand in hand with the processes concentrating power in the agricultural sector and dispossessing many of their lands. The Green Revolution, spanning roughly the 1940s to the 1970s, fundamentally changed agricultural practices around the globe when agronomists from the United States popularized high-yielding grain varieties along with synthetic pesticides and fertilizers, and international development agencies promoted debt-inducing, large-scale irrigation infrastructure.[2] Although input-intensive farming methods succeeded in greatly increasing food production globally, they are also widely critiqued for exacerbating socioeconomic inequalities, deepening the dependence of farmers on multinational corporations, shrinking agrobiodiversity, and upending ecologically suited agricultural practices (Weis 2007; Patel 2013). In the United States, Canada, and Europe, farmers found themselves on a "technological treadmill" that drove agricultural production so high that crop prices, and thus farm incomes, fell precipitously, driving the majority of farmers out of agriculture altogether (Cochrane 1993). The treadmill that accelerated environmental impacts simultaneously deepened inequalities. From an agrarian change perspective, the mid-twentieth-century accelerations reflect a moment when control over the land quickly slipped away from its primary stewards, making the questions of who is responsible for environmental change more complex than simple geography.

Agri-food systems provide a through line that can make prominent Anthropocene origin stories more accountable to the Anthropocene's most influential critiques. In each historic transformation, humanity appears as far from a unified force, and domination over Earth systems is far less complete, unidirectional, or assured than it appears at first glance.

Agri-Food Systems as Source of Theoretical Insight

The study of agrarian change, in addition to enabling a more nuanced empirical picture of proposed Anthropocene origins, provides theoretical support for engaging questions of human exceptionalism and uneven power relations in Anthropocene discussions more broadly. Critical agrarian studies has been wrestling with fundamental questions about nature, the economy, and technological change for well over a century. Although scholars working in this tradition, like most Anthropocene advocates, have generally assumed a problematic Cartesian binary between nature and humanity, their conclusions firmly reject the notion of human dominance: for them, the materiality of soils, plants, and other organisms composing the farm actively resist control and profoundly shape human economic activities (Goodman, Sorj, and Wilkinson 1987; Mann 1990). They further demonstrate how the environmental impact of agriculture is inextricably linked to the growing concentration of economic power within the agri-food system (Buttel 1980; Howard 2016), as well as geopolitical arrangements rooted in colonialism (Friedmann and McMichael 1989). Here we trace how critical theories of agrarian change emphasize the limits to human activities imposed by the biophysical characteristics of agriculture, as well as the centrality of power differentials to agriculture's environmental transformations. Highlighting social struggles within the agri-food system and the unruly character of biological processes, this body of scholarship constitutes a clear countercurrent to Anthropocene narratives of effective dominance by an undifferentiated humanity.

Studies of agrarian political economy have long acknowledged biophysical processes as a source of constraint and unpredictability in human initiatives. In the late nineteenth century, Marxist scholar Kautsky (1988) investigated what he called the "agrarian question." Though capitalism was clearly effecting rapid and seismic change within urban manufacturing, capitalist development in rural areas was much slower and less complete, a result he attributed, in part, to agriculture's unique relationship with land. Land's limited quantity, variable fertility, immobility, and fragmented ownership all served to constrain the profits and scalability of capitalist agriculture, allowing peasant production systems to persist. This conclusion resurfaced during the twentieth century, when geographers and sociologists of agriculture observed that climate, biology, and a wide array of other environmental factors also act as checks on agricultural industrialization. Unlike manufacturing, they argued, the dependence on seasonal change, relatively slow processes of plant

and animal reproduction, and perishability of agricultural products constitute major "obstacles to the development of a capitalist agriculture" (Mann and Dickinson 1978, 466). At the same time, the difficulty of hiring a seasonal workforce, the challenge of controlling workers spread across large distances, the slowness of product innovation in the form of plant or animal breeding, and the unpredictable risks associated with weather and pests continually frustrate ambitions to fully transform fields into factories (FitzSimmons 1986; Goodman, Sorj, and Wilkinson 1987; Kloppenburg 1988; Mann 1990). The biophysical properties of agroecological systems, what some would call "nature," put a major hitch in capitalist plans for planetary transformation; through the lens of agriculture, human domination looks a lot less inevitable and a lot more contingent and constrained.[3]

Recent studies of agriculture incorporate insights from science and technology studies to detail how the material specificities of plant, animal, and fungal life on the farm actively shape economic activity (Watts and Scales 2015). The genetic gymnastics of a soil fungus plaguing strawberry fields, for example, deepens growers' dependence on toxic, ozone-depleting, soon-to-be-banned fumigants, rendering the industry's future exceedingly fragile (Guthman 2019). Industrial hog producers rely on a precarious pool of migrant laborers to hand-feed piglets because reproductively optimized sows now birth more offspring than they have nipples (Blanchette 2020). Biology, in these accounts, is a source not only of "obstacles" but also "opportunities and surprises" (Boyd, Prudham, and Schurman 2001, 555). The physical properties of the apple, for instance, place many idiosyncratic constraints on the apple industry but also produce spontaneous mutations that become a basis for novel varieties and profit streams (Legun 2015). These detailed explorations of how diverse life forms influence human activity counteract the tendency of Anthropocene scholarship to render the more-than-human world as a passive substrate for human intervention.

Agrarian change scholarship also provides a corrective to species-level narratives of dominance in Anthropocene discourse, which tend to gloss over the significance of uneven power relations in global ecological change. Agrarian change scholars show that the industrialization of agriculture and its associated environmental impacts are inextricably linked to the distribution of value along agri-food supply chains and to neocolonial trade relations. Over the course of the twentieth century, the profits earned in agriculture have increasingly been diverted away from farmers, accruing instead to both the "upstream" agricultural input businesses that sell seeds, pesticides, fertilizers, and machinery to farmers and the "downstream" food processors, distributors, and retailers that buy farm output (Goodman, Sorj, and Wilkinson 1987; Lewontin 2000; Howard 2016). The search for profits in agri-food supply chains thus incentivizes agribusinesses to sell farmers technologies that have contributed to water pollution, climate change, and biodiversity decline (Buttel 1980). Farmers' dependence on these inputs has driven up their costs (and reliance on debts) while producing high yields that ultimately drive down the prices they receive for their goods, a "cost–price squeeze" that has devastated smaller or otherwise marginalized farmers (Cochrane 1993, 386). At the international level, differential power relations are at the core of "food regime theory," which interprets agri-food system dynamics and their environmental implications historically as a product of colonial legacies and geopolitical struggle (Friedmann and McMichael 1989, 95). In short, for critical scholars of agrarian change, the planetary impact of agriculture—the monocropped landscapes, the greenhouse gas emissions, the nitrogen pollution—is the outcome of highly unequal distributions of profit and power. It is the struggle *between* human groups—not humanity as a whole—that explains the ecological crises of modern-day agriculture.

Critical agrarian scholarship offers stories that counter the most concerning oversights of Anthropocene thinking. In place of totalizing narratives of human transformation of the Earth, these stories show industrializing ambitions thwarted, at times, by the Earth itself; in place of passive, dominated "nature," they show crops whose diverse biological characteristics shape the possibilities of human action; in place of a homogeneously destructive humanity, they show the search for profit and geopolitical power by some people at the expense of others driving agro-environmental change. In agriculture, we see clearly that human transformations of the Earth are halting, piecemeal, and inextricable from inequality.

Agri-Food Systems as Sites of Agency

Agriculture sheds light on both the uneven and incomplete history of human dominion and the multiple possible futures vying to replace it. The agri-food system has become a prominent site of individual and collective agency and a vibrant context for the imagination of more sustainable alternatives: from the individualized soul-searching of consumer movements, to the explicit discussions of race and poverty central to food justice activism, to the radical reconfiguration demanded by the food sovereignty and agroecology movements. Importantly, these plural pathways of socioecological transformation (E. M. Bennett et al. 2016; Scoones et al. 2020) have gained momentum by taking a very different approach than Anthropocene claims to humanity-scale domination would indicate: embracing the inseparability of people from their environments and directly confronting oppressive structures predicated on social difference.

Consumers increasingly ask how and where their food is grown, linking their own wellness and ethics to connectedness with the land. Many food purchasers are "voting with their dollars" for more ecologically and socially responsible agriculture, leading to the explosive growth of the organics industry (Guthman 2014), the proliferation of fair-trade labeling schemes for tropical commodity crops (Jaffee 2014), and the enthusiastic embrace of locally grown food (Martinez et al. 2010). Purchasing decisions are also a vehicle for rejecting agricultural production systems seen as unnatural, unethical, or unjust, with the consumer boycott of milk produced using recombinant bovine growth hormone (DuPuis 2000) and the campaign for labeling of food containing genetically modified ingredients (Klintman 2002) as prominent examples. These consumer movements are limited in their ability to effect systemic change, given that they operate within existing market logics, are fundamentally undemocractic (i.e., those with the most dollars get the most "votes"), and are often motivated by self-protection in the face of environmental threats (DuPuis and Goodman 2005; Shreck 2005; Szasz 2009). Their rapid growth and momentum, however, also reflect widespread attention to the fundamental unity of our bodies, our cultures, and our environments through eating.

Agriculture and food are also at the heart of movements for reparative justice and emancipatory politics. The movements for food justice, labor justice, and land justice see inequality as the root of socioecological crisis, standing in stark contrast to Anthropocene accounts of domination over nature by humanity as a whole. The food justice movement recognizes that the industrial agri-food system disproportionately harms communities of color and those at the economic margins who bear the physical burden of food production while remaining underserved by its distribution (Alkon and Agyeman 2011). Organizers mobilize agri-food activities—including urban gardening, culinary education, community-supported agriculture, and food policy councils—as vehicles for racial and economic justice (Broad 2016; Sbicca 2018; B. J. Smith 2019). Food has also been central to labor justice organizers' efforts to advance a more livable future through dignified agri-food work, as exemplified by the United Farm Workers (Shaw 2011), the Coalition of Immokalee Workers (Marquis 2017), and Restaurant Opportunities Centers United (Jayaraman and Schlosser 2013). The struggle for land justice seeks to remedy the inequities and environmental harm of agro-industrialization through redistribution of resources (Williams and Holt-Giménez 2017), including the stemming of black land loss in the United States (White 2018), the promotion of indigenous land management (Kamal et al. 2015; Wesche et al. 2016), and the articulation of a decolonial future (Bradley and Herrera 2016). By demanding more just relationships to food, labor, and land, activists across diverse contexts critique capitalism, colonialism, and social inequalities as foundational to the socioecological damages of the Anthropocene, rejecting the homogenization of humanity as a planetary force.

Internationally, the food sovereignty and agroecology movements are two prominent examples of how mobilizations around agri-food systems embrace socioecological unity and confront geopolitical inequities. Rather than pointing to humanity's wrongdoings and suggesting a separation of farmland from "natural ecosystems" (Willett et al. 2019), these movements view social reforms and holistic interlinkages with our environment as the gateways to ecological repair. The food sovereignty movement, which emerged in the mid-1990s from the global peasant coalition La Vía Campesina (Wittman, Desmarais, and Wiebe 2010), pursues food and agricultural self-determination. In doing so, movement organizers name economic imperialism, neocolonialism, and patriarchy as drivers of unsustainable agricultural practices and rural disempowerment (Forum for Food Sovereignty 2007). Among food sovereignty's tenets is the adoption of agroecology as a guiding principle of food production. Agroecology "applies ecological concepts and principles to the design and management of sustainable food systems" (Gliessman 2015, 345), highlighting the necessity of integrating across the interconnected ecological, economic, and social dimensions of agri-food systems (Francis et al. 2003). For the food sovereignty and agroecology movements, agriculture's role in global change cannot be captured as a set of impacts inflicted by humanity at the species level. Instead, it is composed of intertwined social and ecological processes that can and must be transformed in tandem.

Conclusion

Addressing the ills of the Anthropocene requires a fundamental change in perspective; from viewing the Anthropocene as an accretion of alarming statistical indicators, we must shift to seeing it as set of socioecological processes requiring transformation. The agri-food system, we have argued, provides a unique vantage point for engaging critical approaches to the Anthropocene concept. Focusing our attention on food and agriculture reveals the unevenness and contingency characterizing each proposed historical starting point for an epoch defined by human dominance. Theories of agrarian change emphasize the frictions and failures that characterize human projects for taming the environments of which we are part and show how the unequal distribution of power along agri-food value chains and between social groups drives agro-environmental change. This very understanding of socioecological unity and world-shaping inequality motivates a wide range of social movements, revealing food and agriculture to be an essential starting point for confronting the intertwined social and ecological harm of our contemporary moment.

From the vantage point of food and agriculture, it is clear that no matter how stratigraphers date the Anthropocene, the changes they seek to mark are always bound up with processes of profound inequality, fraught with frictions and resistance from beings of all kinds, and accompanied by alternative scripts. To reflexively discuss the Anthropocene, we need all of the socioecological complexity of the agri-food system on the table.

Notes

1. *Golden spike* is a term used by stratigraphers indicating a geological characteristic serving as a temporal boundary marker. It makes reference to the

golden railroad spike completing the first transcontinental railroad in North America and thus, however unintentionally, echoes capitalist and settler colonial ambitions.

2. The co-occurrence of the Green Revolution with the stratigraphic marker of atomic radiation is more than mere coincidence, because synthetic nitrogen and new classes of pesticides were both products of World War II–era military research and manufacturing.

3. Studies of other nature-based industries, including mining (Bunker and Ciccantell 2005) and water delivery (Bakker 2003), have similarly found that the insistent materiality of natural resources has the power to block or redirect human action, but the literature on agriculture is uniquely rich and deep.

ORCID

Emily Reisman http://orcid.org/0000-0002-0199-5746

Madeleine Fairbairn http://orcid.org/0000-0003-0168-4179

References

Alkon, A. H., and J. Agyeman, eds. 2011. *Cultivating food justice: Race, class, and sustainability.* Cambridge, MA: MIT Press.

Altvater, E., E. C. Crist, D. Haraway, D. Hartley, C. Parenti, and J. McBrien. 2016. *Anthropocene or Capitalocene? Nature, history, and the crisis of capitalism.* Oakland, CA: PM Press.

Anderson, K. 2013. *Tending the wild: Native American knowledge and the management of California's natural resources.* Berkeley, Calif.: University of California Press.

Bailey, R. 1994. The other side of slavery: Black labor, cotton, and textile industrialization in Great Britain and the United States. *Agricultural History* 68 (2):35–50.

Bakker, K. J. 2003. *An uncooperative commodity: Privatizing water in England and Wales.* Oxford, UK: Oxford University Press.

Beacham, J. 2018. Organising food differently: Towards a more-than-human ethics of care for the Anthropocene. *Organization* 25 (4):533–49. doi: 10.1177/1350508418777893.

Beckert, S. 2014. *Empire of cotton.* New York: Vintage.

Bennett, C. E., R. Thomas, M. Williams, J. Zalasiewicz, M. Edgeworth, H. Miller, B. Coles, A. Foster, E. J. Burton, and U. Marume. 2018. The broiler chicken as a signal of a human reconfigured biosphere. *Royal Society Open Science* 5 (12):1–11. doi: 10.1098/rsos.180325.

Bennett, E. M., M. Solan, R. Biggs, T. McPhearson, A. V. Norström, P. Olsson, L. Pereira, G. D. Peterson, C. Raudsepp-Hearne, F. Biermann, et al. 2016. Bright spots: Seeds of a good Anthropocene. *Frontiers in Ecology and the Environment* 14 (8):441–48. doi: 10.1002/fee.1309.

Biermann, F., X. Bai, N. Bondre, W. Broadgate, C.-T. Arthur Chen, O. P. Dube, J. W. Erisman, M. Glaser, S. van der Hel, M. C. Lemos, et al. 2016. Down to earth: Contextualizing the Anthropocene. *Global Environmental Change* 39:341–50. doi: 10.1016/j.gloenvcha.2015.11.004.

Blanchette, A. 2020. *Porkopolis: American animality, standardized life, and the factory farm.* Durham, NC: Duke University Press.

Blok, V. 2018. Technocratic management versus ethical leadership redefining responsible professionalism in the agri-food sector in the Anthropocene. *Journal of Agricultural and Environmental Ethics* 31 (5):583–91. doi: 10.1007/s10806-018-9747-2.

Boyd, W., S. Prudham, and R. Schurman. 2001. Industrial dynamics and the problem of nature. *Society & Natural Resources* 14 (7):555–70. doi: 10.1080/08941920120686.

Bradley, K., and H. Herrera. 2016. Decolonizing food justice: Naming, resisting, and researching colonizing forces in the movement: Decolonizing food justice. *Antipode* 48 (1):97–114. doi: 10.1111/anti.12165.

Braun, B. 2008. Environmental issues: Inventive life. *Progress in Human Geography* 32 (5):667–79. doi: 10.1177/0309132507088030.

Broad, G. M. 2016. *More than just food: Food justice and community change.* Oakland: University of California Press.

Bunker, S., and P. Ciccantell. 2005. *Globalization and the race for resources.* Baltimore, MD: Johns Hopkins University Press.

Burnard, T. G. 2019. *Planters, merchants, and slaves: plantation societies in British America, 1650-1820.* Chicago, Ill.: University of Chicago Press.

Buttel, F. H. 1980. Agricultural structure and rural ecology: Toward a political economy of rural development. *Sociologia Ruralis* 20 (1–2):44–62. doi: 10.1111/j.1467-9523.1980.tb00697.x.

Castree, N. 2014. Geography and the Anthropocene II: Current contributions. *Geography Compass* 8 (7):450–63. doi: 10.1111/gec3.12140.

Clark, N. 2011. *Inhuman nature: Sociable life on a dynamic planet.* Los Angeles: Sage.

Cochrane, W. W. 1993. *The development of American agriculture: A historical analysis.* 2nd ed. Minneapolis: University of Minnesota Press.

Cohen, M. N. 2009. Introduction: Rethinking the origins of agriculture. *Current Anthropology* 50 (5):591–95. doi: 10.1086/603548.

Crist, E. 2016. On the poverty of our nomenclature. In *Anthropocene or Capitalocene? Nature, history, and the crisis of capitalism,* ed. J. Moore, 14–33. Oakland, CA: PM Press.

Crutzen, P. J. 2002. Geology of mankind: The Anthropocene. *Nature* 415 (6867):23. doi: 10.1038/415023a.

Davis, H., and Z. Todd. 2017. On the importance of a date, or decolonizing the Anthropocene. *ACME: An International Journal for Critical Geographies* 16 (4):761–80.

Davis, J., A. A. Moulton, L. V. Sant, and B. Williams. 2019. Anthropocene, Capitalocene, … Plantationocene?: A Manifesto for Ecological Justice in an Age of Global Crises. *Geography Compass* 13 (5).

DuPuis, E. M. 2000. Not in my body: rBGH and the rise of organic milk. *Agriculture and Human Values* 17 (3):285–95. doi: 10.1023/A:1007604704026.

DuPuis, E. M., and D. Goodman. 2005. Should we go "home" to eat? Toward a reflexive politics of localism. *Journal of Rural Studies* 21 (3):359–71. doi: 10.1016/j.jrurstud.2005.05.011.

FitzSimmons, M. 1986. The new industrial agriculture. *Economic Geography* 62 (4):334–53. doi: 10.2307/143829.

Forum for Food Sovereignty. 2007. Declaration of Nyéléni. https://nyeleni.org/IMG/pdf/DeclNyeleni-en.pdf (last accessed 10 September 2019).

Foster, J. B. 1999. Marx's theory of metabolic rift: Classical foundations for environmental sociology. *American Journal of Sociology* 105 (2):366–405. doi: 10.1086/210315.

Francis, C., G. Lieblein, S. Gliessman, T. A. Breland, N. Creamer, R. Harwood, L. Salomonsson, J. Helenius, D. Rickerl, R. Salvador, et al. 2003. Agroecology: The ecology of food systems. *Journal of Sustainable Agriculture* 22 (3):99–118. doi: 10.1300/J064v22n03_10.

Friedmann, H., and P. McMichael. 1989. Agriculture and the state system: The rise and decline of national agricultures, 1870 to the present. *Sociologia Ruralis* 29 (2):93–117. doi: 10.1111/j.1467-9523.1989.tb00360.x.

Gibson-Graham, J. K. 2006. *The end of capitalism (as we knew it): A feminist critique of political economy*. Minneapolis: University of Minnesota Press.

Gliessman, S. 2015. *Agroecology: The ecology of sustainable food systems*. 3rd ed. Boca Raton, FL: CRC.

Goodman, D., B. Sorj, and J. Wilkinson. 1987. *From farming to biotechnology: A theory of agro-industrial development*. Oxford, UK: Basil Blackwell.

Griffin, C. J. 2015. *Rural war—Captain Swing and the politics of protest*.

Guthman, J. 2014. *Agrarian dreams: The paradox of organic farming in California*. 2nd ed. Oakland: University of California Press.

Guthman, J. 2019. *Wilted: Pathogens, chemicals, and the fragile future of the strawberry industry*. Oakland: University of California Press.

Haraway, D. 2015. Anthropocene, Capitalocene, Plantationocene, Chthulucene: Making kin. *Environmental Humanities* 6 (1):159–65. doi: 10.1215/22011919-3615934.

Head, L. 2014. Contingencies of the Anthropocene: Lessons from the "Neolithic." *The Anthropocene Review* 1 (2):113–25. doi: 10.1177/2053019614529745.

Howard, P. H. 2016. *Concentration and power in the food system: Who controls what we eat?* New York: Bloomsbury Academic.

Jaffee, D. 2014. *Brewing justice: Fair trade coffee, sustainability, and survival*. 2nd ed. Oakland: University of California Press.

Jayaraman, S., and E. Schlosser. 2013. *Behind the kitchen door*. Ithaca, NY: ILR Press.

Jones, N. 2019. "It tastes like heaven": Critical and embodied food pedagogy with black youth in the Anthropocene. *Policy Futures in Education* 17 (7):905–23. doi: 10.1177/1478210318810614.

Jordan, P. 2014. *The ethnohistory and anthropology of "modern" hunter gatherers*. Oxford University Press. Accessed August 16, 2019. http://oxfordhandbooks.com/view/10.1093/oxfordhb/9780199551224.001.0001/oxfordhb-9780199551224-e-030.

Kamal, A. G., R. Linklater, S. Thompson, J. Dipple, and Ithinto Mechisowin Committee. 2015. A recipe for change: Reclamation of indigenous food sovereignty in O-Pipon-Na-Piwin Cree nation for decolonization, resource sharing, and cultural restoration. *Globalizations* 12 (4):559–75. doi: 10.1080/14747731.2015.1039761.

Kautsky, K. 1988. *The agrarian question: In two volumes*. London: Zwan.

Klintman, M. 2002. The genetically modified (GM) food labelling controversy: Ideological and epistemic crossovers. *Social Studies of Science* 32 (1):71–91. doi: 10.1177/0306312702032001004.

Kloppenburg, J. 1988. *First the seed: The political economy of plant biotechnology*. Cambridge, UK: Cambridge University Press.

Legun, K. A. 2015. Club apples: A biology of markets built on the social life of variety. *Economy and Society* 44 (2):293–315. doi: 10.1080/03085147.2015.1013743.

Lewis, S. L., and M. A. Maslin. 2015. Defining the Anthropocene. *Nature* 519 (7542):171–80. doi: 10.1038/nature14258.

Lewontin, R. C. 2000. The maturing of capitalist agriculture: Farmer as proletarian. In *Hungry for profit: The agribusiness threat to farmers, food, and the environment*, ed. F. Magdoff, J. B. Foster, and F. Buttel, 93–106. New York: Monthly Review Press.

Lynas, M. 2011. *The god species: Saving the planet in the age of humans*. Washington, DC: National Geographic.

Malm, A., and A. Hornborg. 2014. The geology of mankind? A critique of the Anthropocene narrative. *The Anthropocene Review* 1 (1):62–69. doi: 10.1177/2053019613516291.

Mann, S. A. 1990. *Agrarian capitalism in theory and practice*. Chapel Hill: University of North Carolina Press.

Mann, S. A., and J. M. Dickinson. 1978. Obstacles to the development of a capitalist agriculture. *The Journal of Peasant Studies* 5 (4):466–81. doi: 10.1080/03066157808438058.

Marquis, S. L. 2017. *I am not a tractor! How Florida farmworkers took on the fast food giants and won*. Ithaca, NY: ILR Press.

Martinez, S., M. Hand, M. Da Pra, S. Pollack, K. Ralston, T. Smith, S. Vogel, S. Clark, L. Lohr, S. Low, et al. 2010. *Local food systems: Concepts, impacts, and issues*. U.S. Department of Agriculture, Economic Research Service.

Merriam-Webster. 2020. Through line. Accessed August 4, 2020. https://www.merriam-webster.com/dictionary/through%20line.

Mintz, S. W. 1986. *Sweetness and power: The place of sugar in modern history*. New York: Penguin.

Moore, J. W. 2017. The Capitalocene, part I: On the nature and origins of our ecological crisis. *The Journal of Peasant Studies* 44 (3):594–630. doi: 10.1080/03066150.2016.1235036.

Moore, J. W. 2018. The Capitalocene, part II: Accumulation by appropriation and the centrality of unpaid work/energy. *The Journal of Peasant Studies* 45 (2):237–79. doi: 10.1080/03066150.2016.1272587.

Patel, R. 2013. The long green revolution. *Journal of Peasant Studies* 40 (1):1–63. doi: 10.1080/03066150. 2012.719224.

Perelman, M. 2000. *The invention of capitalism: Classical political economy and the secret history of primitive accumulation.* Durham, NC: Duke University Press.

Preiser, R., L. M. Pereira, and R. Biggs. 2017. Navigating alternative framings of human–environment interactions: Variations on the theme of "Finding Nemo." *Anthropocene* 20:83–87. doi: 10.1016/j.ancene.2017. 10.003.

Sbicca, J. 2018. Food justice and the fight for global human flourishing. *Local Environment* 23 (11):1098–102. doi: 10.1080/13549839.2018.1528444.

Scoones, I., A. Stirling, D. Abrol, J. Atela, L. Charli-Joseph, H. Eakin, A. Ely, P. Olsson, L. Pereira, R. Priya, et al. 2020. Transformations to sustainability: Combining structural, systemic and enabling approaches. *Current Opinion in Environmental Sustainability* 42:65–75. doi: 10. 1016/j.cosust.2019.12.004.

Scott, J. C. 2017. Population control: Bondage and war. In *Against the grain: A deep history of the earliest states.* New Haven, CT: Yale University Press.

Sexton, A. E. 2018. Eating for the post-Anthropocene: Alternative proteins and the biopolitics of edibility. *Transactions of the Institute of British Geographers* 43 (4):586–600. doi: 10.1111/tran.12253.

Shaw, R. 2011. *Beyond the fields: Cesar Chavez, the UFW, and the struggle for justice in the 21st century.* Berkeley: University of California Press.

Shreck, A. 2005. Resistance, redistribution, and power in the fair trade banana initiative. *Agriculture and Human Values* 22 (1):17–29. doi: 10.1007/s10460-004-7227-y.

Simpson, M. 2020. The Anthropocene as colonial discourse. *Environment and Planning D: Society and Space* 38 (1):53–71. doi: 10.1177/0263775818764679.

Smith, B. D. 2001. Low-level food production. *Journal of Archaeological Research* 9 (1):1–43. doi: 10.1023/A:1009436110049.

Smith, B. D., and M. A. Zeder. 2013. The onset of the Anthropocene. *Anthropocene* 4:8–13. doi: 10.1016/j. ancene.2013.05.001.

Smith, B. J. 2019. Food justice, intersectional agriculture, and the triple food movement. *Agriculture and Human Values* 36 (4):825–35. doi: 10.1007/s10460-019-09945-y.

Steffen, W., W. Broadgate, L. Deutsch, O. Gaffney, and C. Ludwig. 2015. The trajectory of the Anthropocene: The Great Acceleration. *The Anthropocene Review* 2 (1):81–98. doi: 10.1177/2053019614564785.

Steffen, W., P. J. Crutzen, and J. R. McNeill. 2007. The Anthropocene: Are humans now overwhelming the great forces of nature? *AMBIO: A Journal of the Human Environment* 36 (8):614–621.

Swyngedouw, E., and H. Ernstson. 2018. Interrupting the Anthropo-obScene: Immuno-biopolitics and depoliticizing ontologies in the Anthropocene. *Theory, Culture & Society* 35 (6):3–30. doi: 10.1177/0263276418757314.

Szasz, A. 2009. *Shopping our way to safety: How we changed from protecting the environment to protecting ourselves.* Minneapolis: University of Minnesota Press.

Tsing, A. L. 2015. *The mushroom at the end of the world: On the possibility of life in capitalist ruins.* Princeton, NJ: Princeton University Press.

Veland, S., and A. H. Lynch. 2016. Scaling the Anthropocene: How the stories we tell matter. *Geoforum* 72:1–5. doi: 10.1016/j.geoforum.2016.03.006.

Watts, N., and I. R. Scales. 2015. Seeds, agricultural systems and socio-natures: Towards an actor-network theory informed political ecology of agriculture: Actor-networks and the political ecology of agriculture. *Geography Compass* 9 (5):225–36. doi: 10.1111/gec3.12212.

Weheliye, A. G. 2014. *Habeas viscus: Racializing assemblages, biopolitics, and black feminist theories of the human.* Durham, NC: Duke University Press.

Weis, A. J. 2007. *The global food economy: The battle for the future of farming.* London: Zed.

Wesche, S. D., M. A. F. O'Hare-Gordon, M. A. Robidoux, and C. W. Mason. 2016. Land-based programs in the Northwest Territories: Building Indigenous food security and well-being from the ground up. *Canadian Food Studies/La Revue Canadienne des Études Sur L'alimentation* 3 (2):23–48. doi: 10.15353/cfs-rcea.v3i2.161.

White, M. M. 2018. *Freedom farmers: Agricultural resistance and the black freedom movement.* Chapel Hill: University of North Carolina Press.

Whyte, K. P. 2018. *Food sovereignty, justice, and indigenous peoples.* Oxford, UK: Oxford University Press. Accessed September 5, 2020. http://oxfordhandbooks. com/view/10.1093/oxfordhb/9780199372263.001.0001/oxfordhb-9780199372263-e-34.

Willett, W., J. Rockström, B. Loken, M. Springmann, T. Lang, S. Vermeulen, T. Garnett, D. Tilman, F. DeClerck, A. Wood, et al. 2019. Food in the Anthropocene: The EAT–*Lancet* Commission on healthy diets from sustainable food systems. *Lancet* 393 (10170):447–92. doi: 10.1016/S0140-6736(18)31788-4.

Williams, J. M., and E. Holt-Giménez, eds. 2017. *Land justice: Re-imagining land, food, and the commons in the United States.* Oakland, CA: Food First Books, Institute for Food and Development Policy.

Wittman, H., A. Desmarais, and N. Wiebe. 2010. The origins and potential of food sovereignty. In *Food sovereignty: Reconnecting food, nature and community*, ed. H. Wittman, A. Desmarais, and N. Wiebe, 1–14. Halifax, NS, Canada: Fernwood.

Yusoff, K. 2018. *A billion black Anthropocenes or none.*

Zalasiewicz, J., C. N. Waters, J. A. Ivar do Sul, P. L. Corcoran, A. D. Barnosky, A. Cearreta, M. Edgeworth, A. Gałuszka, C. Jeandel, R. Leinfelder, et al. 2016. The geological cycle of plastics and their use as a stratigraphic indicator of the Anthropocene. *Anthropocene* 13:4–17. doi: 10.1016/j.ancene.2016.01.002.

Zalasiewicz, J., C. N. Waters, M. Williams, A. D. Barnosky, A. Cearreta, P. Crutzen, E. Ellis, M. A. Ellis, I. J. Fairchild, J. Grinevald, et al. 2015. When did the Anthropocene begin? A mid-twentieth century boundary level is stratigraphically optimal. *Quaternary International* 383:196–203. doi: 10.1016/j. quaint.2014.11.045.

Zimmerer, K. S., S. de Haan, A. D. Jones, H. Creed-Kanashiro, M. Tello, M. Carrasco, K. Meza, F. Plasencia Amaya, G. S. Cruz-Garcia, R. Tubbeh, et al. 2019. The biodiversity of food and agriculture (agrobiodiversity) in the anthropocene: Research advances and conceptual framework. *Anthropocene* 25:1–15. doi: 10.1016/j.ancene.2019.100192.

On Decolonizing the Anthropocene: Disobedience via Plural Constitutions

Mark Jackson

This article mobilizes a decolonial critique of the Anthropocene. It argues for a certain epistemic disobedience to what, conceptually and politically, the Anthropocene seeks to legitimate. The article counterposes recent critical and global governance epistemologies, which summon the Anthropocene as a new humanist and statist moment for universal politics, against plural, parochial forms of relational, nonstatist affirmation. Hegemonic governance imaginaries that invoke universalist and naturalizing rationales are shown to reproduce colonial logics. The article argues for marginalized and systematically ignored forms of earthbound relationality that evidence long-standing political and ontological means for responding to modernity's ecological and social harms. Earthbound and rooted life worlds can affirm ecological responsibility and coconstitution otherwise. Two examples are presented: one from Afro-Caribbean geographies and another from Anishinaabe legal scholarship. Together they evidence enduring ecological reciprocities that unsettle and refuse the totalizing rationalities invoked by Anthropocene horizons. *Key Words: Afro-Caribbean, Anthropocene, decoloniality, global environmental governance, Indigenous law, ontology.*

Two recent, influential arguments in the social sciences and humanities propose that the Anthropocene inaugurates a new contradictory moment for contemporary thought and politics. The Anthropocene, it is reasoned, constitutes a contradictory turning point in history, one that ostensibly necessitates nothing less than a "new humanities" (Chakrabarty 2016, 394) "ground[ed] … in a new philosophical anthropology, that is a new understanding of the changing place of humans in the web of life" (Chakrabarty 2019, 30). Postcolonial historian Dipesh Chakrabarty, one proponent of the argument for anthropocenic divergence, famously argued in numerous formative lectures and papers over the past decade that the Anthropocene initiates an inexorable dilemma for modern thought. "[H]uman beings," he maintained, "have tumbled into being a geological agent through our own decisions. The Anthropocene … has been an unintended consequence of human choices" (Chakrabarty 2009, 210). "The climate crisis," he continued, "calls for thinking simultaneously … the immiscible chronologies of capital and species history" (Chakrabarty 2009, 220). Chakrabarty (2012) put the contradiction plainly: "The need arises to view the human simultaneously on contradictory registers: as a geophysical force and as a political agent … belonging at once to differently-scaled histories of the planet, of life and species, and of human societies" (14).

Latour (2018), another such proponent of the Anthropocene's inaugural demand, argued that, in response to an ostensibly new and ineluctable contradiction, what is needed is an "earth-bound," "terrestrial" politics. He wrote that "terrestrials," who he defined as those knowing themselves to be living in the Anthropocene and who seek common cohabitation with others, face an inescapable imperative, the likes of which "modernization has made contradictory: *attaching oneself* to soil on the one hand, *becoming attached to the world* on the other" (Latour 2018, 92, italics in original). Chakrabarty (2016) continued the thought, explicitly invoking Latour as interlocutor: humans must now be "responsible stewards" (390) to both microbial life ("soil") and people ("world"), even when those microbia and people are pathogenic. The question that preoccupies both is how.

Both scholars posited, in response, the need for "an emergent, new universal history of humans" (Chakrabarty 2009, 221), a "new universality" (Latour 2018, 9) in which social difference and natural history must now be thought together. Political modernity requires "alternative descriptions" (Latour 2018, 94) to meet the Anthropocene's constitutive dilemma and so address its hazards. As Latour (2018) intoned further: "Smallness is not an option" (102); "[o]ur only way out is discovering in common what land is inhabitable and with whom to share it" (9).

These arguments are neither unique nor inconsequential, nor are they unchallenged (e.g., Jackson 2014; Schmidt 2019; Boscov-Ellen 2020). Importantly, however, their articulation echoes, legitimates, and reproduces a widespread epistemic privileging of key modern, humanist, and Eurocentric narratives—human-as-anthropos, universalized space, and linearized time—that together frame, and arguably naturalize, a horizon of possible politics for the present. This Anthropocene narrative establishes human species-being as an exceptional, geologic agent obliged to its interdependencies in singular moments of globalized crisis (DeLoughrey 2015). As Lövbrand, Stripple, and Wiman (2009) argued, such privileging also becomes problematic because it constitutes itself as a "taken-for-granted starting point for future … research" (8). Indeed, for humanists, this "idea of global anthropogenic agency is particularly new and exciting," as much as it is also useful for physical scientists "who have finally managed to naturalise human social relations into determinative models" (Morrison 2019, 3).

The taken-for-grantedness of what the Anthropocene naturalizes as human universal agency, and hence governability, is especially prominent, and more instrumentally so than either Latour or Chakrabarty entertained, in global environmental governance agendas, which offer themselves as a "newly emerging paradigm" (Biermann 2014a, 59) for "the human species … the defining element of [the] Anthropocene" (57). Global governance agendas have become progressively hegemonic rationales for environmental action. Proponents from one such influential research program, "Earth System Governance," were, for instance, invited to present to the Fourth Interactive Dialogue of the United Nations General Assembly on Harmony with Nature in 2014. Yet, despite appealing for the need to parse human difference, global environmental governance politics deploy the Anthropocene to naturalize human exceptionalism. Consider the following reasoning expressed by Biermann (2014a), a chief exponent of Earth Systems Governance: "Humans are … a 'political animal' that distinguishes itself from other species by its capacity to collectively organize … through joint institutions. This political characteristic … is fundamental [to] the Anthropocene [which] has to be understood as a global political phenomenon"

(57). Moreover, politics is imagined here through the need to reinforce the status quo: "The Anthropocene creates a new dependence on states … that requires an *effective institutional framework for global co-operation*" (Biermann 2014a, 58, italics in original). Universalizable governance and naturalized human exception work hand-in-glove to materialize an avowedly modern, humanist regulation of power for an ostensibly unique Anthropocene.

This article argues that we do not need to accept these narratives as givens. It argues, instead, for a certain "epistemic disobedience" (Mignolo 2009) to what these and related legitimations erect as a new universal "Anthropos" moment and to the work the Anthropocene concept does in general. This disobedience is not directed at what the Anthropocene seeks to name: geologically detectable planetary environmental impacts precipitated by certain harmful human activities. Specific human activities, structures, and processes are undoubtedly ecocidal. Rather, disobeying the epistemic framing of the Anthropocene and what it seeks to inaugurate, conceptually and politically, reveals much about the coloniality of the concept itself and potential remediations sought with its articulation. This argument builds from recent critical scholarship about the Anthropocene, specifically critical race, feminist, decolonial, black, and Indigenous geographies, and their cognate disciplines, to suggest that although, as is widely argued, the Anthropocene naturalizes and universalizes spatiality and temporality in line with modern and colonial hegemonies (e.g., Yusoff 2018; Fagan 2019), decolonizing the term and its dominant effects requires attending less to an epistemic commensuration of diverse alternatives "under the authority of a power that as yet lacks any political institution" (Latour 2018, 90) and more to the plural, constitutive conditions that are "already here, if hidden from view" (Martineau 2015, 81).

Do we really need a new universal philosophical anthropology of humans, the impulse for which has been so problematic a part of modernity? If an anthropology is necessary, let it be a radically decentering one (e.g., Kohn 2013). What we need is to learn to listen to the many knowledges, human and nonhuman, whose grounds emerge from a sensitively honed millennial attention to constitutive webs of life. Many peoples and places already here, if elsewhere to modernity's hegemon, show how this can be done. As Todd (2016) argued, numerous generations of careful thought and practice have attended, in diverse ways, to affirming common inhabitabilities and their sharing. Critical scholarship just needs to learn to listen to these "already here" worlds, something universalizing Anthropocenic arguments rarely do or, perhaps, are constitutively unable to do. Apparent new contradictions do not discover or reveal previously unrecognized disjunctive imperatives, because the purported problematic of simultaneous universality and particularity is *itself* a product of an anthropocene narrative already "embedded in the regional [i.e., Euro-centric] epistemic frame of modernity," which privileges universality over plural, political possibilities (Mignolo and Walsh 2018, 214).

Indeed, "the Anthropocene ... appears," as Schmidt (2019) wrote, "to create conditions where the social alteration of how the Earth system functions overdetermines 'other' forms of belonging to space, place, or landscapes" (723). If critical scholarship wants to contribute to changing normative ecological and political perceptions, it must disobey the modern epistemic framing of the Anthropocene and its self-determining solutions. Were a "new humanities" necessary, then its newness must reside in looking to extant pluralities wherein "other" forms of life already articulate relational belonging quite outside contradiction or disjunction, where "soil" has been worlding belonging for generations and where the ontological conditions of belonging inhabit, not by way of exclusion but toward coconstitutive, mutual flourishing. Ontology is only a problem, politically, when the expectation of universality reinforces epistemic totality. Decolonizing the Anthropocene entails reading modernist, universalizing imperatives in terms of their local emergence (Mignolo and Walsh 2018), thereby restoring their local scope and, crucially, enabling pluriversal, parochial, and ordinary worlds of relational responsibility to enact ecological belonging (DeLoughrey 2015; Escobar 2018).

The argument proceeds in three steps. First, the article surveys critical and decolonizing accounts of the Anthropocene from feminist, black, and Indigenous perspectives. It builds from these accounts toward the claim that an Anthropocene epistemology struggles to recognize its constitutive outside. This is illustrated with an analysis of the discourse surrounding Earth System Governance, a prominent geographical exemplar of global environmental governance. In its place, the article suggests

border thought and gnosis—knowledge emergent from the exteriorized borders of the modern/colonial world system (Mignolo 2012)—as formative for relational approaches to politics. It concludes with indicative examples of affirmative, enduring practices drawn from borders created by the modern episteme—in this case Afro-Caribbean "geomorphisms" and Anishinaabe law. I have chosen these for their perspicacious articulations of earthbound attention and for the fact that they are relatively underrepresented in the geographical literature. Numerous other geographies could have been offered. Important anticolonial critiques of the Anthropocene's homogenizing force have, for instance, focused on and through Asia (e.g., Chatterjee 2020). The examples mobilized here are offered as "alternative descriptions" (Latour 2018, 94) but ones whose earthbound constitution radically unsettles humanist exceptionalism and its universalizing Anthropocenic logics. The article responds, herein, to Colebrook's (2017) appeal to think with "pre- or non-Anthropocene humans who [do] not ... define themselves as a species by way of climate change" (11); as such, it seeks to enable ways to think "how 'we' might have been otherwise" (11).

Decolonizing Critiques of the Anthropocene

Led by critical race, feminist, black, and Indigenous geographies, and their cognate decolonial scholarships, decolonizing critiques of the Anthropocene suggest themselves as windows onto being-otherwise-than-Anthropocenic. Central to many such critiques launched against the Anthropocene is a focus on the "human-as-anthropos." Who is this human that is now so newly enabled by the concept's paradoxes and its purported geo-illumination? Anthropocene means, after all, "new Man time." For, although the Anthropocene, as a name, claims a generalized human agency responsible for the myriad ecological crises gathered under its auspices, it is simply not the case that, as Ghosh (2016) argued, "every human being, past and present, has contributed to the present cycle of climate change" (115). Such a view totalizes human agency and obscures the particularities of modern/colonial power causal to ecological harm. The human-as-anthropos narrative is reductive; humans are not a "mono-cultural imaginary" (Hayman 2018), nor are they only selfish and destructive (Chernilo 2017). For millennia, most human–animal communities have lived—and many try to do so today—in relationships of dialogue, mutuality, symbiosis, kin, and reciprocity as and with their constitutive ecologies (e.g., Rose 2004; Suchet-Pearson et al. 2013; Haraway 2016; Pascoe 2018; Bawaka Country et al. 2018; Borrows 2019).

It is also the case that many human communities, typically racialized as Indigenous and black, have, for some time, lived with and been made responsible to modernity's Anthropocenic catastrophes and its unfolding legacies. These legacies, far from new, continue to take the form of forced dispossession together with its consequent ecological and cultural devastations (Lear 2010): rampant extractivism (Gómez-Barris 2017; Yusoff 2018), settler colonialisms (e.g., Whyte 2016, 2017; H. Davis and Todd 2017), enslavements (e.g., McKittrick 2013; Wynter and McKittrick 2015; Sharpe 2016), white supremacies (Di Chiro 2017; Mirzoeff 2018), and racialized capitalisms (Vergès 2017; Saldhana 2020). Further, many of these racialized and often subaltern communities, including their wider kinship relations sometimes termed more-than-human persons and multispecies entanglements (e.g., Haraway 2016), have never claimed mastery nor separation from these constitutive relations. Indeed, appeals to the Anthropocene, as both enunciator and enunciated (i.e., as both name and what it names), silence the plurality of extant relational worlds, their histories, and the possibility of them being recognized in our planetary present. As Baskin (2015) wrote, "The term 'Anthropocene' reveals the power of humans, but it conceals who and what is powerful, and how that power is enacted" (16).

Decolonizing critiques of the Anthropocene broadly argue, therefore, that the conceptual narrative invents and generalizes an "anthropos" that, as Yusoff (2018) wrote, giving specificity to that power as coloniality, "universalizes as a geologic commons, thereby neatly erasing, and so denying, the histories of racism and exploitation incubated through its regulatory and legitimating structures" (2). Particular social, political, economic, and cultural structures—epistemologies and geographies—are responsible for this historically unfolding tragedy, storied misleadingly over the past twenty years as the Anthropocene. Specifically, modernity's Eurocentric, and hence local yet globalized epistemic structures, including racial and gender hierarchies; epistemic negations of alterity; authoritative claims on universal validity; property, resource extraction, and commoditization; industrialization; and accumulation—these intersecting "regulatory and legitimating" structures, these powers—have led and, crucially, continue to lead to systematic exploitations and exterminations of life.

Precisely these colonial epistemologies, logics, and structures of practice are argued detectable as geologic signals of what the Anthropocene names. With their sobering "Orbis Spike" thesis, geographers Lewis and Maslin (2015) argued that the beginning of the Anthropocene should be located around 1610, when, following the genocide of an estimated 50 million Indigenous and enslaved black peoples in what is now the Americas, the Caribbean, and Africa, atmospheric carbon dioxide concentrations dropped due to "vegetation regeneration" following "agricultural abandonment" (176). Tens of millions of people and their civilizations were exterminated, land cover returned, and the change is now detectable in the physical record. The Orbis Spike thesis is potentially politically significant because identifying the beginning of modern planetary ecodical impacts with colonialism "names the problem of colonialism as responsible for contemporary environmental crisis," thereby asserting that "the logics that govern our world are not inevitable or 'human nature' but are the result of ... decisions ... that have their origins ... in colonization" (H. Davis and Todd 2017, 763). Beginning with this recognition works to decolonize contemporary geologics, their designs on power, and their naturalizing silences.

Decolonizing critiques interrogate the enunciation as much, if not more, than the enunciated to highlight how responses to what is understood and named as causing harmful changes enables and implements power. Decolonizing critiques reveal that, by presenting itself as an unfolding event of universalizable geologic time, the Anthropocene naturalizes a "global design story" (Vàzquez 2017, 6), thereby denying the local history from which it emanates, and so legitimizing, under the auspices of political improvement (i.e., anthropogenic harm reduction), "a particularly modern/colonial way of living on earth and of worlding the world" (Vàzquez 2017, 6). Universal, naturalizing, improving, global: Through these and related concepts and practices, the Anthropocene reproduces coloniality through and through (e.g., Erickson 2020; Simpson 2020).

Urgency as Coloniality

In few other arenas is this legacy more noticeable than in the ways in which the Anthropocene has been mobilized by contemporary global environmental governance agendas. Discourses like global environmental governance (e.g., Pattberg and Zelli 2016) and Earth Systems Governance (e.g., Biermann 2014b; Burch et al. 2019) have emerged as increasingly hegemonic, analytic, and normative approaches to theorizing a planetarity whose urgent necessity, it is thought, demands universalized political and epistemic response. Urgency demands, wrote proponents of Earth System Governance, "a new 'constitutional moment' in world politics and global governance" (Biermann et al. 2012, 1307). These discourses explicitly invoke Anthropocenic emergency (Biermann 2014a) to call for a "fundamental reorientation and restructuring of national and international institutions toward more effective Earth system governance and planetary stewardship" (Biermann et al. 2012, 1306). The language of stewardship and its institutionalizing legitimacy itself derives from appeals by the scientific popularizers, if not the originators, of the Anthropocene concept for "[e]ffective planetary stewardship ... built around scientifically developed boundaries for critical Earth System processes ... including an architecture of a governance system for planetary stewardship" (Steffen et al. 2011, 757). Stewardship governance is imagined by these proponents as "polycentric and multi-level rather than centralized and hierarchical" (Steffen et al. 2011, 757). It also appeals for the importance of "diversity in norms, worldviews, and knowledge systems" to identify, understand, and theorize governance practices (Burch et al. 2019, 6) and makes explicit that "the focus of Earth System governance is not 'governing the Earth' or managing the entire process of planetary evolution" (Biermann 2014a, 59).

Although global governance aims to include "actors, voices, and knowledge systems that are traditionally excluded" (Biermann 2014a, 59) and attenuate global government, "diversity"—the banner under which difference is recognized—is itself attenuated by key modernist criteria: universality, functionality, and efficiency. Such criteria frame and legitimize the authority of given epistemic and institutional imaginaries, actors, or, in the language of property governance, "stakeholders" (Biermann et al. 2012, 1307): nation-states, policymakers,

governments, corporations, consumers, civil society, and so on. "Environmental goals," global governance advocates, wrote, "must be mainstreamed into global trade, investment and financial regimes" (Biermann et al. 2012, 1307). Thus, the standards and structures with which difference must reconcile are already enclosed and determined by extant analytic, normative, and political horizons. Difference, it seems, must accommodate itself to existing powers and epistemic structures. The scope for fundamental political and economic reorientation, something demanded by the Anthropocene's enunciation, is profoundly—impossibly, even—curtailed.

One might argue that Earth System Governance appeals for an "Earth Alliance" (Biermann 2014b); that is, an integrated and reformed intergovernmental coalition of regulatory institutions whose powers, modeled after the precedents set by global economic and finance governance like the World Trade Organization, World Bank, and International Monetary Fund (Biermann 2014b) but applied to environmental and sustainability concerns, would be a step in the right direction of protecting "vital planetary systems"; these are recognizable, perhaps strategically essential goals. Global cooperation is a must, and enforced environmental regulation works to mitigate some harms. The challenge for global environmental governance approaches, like Earth Systems Governance, however, is to recognize and articulate forms of epistemic and ontological difference that can adapt and change the institutions of political governance themselves and thereby address the underlying causal forces driving planetary ecocide. Otherwise, the conceptual and institutional apparatuses already responsible for reproducing modernity or coloniality will simply be reinforced under greater and more pervasive articulations of integrated governance. If the stated aim of Earth System Governance is to "strengthen the overall framework for effective institutional governance of the interaction of human societies with the planetary system" (Biermann 2014b, 212), it is difficult to see where the scope for transformative difference lies in its universalist, status quo approach.

Worryingly, urgency has also become a key determinant for adjudicating the extent to which recognizing difference might bring about epistemic and political change. As Burch et al. (2019) wrote, "The challenge is to balance the breadth and depth of various forms of knowledge ... with a desire to present timely answers to pressing environmental problems" (6). Recognizing diversity, they continued, "is unlikely to result in an efficient process of decision-making or knowledge creation ... diversity in norms and knowledge systems can be an asset, it can also hamper ... ecologically sound governance: (Burch et al. 2019). Modern environmental harm does require immediate attention, but one of the great dangers in any response to a declared emergency is a certain unreflexive obedience, an obedience that takes the epistemic terms of engagement as given due to claims of imminence, utility, and efficiency. Witness slavery, walls, holocausts, and pandemic power grabs.

Now, not all emergencies are equivalent, but urgency here asserts an organizing principle of obedience, a *force majeure*, for epistemic and political legitimacy that linearizes time. It negates the importance of historical accountability and precedent and "reduces the present to the time of the now" (Vàzquez 2017, 11). Complexities of relational experience—we are alive because of what comes before us, materially and conceptually—are thereby confined to the urgent presence of "harnessing ... earth and all beings to modernity's field of domination" (Vàzquez 2017, 11). Any epistemic disobedience will be, by definition, a threat to the globalized responsibility of species *mastery*, a term explicitly invoked by the Anthropocene's authors: "We ... decide what nature is and what it will be. To master this huge shift, we must change the way we perceive ourselves and our role in the world" (Crutzen and Schwägerl 2011). Invoking mastery and universality, legitimizing architectures of already unequal institutional power, and obscuring differential histories behind naturalized agency do little to change modernist perceptions; it reinforces them. Epistemic refusal in the face of this progressively global coordinating authority thus becomes, in a classic colonial logic, at least, petulance; worse, irrationality or barbarism; or, worst, irrelevant.

Yet, without refusal, what conceptual space is there, under such epistemic and political rubrics, for "changing the way we perceive ourselves and our role in the world"? Where can new possibilities of life come from unless they have a certain "opacity" (Glissant 1997) to dominant orderings? Global design discourses, however well-meaning, repeat qua soundness, the epistemic and environmental problematics inherent to the conditions they seek to overcome: modernization, systematicity, universality, homogeneity, globality, efficiency, control, and technocratic solutions. "[S]alvation cannot come from the same epistemology that created the need for salvation" (Mignolo 2012, xxi). This is why we need epistemic disobedience and to think from the borders and opaque elsewheres created by the continuity of dominant and hegemonic narratives. These border spaces are means that enable diverse forms of life to speak to our modern/colonial present and so enable "changing the way we perceive ourselves," "we" being the "deep mentality" or *imaginaire* of Eurocentric epistemology (Glissant and Hiepko 2011, 255). This can only come about by carefully refusing the reproduction of epistemic obedience the Anthropocene increasingly invokes and by affirming generative practices otherwise.

Earthbound Constitutions

Creative, disobedient responses to the Anthropocene as an organizing metanarrative come from the many exteriorized worlds sustaining decolonizing critiques. They come from "border thinking," or what Mignolo (2012) termed "border gnosis" (13), those forms of "subaltern reason striving to bring to the foreground the force and creativity of knowledges subalternized during a long process of colonization of the planet, which was at the same time the process in which modernity and the modern Reason were constructed." *Gnosis*, a term employed by Mudimbe (1988), counters a modern epistemological emphasis on naturalized systematicity and universality. Gnoseology is conceivably parallel, a precursor, even, to the more contemporary "cosmopolitics" (Stengers 2005; de la Cadena 2010), which is perhaps more familiar to geographers (e.g., Hinchliffe et al. 2005; Last 2017; Jackson 2018). Both concepts designate an attempt to "'slow down' reasoning and create an opportunity to arouse a slightly different awareness of the problems and situations mobilizing us" (Stengers 2005, 995) without resorting to modern or colonial binaries like nature and culture or narratives like species-being, universalized space, and linear time. Border gnosis, instead, locates capacities for different ontological embodiments of problems and solutions in "the exterior[-ized] borders of the modern/colonial world system ... at the conflictive intersection of the knowledge produced from the perspective of modern colonialisms" (Mignolo 2012, 13). These capacities refuse colonizing forms of being made or rendered visible because they affirm grounded, normative, often incommensurable creations of collective and embodied relation that constitute thinking and being otherwise; they are forms of "*affirmative refusal* [or] a positioning/articulation of *becoming other/wise* to power" (Martineau 2015, 43, italics in original).

What if, in other words, the criteria for environmental politics and justice shifted from inaccurately generalized species blame and designs on universality, institutional efficiency, technicity, or singular authority in search of an institution, and so on, and instead became simply care and flourishing? Latour's and Chakrabarty's appeals for a new earthbound terrestrial politics, with which we began, intimate in this direction but at the expense of recognizing the many voices that have, for a very long time, already been living and, hence, articulating alternative descriptions of common, land-based inhabitation. Global governance agendas largely struggle even to recognize such possibilities, given their commitment to strengthening statist and status quo political institutions.

By way of conclusion, I gesture toward—space prohibits further—forms of careful, earthbound constitution, which, as "elsewhere[s] … already here" (Martineau 2015, 81) emerge from constitutive outsides produced by the modern or colonial system, now rendered "geo-logical." These praxes, remarked by decolonizing geographies, embody forms of relational worlding outside dominant epistemic apparatuses. Recognizing how they exceed such norms means beginning not with contradictory species mastery but from care, attention to flourishing, and a certain disobedience to hegemony.

"Plantationocene" emerged recently (e.g., Haraway 2016) as an alternative counterconcept to name planetary ecological crisis. The name suggested itself via the causal modernizing logics (racism, homogeneity, efficiency, mastery, control, industrialization, accumulation) developed first on historical plantations and sustained today in plantations and industrialized agriculture. Literally beside plantation spaces, however, other smaller growing spaces, "plots," also emerged as alternative places of modest flourishing and care (J. Davis et al. 2019). Plots were small gardens outside the global enclosures of plantations with which enslaved peoples grew food to sustain themselves. Plots, wrote Wynter (1971), were soils where once peasants "transplanted all the

structure of values that had been created by traditional societies of Africa, the land remained the Earth—and the Earth was a goddess" (100), which fed and sustained and to whose constitution people returned in death. By growing traditional foodstuffs, a "folk culture as the basis for a social order" emerged that "resist[ed] the market system and market values" (Wynter 1971, 100). Plots, "nutur[ed] an oppositional mode of Black life … as an ethico-political project … where intimately sharing oneself with others has enormous creative potential for the cultivation of new forms of cooperation necessary for a just and sustainable future" (J. Davis et al. 2019, 7, 15). Archetypal border spaces, plots are indicative "point[s] outside the system where the traditional values can give us a focus of criticism against the impossible reality in which we are enmeshed" (Wynter 1971, 101).

The point of plots as real, experimental, and illustrative political and cultural spaces is not that they focus evenly nor as a necessarily networked totality. It is that they existed and acted differentially, as small polysemic worlds of possibility whose responsibility from and to the land and its shared relationships embody and enact multiplicities of creative difference. Not institutionalized designs, nor seeking to be, these were small worlds where material and cultural possibilities returned from the cotemporal African "before" to affirm, with the land, ways and means of "survivance" (Vizenor 1999). Plots materialized pre-enslaved, ancestral relations of space and time wherein the earth, land, and shared inhabitation were not rendered "a vile and base matter" (Wynter 2003, 267) to be extracted for circulation but as affirmative, literal "roots of culture" (Wynter 1971, 101).

Rootedness also articulates ontologies of reciprocity with and through earthbound others in Indigenous legal constitutions (e.g., Borrows 2018; McGregor 2018; Mills 2018). Mills (2018) argued that the language of rootedness in Anishinaabe law grounds stories of reconciling human lifeways, and the constitutional orders it gives rise to, with the deep condition of "earth way." Earthing constitution gives rise to noncolonial, yet diverse, forms of political community. "The vision of harmony," he wrote, "that the earth way sustains isn't one of non-conflict, but of non-disconnection" (Mills 2018, 156). Difference abounds and flourishes but is itself grounded in its own constituting possibility, literally, the earth:

"creation itself, in all its vast complexity" (Mills 2018, 157). Here the constitutional political moment is not commensuration with, or contradiction within, a designed global governance architecture but acting by "affirming our interdependence ... with the earth conditions" that literally subtend (i.e., support and enfold in flourishing) "a range of possibility for what the roots may become" (Mills 2018, 157). Guidance in decision making (i.e., law) derives from shared embodiment, which constitutes the very possibility of learned action. The ability to act does not arise from human exceptionalism, assumed by Western liberal governance tradition, which then leads to contradiction, mastery, and institutionalized design, but from the material and conceptual reciprocities that condition, even exceptionalist conceits, as learned possibility.

Indigenous legal scholar Borrows (2018, 66) wrote that the Anishinaabe word *akinoomaagewin* communicates this "earthbound" sensibility of learning. From *aki* meaning earth and *noomaage* meaning to point toward and take direction from, "teaching and learning literally means the lessons we learn from looking to the earth ... we draw analogies from our surroundings and appropriately apply or distinguish what we see, we learn about how we should live in our surroundings. The earth has a culture and we can learn from it" (Borrows 2018, 66). Grounding earth way relations as the source of just living is referred to in the Anishinaabe legal tradition as *minobimaatisiwin*, the idea and virtue of living a good life. The prefix morpheme *mino-* lends an affirmative motive in the engaged living of embodied relational attention to healthy interdependence. Freedom as the possibility for action, in this context, unlike the Western notion of a reasoned political self-determination that transcends nature, entails acting with interdependencies that enable a social ecology of flourishing (Borrows 2018). *Minobimaatisiwin* is not just acting with but is an embodied relation of, as the Anishinaabe legal scholar and geographer McGregor (2018) wrote, "reciprocal responsibilities and obligations that are to be met in order to ensure harmonious relations" (15). Again, the ground of conceptual reflexivity here is a lived attention to earthbound care rather than an institutionalized governance that begins in epistemic contradiction necessitating technicity, mastery, or a new humanist anthropology. Laws, discrimination, reason, and action are learned capacities that derive not from subjective exceptionalism but toward commitments within which reflexive life is always already constituted. The constitutional moment of politics and action comes not in a designed, global ability to institutionalize right action but from the relations that constitute the very possibility for discrimination's reason: love, care, flourishing, harmony, and so on. The "original position" (Rawls 1993) for just environmental decision is, hence, not an exceptionalist politics of so-called rational separation in decision, which global designs seek to strengthen or transform in appeals for new universals or new planetary anthropologies. They are instead the actual relational, situated, positions of living, breathing, eating, loving, disputing, and so on that come from constitutive earthbound ecologies of care already entangled in ways transformative to Anthropocenic assumptions.

These examples of plots and *akinoomaagewin*—there are, of course, innumerable others—come from and exemplify border gnosis. They speak from outside the registers of statist political recognition reproduced by an "anthropos" whose dominance conditions contemporary global responses. Theirs are careful, disobedient, but affirming "sense-abilities"—aesthetics of touch, taste, smell, and so on but also laws, memories, stories, rituals, and feelings connected through earthbound relationalities. These geographies are gestured to here as instances wherein subjugated people in seemingly impossible situations constitute themselves complicitly and differently. The earth in these stories is the generative relation of what Vàzquez (2017) termed "precedence," the before, which, in this case, was before colonization but that is also earth as "always ahead and always already there in grounding projection ... anterior and foregrounding in its being before the before" (11). Radically antithetical to an anthropocentric earth, these forms of precedence give voice—formally, prosaically, ritually, poetically, ceremonially, sonically, gustatorily, artistically—to a "geomorphism" (Glissant 2006, 176) where geographies, land, and geologies—earth—always already prefigure and prevail human intervention. Indeed, it makes such invention possible.

Many life processes will survive the modern or colonial moment. Listening to how the earth precedes and therefore anticipates, as many Indigenous and Afro-descendent struggles do via these examples of ancestrality and law, might enable political possibilities other than mastery. They are suggested here as means and ways that people have already been responding to, within other earthbound political constitutions and for a very long time, what the Anthropocene seeks to name.

References

Baskin, J. 2015. Paradigm dressed as epoch: The ideology of the Anthropocene. *Environmental Values* 24 (1):9–29. doi: 10.3197/096327115X14183182353746.

Bawaka Country, S. Suchet-Pearson, S. Wright, K. Lloyd, M. Tofa, J. Sweeney, L. Burarrwanga, R. Ganambarr, M. Ganambarr-Stubbs, B. Ganambarr, et al. 2018. Gon Gurtha: Enacting responsibilities as situated co-becoming. *Environment and Planning D: Society and Space* 37 (4):682–702. doi: 10.1177/0263775818799749.

Biermann, F. 2014a. The Anthropocene: A governance perspective. *The Anthropocene Review* 1 (1):57–61. doi: 10.1177/2053019613516289.

Biermann, F. 2014b. *Earth system governance: World politics in the Anthropocene.* Cambridge, MA: MIT Press.

Biermann, F., K. Abbott, S. Andresen, K. Bäckstrand, S. Bernstein, M. M. Betsill, H. Bulkeley, B. Cashore, J. Clapp, C. Folke, et al. 2012. Navigating the Anthropocene: Improving Earth system governance. *Science* 335 (6074):1306–7. doi: 10.1126/science.1217255.

Borrows, J. 2018. Earth-bound: Indigenous resurgence and environmental reconciliation. In *Resurgence and reconciliation: Indigenous-settler relations and earth teachings*, ed. M. Asch, J. Borrows, and J. Tully, 49–82. Toronto: University of Toronto Press.

Borrows, J. 2019. *Law's Indigenous ethics.* Toronto: University of Toronto Press.

Boscov-Ellen, D. 2020. Whose universalism? Dipesh Chakrabarty and the Anthropocene. *Capitalism Nature Socialism* 31 (1):70–83. doi: 10.1080/10455752.2018.1514060.

Burch, S., A. Gupta, C. Y. A. Inoue, A. Kalfagianni, Å. Persson, A. K. Gerlak, A. Ishii, J. Patterson, J. Pickering, M. Scobie, et al. 2019. New directions in earth system governance research. *Earth System Governance* 1 (1). doi: 10.1016/j.esg.2019.100006.

Chakrabarty, D. 2009. The climate of history: Four theses. *Critical Inquiry* 35 (2):197–222. doi: 10.1086/596640.

Chakrabarty, D. 2012. Postcolonial studies and the challenge of climate change. *New Literary History* 43 (1):1–18. doi: 10.1353/nlh.2012.0007.

Chakrabarty, D. 2016. Humanities in the Anthropocene: The crisis of an enduring Kantian fable. *New Literary History* 47 (2–3):377–97.

Chakrabarty, D. 2019. The planet: An emergent humanist category. *Critical Inquiry* 46 (1):1–31. doi: 10.1086/705298.

Chatterjee, E. 2020. The Asian Anthropocene: Electricity and fossil developmentalism. *The Journal of Asian Studies* 79 (1):3–24. doi: 10.1017/S0021911819000573.

Chernilo, D. 2017. The question of the human in the Anthropocene debate. *European Journal of Social Theory* 20 (1):44–60. doi: 10.1177/1368431016651874.

Colebrook, C. 2017. We have always been post-anthropocene: The anthropocene counterfactual. In *Anthropocene feminism*, ed. M. Grusin, 1–20. Minneapolis: University of Minnesota Press.

Crutzen, P., and C. Schwägerl. 2011. Living in the Anthropocene: Towards a new global ethos. Accessed June 26, 2020. https://e360.yale.edu/features/living_in_the_anthropocene_toward_a_new_global_ethos.

Davis, H., and Z. Todd. 2017. On the importance of a date, or decolonising the Anthropocene. *ACME: An International e-Journal for Critical Geographies* 16 (4):761–80.

Davis, J., A. A. Moulton, L. Van Sant, and B. Williams. 2019. Anthropocene, capitalocene, … plantationocene? A manifesto for ecological justice in an age of global crisis. *Geography Compass* 13 (5). doi: 10.1111/gec3.12438.

de la Cadena, M. 2010. Indigenous cosmopolitics in the Andes: Conceptual reflections beyond "politics." *Cultural Anthropology* 25 (2):334–70. doi: 10.1111/j.1548-1360.2010.01061.x.

DeLoughrey, E. 2015. Ordinary futures: Interspecies worldings in the Anthropocene. In *Global ecologies and environmental humanities: Postcolonial approaches*, ed. E. DeLoughrey, J. Didur, and A. Carrigan, 352–72. London and New York: Routledge.

Di Chiro, G. 2017. Welcome to the white (m)anthropocene: A feminist-environmentalist critique. In *Routledge handbook of gender and environment*, ed. S. MacGregor, 487–505. London and New York: Routledge.

Erickson, B. 2020. Anthropocene futures: Linking colonialism and environmentalism in an age of crisis. *Environment and Planning D: Society and Space* 38 (1):111–28. doi: 10.1177/0263775818806514.

Escobar, A. 2018. *Designs for the pluriverse: Radical interdependence, autonomy, and the making of worlds.* Durham, NC: Duke University Press.

Fagan, M. 2019. On the dangers of an Anthropocene epoch: Geological time, political time and post-human politics. *Political Geography* 70:55–63. doi: 10.1016/j.polgeo.2019.01.008.

Ghosh, A. 2016. *The great derangement: Climate change and the unthinkable.* Chicago: University of Chicago Press.

Glissant, É. 1997. *Poetics of relation.* Trans. B. Wing. Ann Arbor: University of Michigan Press.

Glissant, E. 2006. *Une nouvelle région du monde* [A new region of the world]. Paris: Editions Gallimard.

Glissant, E., and A. S. Hiepko. 2011. Europe and the Antilles: An interview with Édouard Glissant. In *The creolization of theory*, ed. F. Lionnet and S.-M. Shih, 255–61. Durham, NC: Duke University Press.

Gómez-Barris, M. 2017. *The extractive zone: Social ecologies and decolonial perspectives.* Durham, NC: Duke University Press.

Haraway, D. 2016. *Staying with the trouble: Making kin in the Chthulucene.* Durham, NC: Duke University Press.

Haraway, D., N. Ishikawa, S. F. Gilbert, K. Olwig, A. L. Tsing, and N. Bubandt. 2016. Anthropologists are talking—About the anthropocene. *Ethnos* 81 (3):535–64. doi: 10.1080/00141844.2015.1105838.

Hayman, E. 2018. Future rivers of the Anthropocene or whose Anthropocene is it? Decolonising the Anthropocene! *Decolonization: Indigeneity, Education and Society* 6 (2):77–92.

Hinchliffe, S., M. B. Kearnes, M. Degen, and S. Whatmore. 2005. Urban wild things: A cosmopolitical experiment. *Environment and Planning D: Society and Space* 23 (5):643–58. doi: 10.1068/d351t.

Jackson, M. 2014. Composing postcolonial geographies: Postconstructivism, ecology and overcoming ontologies of critique. *Singapore Journal of Tropical Geography* 35 (1):72–87. doi: 10.1111/sjtg.12052.

Jackson, M., ed. 2018. *Coloniality, ontology, and the problem of the posthuman*. London and New York: Routledge.

Kohn, E. 2013. *How forests think: Toward an anthropology beyond the human*. Berkeley: University of California Press.

Last, A. 2017. We are the world? Anthropocene cultural production between geopoetics and geopolitics. *Theory, Culture and Society* 34 (2–3):147–68. doi: 10.1177/0263276415598626.

Latour, B. 2018. *Down to Earth: Politics in the new climatic regime*. Cambridge, UK: Polity.

Lear, J. 2010. *Radical hope: Ethics in the face of cultural devastation*. Chicago: University of Chicago Press.

Lewis, S. L., and M. A. Maslin. 2015. Defining the Anthropocene. *Nature* 519 (7542):171–80. doi: 10.1038/nature14258.

Lövbrand, E., J. Stripple, and B. Wiman. 2009. Earth system governmentality: Reflections on science in the Anthropocene. *Global Environmental Change* 19 (1):7–13. doi: 10.1016/j.gloenvcha.2008.10.002.

Martineau, J. 2015. Creative combat: Indigenous art, resurgence, and decolonization. PhD dissertation, University of Victoria.

McGregor, D. 2018. Mino-mnaamodzawin: Achieving Indigenous environmental justice in Canada. *Environment and Society* 9 (1):7–24. doi: 10.3167/ares.2018.090102.

McKittrick, K. 2013. Plantation futures. *Small Axe: A Caribbean Journal of Criticism* 17 (3):1–15. doi: 10.1215/07990537-2378892.

Mignolo, W. D. 2009. Epistemic dis-obedience, independent thought and de-colonial freedom. *Theory, Culture & Society* 26 (7–8):1–23. doi: 10.1177/0263276409349275.

Mignolo, W. D. 2012. *Local histories/global designs: Coloniality, subaltern knowledges and border thinking*. Princeton, NJ: Princeton University Press.

Mignolo, W. D., and C. Walsh. 2018. *On decoloniality: Concepts, analytics, praxis*. Durham, NC: Duke University Press.

Mills, A. 2018. Rooted constitutionalism: Growing political community. In *Resurgence and reconciliation: Indigenous–settler relations and earth teachings*, ed. M. Asch, J. Borrows, and J. Tully, 133–74. Toronto: University of Toronto Press.

Mirzoeff, N. 2018. It's not the Anthropocene, it's the white supremacy scene; or, the geological color line. In *After extinction*, ed. R. Grusin, 123–49. Minneapolis: University of Minnesota Press.

Morrison, K. D. 2019. Provincializing the Anthropocene: Eurocentrism in the Earth system. In *At nature's edge: The global present and long-term history*, ed. G. Cederlöf and M. Rangarajan, 1–16. Delhi: Oxford Scholarship Online. https://oxford.universitypress-scholarship.com/view/10.1093/oso/9780199489077.001.0001/oso-9780199489077.

Mudimbe, V. 1988. *The invention of Africa: Gnosis, philosophy and the order of knowledge*. Bloomington: Indiana University Press.

Pascoe, B. 2018. *Dark emu: Aboriginal Australia and the birth of agriculture*. London: Scribe.

Pattberg, P., and F. Zelli, eds. 2016. *Environmental politics and governance in the Anthropocene: Institutions and legitimacy in a complex world*. London and New York: Routledge.

Rawls, J. 1993. *Political liberalism*. New York: Columbia University Press.

Rose, D. B. 2004. *Reports from a wild country: Ethics for a decolonisation*. Sydney, Australia: University of New South Wales Press.

Saldhana, A. 2020. A date with destiny: Racial capitalism and the beginnings of the Anthropocene. *Environment and Planning D: Society and Space* 38 (1):12–34.

Schmidt, J. J. 2019. The moral geography of the Earth system. *Transactions of the Institute of British Geographers* 44 (4):721–34. doi: 10.1111/tran.12308.

Sharpe, C. 2016. *In the wake: On blackness and being*. Durham, NC: Duke University Press.

Simpson, M. 2020. The Anthropocene as colonial discourse. *Environment and Planning D: Society and Space* 38 (1):53–71. doi: 10.1177/0263775818764679.

Steffen, W., A. Persson, L. Deutsch, J. Zalasiewicz, M. Williams, K. Richardson, C. Crumley, P. Crutzen, C. Folke, L. Gordon, et al. 2011. The Anthropocene: From global change to planetary stewardship. *Ambio* 40 (7):739–61. doi: 10.1007/s13280-011-0185-x.

Stengers, I. 2005. The cosmopolitical proposal. In *Making things public: Atmospheres of democracy*, ed. B. Latour and P. Weibel, 994–1004. Cambridge, MA: MIT Press.

Suchet-Pearson, S., S. Wright, K. Lloyd, and L. Burarrwanga, on behalf of the Bawaka Country. 2013. Caring as country: Towards an ontology of co-becoming in natural resource management. *Asia Pacific Viewpoint* 54 (2):185–97.

Todd, Z. 2016. An Indigenous feminist's take on the ontological turn: "Ontology" is just another word for colonialism. *Journal of Historical Sociology* 29 (1):4–22. doi: 10.1111/johs.12124.

Vàzquez, R. 2017. Precedence, Earth and the Anthropocene: Decolonizing design. *Design Philosophy Papers* 15 (1):77–91. doi: 10.1080/14487136.2017.1303130.

Vergès, F. 2017. Racial capitalocene. In *Futures of black radicalism*, ed. G. T. Johnson and A. Lubin, 72–82. London: Verso.

Vizenor, G. 1999. *Manifest manners: Narratives on postindian survivance*. Lincoln: University of Nebraska Press.

Whyte, K. P. 2016. Is it colonial déjà-vu? Indigenous peoples and climate justice. In *Humanities for the environment: Integrating knowledges, forging new constellations of practice*, ed. J. Adamson, M. Davis, and H. Huang, 88–104. London: Earthscan.

Whyte, K. P. 2017. Indigenous climate change studies: Indigenizing futures, decolonising the Anthropocene. *English Language Notes* 55 (1–2):153–62. doi: 10.1215/00138282-55.1-2.153.

Wynter, S. 1971. Novel and history, plot and plantation. *Savacou* 5:95–102.

Wynter, S. 2003. Unsettling the coloniality of being/power/truth/freedom: Toward the human, after Man, its overrepresentation—An argument. *CR: The New Centennial Review* 3 (3):257–337. doi: 10.1353/ncr.2004.0015.

Wynter, S., and K. McKittrick. 2015. Unparalleled catastrophe for our species? Or, to give humanness a different future: Conversations. In *Sylvia Wynter: On being human as praxis*, ed. K. McKittrick, 9–89. Durham, NC: Duke University Press.

Yusoff, K. 2018. *A billion black Anthropocenes or none*. Minneapolis: University of Minnesota Press.

Part 2

Historical Perspectives on the Anthropocene

Nothing New under the Sun? George Perkins Marsh and Roots of U.S. Physical Geography

Jacob Bendix and Michael A. Urban

U.S. geomorphologists and biogeographers often cite early theoretical roots dating back to late nineteenth- and early twentieth-century exemplars such as Powell, Gilbert, Cowles, and Clements, or earlier European contributors like Hutton, Lyell, von Humboldt, and, of course, Darwin. Yet reviews of our intellectual roots often overlook an early and important U.S. contributor: George Perkins Marsh. Marsh's work on *Man and Nature* is more often cited in the field of environmental history, where it is appropriately noted as a prescient review of human impacts on the landscape. We suggest, however, that his significance extends beyond early environmental activism and that in fact Marsh describes many concepts and analytical approaches that continue to underlie modern geomorphology and biogeography. Moreover, Marsh's ideas and approach presaged fundamental concepts central to our current study of the Anthropocene and coupled human–environment systems, as he emphasized interconnections among biotic, geomorphic and human elements, perhaps most notably with regard to impacts of deforestation on flood regimes. There is, therefore, much to learn from Marsh—both about early thinking in physical geography and about the depth of scientific analysis underlying our discipline's early interest in human impacts. *Key Words: Anthropocene, biogeography, environmental science, geomorphology, human impacts.*

The work of George Perkins Marsh is familiar to many geographers. His classic book *Man and Nature; or, Physical Geography as Modified by Human Action* (Marsh 1864) has a prominent place in U.S. environmental thought and in the genesis of the idea that such a thing as the Anthropocene might even be possible. For Lowenthal (2000b), *Man and Nature* "ushered in a revolution in how people conceived their relations with the earth" (3). As such, it has been described as "a catalyst to the development of the late nineteenth century conservation movement" (Miller 2002, 44) and more famously as "the fountainhead of the conservation movement" in the United States (Mumford 1931, 78). Indeed, Marsh's contributions to environmental thought are sufficiently profound that there is debate as to who should get "credit" for first recognizing his importance (Koelsch 2012).

Notably absent from both accolades and critical analyses of Marsh's work are discussions of how he shaped physical geography, the heart of *Man and Nature*, as specified by its subtitle. Whereas environmental geographers and historians have focused on celebrating Marsh's role in "invert[ing] the historic insights of Ritter (and Guyot and Hegel), to raise the question of human domination of earth" (Clark and Foster 2002, 166), or on debating the originality and universality of his insights (Hall 2004; Judd 2004), physical geographers have largely ignored the detailed environmental science constituting the bulk of *Man and Nature*. Chorley et al.'s comprehensive history of geomorphology (Chorley, Dunn, and Beckinsale 1964; Chorley, Beckinsale, and Dunn 1973), for example, includes but one sentence about Marsh—and that is to note how little influence he had on William Morris Davis.

Within the broader field of geography, Marsh's ideas concerning the centrality of the human–environment interface served as the benchmark guiding generations of scholars. In 1955, a landmark symposium convened and cochaired by Carl Sauer, Marston Bates, and Lewis Mumford at Princeton University brought together seventy of the most widely known and influential scholars of the day to discuss the legacy of Marsh and address the question of how humans shape the surface of the Earth. It resulted in a 1,000-page text containing contributions from the three cochairs as well as luminaries such as C. W. Thornthwaite, Arthur Strahler, Luna Leopold, and Abel Wolman, who focused much of their attention on surficial processes, climatic alterations, and biotic communities (Thomas 1956). The concept underlying the symposium directly echoed the implicit imperative of *Man and Nature*: "scholars move toward theoretical and conceptual ordering of man's knowledge of himself and his world … by synthesis, transcending the limits of present disciplines or branches of science" (Fejos 1956, vii). For Marsh, understanding this relationship between humans and the environment was the central question at the root of his work. Echoing Huxley before him, the purpose was to determine whether "man is of nature or above her" (Lowenthal 2000a, 309).

Yet despite these integrated visions of human–environment interaction, in many ways, geography in the twentieth century disassembled Marsh as it specialized. Marsh and his writings remained well known but unrecognizable, as modern geography became less synthetic and more divergent, leading Trimble (1992) to refer to the human–environment interface antithetically as both a major strength of geography and as *terra incognita*. In practice, physical geography shifted its focus to structural changes in environmental systems and became more reductionist as scientific methods from biology and physics were adopted as the appropriate means of inquiry (Sack 1992). The focus of investigations largely shifted to trying to understand underlying biophysical systems and how they function in isolation from human intervention. At the same time, U.S. environmental thought was increasingly influenced by an exclusionary conception of nature outside of human pollution (Leopold 1949; Carson 1962). Whereas Marsh offered a vision of integrative synergy between biotic, social, and environmental dynamics and cited Stoppani's proposal of the Anthropozoic Era to bolster the centrality of human action, twentieth-century physical geography was largely defined by a fundamental separation between humans and environmental systems (Urban 2018).

In this article, we focus on the specific legacy of the physical geography Marsh discussed in *Man and Nature*, highlighting the extent to which the topics he addressed framed much of the conception and research pursued by "modern" physical geographers. Such a review demonstrates that although Marsh was (demonstrably) not omniscient, he was certainly prescient. We further suggest that his integrative approach to physical geography and the role of humans in effecting kinetic change within environmental systems similarly foreshadowed a physical geography central to addressing the challenges of the Anthropocene, antecedent to the very concept itself.

Marsh's Physical Geography

It should be noted at the outset that most of Marsh's content was not based on his own empirical research. He certainly drew on his own anecdotal observations in New England and the Old World (Lowenthal 2000a), but Marsh compiled much of his synthesis around the detailed observations of others, with the initial edition drawing on more than 600 sources (Lowenthal 1965). In the tradition of Humboldt, those sources included an extraordinarily thorough compendium of descriptions and measurements from travelers, scientists, and engineers, not just of the eighteenth and nineteenth centuries but reaching back to the Classical and Renaissance periods (Wulf 2015). His ability to compile and unify information from all of those sources into a coherent, analytical whole resulted in a volume that was indeed not "just" a reflection on humans and nature but also, true to his subtitle, a comprehensive physical geography. Marsh assembled a catalog of examples illustrating the efficacy of human agency in shaping environments. Yet although recent literature in physical geography follows Marsh's lead on many of these topics (Table 1), it rarely acknowledges the provenance of these ideas.

Biogeographic Examples

One of the most widely taught, researched, and debated concepts in biogeography and plant ecology is that of succession (Johnson and Miyanishi 2008). The scholars most often cited as the originators of successional theory are Cowles (1899, 1911) and Clements (1904, 1916)—an attribution that itself ignores Clements's own literature reviews in which he cited several (mostly European) authors as having "called attention to … succession" (Clements 1916, 18). Fairly or not, Clements is widely regarded as having provided the "dominant conceptual apparatus for vegetation dynamics studies during the first part of the twentieth century" (Pickett, Cadenasso, and Meiners 2009, 10). A common scenario in Clements's writing (and that of many subsequent authors) was that of hydrarch succession:

Table 1. Partial list of physical geography topics of modern interest addressed in Marsh (1864)

Topic	Page on which topic first appears
Secondary succession	28
Beavers as ecosystem engineers	29
Hydrarch succession	30
Plant and animal extinctions by humans	36
Cold air drainage	52
Impact of invasive species on native plants	58
Columbian exchange	62
Pleistocene overkill as possible cause of megafauna extinctions	77
Birds as agents of long-distance plant dispersal	88
Bioturbation	100
Introduction of destructive animal species and importance of absence of natural predators	104
Biological pest control	105
Impact of removing dead trees on dependent bird species	109
Impacts of agriculturally induced sedimentation, loss of shade trees, dams, and industrial water pollution on fish spawning	122
Plant macrofossils as evidence of past vegetation	129
Role of wolves in limiting deer (broadly referring to Cervidae) populations and destructive ("injurious") impacts on forests when deer populations exceed carrying capacity	130
Xerarch succession (described, although not named as such)	131
Role of fire and bison grazing in maintaining North American prairies	135
Impacts of Native American use of fire on forest composition and structure	136
Impacts of fire on forest soils	138
Impact of land use/land cover change on albedo	144
Rainfall interception by forests	162
Plant litter impacts on soil moisture (both infiltration and evaporation rates)	165
Contribution of vegetation to fog drip	185
Impacts of deforestation and reforestation on hydrologic lag time	209
Hydrologic and geomorphic impacts of deforestation	215
Geographic variability of hydrogeomorphic response to environmental change	216
Role of forests in favoring "infiltration and percolation" over "flow of water over the surface" and implications for flood frequency and severity	224
Contribution of tree roots to rock weathering	229
Use of check dams to induce sedimentation (a practice that he in turn attributed to the ancient Romans)	237
Abrasion and downstream fining	251
Sediment mobilization following dam failure	253
Fluvial sediment budgets	257

(Continued)

Table 1. (*Continued*).

Topic	Page on which topic first appears
Weathering by frost shatter	265
Contribution of weathered marl strata to rockslides	269
Impact of trees on reducing landslides	270
Windthrow at forest edges created by logging	272
Postfire slope erosion	273
Seedbank vs. long-distance dispersal as sources for postdisturbance revegetation	287
Fire- and/or char-stimulated germination	287
Potential loss of plants with as-yet unknown medicinal value due to deforestation	290
Role of plant litter in nutrient biocycling	323
Wetland drainage and land reclamation	333
Flood control levees (embankments)	334
Impact of drainage and irrigation on soil, climate, and vegetation	360
Salinization caused by irrigation	382
Timing of scour and fill relative to rising and falling limbs of the flood hydrograph	391
Construction of flood control reservoirs	397
Cost–benefit (and risk) analysis of flood control dams	398
Occurrence and impacts of *jökulhlaups*	403
Observations (and speculation) regarding groundwater hydrology	434
Comparative weathering rates of minerals	451
Sedimentology of sand dunes	463
Potential impacts of Volga River diversions on Caspian sea levels	531

Note: See text for discussion of examples from the list.

The area once covered by deep water becomes transformed into forest, a phenomenon clearly conceivable when one follows the actual processes of development. The various stages—submerged plant, floating plant, reed swamp, sedge meadow, woodland, and climax forest—are merely cross-sections of a continuous development. (Weaver and Clements 1929, 60)

Although this notion is attributed to Clements and the ecologists who followed him, a decade before Clements was born, Marsh (1864) had graphically described just such a sequence:

Aquatic and semiaquatic plants propagate themselves, and spread until they more or less completely fill up the space occupied by the water, and the surface is gradually converted from a pond to a quaking morass. The morass is slowly solidified by vegetable production and deposit, then very often restored to the forest condition by the growth of black ashes, cedars, or, in southern latitudes, cypresses, and other trees suited to such a soil, and thus the interrupted harmony of nature is at last reestablished. (30–31)

Similarly, Marsh's assertion of sequential forest recovery from clearing (albeit substituting abandoned Native American fields for those of European-American farmers) could, if published a century or more later, have served as a simple summary of the twentieth-century old-field succession studies (e.g., Oosting 1942) that gave rise to the simplistic diagrams of secondary succession now found in most introductory biogeography and ecology texts:

> Whenever the Indian, in consequence of war or the exhaustion of the beasts of the chase, abandoned the narrow fields he had planted and the woods he had burned over, they speedily returned, by a succession of herbaceous, arborescent, and arboreal growths, to their original state. (Marsh 1864, 28)

He repeated the term later on the same page and elsewhere repeatedly referred to "the order of succession" (139, 287, 324). Marsh's biogeographic insights were hardly limited to succession (Table 1). To add just one more example, modern research on the role of fire in stimulating germination, whether due to heat (Keeley 1987) or to the addition of charred wood to the substrate (Thanos and Rundel 1995), could hardly surprise Marsh (1864), who asserted that some

> … seeds, whether the fruit of an ancient vegetation, or newly sown by winds or birds require either a quickening by a heat which raises to a certain high point the temperature of the stratum where they lie buried, or a special pabulum furnished only by the combustion of the vegetable remains that cover the ground in the woods. (287)

Geomorphic Examples

Marsh similarly presented many themes that are familiar in various branches of geomorphology. At the network level, his description is strikingly reminiscent of the Zone 1 (drainage basin, sediment-source area) and Zone 3 (depositional: alluvial fan, alluvial plain, delta) described by Schumm (1977) in his idealized fluvial system.

> It has been contended that all rivers which take their rise in mountains originated in torrents. These, it is said, have lowered the summits by gradual erosion, and, with the material thus derived, have formed shoals in the sea

which once beat against the cliffs; then, by successive deposits, gradually raised them above the surface, and finally expanded them into broad plains traversed by gently flowing streams. (Marsh 1864, 262)

Marsh was much interested in fluvial geomorphology, devoting several pages to what would today be described as "back of the envelope" calculations of the impact of deforestation on the sediment budget of the Po River over a span of centuries. His description of sediment redistribution and channel form adjustments following the washout of a mill dam in New England could be seen as a qualitative precursor to contemporary studies of dam removal impacts in the same region (e.g., Magilligan et al. 2016).

Notwithstanding his emphasis on hydrological and fluvial topics, Marsh addressed a range of geomorphic subjects (Table 1). On the topic of mass movements, for example, he explained the 1771 rockslide that created Lake Alleghe and killed more than fifty people (Bonnard 2006) as follows:

> The [mountainside] rested on a steeply inclined stratum of limestone, with a thin layer of calcareous marl intervening, which, by long exposure to frost and the infiltration of water, had lost its original consistence, and become a loose and slippery mass instead of a cohesive and tenacious bed. (Marsh 1864, 268–69)

That description is entirely congruent with the explanations that modern researchers have arrived at in similar settings via both field observations and modeling efforts (Eberhardt, Thuro, and Luginbuehl 2005).

Disclaimer: Insightful, but Not Omniscient

We do not suggest that all of Marsh's conclusions proved in the long run to be accurate. His belief that rows of trees provide protection from malaria by blocking "miasmatic vapors" (Marsh 1864, 154) or that forests would soon cover many parts of the Arabian and African deserts, if man and domestic animals, especially the goat and the camel, were banished from them (Marsh 1864) find no support in modern science. And Marsh's repeated references to the positive impacts of coal usage due to his concern for fuelwood contributing to deforestation (295, 298), although logical in the context of what was known at the time, seem far more problematic in light of the accelerating reality of anthropogenic climate change. These and other erroneous assertions make clear that *Man and Nature* is not a textbook for modern times.

Rather, it introduces many of the topics that still intrigue us today. Moreover, and crucially, it provides an integrative approach that is central to physical geography, perhaps more now than ever.

Integrative Physical Geography

Although the examples in Table 1 show Marsh's interest in various physical geography topics, the whole of his interest was greater than the sum of its parts. That is because he was keenly aware of the manner in which the components of the environmental system he was studying interacted dynamically with each other. This awareness is perhaps best illustrated by his discussion of the impacts of deforestation on, sequentially, hydrology, hillslope erosion, and valley bottom alluviation:

> The soil is bared of its covering of leaves, broken and loosened by the plough, deprived of the fibrous rootlets which held it together, dried and pulverized by sun and wind, and at last exhausted by new combinations. The face of the earth is no longer a sponge, but a dust heap, and the floods which the waters of the sky pour over it hurry swiftly along its slopes, carrying in suspension vast quantities of earthy particles. …

> From these causes, there is a constant degradation of the uplands, and a consequent elevation of the beds of watercourses and of lakes by the deposition of the mineral and vegetable matter carried down by the waters. (Marsh 1864, 215)

This description bears marked similarity to the themes in Knox's (1972, 1977, 1987) classic studies of human impacts in Wisconsin watersheds, studies that themselves have been noted for their significance in influencing how contemporary physical geographers regard river systems (Graf 2013; Bendix and Vale 2014). That similarity, in turn, serves not only as a reminder of the prescience of Marsh but of the importance of such integrative research as physical geographers confront the challenges of the Anthropocene.

Marsh in the Anthropocene

At a basic level, Marsh offers physical geography a conceptual guide for what is possible, ideas that describe how our science can uniquely contribute to explaining the transmutation of a world influenced by humans. Scholars such as von Humboldt had previously highlighted the inherent complexity and interconnections governing environmental systems; Stoppani saw humans as limitless and powerful; Babbage asserted human behavior, intellect, and even emotions to become etched into the landscape through material interactions; and philosophers going back to Kant struggled with the role of humans in physical geography (Lowenthal 2000a; Turpin and Federighi 2012; Wulf 2015; Urban 2018).

For Marsh, "the great question" of whether people are part of nature or set apart was central to how he understood human–environment interactions and in many ways was never resolved. This is no slight to Marsh's intellectual capacity or the contribution of his work; the question remains contentious and is central to discussions about the scope of geomorphological and biogeographical science and environmental management today. Underlying *Man and Nature* is a set of assumptions delineating the complexity of the actions and reactions constantly taking place between people and the physical world around them (Hall 2005). First, people have the capacity to significantly modify environments and those underlying processes creating structure and form. These modifications are not inherently all negative; he also asserted throughout that certain interventions can be beneficial. Second, nature heals damage and degradation slowly, in part because many human actions in such systems cause multiple compounding disruptions. Finally, because of this, any repair needs to be an intentional intervention and cannot solely rely on background rates of change or benign neglect.

The lessons derived from Marsh are in some ways as fluid as Wittgenstein's (1953) duckrabbit. *Man and Nature* has been read and understood in different ways, contingent on the inherent perspective of the audience. As Hall (2005) suggested, it is why Marsh has influenced different traditions in antithetical ways. In Italy, Marsh was understood to advocate for active measures of reconstructing or mending environments that influenced the Bonifica movement in the late nineteenth century (Gachelin 2013). In the United States, Marsh's impact largely derived from his first assumption, that human action can lead to significant environmental damage. Conservationists saw this as mostly negative, in opposition to an idealized, wild nature that should be left alone or isolated from human interference. As we described earlier, many physical scientists instead sought to understand human disturbance by isolating the mechanics involved in driving systemic change. People and their impacts on the environment were not integrated into a systemic understanding of landscape change; they were the jumping-off point from which underlying biophysical systems and their function could be understood in light of disturbance events. In the United States, Marsh's legacy was focusing attention on the ways humans can generate environmental damage, whereas in Europe, the message that resonated most throughout the twentieth century was the need to repair such damage (Hall 2005). Each of these disparate approaches owes a debt to Marsh, yet also simplifies and disassembles the overall enterprise in which he was engaged.

Recent concepts reframing physical geography have begun to analytically break down categorical distinctions separating or isolating humans from environmental function as well as disparate systems from each other. Such distinctions largely arose during the reductionism of physical geography as it grew more expressly scientific and perhaps even undermined the creativity and limited the imaginative capacity of researchers to see connections (Baker and Twidale 1991). Growing recognition of the "wicked" nature of intractable, complex, multifarious environmental problems initially led the National Science Foundation to encourage research into coupled human–environment systems. This evolved into integrated socioenvironmental systems as the original concept was seen as too modular to fully capture the truly integrated nature of socioenvironmental systems versus two discrete sets of natural and human systems considered in parallel.

Critical physical geography is another recent theoretical approach with the goal of bringing together the analytical strengths of social geography with the fundamental understanding of the mechanics driving biophysical environments for the express purpose of proactively addressing environmental problems in ways that can also mitigate social ills (Lave, Biermann, and Lane 2018). By breaking down categorical distinctions, "social elements of perception, valuation, power, politics, and scale become forcing mechanisms for the biophysical landscape" (Urban 2018, 57).

Broadly understood, the Anthropocene is another concept that affects how practitioners define the scope and methodological complexity of physical geography. Casting human action as more than external disturbances to biophysical systems, the Anthropocene instead represents a "fundamental rupture" from our historical understanding of limited human efficacy (Hamilton and Grinevald 2015). The assumption that the net magnitude of human behavior has grown beyond localized impacts and is now causing systemic changes at global scales disorients conceptions that have grounded physical geography throughout much of the last century. Human agency is of such magnitude that it compromises or destroys the resilient capacity of biophysical systems.

Throughout *Man and Nature*, Marsh focused explicitly on the efficacy of human agency in generating landscape change, and the critical interconnections among biotic, geomorphic, and human elements. Writing before the reification of academic disciplines, he transcended what might be seen by modern eyes as systemic boundaries and concentrated instead on linkages and interconnections. Perhaps all along, this very difference in perspective and universality is what attracts many geographers to Marsh. Marsh has inspired by imagining the gross scale interactions that cause simple environmental processes to cascade into other states. Throughout the twentieth century, as physical geographers became more scientific, we began to examine such interactions on a more restricted and reductionistic scale. The strength of environmental interactions, however, is generally more apparent at coarse spatial and temporal scales than at fine ones (Vale 2003), so as we focused our studies and increased our technical competence, we too often lost sight of those interconnections and broad-scale interrogations that inspired us in the first place. Of course, the disassembling of Marsh that took place over the last century properly illustrates that there is no guarantee that synthetic ideas such as the Anthropocene will, now that they have been articulated, become the new theoretical frame through which we understand physical geography. The levels of complexity and uncertainty characteristic of "wicked" problems make it especially challenging to integrate physical, biological, and social elements (Ludwig 2001; Allen et al. 2011). The Anthropocene is an unwieldy conceptual beast that might be of only rhetorical use unless physical geography can embrace the untidiness and uncertainty that come with complex syntheses.

It would be simplistic to say that Marsh presaged the breadth of technical advances in biogeography and geomorphology that has taken place since he wrote *Man and Nature*. It would also be misleading to suggest that integrative concepts such as coupled human–environment systems, critical physical geography, and the Anthropocene originated with Marsh. We do assert, however, that these concepts build on and reflect the totality of Marsh's original enterprise with more fidelity than earlier work that was more fragmented and piecemeal. He ultimately influenced not only our practice but also our underlying assumptions of how the biophysical world works. This legacy of wedding practice and utility with theoretical ideas has invariably led us toward synthetic concepts such as the Anthropocene. In Marsh, we not only see a roadmap guiding physical geography through topics of perpetual interest; we also see deep-seated unresolved implications for how we theorize our field.

Acknowledgment

We are indebted to two anonymous reviewers, whose comments resulted in substantive improvements to the article.

References

Allen, C. R., J. J. Fontaine, K. L. Pope, and A. S. Garmestani. 2011. Adaptive management for a turbulent future. *Journal of Environmental Management* 92 (5):1339–45. doi: 10.1016/j.jenvman.2010.11.019.

Baker, V., and C. R. Twidale. 1991. The reenchantment of geomorphology. *Geomorphology* 4 (2):73–100. doi: 10.1016/0169-555X(91)90021-2.

Bendix, J., and T. R. Vale. 2014. Placing the river in context: James C. Knox, fluvial geomorphology, and physical geography. *Geography Compass* 8 (5):325–35. doi: 10.1111/gec3.12129.

Bonnard, C. 2006. Technical and human aspects of historic rockslide dammed lakes and landslide dam breaches. *Italian Journal of Engineering Geology and Environment* (Special Issue 1):21–30. http://www.ijege.uniroma1.it/rivista/special-2006/special-2006/technical-and-human-aspects-of-historic-rockslide-dammed-lakes-and-landslide-dam-breaches/ijege-special-06-bonnard.pdf

Carson, R. 1962. *Silent spring*. Boston: Houghton Mifflin.

Chorley, R. J., R. P. Beckinsale, and A. J. Dunn. 1973. *The history of the study of landforms: Or the development of geomorphology: Volume Two. The life and work of William Morris Davis*. London: Methuen.

Chorley, R. J., A. J. Dunn, and R. P. Beckinsale. 1964. *The history of the study of landforms: or, the development of geomorphology: Volume One. Geomorphology before Davis*. London: Methuen.

Clark, B., and J. B. Foster. 2002. George Perkins Marsh and the transformation of Earth: An introduction to Marsh's *Man and nature*. *Organization & Environment* 15 (2):164–69. doi: 10.1177/10826602015002003.

Clements, F. E. 1904. *The development and structure of vegetation*. Lincoln: University of Nebraska-Botanical Seminar.

Clements, F. E. 1916. *Plant succession: An analysis of the development of vegetation*. Washington, DC: Carnegie Institution.

Cowles, H. C. 1899. The ecological relations of the vegetation on the sand dunes of Lake Michigan: Part I. Geographical relations of the dune floras. *Botanical Gazette* 27 (2):95–117. doi: 10.1086/327796.

Cowles, H. C. 1911. The causes of vegetational cycles. *Annals of the Association of American Geographers* 1 (1):3–20. doi: 10.1080/00045601109357004.

Eberhardt, E., K. Thuro, and M. Luginbuehl. 2005. Slope instability mechanisms in dipping interbedded conglomerates and weathered marls—The 1999 Rufi landslide. *Engineering Geology* 77 (1–2):35–56. doi: 10.1016/j.enggeo.2004.08.004.

Fejos, P. 1956. Foreword. In *Man's role in changing the face of the earth*, ed. W. L. Thomas, Jr., vii–viii. Chicago: University of Chicago Press.

Gachelin, G. 2013. The interaction of scientific evidence and politics in debates about preventing malaria in 1925. *Journal of the Royal Society of Medicine* 106 (10):415–20. doi: 10.1177/0141076813501743.

Graf, W. L. 2013. James C. Knox (1977) Human impacts on Wisconsin stream channels. *Progress in Physical Geography: Earth and Environment* 37 (3):422–31. doi: 10.1177/0309133313490008.

Hall, M. 2004. The provincial nature of George Perkins Marsh. *Environment and History* 10 (2):191–204. doi: 10.3197/0967340041159803.

Hall, M. 2005. *Earth repair: A transatlantic history of environmental restoration*. Charlottesville: University of Virginia Press.

Hamilton, C., and J. Grinevald. 2015. Was the Anthropocene anticipated? *The Anthropocene Review* 2 (1):59–72. doi: 10.1177/2053019614567155.

Johnson, E. A., and K. Miyanishi. 2008. Testing the assumptions of chronosequences in succession. *Ecology Letters* 11 (5):419–31. doi: 10.1111/j.1461-0248.2008.01173.x.

Judd, R. W. 2004. George Perkins Marsh: The times and their man. *Environment and History* 10 (2):169–90. doi: 10.3197/0967340041159821.

Keeley, J. E. 1987. Role of fire in seed germination of woody taxa in California chaparral. *Ecology* 68 (2):434–43. doi: 10.2307/1939275.

Knox, J. C. 1972. Valley alluviation in southwestern Wisconsin. *Annals of the Association of American Geographers* 62 (3):401–10. doi: 10.1111/j.1467-8306.1972.tb00872.x.

Knox, J. C. 1977. Human impacts on Wisconsin stream channels. *Annals of the Association of American Geographers* 67 (3):323–42. doi: 10.1111/j.1467-8306.1977.tb01145.x.

Knox, J. C. 1987. Historical valley floor sedimentation in the Upper Mississippi Valley. *Annals of the Association of American Geographers* 77 (2):224–44. doi: 10.1111/j.1467-8306.1987.tb00155.x.

Koelsch, W. A. 2012. The legendary "rediscovery" of George Perkins Marsh. *Geographical Review* 102 (4):510–24. doi: 10.1111/j.1931-0846.2012.00172.x.

Lave, R., C. Biermann, and S. N. Lane, eds. 2018. *The Palgrave handbook of critical physical geography*. London: Palgrave.

Leopold, A. 1949. *A Sand County almanac, and sketches here and there*. New York: Oxford University Press.

Lowenthal, D. 1965. Introduction. In *Man and nature; or, physical geography as modified by human action*, ed. D. Lowenthal, xv–xxix. Cambridge, MA: Harvard University Press.

Lowenthal, D., ed. 2000a. *George Perkins Marsh: Prophet of conservation*. Seattle: University of Washington Press.

Lowenthal, D. 2000b. Nature and morality from George Perkins Marsh to the millennium. *Journal of Historical Geography* 26 (1):3–23. doi: 10.1006/jhge.1999.0188.

Ludwig, D. 2001. The era of management is over. *Ecosystems* 4 (8):758–64. doi: 10.1007/s10021-001-0044-x.

Magilligan, F. J., K. H. Nislow, B. E. Kynard, and A. M. Hackman. 2016. Immediate changes in stream channel geomorphology, aquatic habitat, and fish assemblages following dam removal in a small upland catchment. *Geomorphology* 252:158–70. doi: 10.1016/j.geomorph.2015.07.027.

Marsh, G. P. 1864. *Man and nature; or, physical geography as modified by human action*. New York: Scribner.

Miller, C. 2002. Thinking like a conservationist. *Journal of Forestry* 100 (8):42–45. doi: 10.1093/jof/100.8.42.

Mumford, L. 1931. *The brown decades: A study of the arts in America, 1865–1895*. New York: Harcourt, Brace.

Oosting, H. J. 1942. An ecological analysis of the plant communities of Piedmont, North Carolina. *American Midland Naturalist* 28 (1):1–126. doi: 10.2307/2420696.

Pickett, S. T. A., M. L. Cadenasso, and S. J. Meiners. 2009. Ever since Clements: From succession to vegetation dynamics and understanding to intervention. *Applied Vegetation Science* 12 (1):9–21. doi: 10.1111/j.1654-109X.2009.01019.x.

Sack, D. 1992. New wine in old bottles: The historiography of a paradigm change. *Geomorphology* 5 (3–5):251–63. doi: 10.1016/0169-555X(92)90007-B.

Schumm, S. A. 1977. *The fluvial system*. New York: Wiley.

Thanos, C. A., and P. W. Rundel. 1995. Fire-followers in chaparral: Nitrogenous compounds trigger seed germination. *The Journal of Ecology* 83 (2):207–16. doi: 10.2307/2261559.

Thomas, W. L., Jr., ed. 1956. *Man's role in changing the face of the earth*. Chicago: University of Chicago Press.

Trimble, S. W. 1992. Preface. In *The American environment: Interpretations of past geographies*, ed. L. M. Dilsaver and C. E. Colten, xv–xxii. Totowa, NJ: Rowman and Littlefield.

Turpin, E., and V. Federighi. 2012. A new element, a new force, a new input: Antonio Stoppani's Anthropozoic. In *Making the geologic now*, ed. E. Ellsworth and J. Kruse, 34–41. Brooklyn, NY: Punctum.

Urban, M. A. 2018. In defense of crappy landscapes (Core Tenet #1). In *The Palgrave handbook of critical physical geography*, ed. R. Lave, C. Biermann, and S. Lane, 49–66. London: Palgrave.

Vale, T. R. 2003. Scales and explanations, balances and histories: Musings of a physical geography teacher. *Physical Geography* 24 (3):248–70. doi: 10.2747/0272-3646.24.3.248.

Weaver, J. E., and F. E. Clements. 1929. *Plant ecology*. New York: McGraw-Hill.

Wittgenstein, L. 1953. *Philosophical investigations*. New York: Macmillan.

Wulf, A. 2015. *The invention of nature*. New York: Knopf.

Synchronizing Earthly Timescales: Ice, Pollen, and the Making of Proto-Anthropocene Knowledge in the North Atlantic Region

Sverker Sörlin and Erik Isberg

The Anthropocene concept frames an emerging new understanding of the human–Earth relationship. It represents a profound temporal integration that brings historical periodization on a par with geological time and creates entanglements between timescales that were previously seen as detached. Because the Anthropocene gets this role of a unifying planetary concept, the ways in which vast geological timescales were incorporated into human history are often taken for granted. By tracing the early history of the processes of synchronizing human and geological timescales, this article aims to historicize the Anthropocene concept. The work of bridging divides between human and geological time was renegotiated and took new directions in physical geography and cognate sciences from the middle decades of the twentieth century. Through researchers such as Ahlmann (Sweden), Seligman (United Kingdom), and Dansgaard (Denmark) in geography and glaciology and Davis (United States) and Iversen (Denmark) in palynology and biogeography, methodologies that became used in synchronizing planetary timescales were discussed and practiced for integrative understanding well before the Anthropocene concept emerged. This article shows through studies of their theoretical assumptions and research practices that the Anthropocene could be conceived as a result of a longer history of production of integrative geo-anthropological time. It also shows the embedding of concepts and methodologies from neighboring fields of significance for geography. By situating and historicizing spaces and actors, texture is added to the Anthropocene, a concept that has hitherto often been detached from the specific contexts and geographies of the scientific work that enabled its emergence. *Key Words: environmental object, geo-anthropology, glaciology, palynology, proto-Anthropocene.*

The present has proven correct what German historian Koselleck (1979) suggested: There is a coexistence of multiple times. The full implication of this statement has become abundantly clear in the last two decades with the arrival of the Anthropocene concept (Crutzen and Stoermer 2000; Steffen et al. 2015; Zalasiewicz et al. 2018), suggesting new, geo-anthropological temporalities for the human–earth relationship. Proposed starting points of the Anthropocene epoch have ranged from the early Holocene up to almost the present day (Swanson 2016; Warde, Robin, and Sörlin 2017).

Many of the propositions are familiar to geography, where patterns of space have been unavoidably linked to reconfigurations in the understanding of time and the waves of human expansion through agriculture, empires, resource colonialism, and techno-scientific dominance (Ruddiman 2003; Lewis and Maslin 2015; Waters et al. 2016). The potential chronologies following from these by now comprehensive exercises of Anthropocene time-making— "seemingly banal time charts" (Swanson 2016, 172)—have become such a hotbed of arguments because they are, on the contrary, of utmost importance, functioning as "a form of infrastructure that shapes environmental management practices, research agendas, and policy negotiations" (Swanson 2016, 172). For some practitioners of geography, the Anthropocene might even herald a "rediscovery" of its "aspirations to be a 'world discipline' about the human–environment relationship, extending to the largest spatio-temporal scales" (Castree 2014, 449).

The interest in time periods and temporal dimensions has grown not just in disciplines, like the geosciences, that were always preoccupied with the record of time but increasingly across the environmental sciences and in the social sciences and humanities. Through recent research within the latter, timescales are widening beyond the confines of historical times to engage with science-based chronologies of geological "deep time," the biological past, and archaeology (Rudwick 2005; Chakrabarty 2009, 2019; Shryock, Smail, and Earle 2011; McGrath and Jebb 2015). This *temporal turn*—the term itself becoming increasingly used in the previous decade (e.g., Hassan 2010; Corfield 2015)—has led to a burgeoning interdisciplinary collaboration in which geographers have provided core articulations (Castree et al. 2014; Yusoff 2016). The Anthropocene soon became the key concept to crystallize the raised sensibilities of this critical interface, undoubtedly because of its major claim that human impacts had reached multiple critical thresholds and scaled to the planetary level in ways that would merit an acknowledged addition to the chrono-stratigraphic record. Even though it is evident that integrative human–geological timescales have permeated recent debates on what impact was major enough to warrant the use of the concept, with the proposed mid-twentieth-century rise of the Great Acceleration as a strong candidate (Zalasiewicz et al. 2015), it is less evident where these timescales came from. Engagement in these debates, including among geographers, who in bibliometric terms are leading users of the concept (Knitter et al. 2019), has somewhat obscured how the Anthropocene belongs in wider processes of temporalization that date further back into the twentieth century.

By following these processes, as we do in this article, we discover not only geographers at work but also interesting fault lines relating to *whether* and *how* different timescales—geological, geophysical, biological, and social—should be aligned. Some researchers, chiefly anchored in geological and geophysical time records, argued for strong alignment in line with what later appeared in Earth system science and eventually crystallized in the Anthropocene mode of thought. Others, typically working on organic time layers of the late Quaternary, tended to prefer a more cautious 'individualistic response,' arguing for less alignment, along the lines of latter-day humanist critiques against the omnibus character of the Anthropocene concept (e.g., Crist 2013; Pálsson et al. 2013). The production of planetary knowledge was always situated in political and scientific geographies that could be scaled up to speak for the entire planet. Lehman (2020), for example, showed how the oceans became an object of knowledge on a planetary scale through geographical work in the field. These kinds of processes, we argue, were not solely about planetary spatiality but also about timescales rendered visible through materialized proxy records and made to speak to an aggregated human–Earth relationship. In this article, the term we use for this kind of complex integrative work on multiple timescales is *synchronization* (Jordheim 2014, 2017), because it denotes the integration of different temporalities across the entire disciplinary spectrum into an emerging geo-anthropological temporality.

Environmental Objects and the Proto-Anthropocene

Our focus is on the North Atlantic region, a limited but essential context where the origination of synchronized timescales cross-pollinated physical geography with neighboring knowledge fields in climate history, earth history, and paleoecology. A key material element in our story is ice and how it turned from a strictly geophysical entity into an *environmental object*. Environmental objects appear as objects of knowledge on different scales through interconnections with their surroundings on both local and, increasingly, planetary levels and with the rate and direction of change in these surroundings (Aronowsky 2018). Ice, in this capacity, was used in stories about the historical geography of the region, linking timescales of ice cores—assembled vertical time—with those of human settlements and impacts in conventional landscapes of time and agency. A second key material is the organic paleolayer where pollen, often distributed and relocated in sediments, contributed further chronogeographical detail and nuances of human agency, hence also problematizing some of the grand narratives that arose from the historical geography of ice.

Geography and geographical practices were present in this research in ways that merit further attention. In the following sections, we focus on a set of loosely connected scholars from Denmark, Sweden, and the United States, whose work and research styles we identify as essential for the early synchronization of disparate timescales in or related to physical geography. These scholars would all become international leaders with defining influence in their fields over long periods of time. That is not the only reason we find them interesting, though. They are selected here because they stand out as ideal types, quintessential representatives, of broader patterns in the intellectual history of the Anthropocene and of an evolution and calibration of methods and ideas that demonstrate how key features of the Anthropocene outlook took shape several decades before the concept was articulated. We call this a *proto-Anthropocene* understanding. We further posit that their work represents several significant contributions in a succession whereby entanglements between human and societal timescales and those of geo-chronologies were gradually established and affected globally through multiple disciplines and subsequently by policy.

Their work covered a half-century, from the 1920s through to the 1970s, with a concentration around the peak of the Cold War in the 1950s and the 1957–1958 International Geophysical Year. This was a period when Fennoscandia and the North Atlantic, including Greenland, attracted a great deal of interest for a combination of strategic and scientific reasons (Doel et al. 2014; Doel, Harper, and Heymann 2016). A cross-cutting theme of the research was climate related, aimed at explaining and describing changes in climate and its impact. The theme ran through issues as far apart as the geographical distribution of vegetation, the retraction of glaciers, the patterns of sea ice, and the causes of the demise of the Norse colony on Greenland in the Middle Ages. All of these questions had strong temporal components, none of which were even remotely known in the early decades of the twentieth century. Geographers and paleoscientists working in the North Atlantic region shared interests in geophysical and biochemical processes and their implications for what was in the course of the postwar decades identified as anthropogenic "environmental change" (Warde, Robin, and Sörlin 2018, 112, 121, 221). They found in the far north attractive sites of inquiry that were accessible despite their remote location. As it turned out, they also produced scientific work that in various ways informed conceptions of anthropogenic transformations on geophysical scales.

Horizontal Gradualism: Glacial Orthodoxy

Scientific curiosity in the region was older (Bravo and Sörlin 2002; McCannon 2012) but received a substantial boost because of the early twentieth-century Arctic warming trend, which saw temperatures between 1919 and 1939 rise several degrees above the earlier average (Yamanouchi 2011). It was during that period that glaciologist Ahlmann initiated his comprehensive studies of glacier reduction across the entire region, from Scandinavian glaciers in the 1920s through to ice sheets on Svalbard and Greenland in the 1930s. A professor of geography at Stockholms Högskola (now Stockholm University), Ahlmann rose to become a prominent international leader of the discipline, a tireless networker and science diplomat, serving as president of the International Geographical Union from 1956 to 1960. After documenting massive glacial retreat across the North Atlantic region, he nonetheless remained stubbornly skeptical to the idea of human climate forcing, reflecting a stand among many field scientists, glaciologists not the least (Sörlin 2009).

His predominantly empirical work style prioritized site-specific field research to monitor glacial change in real time, with digs in the upper snow layers to uncover change over years, possibly decades, but rarely further (Sörlin 2011). Ever anxious to measure ice extent and volume properly, his methodological rigor was impeccable. He used teams of students and close colleagues, supported by assistants who were by trade and tradition bound to field sites that they measured and recorded repeatedly over the course of years. He was assisted by Sami in the Scandinavian North, farmers in Iceland, local Inuit and Norwegian hunters in Greenland, and often by local scientific expertise. He ultimately established a glaciological research station at Tarfala in northern Sweden in 1945, a "microgeography of authority" based on his site-specific monitoring approach (Sörlin 2018).

Ahlmann was well respected, and his meticulous, data-centered work was admired by his international geography peers, such as Seligman and Manley (1944; Endfield, Veale, and Hall 2015). Seligman, himself an avid glaciologist, president of the International Glaciological Society, and author of the much-acclaimed *Snow Structures and Ski Fields* (Seligman 1936), was a loyal supporter of Ahlmann's climate change skepticism (Seligman 1944). Their work in geographical glaciology represented inter- and immediate postwar orthodoxy, just before it was about to be questioned. As much as Ahlmann was an innovative physical geographer, his multisite comparative "horizontal" method—favoring linear and gradualist interpretations of natural

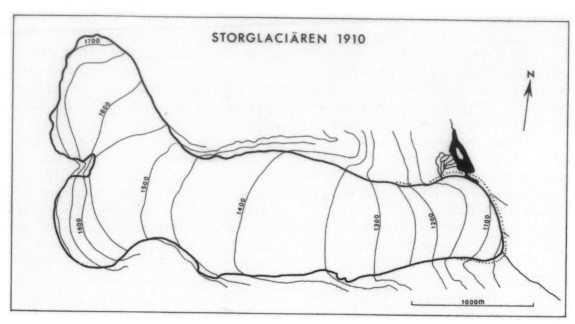

Figure 1.   Stylized map of the Storglaciären in North Sweden in 1910 with no timescale. *Source:* Bolin Center for Climate Research, Swedish Glaciers (https://bolin.su.se/data/svenskaglaciarer/glacier.php?g=69). Map based on a photograph by Fredrik Enquist (1910).

variability to explain climate "fluctuations" (his desired term)—kept him outside of the wider systemic approaches to climate and geophysical dynamics that grew among his international contemporaries working closer to fields such as meteorology and climate physics (Fleming 2016; see Figure 1). As a result, he lacked conceptual and theoretical tools to propose explanations or hypotheses that took nonlinear or anthropogenic drivers into account. His research established the reduction of glaciers as a scientific fact, but he spoke timidly on theories of climate fluctuations and their causes (e.g., Ahlmann 1948, 1953).

This cautious outlook did not allow for human agency or timescales in explaining the causes of change. In Ahlmann's rendering, glacial change had little connection with deep time chronologies and it was only loosely related to earlier time-organizing efforts for the late Quaternary period such as the sediment-based "varve" geochronology, proposed by the Stockholm geologist De Geer (De Geer 1912; Bergwik 2014). This succession of dating techniques aspired to overcome the basic shortcoming of geology, that "the history of the earth" had "hitherto been a history without years" (De Geer 1912, 241), but it did so without any deeper reflection on the possibility that this "history" might include the presence and agency of humans. The idea that geological time manifested itself with a high frequency, on the

annual scale in both clay and ice, however, opened the door to further reflection on human–geological interaction.

Ahlmann and his field-based contemporaries can be seen as standing on the threshold of a proto-Anthropocene understanding. Based on fledgling research on what had just been termed the *cryosphere* by Polish glaciologist Dobrowolski (1923)—a term resolutely refuted by Seligman and the British establishment (Barry, Jania, and Birkenmajer 2011)—Ahlmann identified ice as an element of geographical change. His epistemic project did not include ice as part of "the environment," a concept he did not use, however, and his work on glacial change contained few ideas about temporalization or the combination of multiple temporalities.

Scale Jumping through Field Jumping: The Portable Cryosphere

In 1966, just about two decades after Ahlmann founded his research station in Tarfala, the Camp Century ice core drill penetrated the thick north Greenland ice sheet to reach solid ground underneath. The drilling produced the deepest ice core— 1,387 m—to date. It also served as a validation of the possibilities of ice core drilling that had been expressed by hopeful scientists for more than a decade (Lolck 2004; Langway 2008; Martin-Nielsen

2013; Nielsen, Nielsen, and Martin-Nielsen 2014). Compared to previous ways of measuring change in the cryosphere, the Camp Century ice core offered a drastic temporal expansion, allowing for a climatic record reaching 100,000 years back in time.

This jump in scales, from annual measurements of incremental change in glaciers to millennia of climatic shifts stored inside a single object, allowed new temporalities of ice to materialize and new domains of expertise to take form. In the years following the Camp Century ice core, a vertical geography of the cryosphere (a concept now embraced) emerged, and with it came an unsettling of disciplinary boundaries. Dansgaard, a professor at the University of Copenhagen and a key figure at Camp Century, used his expertise in ice core drilling to assert himself as an expert in fields well beyond his own in geophysics.

By replacing Ahlmann's horizontal outlook with a vertical one, Dansgaard and his colleagues apprehended a radically different cryosphere: A richly textured history became visible through the layers in the ice cores. Ice appeared as a proxy for thousands of years of accumulated data rather than a research object in itself, a 'cold species' that should be monitored in real time. This turn, from horizontal to vertical and from research object to proxy, altered the qualities of ice, transforming it into an environmental object of a different kind, interesting not just in itself but because of the deep time it rendered visible (Carey 2010; Chu 2015, 2020). Ice as proxy, rather than ice as an empirical object in itself, enabled it to move beyond the cryosphere (Isberg and Paglia forthcoming) and become part of larger planetary dynamics, thereby also broadening the scope of what kind of knowledge could be produced by studying it. "The scope of ice core studies reaches far beyond glaciology itself," as Dansgaard himself put it, mentioning climatology, meteorology, geology, and solar physics as possible disciplines with which ice cores could engage (Dansgaard et al. 1973, 5).

Dansgaard's publications after 1966 indicate that the jump in temporal scales also allowed for jumps between disciplines, including as far afield as archaeology and history. In a 1975 article in *Nature*, Dansgaard set out to explain the fifteenth-century demise of a Norse settlement on Greenland by using ice core data to track climatic changes, thereby venturing into the field of historical geography. He went on to connect the fall of the medieval settlement—a proto-Anthropocene event in itself—with contemporary environmental problems

on a planetary scale (Dansgaard et al. 1975). The temporal expansion rendered possible through ice core drilling enabled a temporal compression as well, in which the fifteenth century and contemporary politics appeared connected across time by the similarities in environmental degradation. Through the ice cores, Dansgaard emerged as a versatile expert who could make such connections visible. The future also became a domain of his expertise, and Dansgaard developed a preoccupation with projecting climate futures through his "frozen annals" (Dansgaard 2005; see also Dansgaard et al. 1972), as he called them.

In the evolution of a proto-Anthropocene understanding of the relations between historical time and geochronology, Dansgaard's publications in the early 1970s show how they were growing increasingly entangled. The new vertical spatiality of the cryosphere made visible vastly longer timescales, enabling connections between local environments, glaciers, and the dynamics of the entire planet (Figure 2). If Ahlmann was standing on the threshold of a proto-Anthropocene understanding, Dansgaard entered in, albeit without conceptualizing it in this manner and, as his jumps between scales and fields indicate, productively using the lack of a strict disciplinary framework within which to operate.

Another shift that gradually appeared with the emergence of ice as a new kind of environmental object was the geography of the fieldwork itself. The ice cores were no longer bound to local scientific work in the "field cryosphere"; they could be recovered and circulated in a larger scientific infrastructure of ice core repositories that began coming online in Europe and the United States. They became a *portable cryosphere*. Even though ice core drilling still involved a great deal of fieldwork (O'Reilly 2016), much of the scientific work on ice and environmental change could now be conducted in laboratories far away from the site at which the ice originated. Not only the environmental object had changed but its geographical boundaries had changed as well.

After 1966, ice core drilling evolved from being a fairly marginal scientific activity to expand its scope into several scientific disciplines and increasingly into the environmental debate and attempts to project future trajectories of the planet's climate (Carey and Antonello 2017; Elzinga 2017). The ice core brought the glaciers with it to new epistemic geographies of climate knowledge (Mahony and Hulme 2018), altering both the environmental object itself

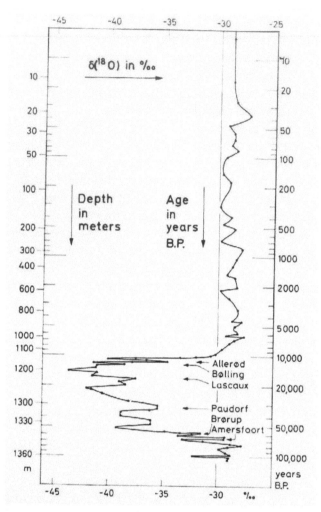

Figure 2. Representation of the 1966 Camp Century ice core with a 100,000-year timescale. *Source:* Dansgaard and Johnsen (1969, 221) (Published with permission by International Glaciological Society, Cambridge, UK.).

and the manner in which to study it. Ice became an environmental object that could be translated into quantifiable data through large-scale computerized climate modeling (Heymann, Gramelsberger, and Mahony 2017), enabling it to move from a remote, singular existence in the cryosphere into a data point in the bourgeoning understanding of an inter-connected earth system.

Seeing Scales in Seeds

By the middle of the 1970s, ice core data began to appear in scientific work beyond glaciology in fields such as climatology and oceanography (e.g., Broecker 1975). Rather than watching glacier ice transforming in real time like Ahlmann, or drilling into the ice to uncover the past of a piece of Arctic geography as Dansgaard did,

the scientific object that now appeared was ice in a quantified and universalized form. As such, it was some-thing that could operate not just in a North Atlantic context but on a planetary scale. Ice cores, along with other climate proxies such as deep-sea cores, corals, and tree rings, became objects of the earth system, rather than of the geographies from which they were retrieved. They appeared as a new kind of environmental object that could function as proxies for environmental changes well beyond the materiality and location of the objects themselves. In early Earth system science (ESS), these material time records functioned as "archives" that could underpin projections and calculations of the Earth system and were completely detached from the local envi-ronments that enabled their appearance in the first place (e.g., Oeschger 1985; NASA Earth System Sciences Committee 1986). With this increasing detachment from the field, concerns were also raised regarding the reductionist dangers in the ESS approach. What were the real-time relationships between aggregated as well as despatialized chronologies and the rates of change and related risks on local or regional levels? The negotiation between different spatial and temporal scales and the materialities that rendered them visible was itself a pro-cess unfolding over time.

A case in which this process can be seen is paly-nology. The study of pollen, spores, and microscopic planktonic organisms began as a scientific field in the early 1900s through the work of the Swedish sci-entist von Post but later expanded to the United States, and in the 1970s became increasingly enmeshed within the emerging ESS field (von Post 1916; Nordlund 2014; Birks and Berglund 2018; Edwards 2018). Despite the differences in their respective objects of study, ice and pollen, ice core drilling and palynology shared early institutional for-mation, particularly in a Scandinavian context. In 1951, Dansgaard, together with colleagues at the University of Copenhagen, among them the promi-nent palynologist Iversen, attended the first "isotope colloquium," in which interdisciplinary approaches to dating methods and the establishment of a special dating laboratory was discussed (Lolck 2004).

It was, to put it differently, a colloquium for dif-ferent earthly timescales, which shared some key features—being able to interpret through isotope analysis, stratification, and sedimentation—but also differed in some key ways: temporal scope, level of detail, local variabilities, and spatial distribution. Both Dansgaard and Iversen were interested in

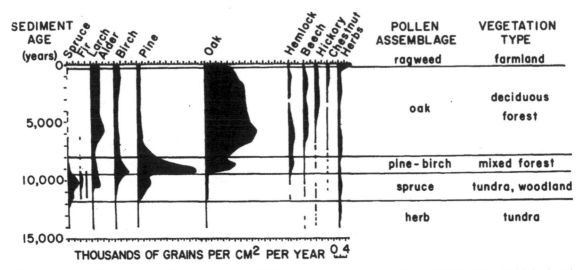

Figure 3. Pollen diagram from southern England with a 15,000-year timescale. *Source:* Davis (1969, 325) (Published with permission by American Scientist.).

climates and environments of the past, but the different materialities and technologies of their respective fields rendered results and representations of historical change that appeared in different ways (Iversen 1953; Dansgaard and Johnsen 1969). The shared institutional framework, yet separate methodologies and results, of Iversen and Dansgaard can serve as an indicative example of the multiple timescales that have coexisted in the making of a synchronized earth system. Particularly the practice of coring, which allowed for analysis of sedimentation and vertical temporal divisions, was common in both fields and rendered it possible to find common ground despite disparate materialities, timescales, and geographies (Figures 2 and 3). Their work, however, also illustrates the nonsynchronicities and the difficulties in reconciling local differences with an all-encompassing idea of the planet as an integrated entity and with the latter becoming an environmental object in its own right.

Three years after the "isotope colloquium," a young Davis, then a Harvard graduate student, went to Copenhagen to study under Iversen and work with the interglacial pollen spectra in western Greenland (Davis 1954). She would later become the head of the Department of Ecology and Behavioral Biology at the University of Minnesota and a leading figure in palynology. Rather than researching the ice sheet itself, which Dansgaard began sampling in the early 1950s, Davis was preoccupied with studying the remains of vegetation that were stored in Greenland's soil and ice. Even though their work in Greenland shared many similarities, and occurred in parallel,

their respective approaches to generalization and large-scale modeling came to differ over time. In this way, Davis can serve as an example of the "individualistic response" approach to oversimplified models that threatened to obscure local variation.

From the 1980s, she became increasingly concerned with human environmental impact, particularly emphasizing the relationship between local environments and global climatic changes (Davis 1986a, 1986b). The complex dynamics of local or regional vegetation histories did not self-evidently translate into the models of Earth systems scientists, however (Figure 3). In a 1989 presidential address to the Ecological Society of America, she displayed a cautious stance toward the shifting biome approach visible in generalized models: "This result from the past means that species can be expected to respond individualistically to climatic changes in the future. We must not build models that predict the future by shifting existing communities or biomes around on the surface of the globe" (Davis 1989, 222).

The following year she reiterated the same point, explicitly addressing the International Geosphere-Biosphere Program, and cautioned against oversimplifying ecological processes. Asserting that "the paleorecord argues against such simplification" (Davis 1990, 269), she highlighted a tension between scale jumping and the need to ground scale practices in material and situated environments. Similar concern was voiced by glaciologists working in the Ahlmann–Seligman gradualist tradition, assembling local data that did not fit the modeled reality (Sörlin 2018).

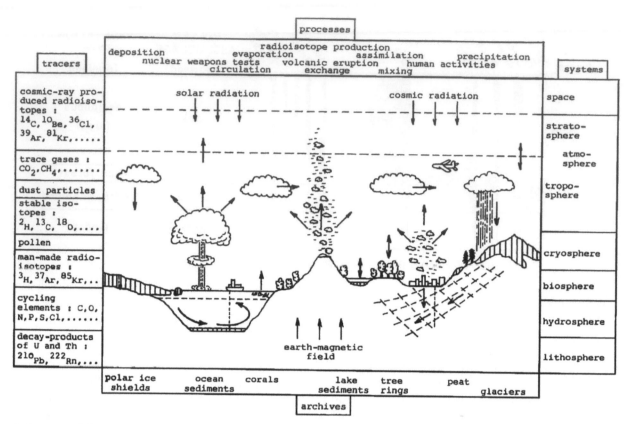

Figure 4. Diagram of the environmental system concept with multiple stylized temporal dimensions. *Source:* Oeschger (1985) (Published with permission by the American Geophysical Union.).

This does not suggest that Davis wanted to distance herself from the integrative ambitions of the early ESS. Rather, their difference in outlook could be understood as a difference in emphasis linked to historically developed materialities and practices. ESS favored big data and large spatial scales, which facilitated synchronizing across timescales. The geographical approaches (gradualist glaciology, case-based palynology), rooted in the field and in comparative data sets and real-time observation over shorter time spans, tended, on the contrary, to produce skepticism toward both large scales and cross-field synchronization. Thus, contemporary critique of the reductionist tendencies of the Anthropocene concept (e.g., Haraway 2015; Yusoff 2019) can be seen in the light of a longer epistemological split, which dates back to the scientific practices that underpin the concept itself.

In the depiction of the 1910 surface extension of Storglaciären in northern Sweden (Figure 1), the *horizontal way of temporalizing* the glacier produced a "landscape of recorded change" (Sörlin 2018, 257) that traced environmental change in real time as a shrinking local area of ice subject to natural climate

variability. Photographic documentation, maps, and mass balance measurements made annually were spatially limited to the boundaries of the body of the glacier. The *vertical temporalization* visible in ice core drilling (Figure 2) created a spatially and temporally *different environmental object*: The timescale is vastly increased and the spatial dimensions are not limited to a single glacier or local environment but comparable with timescales from other environmental objects, such as pollen. Palynology, here exemplified by the work of Davis (Figure 3), shares some similarities with the ice core diagram—vertical temporality and a vast timescale—but is also geographically bounded (to New England in this case) and accounts for other environmental factors such as biodiversity. These differences are, however, not visible in the 1985 illustration of the Earth system (Figure 4). Here, different paleochronologies are put under the headline "archives" and are visualized as seemingly unified entities into a larger planetary "environmental system concept." The diagram marks an integration of human agency and politics with geological phenomena—nuclear weapons tests and volcanic eruptions are presented as similar kinds of events—and the synchronized archives are used to underpin this understanding, characteristic of the proto-Anthropocene.

Grounding Earthly Times: Concluding Discussion

Through empirical study of research practices in physical geography and cognate fields, we have tried to demonstrate that the emergence of the Anthropocene can be usefully understood as the result of a longer history of production of integrative human–geological timescales. This work of synchronization, taking place in the second half of the twentieth century, occurred within a broader process of configuring planetary knowledge that would later crystallize in the concept of the Anthropocene. The new approach to geo-chronologies differed fundamentally from earlier understandings when both dating techniques and Quaternary time layers were detached from human agency on large geographical scales. This process of configuration was not the outcome of abstract, intellectual work but rather a product of material and situated scientific practice. When the elements of study, such as ice and pollen, were researched with new methods from neighboring fields (field jumping) and thereby provided new sets of data, scale jumping became possible and eventually resulted in a planetary-scale environmental object with its own embedded temporality.

What can explain the rapid decline of the detached temporalizations of the Earth and human history and the equally swift ascent of a human-agential regime linked to geological timescales? There was obviously nothing wrong with the field data of glaciology, nor with previous data in Arctic plants. Using the North Atlantic as a lens, we have been able to identify distinct moments where new temporal understandings of established elements of study (ice and pollen) emerged at about the same time in the 1950s. Our story, admittedly brief and stylized, suggests that integrative combinations of knowledge were essential in making scale jumping possible, which in turn was necessary to quantify and assess anthropogenic impacts on the global scale.

Further impetus was provided by intellectual forerunners, some of them transformed fieldworkers in possession of strong environmental objects, like Dansgaard. Others were more full-fledged sedentary "meta-specialists" (Warde, Robin, and Sörlin 2018, 63), like oceanographer Broecker, who played a key role in allowing ice core data to travel from their local Arctic confines to become temporal objects with implications for multiple properties of the Earth system (Broecker 1975). In this latter lineage, we can place Crutzen himself, who in 2000 first articulated the Anthropocene concept in its modern form (Steffen, McNeill, and Crutzen 2007). The conceptual innovation marks the beginning of a period of organized scholarly, cultural, and political thinking around the idea of the Anthropocene. Apart from the concept itself, however, very little differentiates Crutzen from his North Atlantic postwar forerunners; he was a product of long-standing work in scaling and synchronizing.

It is therefore useful to identify a proto-Anthropocene understanding beginning before the institutionalization of ESS and the emergence of the Anthropocene concept. This period, largely coinciding with the Great Acceleration since approximately 1950, speaks to the strengths and weaknesses of, respectively, scientific disciplines and the work in those integrative research fields we have discussed here. The proto-Anthropocene phase highlights how the appearance of standardized planetary temporalities, visible in ESS, was the outcome of elaborate work of synchronization and processes of standardizing a multiplicity of proxy records into unified ideas about human–geological timescales. Thus, contemporary debates on the reductionist, and possibly obscuring, tendencies of the Anthropocene should also be seen in light of a longer history of negotiating local variabilities with aggregated, planetary-scale temporalizations.

The production of integrative timescales, and the scientific work that enabled their appearance, speaks to larger questions regarding the ways in which planetary knowledge was produced (Figure 4). The challenge was the combination of vertical and horizontal geographies with the synchronization work that could allow enough explanatory power to human agency. ESS managed to mobilize enough intellectual, institutional, and infrastructural resources to provide the essential building blocks of a new understanding of the human–Earth relationship (Seitzinger et al. 2015). By the mid-1980s, integrative representations of the "environmental system concept" were already in circulation, complete with "tracers" such as pollen, "archives" such as glaciers and peat, "systems" such as cryosphere and biosphere, and "processes" including human agency (Oeschger 1985). Ice and pollen had eventually become input to computerized modeling of planetary dynamics. The innovations that we have identified as instrumental in reaching the new understanding—scale jumping, portability of data, systems approaches, and synchronized, integrative timescales—were all present in ESS and facilitated the formation of a particular epistemology of the human–Earth relationship. This interdisciplinary framework later surfaced as the Anthropocene concept, about a half-century after important work on timescales took place in the North Atlantic region.

Acknowledgments

We thank our colleagues in the European Research Council–funded project SPHERE—Study of the Planetary Human Earth Relationship: The Rise of Global Environmental Governance at KTH in Stockholm, Sweden, with partners at the University of Cambridge, the European University Institute in Florence, and the University of California, Berkeley, for their comments on an earlier draft and two anonymous reviewers for helpful suggestions.

References

Ahlmann, H. W. 1948. The present climate fluctuation. *The Geographical Journal* 112 (4–6):165–93. doi: 10. 2307/1789696.

Ahlmann, H. W. 1953. *Glacier variations and climate fluctuations.* New York: The American Geographical Society. doi: 10.1017/S0022143000025338.

Aronowsky, L. V. 2018. The planet as self-regulating system: Configuring the biosphere as an object of knowledge, 1940–1990. PhD diss., Harvard University, Graduate School of Arts & Sciences.

Barry, R. G., J. Jania, and K. Birkenmajer. 2011. A. B. Dobrowolski—The first cryospheric scientist and the subsequent development of cryospheric science. *History of Geo- and Space Sciences* 2 (1):75–79. doi: 10.5194/hgss-2-75-2011.

Bergwik, S. 2014. A fractured position in a stable partnership: Ebba Hult De Geer, Gerard De Geer and early twentieth century Swedish geology. *Science in Context* 27 (3):423–51. doi: 10.1017/S0269889714000131.

Birks, H. J. B., and B. E. Berglund. 2018. One hundred years of Quaternary pollen analysis 1916–2016. *Vegetation History and Archaeobotany* 27 (2):271–309. doi: 10.1007/s00334-017-0630-2.

Bolin Centre for Climate Research. n.d. Swedish glaciers. Accessed June, 29 2020. https://bolin.su.se/data/sven-skaglaciarer/glacier.php?g=69.

Bravo, M., and S. Sörlin, eds. 2002. *Narrating the Arctic: A cultural history of Nordic scientific practices.* Canton, MA: Science History Publications.

Broecker, W. 1975. Climatic change: Are we on the brink of a pronounced global warming? *Science* 189 (4201):460–63. doi: 10.1126/science.189.4201.460.

Carey, M. 2010. In the Shadow of Melting Glaciers: Climate Change and Andean Society. New York: Oxford University Press.

Carey, M., and A. Antonello. 2017. Ice cores and the temporalities of the global environment. *Environmental Humanities* 9 (2):181–203. doi: 10.1215/22011919-4215202.

Castree, N. 2014. The Anthropocene and Geography I: The back story. *Geography Compass* 8 (7):436–49. doi: 10.1111/gec3.12141.

Castree, N., W. M. Adams, J. Barry, D. Brockington, B. Büscher, E. Corbera, D. Demeritt, R. Duffy, U. Felt, K. Neves, et al. 2014. Changing the intellectual climate. *Nature Climate Change* 4 (9):763–68. doi: 10.1038/nclimate2339.

Chakrabarty, D. 2009. The climate of history: Four theses. *Critical Inquiry* 35 (2):197–222. doi: 10.1086/596640.

Chakrabarty, D. 2019. The planet: An emergent humanist category. *Critical Inquiry* 46 (1):1–31. doi: 10.1086/705298.

Chu, P.-Y. 2015. Mapping permafrost country: Creating an environmental object in the Soviet Union, 1920s–1940s. *Environmental History* 20 (3):396–421. doi: 10.1093/envhis/emv050.

Chu, P.-Y. 2020. *The life of permafrost: A history of frozen Earth in Russian and Soviet science.* Toronto, Canada: University of Toronto Press.

Corfield, P. J., ed. 2015. History and the temporal turn: Returning to causes, effects and diachronic trends. In *Les âges de Britannia: Repenser l'histoire des mondes Britanniques: (Moyen Âge-XXIe siècle):* [The ages of Britannia: Rethink the history of Britannique worlds, Middle Ages through 21st century.] ed. J.-F. Dunyach and A. Mairey, 259–73. Rennes, France: Presses Universitaires de Rennes.

Crist, E. 2013. On the poverty of our nomenclature. *Environmental Humanities* 3 (1):129–47. doi: 10.1215/22011919-3611266.

Crutzen, P., and E. Stoermer. 2000. The Anthropocene. *Global Change Newsletter* 41 (May):17–18. http://www.igbp.net/download/18.316f18321323470177580001401/1376383088452/NL41.pdf.

Dansgaard, W. 2005. *Frozen annals: Greenland ice cap research.* Copenhagen: Niels Bohr Institute.

Dansgaard, W., and S. J. Johnsen. 1969. Flow model and a time scale for the ice core from Camp Century, Greenland. *Journal of Glaciology* 8 (53):215–23. doi: 10.1017/S0022143000031208.

Dansgaard, W., S. J. Johnsen, H. B. Clausen, and N. Gundestrup. 1973. Stable isotope glaciology. *Meddelelser om Grønland* 197 (2):1–53.

Dansgaard, W., S. J. Johnsen, H. B. Clausen, and C. C. Langway. 1972. Speculations about the next glaciation. *Quaternary Research* 2 (3):396–98. doi: 10.1016/0033-5894(72)90063-4.

Dansgaard, W., S. J. Johnsen, N. Reeh, N. Gundestrup, H. B. Clausen, and C. U. Hammer. 1975. Climatic changes, Norsemen, and modern man. *Nature* 255 (5503):24–28. doi: 10.1038/255024a0.

Davis, M. B. 1954. Interglacial pollen spectra from Greenland. In *Danmarks Geologiske Undersøgelse. II Raekke* [Studies in vegetational history in honour of Knud Jessen], ed. J. Iversen, 65–72. Copenhagen: C.A. Reitzels Forlag.

Davis, M. B. 1969. Palynology and environmental history during the Quaternary period. *American Scientist* 57 (3):317–32.

Davis, M. B. 1986a. Climatic instability, time lags, and community disequilibrium. In *Community ecology*, ed. J. M. Diamond and T. J. Case, 269–84. New York: Harper and Row.

Davis, M. B. 1986b. Foreword: Symposium on vegetation–climate equilibrium. VIth International Palynological Conference, Calgary. *Vegetatio* 67 (2):64.

Davis, M. B. 1989. Address of the president: Insights from paleoecology on global change. *Bulletin of the Ecological Society of America* 70 (4):222–28.

Davis, M. B. 1990. Biology and paleobiology of global climate change: Introduction. *Trends in Ecology & Evolution* 5 (9):269–70. doi: 10.1016/0169-5347(90)90078-r.

De Geer, G. 1912. A geochronology of the last 12,000 years. In *Compte Rendu 11 Congrès Géologique International, 11th International Geological Congress (1910)*, Vol. 1, 241–53. Stockholm, Sweden: P.A. Norstedt & Söner. doi: 10.1007/BF01802565.

Dobrowolski, A. B. 1923. *Historia naturalna lodu* [The natural history of ice]. Warsaw, Poland: Kasa Pomocy im. Dr. J. Mianowskiego.

Doel, R. E., R. M. Friedman, J. Lajus, S. Sörlin, and U. Wråkberg. 2014. Strategic Arctic science: National interests in building natural knowledge—Interwar era through the Cold War. *Journal of Historical Geography* 44 (April):60–80. doi: 10.1016/j.jhg.2013.12.004.

Doel, R. E., K. C. Harper, and M. Heymann. 2016. *Exploring Greenland: Cold War science and technology on ice.* New York: Palgrave.

Edwards, K. 2018. Pollen, women, war and other things: Reflections on the history of palynology. *Vegetation History and Archaeobotany* 27 (2):319–35. doi: 10.1007/s00334-017-0629-8.

Elzinga, A. 2017. Polar ice cores: Climate change messengers. In *Research objects in their technological setting*, ed. B. B. Vincent, S. Loeve, A. Nordmann, and A. Schwarz, 215–31. London and New York: Routledge.

Endfield, G. H., L. Veale, and A. Hall. 2015. Gordon Valentine Manley and his contribution to the study of climate change: A review of his life and work. *Wiley Interdisciplinary Reviews: Climate Change* 6 (3):287–99. doi: 10.1002/wcc.334.

Fleming, J. R. 2016. *Inventing atmospheric science: Bjerknes, Rossby, Wexler, and the foundations of modern meteorology.* Cambridge, MA: MIT Press.

Haraway, D. 2015. Anthropocene, Capitalocene, Plantationocene, Chthulucene: Making kin. *Environmental Humanities* 6 (1):159–65. doi: 10.1215/22011919-3615934.

Hassan, R. 2010. Globalization and the "temporal turn": Recent trends and issues in time studies. *The Korean Journal of Policy Studies* 25 (2):83–102.

Heymann, M., G. Gramelsberger, and M. Mahony, eds. 2017. *Cultures of prediction in atmospheric and climate science.* London and New York: Routledge.

Isberg, E., and E. Paglia. Forthcoming. On record: Political temperature and the temporalities of climate change. In *Times of history, times of nature: Temporalization and the limits of modern knowledge*, ed. A. Ekström and S. Bergwik. New York: Berghahn.

Iversen, J. 1953. Radiocarbon dating of the Alleröd period. *Science* 118 (3053):9–11. doi: 10.1126/science.118.3053.9.

Jordheim, H. 2014. Introduction: Multiple times and the work of synchronization. *History and Theory* 53 (4):498–518. doi: 10.1111/hith.10728.

Jordheim, H. 2017. Synchronizing the world: Synchronism as historiographical practice, then and now. *History of the Present* 7 (1):59–95. doi: 10.5406/historypresent.7.1.0059.

Knitter, D., K. Augustin, E. Biniyaz, W. Hamer, M. Kuhwald, M. Schwanebeck, and R. Duttmann. 2019. Geography and the Anthropocene: Critical approaches needed. *Progress in Physical Geography: Earth and Environment* 43 (3):451–61. doi: 10.1177/0309133319829395.

Koselleck, R. 1979. *Vergangene Zukunft: Zur Semantik geschichtlicher Zeiten.* Frankfurt am Main, Germany: Suhrkamp.

Langway, C. C. 2008. *The history of early polar ice cores.* Buffalo, NY: ERDC/CRREL.

Lehman, J. 2020. Making an Anthropocene ocean: Synoptic geographies of the International Geophysical Year (1957–1958). *Annals of the American Association of Geographers* 110 (3):606–22. doi: 10.1080/24694452.2019.1644988.

Lewis, S. L., and M. A. Maslin. 2015. Defining the Anthropocene. *Nature* 519 (7542):171–80. doi: 10.1038/nature14258.

Lolck, M. 2004. *Klima, Kold Krig og Iskerner* [Climate, cold war and ice cores]. Aarhus, Denmark: Steno Instituttet, Aarhus Universitet.

Mahony, M., and M. Hulme. 2018. Epistemic geographies of climate change: Science, space and politics. *Progress in Human Geography* 42 (3):395–424. doi: 10.1177%2F0309132516681485.

Manley, G. 1944. Some recent contributions to the study of climatic change. *Quarterly Journal of the Royal Meteorological Society* 70 (305):197–219. doi: 10.1002/qj.49607030508.

Martin-Nielsen, J. 2013. "The deepest and most rewarding hole ever drilled": Ice cores and the Cold War in Greenland. *Annals of Science* 70 (1):47–70. doi: 10.1080/00033790.2012.721123.

McCannon, J. 2012. *A history of the Arctic: Nature, exploration, exploitation.* London: Reaktion.

McGrath, A., and M. Jebb, eds. 2015. *Long history, deep time: Deepening histories of place.* Canberra, ACT, Australia: ANU Press.

NASA Earth System Sciences Committee. 1986. *Earth System Science overview: A program for global change.* Washington, DC: National Aeronautics and Space Administration.

Nielsen, K., H. Nielsen, and J. Martin-Nielsen. 2014. City under the ice: The closed world of Camp Century in Cold War science. *Science as Culture* 23 (4):443–64. doi: 10.1080/09505431.2014.884063.

Nordlund, C. 2014. Peat bogs as geological archives: Lennart von Post and the development of quantitative pollen analysis during World War I. *Earth Sciences History* 33 (2):187–200. doi: 10.17704/eshi.33.2.b905813153638715.

Oeschger, H. 1985. The contribution of ice core studies to the understanding of environmental process. In *Greenland ice cores: Geophysics, geochemistry and the environment*, ed. C. C. Langway, Jr., H. Oeschger, and W. Dansgaard, 9–19. Washington, DC: American Geophysical Union.

O'Reilly, J. 2016. Sensing the ice: Field science, models, and expert intimacy with knowledge. *Journal of the Royal Anthropological Institute* 22 (1):27–45. doi: 10.1111/1467-9655.12392.

Pálsson, G., B. Szerszynski, S. Sörlin, J. Marks, B. Avril, C. Crumley, H. Hackmann, P. Holm, J. Ingram, A. Kirman, et al. 2013. Reconceptualizing the "Anthropos" in the Anthropocene: Integrating the social sciences and humanities in global environmental change research. *Environmental Science & Policy* 28 (1):3–13. doi: 10.1016/j.envsci.2012.11.004.

Ruddiman, W. F. 2003. The Anthropogenic greenhouse era began thousands of years ago. *Climatic Change* 61 (3):261–93. doi: 10.1023/B:CLIM.0000004577.17928.fa.

Rudwick, M. 2005. *Bursting the limits of time: The reconstruction of geohistory in the age of revolution.* Chicago: Chicago University Press.

Seitzinger, S., O. Gaffney, G. Brasseur, W. Broadgate, P. Ciais, M. Claussen, J. W. Erisman, et al. 2015. International Geosphere–Biosphere Programme and Earth system science: Three decades of co-evolution. *Anthropocene* 12 (December):3–16. doi: 10.1016/j.ancene.2016.01.001.

Seligman, G. 1936. *Snow structure and ski fields: Being an account of snow and ice forms met with in nature, and a study on avalanches and snowcraft.* London: Macmillan.

Seligman, G. 1944. Glacier fluctuation. *Quarterly Journal of the Royal Meteorological Society* 70 (303):22. doi: 10.1002/qj.49707030305.

Shryock, A., D. L. Smail, and T. Earle. 2011. *Deep history: The architecture of past and present.* Berkeley: University of California Press.

Sörlin, S. 2009. Narratives and counter-narratives of climate change: North Atlantic glaciology and meteorology, c. 1930–1955. *Journal of Historical Geography* 35 (2):237–255. doi: 10.1016/j.jhg.2008.09.003.

Sörlin, S. 2011. The anxieties of a science diplomat: Field co-production of climate knowledge and the rise and fall of Hans Ahlmann's "Polar Warming." *Osiris* 26 (1):66–88. doi: 10.1086/661265.

Sörlin, S. 2018. A microgeography of authority: Glaciology and climate change at the Tarfala Station 1945–1980. In *Understanding field science institutions*, ed. H. Ekerholm, K. Grandin, C. Nordlund, and P. A. Schell, 255–85. Sagamore Beach, MA: Science History Publications.

Steffen, W., W. Broadgate, L. Deutsch, O. Gaffney, and C. Ludwig. 2015. The trajectory of the Anthropocene: The great acceleration. *The Anthropocene Review* 2 (1):81–98. doi: 10.1177/2053019614564785.

Steffen, W., J. R. McNeill, and P. J. Crutzen. 2007. The Anthropocene: Are humans now overwhelming the great forces of nature? *Ambio* 36 (8):614–621. doi: 10.1579/0044-7447(2007)36[614:TAAHNO]2.0.CO;2.

Swanson. H. A. 2016. Review of: Anthropocene as Political Geology: Current Debates over how to tell Time. *Science as Culture* 25 (1):157–163.

von Post, L. 1916. Einige südschwedischen Quellmoore (trans. Some South Swedish Spring Mires). *Bulletin of the Geological Institution of the University of Uppsala* 15:219–78.

Warde, P., L. Robin, and S. Sörlin. 2017. Stratigraphy for the Renaissance: Questions of expertise for "the environment" and "the Anthropocene." *The Anthropocene Review* 4 (3):246–58. doi: 10.1177/2053019617738803.

Warde, P., L. Robin, and S. Sörlin. 2018. *The environment: A history of the idea.* Baltimore, MD: Johns Hopkins University Press.

Waters, C. N., J. Zalasiewicz, C. Summerhayes, A. D. Barnosky, C. Poirier, A. Cearreta, A. Gałuszka, M-. Edgeworth, E. C. Ellis, M. Ellis, et al. 2016. The Anthropocene is functionally and stratigraphically distinct from the Holocene. *Science* 351 (6269):aad2622. doi: 10.1126/science.aad2622.

Yamanouchi, T. 2011. Early 20th century warming in the Arctic: A review. *Polar Science* 5 (1):53–71. doi: 10.1016/j.polar.2010.10.002.

Yusoff, K. 2016. Anthropogenesis: Origins and endings in the Anthropocene. *Theory, Culture & Society* 33 (2):3–28. doi: 10.1177%2F0263276415581021.

Yusoff, K. 2019. *A billion black Anthropocenes or none.* Minneapolis: University of Minnesota Press.

Zalasiewicz, J., C. N. Waters, M. Williams, A. D. Barnosky, A. Cearreta, P. Crutzen, E. Ellis, M. A. Ellis, I. J. Fairchild, J. Grinevald, et al. 2015. When did the Anthropocene begin? A mid-twentieth century boundary level is stratigraphically optimal. *Quaternary International* 383:196–203. doi: 10.1016/j.quaint.2014.11.045.

Zalasiewicz, J., C. N. Waters, M. Williams, and C. Summerhayes, eds. 2018. *The Anthropocene as a geological time unit: A guide to the scientific evidence and current debate.* Cambridge, MA: Cambridge University Press.

Geographic Thought and the Anthropocene: What Geographers Have Said and Have to Say

Thomas Barclay Larsen and John Harrington, Jr.

Drawing from early modern and contemporary geographic thought, this article explores how the premise of an Anthropocene (Age of Humans) can be used to reinforce enduring modes of human–environment thinking. Anthropocene dialogues build on insights posed by geographers of the eighteenth and early nineteenth centuries: unity of nature, humans as nature made conscious, humans as nature's conscience, and time periods as devices for thinking about human–environment relations. Complementing these ideas, contemporary geographers are making compelling statements about the Anthropocene, affirming that interpretations of the proposed geologic time period differ according to socioenvironmental variables, geographic imaginations, local contexts, and critical perspectives. Three forms of human–environment thinking emerge from examining links between early modern geographers and current geographers addressing the Anthropocene: synthesis thinking, epistemological thinking, and ethical thinking. Connections across ideas concerning the Anthropocene and geographic thought will be strengthened by developing systematic chronologies of the human–environment relationship. *Key Words: Anthropocene, consilience, geographic thought, human–environment, synthesis.*

On 23 June 1802, Alexander von Humboldt neared the 6,263 m summit of Chimborazo, in the Andes Mountains south of Quito, Ecuador. At around 5,917 m, he was the first documented European to climb so high. Sucking the thin air and peering at landscapes below, the geographer fully grasped his eminent impression of nature's togetherness and interdependence (Wulf 2016). Humboldt's *eutierra*—a feeling of oneness with Earth coined by philosopher Glenn Albrecht—was the 1800s version of the Blue Marble, a first full portrait of the sunlit "Spaceship Earth," photographed during the 1972 Apollo 17 expedition (Louv 2011; Reinert 2011; National Aeronautics and Space Administration 2017). A sense of wholeness captured by the Blue Marble image expanded the public's outer-spatial perception of Earth and humbled individuals preoccupied by anxieties transpiring in cities, war zones, coastlines, farmlands, and elsewhere (R. Fuller 1969; Reinert 2011; Zurita, Munro, and Houston 2018).

To glimpse the Chimborazo eutierra or the Blue Marble of the twenty-first century, one must descend from the aerial view and direct attention among the rock strata exposed at Earth's surface (see Zurita, Munro, and Houston 2018; Jackson 2019). From that stratigraphic perspective reemerged an idea

when Nobel chemist Paul Crutzen uttered the word *Anthropocene* (Age of Humans) as a potential geologic progression from the Holocene (New Age). The insight happened during a February 2000 conference in Cuernavaca, Mexico, sponsored by the Scientific Committee of the International Geosphere-Biosphere Program (Crutzen 2002). The Anthropocene term was first employed by Russian geologist Aleksei Pavlov in 1922 (Lewis and Maslin 2015).[1] One word expressed the inadequacy of the Holocene to describe today's distinctive socioecological situation. Scholars gathered enough evidence to back claims that humans were more than just part of nature; they had irreversibly rewired Earth's system and were overriding its planetary boundaries (Crutzen and Stoermer 2000; Crutzen 2002; see also Rockström et al. 2009).

Among academic circles, the Anthropocene idea went viral (Castree 2014a, 2014b). Many scholars believe that the label, Anthropocene, ought to become a formal epoch in the official geologic timeline (Crutzen and Stoermer 2000; Waters et al. 2016). Advancing the cause is the Anthropocene Working Group, a coalition of prominent academics in the realm of human–environment relations (Zalasiewicz et al. 2015; Ellis et al. 2016; Ellis 2017). Such an addition to the geologic timeline would secure the significance of humans as geological agents, transforming Earth's system and surface to the point where the human impress can be globally excavated in layers of rock and sediment (Crutzen and Stoermer 2000; Crutzen 2002; Häusler 2018). Scholars of different disciplines have defended and criticized the proposed geologic epoch (Lepori 2015; Toivanen et al. 2017; Edgeworth et al. 2019; Ruddiman 2019; Zalasiewicz et al. 2019). According to Bjornerud (2018), "In many ways, the exact start of the Anthropocene matters less than the concept behind it" (131).

Geographers are well practiced in contributing to and evaluating planetary knowledge and, thus, have important things to say about the Anthropocene concept (Lehman 2020). Historiographical accounts document a discipline defined by its expertise in human–environment relations (Zimmerer 2010; Harden 2012). How does the premise of an Anthropocene influence interpretations of geography's disciplinary history? How can the theories and ideas of geography advance thinking about the Anthropocene and human–environment relations? A revisit of classic and contemporary contributions confirms that geographers have been studying aspects of the so-called Anthropocene long before the idea's rebirth at the turn of the new millennium (Ellis 2017; Butler 2018).

At first, contemporary geographers who addressed aspects of the Anthropocene "made little of Geography's wider intellectual resources in advancing their particular Anthropocene propositions" (Castree 2014b, 460). Anthropocene dialogues now hold court in the geography community, as more geographers use the concept as a medium for grappling with global change and human–environment relations. Exploration of the literature on geographic thought and the Anthropocene culminates in at least three forms of human–environment thinking: synthesis, epistemological, and ethical thinking. In a "complex, interconnected, rapidly changing world," a broad perspective and range of skills benefit individual and group responses to twenty-first-century issues and beyond (Epstein 2019, 51). Geography fits the bill perfectly, much more so than any other academic discipline (Murphy 2018).

A Revisionist Historiography of Geographic Thought?

The Anthropocene inspires a retrospective look at geography's disciplinary history. In the midst of a very short issue-attention cycle, an inattentive public and ecologically anxious academy appear to zigzag from issue to issue, placing society at risk of ignoring procedural knowledge and past wisdoms (Downs 1972; Robbins and Moore 2013). Accounts of the Anthropocene's intellectual roots credit "classic efforts to illustrate the importance of humans as agents of landscape change," many of which deserve more than a brief nod or passing citation (Butler 2018, 530). Human–environment aspects of the Anthropocene provide chances for a revisionist historiography that recasts geography's disciplinary history according to emerging ideas and circumstances (cf. Domosh 1991).

Acquiring renewed significance in the Anthropocene are ideas from geography's past, a number of which become apparent in the nineteenth and early twentieth centuries. Humboldt and Marsh envisaged nature as an integrated whole with humans affecting it. Humboldt (1849) conceptually braided Earth's systems together through *naturgemälde* (the unity of nature; Wulf 2016; Jackson 2019). *Man and nature*, authored by Marsh ([1864] 2003), complemented Humboldt's conclusions by describing human impacts on the environment in various contexts around the world. Integrating humans with the environment represents one important contribution of early modern geographers toward the Anthropocene's eventual conception (Bendix and Urban 2020).

Nature's unity meshes with Reclus's ([1866] 2013, 110) argument that humans are nature made conscious and that "[h]umanity's development is most intimately connected with the nature that surrounds it." Studying nature along with humanity's past downfalls—tales of "tribes, peoples, cities, and empires that have perished miserably as the consequence of slow evolutions"—can instruct future adaptations (Reclus [1898] 2013, 139). With environmental awareness comes environmental responsibility. Kropotkin (1902) proclaimed the primacy of mutual aid, an ethical duty of humans to cooperate with other human and nonhuman life. Therefore, humans are nature's conscience just as much as they are nature made conscious. Correspondingly, ideas like the Anthropocene shape and are molded by a need to consider how history meets accountability (Bartel 2018; Schmidt 2019).

Time periods represent one device employed by early modern geographers to think about human–environment relations. In the manner of the *Annales* School of geographers and historians, de la Blache and Reclus established grand narratives of human history, positing that slow change is true change when examining human interaction with the world (Ferretti 2015). To document these changes, de la Blache and Reclus synthesized history akin to generalizing places into regions: by aggregating similar events into time periods of long durations (*longue durée*; Reclus 1866; Wishart 2004; Ferretti 2015). Geographers probing the Anthropocene restore a bygone geographic practice of scaffolding human–environment relations according to time periods (Sauer 1974).

Humboldt's naturgemälde, Marsh's human impacts, Reclus's consciousness, Kropotkin's mutual aid, and the *Annales* School's longue durée coalesce to produce a brief historiography of connections between the Anthropocene and geographic thought. These ideas, along with others, exemplify how fundamental components of the Anthropocene concept existed in one form or another long before 2000. Anthropocene discourses compel contemporary geographers to piece together ideas that challenge, reconcile, and build on geography's past.

Geographic Contributions to Anthropocene Dialogues

Timely statements have surfaced about the Anthropocene and its affiliation with geographic knowledge (Ziegler and Kaplan 2019). Links to Anthropocene scholarly dialogue tend to be made by emplacing [insert topic] in the Anthropocene, rethinking [insert epistemology] of the Anthropocene, and venturing "toward" or "beyond" [insert point(s) of contention] about the Anthropocene. Language conveys a multitude of contexts, directions, and ideations. What do geographers have to say about the Anthropocene and how do they support their statements with disciplinary concepts, skills, and perspectives? Several statements arise from examining Anthropocene literature in geography.[2]

The Anthropocene Is More Than Stratigraphic

Debating the Anthropocene's existence is an exercise in geographic inquiry: Scholars inspect socioenvironmental systems at graduating spatial and temporal scales to reach a global geologic conclusion.

Despite the epoch's lithic theme, the structure and makeup of the Anthropocene prove difficult to define, and these ambiguities muddle the placement of a temporal boundary or boundaries (Steffen et al. 2011). Some interlocutors claim that the Anthropocene reaches back to the human discovery of fire, whereas others suggest that the geologic epoch begins around 1945 with developing nuclear technology and weaponry (Lewis and Maslin 2015). One faction of geographers and geography-adjacent scholars comments directly on the Anthropocene's proposed formalization in the geologic timeline (Edgeworth et al. 2019). *Progress in Physical Geography*, in particular, has fielded critiques and replies regarding the Anthropocene Working Group's geologic methods for establishing time periods (Edgeworth et al. 2019; Knitter et al. 2019; Ruddiman 2019).

Castree (2015) introduced the Anthropo(s)cene to characterize the epoch's relevant networks, publications, and institutions. Anthropocene ideas have geography journal editors coordinating provocative papers and commentaries (E. Johnson et al. 2014; Bauer and Ellis 2018; Edgeworth et al. 2019; Ziegler and Kaplan 2019). Characterizing this collective commentary is the view of the Anthropocene as a "rough place-holder" for an "unprecedented historical condition" riddled with complexity and uncertainty, that demands "critical assessments of how material engagements take form, hold fast, and/or break apart in space and through time" (E. Johnson et al. 2014, 440). Applying the geographic perspective pushes the geologic limits of defining the Anthropocene (Knitter et al. 2019). Geography straddles the social and physical sciences, along with the humanities, in its search for synthesis or consilience (Wilson 1998; Gober 2000). Calls to integrate geographers of all pedigrees complicate a logical positivist process for determining where the Age of Humans begins in the geologic record (Castree 2015; Ellis 2017).

Acknowledgment that Earth's system encompasses social and physical processes necessitates a more-than-stratigraphic Anthropocene timeline (Edgeworth et al. 2019). For instance, biogeographers might arrive at an alternative chronology by prioritizing biospheric as opposed to lithospheric processes. Plant and animal domestications, accompanied by human-induced species extinctions, signal systemic changes to the biogeographer in the intensity and timing of human impacts (Young 2014, 2016). Political and economic geographers uncover answers in milieus—the social world of ideas, practices, and interpersonal relations—depicting the Anthropocene less as a geologic artifact and more as a social outcome of production and consumption (Castree 2015; Millar and Mitchell 2017; Sexton 2018; Gibson-Graham 2019). The character, language, and types of strata differ greatly across geography's subfields.

Transdisciplinary researchers outside geography have reached similar conclusions, ascribing a pluralized nature to the Anthropocene idea(s) (Toivanen et al. 2017; Stallins 2020). Multiple Anthropocenes exist and can be linked to the identification and visualization of geographic phenomena through time. Much like a zone of transition from one geographic region to an adjacent one, a temporal boundary can be expressed in a number of ways.

- The Anthropocene can be viewed as a *gradient* of increasing human transformation of the environment, a thin Anthropocene beginning with fire discovery and intensifying as innovations took place in agriculture, technology, and the development of nation-states (Ellis et al. 2016).
- A *transition* can be represented as a golden spike, a specific date and layer in Earth's stratigraphy, such as 1945 when the Great Acceleration exhibited rapid socioenvironmental change and above-ground nuclear bomb testing altering Earth's radiative composition of soils (Steffen et al. 2015).
- An *emphasis* on humans as agents of change indicates shifts from traditional agricultural and societal transformations of Earth to industrial and postindustrial transformations (Smith and Zeder 2013).
- *Plurality*, with a series of overlapping variables, can represent the Anthropocene, with contemporaneous increases in methane or carbon dioxide, intensification of deforestation and soil erosion, and mass extinction of species (Ruddiman et al. 2015).

Embracing a more-than-stratigraphic perspective yields imaginative inquiry into the changing character of human–environment relations.

Anthropocene Studies Stimulate the Geographic Imagination

Whereas some geographers engage in the Anthropocene's officiation, others reimagine the epoch in ways that traverse the sciences and humanities (Matless 2017). Patterns across space dominate the spatial-chorological identity within geographic studies; the Anthropocene brings depth to the Blue Marble, exposing humanity's knotted entanglement with the subterranean past (Turner 2002; Zurita, Munro, and Houston 2018). Depth can refer to a literal Geography Underground, a layered world beneath the chorology of Earth's surface. "Skyscrapers, air-space, mountains" and other focal points turn Western minds "up and out, by flight, adventures in outer space and of late, by climate change and other atmospheric preoccupations" (Hawkins 2020, 4). The Anthropocene literally grounds our gaze, disclosing previously unknown physical landforms like *aquaterra*, influenced by continuous advance and retreat of sea level during the last 120,000 years (Dobson 2014; Dobson, Spada, and Galassi 2020). Landscape ideas like aquaterra characterize the shifting Earth system setting and add dynamism that prior typologies lack, such as Humboldt's (1849) elevation-based classes of *tierra caliente* (hot land), *templada* (temperate), *fría* (cold), and *helada* (frozen).

In a humanistic light, depth can refer to an imaginative Geography Underground, a reaction to concerns that the current environmental crisis doubles as a crisis of the geographic imagination (Hawkins 2020). Anthropocene musings have inspired visions of Earth as "the scene of the crime" and a composted poem containing layers of meaning thrown together (Magrane 2020, 8). Applications of these thoughts can rewild human interactions with the environment to promote better health outcomes and more sustainable futures (Leiper 2019).

Geographers and other scholars interpret the Anthropocene using different lenses to identify its wholeness, deconstruct it, and relate it to other ideas. Dividing global time functions more as a telescope than an accurate clock, capable of "looking back through time from the present, adjusting the focus according to temporal distance between the observer and the observed" (Edgeworth et al. 2019, 340). Such a telescopic vision can also be kaleidoscopic, containing a multitude of colors and hues. Despite its origins in the Earth system scientific perspective, the Anthropocene has manifested in a number of forms: a stand-alone idea or concept; stratigraphic boundary, geologic epoch, equation, or factor (Gaffney and Steffen 2017); rupture in Earth system functions (Hamilton 2016); paradigm shift from environmental science to Earth system science (Hamilton 2015); evolving paradigm (Butzer 2015); "differential lens through which disciplines across the academy are reviewing, debating, and reinventing their conceptions of humanity and nature" (Bauer and Ellis 2018, 210); powerful environmental education tool (Braje 2018); "singular trophy to be fought over and won or lost" (Zalasiewicz et al. 2018, 222); and terms with multiple meanings (like countless other Anglophone concepts).

As a result, the Anthropocene has become a palimpsest, in which older versions become replaced by newer translations, yet nonetheless retain the watermark of the original idea. Aiding this palimpsest view is an assortment of collateral concepts, additional ideas that supplement a broader idea (e.g., planetary boundaries, views of nature; Castree 2014c). Perspectives appear as *portmanteaus*, new words gathering "various meanings and associations from bumping into, and off of, other related terms in the same and other languages" (Usher 2016, 57).

Auxiliary "cenes" depict alternative narratives. Known for originating the Gaia hypothesis, Lovelock (2019) claimed that the Anthropocene began in 1712, with the invention of Thomas Newcomen's atmospheric engine. Lovelock's Anthropocene has since ended, followed by a transition to the Novacene, a state of "techno-nature" in which humans are eventually replaced by cyborgs living harmoniously with Earth. Wilson (2016) termed the Eremocene (Age of Loneliness) as "the age of people, our domesticated plants and animals, and our croplands all around the world as far as the eye can see" Wilson (2016, 20). When describing the physics of scale, West (2017) claimed that the Anthropocene ended with the Industrial Revolution. In its place, the Urbanocene emerged as urbanization altered the global reality so that the "future of humanity and the long-term sustainability of the planet are inextricably linked to the fate of our cities" (214). Although coined by nongeographers, the Novacene, Eremocene, and Urbanocene demonstrate how the Anthropocene expands and diversifies scholarly and geographic imaginations.

Geographers who address issues in Earth system science can link back to the Greek philosopher Heraclitus (born 544 BCE), with the observation that humans cannot step in the same river twice, because both the river and humans change (Graham 2019). Similarly, the Anthropocene merges *terrae incognitae* (unexplored places) with *saecula incognita* (unexplored lengths of time). Anthropocene thinking enables philosophers of geography to examine myriad perceptions of human–environment relations among various places and times (Wright 1947).

Global Time Periods Have Varying Implications for Local Places

Studies of the Anthropocene reinvigorate the geographic awareness that global changes manifest differently in local places (Wilbanks and Kates 1999). Grand narratives tend to be myopic, so ground-truthing a local Anthropocene produces finer resolutions of the situation (Pawson 2015). Fieldwork is being conducted in localities with the Anthropocene as a conceptual setting. For example, human impacts on rivers in New Zealand feature a place-based chronology marked by early Polynesian settlement around 1300 CE, followed by the impacts from the late 1800s CE when Europeans began clearing land (I. Fuller, Macklin, and Richardson 2015). Moreover, incipient evidence questions conventional wisdom(s) holding that early hunter-gatherers practiced sustainable land management (Feeney 2019). Scales of such land transformations tend to be localized, with more significant evidence for regional change with early mining and smelting transformations during the Late Bronze Age and Early Iron Age (3500–2800 BP; Wagreich and Draganits 2018).

Local studies of urban geomorphology have also contributed toward understanding human transformations in the Anthropocene. Earthquake-prone cities like Christchurch, New Zealand, face distinctive challenges for disaster recovery and resilience under the weight of multiple Anthropocene-related environmental hazards (Westgate 2020). European cities like Genoa, Rome, Naples, Palermo (Italy), and Patras (Greece) are long-standing relics of human modifications to geomorphic processes: diversion of rivers, quarrying rocks, leveling areas of relief, building high grounds for defense, and amassing artificial ground (Brandolini et al. 2020).

Of additional interest are local changes between human and nonhuman species in the Anthropocene. Climate change and marine degradation prompt inquiry into co-evolution between human and nonhuman species among the Galapagos Islands (Salinas-de-León et al. 2020). In Australia, geographers navigate uncertainty in the livestock industry through recommendations like making cattle production more efficient, advocating ethical forms of consumption, promoting artificial beef and dairy, and supporting veganism (McGregor and Houston 2018). Evidently, morphologies of landscape matter in the midst of global transformations (Sauer 1925).

The Anthropocene Is Not Neutral

Cartographic representations of the Anthropocene portray the globe in an array of panic-inducing yellows, oranges, and reds, designed to provoke emotion and politicize the issue (Schneider 2016; see also Xu et al. 2020). Recognizing an Anthropocene comes with strong ethical implications (Bartel 2018; Gibson-Graham et al. 2019; Krzywoszynska 2019; Schmidt 2019). Ideas and debates move the contribution of the Anthropocene beyond normal science circles to the "nonorthodoxy" of the late twentieth century (see Skaggs 2004, 448). Bullard (2000, 2018), founder of environmental justice and the Association of American Geographers 2018 Honorary Geographer, affirmed that environmental decisions affect race, ethnicity, and place and that a humane response has only begun. Efforts to define an Anthropocene ethic call attention to geographers' desires to reach conclusions about "why do one thing over another, or why do anything" (Harman, Harrington, and Cerveny 1998, 277). A cadre of gender-focused, feminist, and related geographers engage with problems of othering, racism, capitalocentrism, and colonialism in the Anthropocene (Leichenko and Mahecha 2015; Gibson-Graham 2019; Gibson-Graham et al. 2019).

Critical social theorists interject Anthropocene dialogues by interrogating trajectories of colonialism, neoliberalism, and racism through human–environmental history (Haraway 2015, 2016; Erickson 2020; Joo 2020; Luke 2020; Simpson 2020). Crucial to that inquisition is questioning "the utility of the knowledge we are creating and the reasons *why* and for *whom* this knowledge is produced" (Leichenko and Mahecha 2015, 330). The process entails investigating how shifts in discourse, government regulation, economic practices, and social norms reveal or obscure ethical consequences of environmental decisions (Schmidt 2019). Media can be a powerful tool for downplaying the seriousness of global change, owing to the observation that "humans are not so much rational creatures as they are very good rationalizers" (Orr 2016, 20). Through use of the Anthropocene concept, geographers can educate the public to interpret media with a critical eye, inspire

hope for change, and use the past to map out future human–environment interactions (Pawson 2015).

Along with the Anthropocene, critical scholars have considered portmanteaus like the Plantationocene, bringing slavery to the fore in the human–environment relationship; the Capitalocene, underscoring the Marxist importance of capitalism in transforming environment and society; and the Chthulucene, envisioning a postdisaster union of human and nonhuman species (Haraway 2015, 2016). For critical geographers, these terms contain broader implications for articulating "struggles, strategies to cultivate more just futures, and alternative ways of being" in the context of black geographies, indigenous geographies, and other displaced populations (Davis et al. 2019, 11). Epochs become more than containers of time. They can contain revolutionary ideas.

Modes of Human–Environment Thinking in the Anthropocene

Reviews of geography's human–environment tradition can be programmatic, thematic, biographic, and prescriptive (Kates 1987; Turner 1997; Zimmerer 2010; Harden 2012; J. Harrington and Harrington 2012). Given geography's rich human–environment history and current interest in the Anthropocene, what might the assemblage of ideas look like and how can it inform thinking about human–environment relations? Classifying human–environment thinking strengthens the human–environment identity as a geographic subject and assists geographers' observations and interpretations of Anthropocene phenomena (cf. Golledge 2002; Turner 2002). Running through the statements, topics, and themes in this article are three modes of human–environment thinking: synthesis thinking, epistemological thinking, and ethical thinking (Table 1).

Synthesis Thinking

As geographers address present and future challenges, "human–environment interaction should become a larger part of the discipline" (Malanson 2014, 257), with polymaths building on the detailed work of specialists (see Moran and Lopez 2016; Goudie 2017). Central to being a polymath geographer is linking knowledge together through synthesis thinking (Gober 2000). Consilience, or the convergence of knowledge, conveys an integrated approach to

Table 1. Modes of human–environment thinking and how they relate to statements by early modern geographers, contemporary geographers discussing the Anthropocene, and identified topics and themes of human–environment geography

Mode of human–environment thinking	Context-relevant definition	Early modern geographers discussing human–environment relations	Contemporary geographers discussing the Anthropocene	Zimmerer's (2010, 2017) related topics and themes for human–environment geography[a]
Synthesis thinking	Linking knowledge together to inform understandings of human–environment relations	• Nature as a unified web and humans affecting it (Humboldt 1849; Marsh 1864) • Time periods as devices for mapping human–environmental change (longue durée)	• Global time periods taking different contexts in local places (Wilbanks and Kates 1999; Pawson 2015) • The Anthropocene encompassing more than just geologic indicators (Edgeworth et al. 2019; Knitter et al. 2019)	• Land use, land systems, land change, and biodiversity • Social–ecological and coupled human–environment systems • Resource political economy, management, and politics • Food, health, and bodies in relation to the environment
Epistemological thinking	Applying theoretical lenses to interpret and critique aspects of the human–environment relationship	• Humans as nature made conscious (Reclus 1866)	• Anthropocene studies stimulating the geographic imagination (Matless 2017; Zurita, Munro, and Houston 2018) • Collateral concepts and portmanteaus of the Anthropocene (Castree 2014c) • Revisionist historiographies of geography	• Environmental landscape history and ideas • Knowledge concepts in environmental management and policy
Ethical thinking	Considering broader implications and responsibilities of environmental decisions	• Humans as nature's conscience; Kropotkin's (1902) Mutual Aid	• The Anthropocene as an ethical statement (Schneider 2016; Bartel 2018; Gibson-Graham et al. 2019; Krzywoszynska 2019; Schmidt 2019) • Responsibility of geographers to address broader impacts of human–environment research (Leichenko and Mahecha 2015; Pawson 2015)	• Livelihoods and agricultural landscapes • Environmental hazards, risk, vulnerability, and resilience • Political ecology and environmental governance

Notes: [a]Each of Zimmerer's (2010, 2017) topics incorporates multiple modes of human–environment thinking. Slight adjustments were made in organizing this table.

grasping global phenomena (Wilson 1998; Lowenthal 2019). Geographers collect concepts, skills, and perspectives to help humans comprehend their relationship to the world (Unwin 1992; Sack 1997).

Anthropocene dialogues include ongoing efforts to unify geography in response to contemporary challenges across Earth's critical zones, while also appreciating disciplinary diversity (Minor et al. 2020). Attaching the Anthropocene to geographic concepts can encourage physical geographers, economic geographers, political geographers, cultural geographers, urban geographers, and others to share distinctive takes on nature–society issues (see Lorimer 2012; Moran and Lopez 2016; Ellis 2017; Goudie 2017; Derickson 2018; L. Johnson, Lobo, and Kelly 2019). Multiple routes to geographic synthesis are available, ranging from logical positivist to humanist methods (Turner and Robbins 2008; Seamon 2018). Epistemological thinking assists in exploring how geographers, nongeographers, and the public synthesize aspects of the human–environment relationship.

Epistemological Thinking

Epistemological thinking deploys different ways of knowing to navigate the "bazaar of exchange" within the "cultural heterotopia" of human–environment thought (Mugerauer 1994, 3; Zimmerer 2010). Pragmatic, critical, and empathetic ways of knowing are available to approach a human–environment topic (Mugerauer 1994; Cresswell 2013; Bousquet et al. 2015). Schools of thought and epistemologies advance dynamic understandings of how time periods (like the Anthropocene) are created and interpreted (Wishart 2004; Stallins 2020).

Ideas and knowledge concepts drive epistemologies (Zimmerer 2017). "It matters what ideas we use to think other ideas," wrote anthropologist Strathern (1992, as quoted in Haraway 2016, 12). Geographers offer a number of big ideas to inquire about human interaction with the biogeophysical world. Contributions include sustainability science (Kates et al., 2001; see also L. Harrington 2016), vulnerability science (Cutter, Mitchell, and Scott 2000; Cutter 2003), land change science (Turner, Lambin, and Reenberg 2007), global change in local places (Wilbanks and Kates 1999), anthropogenic biomes (Ellis and Ramankutty 2008), and urbanization science (Solecki, Seto, and Marcotullio 2013). Steeped in geographic thought, epistemological thinking

diversifies planetary knowledge about the Anthropocene and facilitates contemplating broader implications for environmental decisions.

Ethical Thinking

Both synthesis and epistemological thinking have ethical ramifications because they influence how geographers assess vulnerability and risk to environmental hazards, as well as indicate equitable solutions for environmental governance (Zimmerer 2017; Cutter 2020). Righteousness pervades environmental matters, owing to the philosophical observation that humans "understand nothing about ecological mutations if we don't measure the extent to which they throw everyone into a panic. Even if they have several different ways of driving us crazy!" (Latour 2017, 10–11). Amid a "renewed emphasis specifically on [geography's] human/environment tradition," research must reflect broader societal implications and connect "with our needs as humans" (Harman 2003, 421).

By virtue of its breadth, geography has the flexibility to reconcile the tension between nature and culture and offer space for examining environmental ethics (Bartel 2018). A well-defined Anthropocene ethic thus presents a worthwhile endeavor to further shape geography's human–environment identity (Gibson-Graham et al. 2019; Krzywoszynska 2019; Schmidt 2019). Efforts might evolve into an applied Anthropocene geography that fosters ecologically conscious leaders who build communities of proactive planetary residents through environmental citizenship (Chapin and Fernandez 2013).

The Road Still Beckoning for Human–Environment Geography

Synthesis thinking, epistemological thinking, and ethical thinking help characterize the relationship between geographic thought and human–environment topics like the Anthropocene. As geographers and other scholars consider prospects for constructing planetary knowledge, it becomes necessary to identify cross-cutting ideas that join diverse human–environment research (see Gober 2000; Zimmerer 2010; Harden 2012; Lowenthal 2019). To adapt the words of Kates (1987), the road still beckons for human–environment geography in the Age of Humans.

One path forward involves geographers combining synthesis, epistemological, and ethical thinking to address how to assemble human–environment events into time periods. Curiosity about the Anthropocene concept verifies that time periods can contain powerful ideas about human–environment relations that inform understandings and decisions (Bjornerud 2018). Anthropocene dialogues open possibilities for geographers to participate in scientists' calls to develop systematic histories of human–environment interactions (Costanza et al. 2007; Cornell et al. 2010).

Geographers might venture so far as to begin developing, officiating, and interpreting a human–environment timeline, one that accommodates both physical and social scientific qualities of nature–society relations (Larsen and Harrington 2020). As some Anthropocene scholars have hammered away at the rocks exposed at the Blue Marble's surface, it is important for geographers to continue collecting knowledge and perspectives, while aggregating these understandings to create new time periods and ideas through human–environment thinking.

Acknowledgments

The authors thank Dr. David Butler and the two anonymous reviewers for their input on improving this article. They would also like to thank Dr. Alexander Hall for advice on Latin translation.

Notes

1. The Anthropocene also contains conceptual ties to the Noösphere, developed by Ukrainian geochemist Vernadsky in 1945 to describe the interface between the biosphere and human cognition (Lewis and Maslin 2015).
2. Although these declarations overlap with discourses outside of geography, they are imbued with distinct geographic approaches to human–environment thinking.

References

Bartel, R. 2018. Place-speaking: Attending to the relational, material and governance messages of *Silent spring*. *The Geographical Journal* 184 (1):64–74. doi: 10.1111/geoj.12229.

Bauer, A., and E. Ellis. 2018. The Anthropocene divide: Obscuring understanding of social–environmental change. *Current Anthropology* 59 (2):209–27. doi: 10.1086/697198.

Bendix, J., and M. Urban. 2020. Nothing new under the sun? George Perkins Marsh and roots of U.S. physical geography. *Annals of the American Association of Geographers*. doi: 10.1080/24694452.2020.1761769.

Bjornerud, M. 2018. *Timefulness: How thinking like a geologist can help save the world*. Princeton, NJ: Princeton University Press.

Bousquet, F., P. Robbins, C. Peloquin, and O. Bonato. 2015. The PISA grammar decodes diverse human–environment approaches. *Global Environmental Change* 34:159–71. doi: 10.1016/j.gloenvcha.2015.06.013.

Braje, T. 2018. Comments on "The Anthropocene divide: Obscuring understanding of social-ecological change." *Current Anthropology* 59:215–16.

Brandolini, P., C. Cappadonia, G. Luberti, C. Donadio, L. Stamatopoulos, C. D. Maggio, F. Faccini, et al. 2020. Geomorphology of the Anthropocene in Mediterranean urban areas. *Progress in Physical Geography* 44 (4):461–94. doi: 10.1177/0309133319881108.

Bullard, R. 2000. *Dumping in Dixie*. 3rd ed. Boulder, CO: Westview.

Bullard, R. 2018. 2018 Honorary Geographer Address with Dr. Robert Bullard, father of environmental justice. Accessed May 12, 2020. https://www.youtube.com/watch?v=9KtKzldxLAM.

Butler, D. R. 2018. Man as a geological agent: An account of his actions on inanimate nature, by Robert Lionel Sherlock, 1922. *Progress in Physical Geography: Earth and Environment* 42 (4):530–34. doi: 10.1177/0309133318787999.

Butzer, K. 2015. *Anthropocene* as an evolving paradigm. *The Holocene* 25 (10):1539–41. doi: 10.1177/0959683615594471.

Castree, N. 2014a. The Anthropocene and geography I: The back story. *Geography Compass* 8 (7):436–49. doi: 10.1111/gec3.12141.

Castree, N. 2014b. The Anthropocene and geography II: Current contributions. *Geography Compass* 8 (7):450–63. doi: 10.1111/gec3.12140.

Castree, N. 2014c. *Making sense of nature*. London and New York: Routledge.

Castree, N. 2015. Changing the Anthropo(s)cene. *Dialogues in Human Geography* 5 (3):301–16. doi: 10.1177/2043820615613216.

Chapin, F. S., III, and E. Fernandez. 2013. Proactive ecology for the Anthropocene. *Elementa* 1:p.000013. doi: 10.12952/journal.elementa.000013.

Cornell, S., R. Costanza, S. Sörlin, and S. van der Leeuw. 2010. Developing a systematic "science of the past" to create our future. *Global Environmental Change* 20 (3):426–27. doi: 10.1016/j.gloenvcha.2010.01.005.

Costanza, R., L. Graumlich, W. Steffen, C. Crumley, J. Dearing, K. Hibbard, R. Leemans, C. Redman, and D. Schimel. 2007. Sustainability or collapse: What can we learn from integrating the history of humans and the rest of nature? *Ambio* 36 (7):522–27. doi: 10.1579/0044-7447(2007)36[522:socwcw]2.0.co;22.0.co;2].

Cresswell, T. 2013. *Geographic thought: A critical introduction*. West Sussex, UK: Wiley-Blackwell.

Crutzen, P. 2002. Geology of mankind. *Nature* 415 (6867):23. doi: 10.1038/415023a.

Crutzen, P., and E. Stoermer. 2000. The Anthropocene. *Global Change Newsletter* 41:17–18.

Cutter, S. 2003. The vulnerability of science and the science of vulnerability. *Annals of the Association of American Geographers* 93 (1):1–12. doi: 10.1111/1467-8306.93101.

Cutter, S. 2020. The changing nature of hazard and disaster risk in the Anthropocene. *Annals of the American Association of Geographers*. doi: 10.1080/24694452.2020.1744423.

Cutter, S., J. Mitchell, and M. Scott. 2000. Revealing the vulnerability of people and places: A case study of Georgetown County, South Carolina. *Annals of the Association of American Geographers* 90 (4):713–37. doi: 10.1111/0004-5608.00219.

Davis, J., A. Moulton, L. Van Sant, and B. Williams. 2019. Anthropocene, Capitalocene, … Plantationocene? A manifesto for ecological justice in an age of global crises. *Geography Compass* 13 (5):e12438. doi: 10.1111/gec3.12438.

Derickson, K. 2018. Urban geography III: Anthropocene urbanism. *Progress in Human Geography* 42 (3):425–35. doi: 10.1177/0309132516686012.

Dobson, J. E. 2014. *Aquaterra Incognita*: Lost land beneath the sea. *Geographical Review* 104 (2):123–38. doi: 10.1111/j.1931-0846.2014.12013.x.

Dobson, J. E., G. Spada, and G. Galassi. 2020. Global choke points may link sea level and human settlement at the Last Glacial Maximum. *Geographical Review*. doi: 10.1080/00167428.2020.1728195.

Domosh, M. 1991. Toward a feminist historiography of geography. *Transactions of the Institute of British Geographers* 16 (1):95–104. doi: 10.2307/622908.

Downs, A. 1972. Up and down with ecology—The "issue-attention cycle." *The Public Interest* 28:38–50.

Edgeworth, M., E. Ellis, P. Gibbard, C. Neal, and M. Ellis. 2019. The chronostratigraphic method is unsuitable for determining the start of the Anthropocene. *Progress in Physical Geography: Earth and Environment* 43 (3):334–44. doi: 10.1177/0309133319831673.

Ellis, E. 2017. Physical geography in the Anthropocene. *Progress in Physical Geography: Earth and Environment* 41 (5):525–32. doi: 10.1177/0309133317736424.

Ellis, E., M. Maslin, N. Boivin, and A. Bauer. 2016. Involve social scientists in defining the Anthropocene. *Nature* 540 (7632):192–93. doi: 10.1038/540192a.

Ellis, E., and N. Ramankutty. 2008. Putting people in the map: Anthropogenic biomes of the world. *Frontiers in Ecology and the Environment* 6 (8):439–77. doi: 10.1890/070062.

Epstein, D. 2019. *Range: Why generalists triumph in a specialized world*. New York: Riverhead.

Erickson, B. 2020. Anthropocene futures: Linking colonialism and environmentalism in an age of crisis. *Environment and Planning D: Society and Space* 38 (1):111–28. doi: 10.1177/0263775818806514.

Feeney, J. 2019. Hunter-gatherer land management in the human break from ecological sustainability. *The Anthropocene Review* 6 (3):223–42. doi: 10.1177/2053019619864382.

Ferretti, F. 2015. Anarchism, geohistory, and the *Annales*: Rethinking Elisée Reclus's influence on Lucien Febvre. *Environment and Planning D: Society and Space* 33 (2):347–65. doi: 10.1068/d14054p.

Fuller, I., M. Macklin, and J. Richardson. 2015. The geography of the Anthropocene in New Zealand: Differential river catchment response to human impact. *Geographical Research* 53 (3):255–69. doi: 10.1111/1745-5871.12121.

Fuller, R. 1969. *Operating manual for Spaceship Earth*. New York: Simon and Schuster.

Gaffney, O., and W. Steffen. 2017. The Anthropocene equation. *The Anthropocene Review* 4 (1):53–61. doi: 10.1177/2053019616688022.

Gibson-Graham, J. 2019. Reading for difference in the archives of tropical geography: Imagining an(other) economic geography for beyond the Anthropocene. *Antipode* 52:12–35. doi: 10.1111/anti.12594.

Gibson-Graham, J., J. Cameron, S. Healy, and J. McNeill. 2019. Economic geography and ethical action in the Anthropocene: A rejoinder. *Economic Geography* 95 (1):27–29. doi: 10.1080/00130095.2018.1538696.

Gober, P. 2000. In search of synthesis. *Annals of the Association of American Geographers* 90 (1):1–11. doi: 10.1111/0004-5608.00181.

Golledge, R. 2002. The nature of geographic knowledge. *Annals of the Association of American Geographers* 92 (1):1–14. doi: 10.1111/1467-8306.00276.

Goudie, A. S. 2017. The integration of human and physical geography revisited. *The Canadian Geographer/Le Géographe Canadien* 61 (1):19–27. doi: 10.1111/cag.12315.

Graham, D. 2019. Heraclitus. In *Stanford encyclopedia of philosophy*, ed. E. N. Zalta. Accessed November 20, 2019. https://plato.stanford.edu/entries/heraclitus/.

Hamilton, C. 2015. Getting the Anthropocene so wrong. *The Anthropocene Review* 2 (2):102–7. doi: 10.1177/2053019615584974.

Hamilton, C. 2016. The Anthropocene as rupture. *The Anthropocene Review* 3 (2):93–106. doi: 10.1177/2053019616634741.

Haraway, D. 2015. Anthropocene, Capitalocene, Plantationocene, Chthulucene: Making kin. *Environmental Humanities* 6 (1):159–65. doi: 10.1215/22011919-3615934.

Haraway, D. 2016. *Staying with the trouble: Making kin in the Chthulucene*. Durham, NC: Duke University Press.

Harden, C. 2012. Framing and reframing questions of human–environment interactions. *Annals of the Association of American Geographers* 102 (4):737–47. doi: 10.1080/00045608.2012.678035.

Harman, J. 2003. Whither geography? *The Professional Geographer* 55 (4):415–21. doi: 10.1111/0033-0124.5504001.

Harman, J., J. Harrington, Jr., and R. Cerveny. 1998. Science, policy, and ethics: Balancing scientific and ethical values in environmental science. *Annals of the Association of American Geographers* 88 (2):277–86. doi: 10.1111/1467-8306.00094.

Harrington, J., Jr., and L. Harrington. 2012. Global change and geographic thought. In *21st century geography: A reference handbook*, ed. J. P. Stoltman, 50–66. Thousand Oaks, CA: Sage.

Harrington, L. 2016. Sustainability theory and conceptual considerations: A review of key ideas for sustainability, and the rural context. *Papers in Applied Geography* 2 (4):365–82. doi: 10.1080/23754931.2016.1239222.

Häusler, H. 2018. Did anthropogeology anticipate the idea of the Anthropocene? *The Anthropocene Review* 5 (1):69–86. doi: 10.1177/2053019617742169.

Hawkins, H. 2020. Underground imaginations, environmental crisis and subterranean cultural geographies. *Cultural Geographies* 27 (1):3–21. doi: 10.1177/1474474019886832.

Humboldt, A. 1849. *Cosmos: A sketch of a physical description of the universe*, trans. E. Otté. London: Henry G. Bohn.

Jackson, S. 2019. Humboldt for the Anthropocene. *Science* 365 (6458):1074–76. doi: 10.1126/science.aax7212.

Johnson, E., H. Morehouse, S. Dalby, J. Lehman, S. Nelson, R. Rowan, S. Wakefield, and K. Yusoff. 2014. After the Anthropocene: Politics and geographic inquiry for a new epoch. *Progress in Human Geography* 38 (3):439–56. doi: 10.1177/0309132513517065.

Johnson, L., M. Lobo, and D. Kelly. 2019. Encountering naturecultures in the urban Anthropocene. *Geoforum* 106:358–62. doi: 10.1016/j.geoforum.2019.05.005.

Joo, H. 2020. We are the world (but only at the end of the world): Race, disaster, and the Anthropocene. *Environment and Planning D: Society and Space* 38 (1):72–90. doi: 10.1177/0263775818774046.

Kates, R. 1987. The human environment: The road not taken, the road still beckoning. *Annals of the Association of American Geographers* 77 (4):525–34. doi: 10.1111/j.1467-8306.1987.tb00178.x.

Kates, R., W. Clark, R. Corell, J. Hall, C. Jaeger, I. Lowe, J. McCarthy, H. Schellnhuber, B. Bolin, N. Dickson. 2001. Environment and development. Sustainability science. *Science* 292 (5517):641–42. doi: 10.1126/science.1059386.

Knitter, D., K. Augustin, E. Biniyaz, W. Hamer, M. Kuhwald, M. Schwanebeck, and R. Duttmann. 2019. Geography and the Anthropocene: Critical approaches needed. *Progress in Physical Geography: Earth and Environment* 43 (3):451–61. doi: 10.1177/0309133319829395.

Kropotkin, P. 1902. *Mutual aid: A factor of evolution*. New York: McClure Phillips.

Krzywoszynska, A. 2019. Caring for soil life in the Anthropocene: The role of attentiveness in more-than-human ethics. *Transactions of the Institute of British Geographers* 44 (4):661–75. doi: 10.1111/tran.12293.

Larsen, T., and J. Harrington, Jr. 2020. A human–environment timeline. *Geographical Review*. doi: 10.1080/00167428.2020.1760719.

Latour, B. 2017. *Facing Gaia: Eight lectures on the new climatic regime*. Cambridge, UK: Polity.

Lehman, J. 2020. Making an Anthropocene ocean: Synoptic geographies of the International Geophysical Year (1957–1958). *Annals of the American Association of Geographers* 110 (3):606–22. doi: 10.1080/24694452.2019.1644988.

Leichenko, R., and A. Mahecha. 2015. Celebrating geography's place in an inclusive and collaborative Anthropo(s)cene. *Dialogues in Human Geography* 5 (3):327–32. doi: 10.1177/2043820615613252.

Leiper, C. 2019. The Paleo paradox: Re-wilding as a health strategy across scales in the Anthropocene. *Geoforum* 105:122–30. doi: 10.1016/j.geoforum.2019.05.015.

Lepori, M. 2015. There is no Anthropocene: Climate change, species-talk, and political economy. *Telos* 2015 (172):103–24. doi: 10.3817/0915172103.

Lewis, S., and M. Maslin. 2015. Defining the Anthropocene. *Nature* 519 (7542):171–80. doi: 10.1038/nature14258.

Lorimer, J. 2012. Multinatural geographies for the Anthropocene. *Progress in Human Geography* 36 (5):593–612. doi: 10.1177/0309132511435352.

Louv, R. 2011. *The nature principle*. Chapel Hill, NC: Algonquin.

Lovelock, J. 2019. *Novacene: The coming age of hyperintelligence*. Cambridge, MA: MIT Press.

Lowenthal, D. 2019. *Quest for the unity of knowledge*. London and New York: Routledge.

Luke, T. 2020. Tracing race, ethnicity, and civilization in the Anthropocene. *Environment and Planning D: Society and Space* 38 (1):129–46. doi: 10.1177/0263775818798030.

Magrane, E. 2020. Climate geopoetics (the earth is a composted poem). *Dialogues in Human Geography*. doi: 10.1177/2043820620908390.

Malanson, G. 2014. Physical geography on the methodological fence: David Stoddart (1965) Geography and the ecological approach: The ecosystem as a geographic principle and method. *Geography* 50: 242–251. *Progress in Physical Geography* 38 (2):251–258. doi: 10.1177/0309133314525184

Marsh, G. P. [1864] 2003. *Man and nature: Or, physical geography as modified by human action*. Seattle: University of Washington Press.

Matless, D. 2017. The Anthroposcenic. *Transactions of the Institute of British Geographers* 42 (3):363–76. doi: 10.1111/tran.12173.

McGregor, A., and D. Houston. 2018. Cattle in the Anthropocene: Four propositions. *Transactions of the Institute of British Geographers* 43 (1):3–16. doi: 10.1111/tran.12193.

Millar, S., and D. Mitchell. 2017. The tight dialectic: The Anthropocene and the capitalist production of nature. *Antipode* 49:75–93. doi: 10.1111/anti.12188.

Minor, J., J. Pearl, M. Barnes, T. Colella, P. Murphy, S. Mann, and G. Barron-Gafford. 2020. Critical zone science in the Anthropocene: Opportunities for biogeographic and ecological theory and praxis to drive earth science integration. *Progress in Physical Geography: Earth and Environment* 44 (1):50–69. doi: 10.1177/0309133319864268.

Moran, E., and M. Lopez. 2016. Future directions in human–environment research. *Environmental Research* 144 (B):1–7. doi: 10.1016/j.envres.2015.09.019.

Mugerauer, R. 1994. *Interpretations on behalf of place: Environmental displacements and alternative responses*. Albany, NY: SUNY Press.

Murphy, A. 2018. *Geography: Why it matters*. Cambridge, UK: Polity.

National Aeronautics and Space Administration. 2017. Apollo 17: Blue Marble, August 7. Accessed May 12, 2020. https://www.nasa.gov/image-feature/apollo-17-blue-marble.

Orr, D. 2016. *Dangerous years: Climate change, the long emergency, and the way forward*. New Haven, CT: Yale University Press.

Pawson, E. 2015. What sort of geographical education for the Anthropocene? *Geographical Research* 53 (3):306–12. doi: 10.1111/1745-5871.12122.

Reclus, E. 1866. *L'homme et la terre* [Man and earth]. Paris: Librairie Universelle.

Reclus, E. [1866] 2013. The feeling for nature in modern society. In *Anarchy, geography, modernity: Selected writings of Élisée Reclus*, ed. J. Clark and C. Martin, 103–12. Oakland, CA: PM.

Reclus, E. [1898] 2013. Evolution, revolution, and the Anarchist ideal. In *Anarchy, geography, modernity: Selected writings of Élisée Reclus*, ed. J. Clark and C. Martin, 138–55. Oakland, CA: PM.

Reinert, A. 2011. The Blue Marble shot: Our first complete photograph of Earth. *The Atlantic*, April 12. Accessed May 12, 2020. https://www.theatlantic.com/technology/archive/2011/04/the-blue-marble-shot-our-first-complete-photograph-of-earth/237167/.

Robbins, P., and S. Moore. 2013. Ecological anxiety disorder: Diagnosing the politics of the Anthropocene. *Cultural Geographies* 20 (1):3–19. doi: 10.1177/1474474012469887.

Rockström, J., W. Steffen, K. Noone, Å Persson, F. Chapin, E. Lambin, T. M. Lenton, M. Scheffer, C. Folke, H. Schellnhuber. 2009. Planetary boundaries: Exploring the safe operating space for humanity. *Ecology and Society* 14 (2):1–32. doi: 10.5751/ES-03180-140232.

Ruddiman, W. 2019. Reply to Anthropocene Working Group responses. *Progress in Physical Geography: Earth and Environment* 43 (3):345–51. doi: 10.1177/0309133319839926.

Ruddiman, W., E. Ellis, J. Kaplan, and D. Fuller. 2015. Defining the epoch we live in. *Science* 348 (6230):38–39. doi: 10.1126/science.aaa7297.

Sack, R. 1997. *Homo geographicus*. Baltimore, MD: Johns Hopkins.

Salinas-de-León, P., S. Andrade, C. Arnés-Urgellés, J. Bermudez, S. Bucaram, S. Buglass, F. Cerutti, W. Cheung, C. De la Hoz, V. Hickey. 2020. Evolution of the Galapagos in the Anthropocene. *Nature Climate Change* 10 (5):380–82. doi: 10.1038/s41558-020-0761-9.

Sauer, C. 1925. The morphology of landscape. *University of California Publications in Geography* 2:19–53.

Sauer, C. 1974. The fourth dimension of geography. *Annals of the Association of American Geographers* 64 (2):189–92. doi: 10.1111/j.1467-8306.1974.tb00969.x.

Schmidt, J. 2019. The moral geography of the Earth system. *Transactions of the Institute of British Geographers* 44 (4):721–34. doi: 10.1111/tran.12308.

Schneider, B. 2016. Burning worlds of cartography: A critical approach to climate cosmograms of the Anthropocene. *GEO: Geography and Environment* 3 (2):e00027. doi: 10.1002/geo2.27.

Seamon, D. 2018. *Life takes place: Phenomenology, lifeworlds, and place making*. London and New York: Routledge.

Sexton, A. 2018. Eating for the post-Anthropocene: Alternative proteins and the biopolitics of edibility. *Transactions of the Institute of British Geographers* 43 (4):586–600. doi: 10.1111/tran.12253.

Simpson, M. 2020. The Anthropocene as colonial discourse. *Environment and Planning D: Society and Space* 38 (1):53–71. doi: 10.1177/0263775818764679.

Skaggs, R. 2004. Climatology in American geography. *Annals of the Association of American Geographers* 94 (3):446–57. doi: 10.1111/j.1467-8306.2004.00407.x.

Smith, B., and M. Zeder. 2013. The onset of the Anthropocene. *Anthropocene* 4:8–13. doi: 10.1016/j.ancene.2013.05.001.

Solecki, W., K. Seto, and P. Marcotullio. 2013. It's time for an urbanization science. *Environment: Science and Policy for Sustainable Development* 55 (1):12–17. doi: 10.1080/00139157.2013.748387.

Stallins, J. 2020. The Anthropocene: The one, the many, and the topological. *Annals of the American Association of Geographers*. doi: 10.1080/24694452.2020.1760781.

Steffen, W., W. Broadgate, L. Deutsch, O. Gaffney, and C. Ludwig. 2015. The trajectory of the Anthropocene: The great acceleration. *The Anthropocene Review* 2 (1):81–98. doi: 10.1177/2053019614564785.

Steffen, W., J. Grinevald, P. Crutzen, and J. McNeill. 2011. The Anthropocene: Conceptual and historical perspectives. *Philosophical Transactions: Series A. Mathematical, Physical, and Engineering Sciences* 369 (1938):842–67. doi: 10.1098/rsta.2010.0327.

Toivanen, T., K. Lummaa, A. Majava, P. Järvensivu, V. Lähde, T. Vaden, and J. Eronen. 2017. The many Anthropocenes: A transdisciplinary challenge for the Anthropocene research. *The Anthropocene Review* 4 (3):183–98. doi: 10.1177/2053019617738099.

Turner, B. L., II. 1997. Spirals, bridges, and tunnels: Engaging human–environment perspectives in geography. *Ecumene* 4 (2):196–217. doi: 10.1177/147447409700400205.

Turner, B. L., II. 2002. Contested identities: Human–environment geography and disciplinary implications in a restructuring academy. *Annals of the Association of American Geographers* 92 (1):52–74. doi: 10.1111/1467-8306.00279.

Turner, B. L., II, E. Lambin, and A. Reenberg. 2007. The emergence of land change science for global environmental change and sustainability. *Proceedings of the National Academy of Sciences of the United States of America* 104 (52):20666–71. doi: 10.1073/pnas.0704119104.

Turner, B. L., II, and P. Robbins. 2008. Land-change science and political ecology: Similarities, differences, and implications for sustainability science. *Annual Review of Environment and Resources* 33 (1):295–316. doi: 10.1146/annurev.environ.33.022207.104943.

Unwin, T. 1992. *The place of geography*. London and New York: Taylor & Francis.

Usher, P. 2016. Untranslating the Anthropocene. *Diacritics* 44 (3):56–77. doi: 10.1353/dia.2016.0014.

Wagreich, M., and E. Draganits. 2018. Early mining and smelting lead anomalies in geological archives as potential stratigraphic markers for the base of an early Anthropocene. *The Anthropocene Review* 5 (2):177–201. doi: 10.1177/2053019618756682.

Waters, C., J. Zalasiewicz, C. Summerhayes, A. Barnosky, C. Poirier, A. Galuszka, A. Cearreta, M. Edgeworth, E. Ellis, M. Ellis. 2016. The Anthropocene is functionally and stratigraphically distinct from the Holocene. *Science* 351 (6269). doi: 10.1126/science.aad2622.

West, G. 2017. *Scale: The universal laws of growth, innovation, sustainability, and the pace of life in organisms, cities, economies, and companies.* New York: Penguin.

Westgate, J. 2020. Anthropocene dwelling: Lessons from post-disaster Christchurch. *New Zealand Geographer* 76 (1):26–38. doi: 10.1111/nzg.12242.

Wilbanks, T., and R. Kates. 1999. Global change in local places: How scale matters. *Climatic Change* 43 (3):601–28. doi: 10.1023/A:1005418924748.

Wilson, E. 1998. *Consilience: The unity of knowledge.* New York: First Vintage.

Wilson, E. 2016. *Half-Earth: Our planet's fight for life.* New York: Liveright.

Wishart, D. 2004. Period and region. *Progress in Human Geography* 28 (3):305–19. doi: 10.1191/0309132504ph488oa.

Wright, J. K. 1947. *Terrae incognitae*: The place of imagination in geography. *Annals of the Association of American Geographers* 37 (1):1–15. doi: 10.1080/00045604709351940.

Wulf, A. 2016. *The invention of nature: Alexander von Humboldt's New World.* New York: Vintage.

Xu, C., T. A. Kohler, T. M. Lenton, J.-C. Svenning, and M. Scheffer. 2020. Future of the human climate niche. *Proceedings of the National Academy of Sciences of the United States of America* 117 (21):11350–55. doi: 10.1073/pnas.1910114117.

Young, K. 2014. Biogeography of the Anthropocene: Novel species assemblages. *Progress in Physical Geography: Earth and Environment* 38 (5):664–73. doi: 10.1177/0309133314540930.

Young, K. 2016. Biogeography of the Anthropocene: Domestication. *Progress in Physical Geography: Earth and Environment* 40 (1):161–74. doi: 10.1177/0309133315598724.

Zalasiewicz, J., C. Waters, M. Head, C. Poirier, C. Summerhayes, R. Leinfelder, J. Grinevald, et al. 2019. A formal Anthropocene is compatible with but distinct from its diachronous anthropogenic counterparts: A response to W. F. Ruddiman's "Three flaws in defining a formal Anthropocene." *Progress in Physical Geography: Earth and Environment* 43 (3):319–33. doi: 10.1177/0309133319832607.

Zalasiewicz, J., C. Waters, M. Head, C. Poirier, C. Summerhayes, R. Leinfelder, J. Grinevald, W. Steffen, J. Syvitski, P. Haff, et al. 2018. The geological and Earth system reality of the Anthropocene. *Current Anthropology* 59 (2):220–23. doi: 10.1086/697198

Zalasiewicz, J., C. Waters, M. Williams, A. Barnosky, A. Cearreta, P. Crutzen, E. Ellis, M. Ellis, I. Fairchild, J. Grinevald, et al. 2015. When did the Anthropocene begin? A mid-twentieth century boundary level is stratigraphically optimal. *Quaternary International* 383:196–203. doi: 10.1016/j.quaint.2014.11.045.

Ziegler, S., and D. Kaplan. 2019. Forum on the Anthropocene. *Geographical Review* 109 (2):249–51. doi: 10.1111/gere.12336.

Zimmerer, K. 2010. Retrospective on nature–society geography: Tracing trajectories (1911–2010) and reflecting on translations. *Annals of the Association of American Geographers* 100 (5):1076–94. doi: 10.1080/00045608.2010.523343.

Zimmerer, K. 2017. Geography and the study of human–environment relations. In *International encyclopedia of geography: People, the earth, environment, and technology*, ed. D. Richardson, N. Castree, M. F. Goodchild, A. Kobayashi, W. Liu, and R. A. Marston. West Sussex, UK: Wiley. doi: 10.1002/9781118786352.wbieg1028.

Zurita, M., P. Munro, and D. Houston. 2018. Un-earthing the subterranean Anthropocene. *Area* 50 (3):298–305. doi: 10.1111/area.12369.

Part 3

Physical Geography and the Anthropocene

Floodplain and Terrace Legacy Sediment as a Widespread Record of Anthropogenic Geomorphic Change

L. Allan James,⊕ Timothy P. Beach,⊕ and Daniel D. Richter

Anthropogenic erosion and sedimentation are critical components of global change that involve life-sustaining natural resources of soil and water. Many geomorphic systems have responded to intense land use disturbance with episodic erosion and sedimentation, often orders of magnitude greater than background geological rates in the Holocene. Accelerated sedimentation is a metric for land use change and provides evidence of geomorphic change in fans, floodplains, terraces, deltas, lakes, karst sinks, estuaries, and coastal marine deposits. This review describes high variability in the timing of alluvial sedimentation and the value of bottomland stratigraphy with emphasis on records and heterogeneity of anthropogenic change. Although floodplain sedimentary evidence might be ill-suited for defining the proposed Anthropocene epoch boundary based on stratigraphic boundary criteria, sedimentology and stratigraphy provide rich evidence for long-term human activities that measures buildups of human environmental alterations long before the proposed mid-twentieth-century onset of the Anthropocene. Anthropogenic sedimentation is globally widespread but is too time-transgressive to serve as a stratigraphic indicator for onset of the proposed Anthropocene epoch. The emphasis here is on an example of *longue durée*; that is, the long record of anthropogenically accelerated sedimentation and the valuable evidence that it provides of human-induced environmental change. A conceptual tripartite model summarizes the evolution of anthropogeomorphic change. Anthropogenic sediments preserve geoarchaeological evidence and other important contextual information in buried landscapes that include human infrastructure, fields, material culture, contaminants, and paleosols. Geomorphic change can threaten food production, flooding, water quality, and so on. Understanding erosion and sedimentation dynamics is a vital concern for humanity. *Key Words: Anthropocene, floodplain sediment, global geomorphic change.*

The purpose of this review is to recognize the value of alluvial stratigraphy as an enduring record of anthropogenic environmental change and how this record indicates clearly extensive anthropogenic change long before the twentieth century. One objective is to demonstrate how scientists can decipher records of change and use this information to understand rates and processes of long-term global anthropogenic change. A second objective of this article is to emphasize the millennial scale of this anthropogenic change in geomorphology. Specifically, a long transition was involved in developing the present humanized global system, and accelerated changes in the mid-twentieth century are best understood with knowledge of these prior conditions. A third objective is to present a temporal framework that conceptualizes long-term global anthropogeomorphic change. The three objectives are addressed by reviewing the timing of global episodic sedimentation and presenting three case studies from North America as examples of human-caused changes in erosion and sedimentation. The alluvial stratigraphic record provides abundant evidence of complexities and deep-rooted change and is fundamentally important to reconstructing accurate records of global change that lead toward the formal stratigraphic definitions of the Anthropocene.

Accelerated floodplain sedimentation as a response to human land disturbances has left widespread alluvial deposits overlying Holocene discontinuities in six of the seven continents. Sedimentary deposition often followed intensification of agriculture associated with the early great civilizations, such as the Sumerian and Assyrian civilizations in

Mesopotamia, the Harappan civilization in the Indus River Valley, the Xia and Shang Dynasties and earlier Neolithic cultures on the Huang He, the Greek and Roman worlds, and the Maya and others in Mesoamerica (van Andel, Zangger, and Demitrack 1990; Xu 1998; Bintliff 2002; Beach et al. 2009; Anthony, Marriner, and Morhange 2014). Sedimentary deposits resulting from human activities (legacy sediment) also include spoils from mines and construction (Mossa and James 2013). These sediments record the changes to geomorphic systems that have stratigraphic markers such as buried soils, pollen assemblages, geochemical signatures, and other proxies that allow identification, environmental interpretation, and dating of the deposits (Happ, Rittenhouse, and Dobson 1940; Knox 1972, 1987, 2006; Jacobson and Coleman 1986; Nanson 1986; Beach 1994; Leigh 1997; Richter and Markewitz 2001; Beach et al. 2015). On the other hand, key layers in fluvial stratigraphy can be time transgressive; that is, the feature might be of a different age in different locations and might be difficult to distinguish from alluvial deposits generated by natural changes, such as following climate and vegetation change or tectonic activity. Despite difficulties in interpretation, legacy sediment and the proxies contained within it provide invaluable evidence of environmental change generated by human agency. This empirical basis for reconstructing past environments complements and tempers efforts to develop constructs such as defining the Anthropocene or developing a temporal framework for it. It leads to a view of change with an emphasis on extended periods that is to the Anthropocene as *longue durée* has been to the human sciences (Braudel 1949, 1958) based on slowly evolving systems over a plurality of times rather than on single, isolated events.

The amount of sediment moved by agriculture, mining, and construction makes humans the dominant geomorphic agents on Earth (Hooke 2000; Wilkinson and McElroy 2007; Haff 2010; Hooke, Martin-Duque, and Pedraza 2012). Humans also regulate sediment deliveries downstream to the extent that vast volumes of eroded material in large rivers no longer reach the oceans (Vörösmarty et al. 2003; Walling and Fang 2003; Walling 2006). Fluvial sediment in major rivers is often arrested before it can compensate for subsidence and replenish soils in deltas where major human populations and food production centers are located. In many large basins, a chronological sequence of anthropogenic change to sediment budgets can be recognized in which accelerated erosion and downstream sedimentation were followed by dam construction over the past century that now arrests down-valley transport of sediment (Williams and Wolman 1984; Walling and Fang 2003). These processes are often complicated by further anthropogenic changes. For example, levees and channelization narrow channels, deepen flood flows, and encourage transport, whereas channel mining removes or sequesters alluvium.

Anthropogenic bottomland sedimentation is global in scale, although the timing and magnitudes of sedimentation vary greatly among regions. Several studies of anthropogenic sedimentation have been made globally (Macklin and Lewin 2019; Bravard 2019; Jenny et al. 2019), in Europe (Dotterweich 2008; Notebaert and Verstraeten 2010; Macklin, Lewin, and Jones 2014; Brown et al. 2018), in Asia (Xu 1998; Shi, Dian, and You 2002; Zhuang and Kidder 2014), and in Oceania (Wasson et al. 1998; Rustomji and Pietsch 2007; Hughes et al. 2010; Richardson et al. 2014). This article considers the timing of episodic sedimentation globally on a conceptual basis and presents three case studies from North America.

Alluviation and the Proposed Anthropocene Epoch

Widespread controversy both inside and outside of stratigraphy continues over when the proposed Anthropocene epoch begins. Fluvial and soil geomorphologists might be tempted to oppose the stratigraphers' recent movement to define the epoch based on strata from the mid-twentieth century. Some of the most studied and extensive anthropogenic sedimentary deposits are associated with settlement periods from Greek and Roman times and from the late eighteenth through the nineteenth centuries. Two definitions to the Anthropocene can be recognized (Richter 2020), however: definitions for the geological discipline known as stratigraphy and definitions appropriate for other scientific disciplines and scholars. Stratigraphers seek a particular stratum that provides a pragmatic answer to disciplinary questions constrained by the need for synchronicity and global scale. Other scholarly and public debates seek much broader definitions of the Anthropocene that encompass environmental effects of the full development of the human species and cultures. How stratigraphers define the Anthropocene epoch should enrich and not divert attention from questions of the complex timing, nature, and magnitude of human changes and implications about inertia and behavior of environmental systems in the future.

Considerable debate has taken place over when and where the proposed Anthropocene epoch began since the concept was advocated by Crutzen (2002). Arguments for designating the onset at an early stage with the advent of agriculture in the Neolithic (Ruddiman 2003; Ruddiman and Ellis 2009)—that is, the Early Anthropocene or Paleoanthropocene—have been based on changes to ecosystems, land clearance for agriculture, population density increases, carbon production, and changes in atmospheric gas concentrations (Ellis et al. 2013; Foley et al. 2013; Beach et al. 2019; Stephens et al. 2019). The industrial revolution has also been identified for the onset of the Anthropocene epoch with its threshold rises in population densities, resource extraction, fossil fuel combustion, and atmospheric carbon production (Crutzen 2002). A third common onset is the Great Acceleration in the mid-twentieth century, which adds aboveground nuclear bomb testing and widespread distribution of radioactive nuclides on a global scale to the previous threshold rises (Sreffen et al. 2015; Zalasicwicz et al. 2015, 2019; Waters et al. 2016).

Designation of the Anthropocene as a formal unit in the stratigraphic system maintained by the International Commission on Stratigraphy requires careful scrutiny in a rigorous review process. To avoid ambiguity and confusion, the factors used for that purpose are heavily weighted to ensure that evidence of change is ubiquitous and synchronous. As is common in stratigraphy, older boundaries are associated with less well-defined dates than younger boundaries that tend to be more precise. These criteria have favored selection of radionuclides (fallout from nuclear bomb tests) as a globally distributed and synchronous signal that is easily detected in sediment, provides a clear and widespread chronostratigraphic marker, and corresponds with the Great Acceleration in the mid-twentieth century (Waters et al. 2016).

Attempts to define a global stratigraphic section or point or stratotypes (type sections) to define the Anthropocene boundary are being prepared, with the Anthropocene Working Group moving ahead to a mid-twentieth-century proposal. The emphasis on identifying a specific stratum for the onset of the Anthropocene is challenging given the time-transgressive behavior of Earth system change, especially at a global scale. Earth history indicates that even from a stratigraphic perspective, changes rarely occur synchronously on a global scale. Most changes detectable in the stratigraphic record had strong diachronous behaviors; that is, no unique beginning in time that applies on a global scale. Although stratigraphers seek a stratigraphic boundary that answers a disciplinary question constrained by synchronicity and global scale, most other scholarly and public questions about the proposed Anthropocene epoch seek much broader perspectives of anthropogenic effects on the Earth system. These questions encompass the full development of the human species and cultures, questions not constrained by synchronicity or spatial scale.

Timing of Anthropogenic Sedimentation

Understanding sediment dynamics on millennial to decadal timescales is of great practical importance with regard to stratigraphy and global change. Geologic discontinuities in floodplain alluvial sequences underlying anthropogenic sediments typically range widely in dates of occurrence between the mid-Holocene and the twentieth century. In some cases these dates span millennia in a single region (Brown et al. 2013). The diachronous nature of anthropogenic sedimentation is particularly problematic when considering ages of the bottom boundary layer representing a geologic unconformity between pre- and postanthropogenic deposits. Physical, chemical, and paleontological changes across this critical boundary are characteristically distinct, as is the diachronism of these layers (Edgeworth et al. 2015). This boundary is referred to as Boundary A or the JinJi Boundary (in Japanese, *jin* is human being and *ji* is natural). In floodplain alluvial stratigraphy, this surface is often marked by a distinct soil profile indicating a geomorphically stable landscape prior to accelerated sedimentation (Happ, Rittenhouse, and Dobson 1940; Knox 1972, 2006; Wasson et al. 1998; James 2019; Wade et al. 2020). The onset of anthropogenically accelerated alluviation does not always define a particular time or event or process of ubiquitous extent. When details of geomorphic change are chronicled, spatial and temporal complexity often emerges with a variety of factors driving geomorphologic change. Although designation of a specific time for the boundary is important from a practical perspective, the longevity, complexity, and transitional nature of change also should be considered to recognize (1) environmental and cultural processes involved in sediment generation, (2) evidence for soil formation processes, and (3) the physical processes that must be understood for containment or remediation.

Rates of geomorphic change at the global scale could be broadly conceived to have three general stages: (I) early gradualism, (II) rapid change in response to hierarchical civilizations or intensified land use, and (III) postcolonial accelerated growth. These stages are largely time-transgressive and, in some cases, a region could revert back to an earlier stage of erosion and sediment production (Figure 1).

Stage I is characterized by gradualism that involves primitive agricultural techniques, settlements with low population densities, and a lack of large trade surpluses. Gradualism has been a hallmark of geologic theory since Hutton (1795) introduced the concept of uniformitarianism. Recent studies indicate early, widespread anthropic soil erosion and sedimentation by 4,000 BP and even earlier extensive agriculture (Jenny et al. 2019; Stephens et al. 2019). Early land clearance for agriculture often involved a mosaic pattern of shifting fields and settlements, so rapid erosion and sedimentation was discontinuous in time and space. In some regions, especially where the history of shifting agriculture extends back several millennia, initial anthropogeomorphic change might have occurred as a form of punctuated gradualism in which severe erosion at discrete times and places generated sudden local pulses of sediment but the integration of these pulses through time and space resulted in a gradual accumulation of localized storage or delayed propagation of sediment downstream.

Stage II of anthropogeomorphic development is characterized by periods of rapid change and increasing land use intensity with developments in mechanical and agricultural technologies and surplus trade. Land use technological innovations include improvements in large draft animal power (e.g., heavy harnesses and wheeled plows) and water power for irrigation and grinding grain but limited availability of other mechanization. Stage IIa occurred where

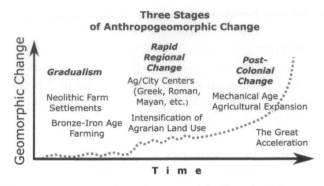

Figure 1. Conceptual tripartite model of geomorphic rates having (I) early periods of slow, gradual change by localized land clearance; (II) early periods of rapid change; and (III) intense postcolonial changes, including the Great Acceleration. Dashed line shows schematically the generalized rates of global geomorphic change.

hierarchical societies developed with high population densities, enslaved labor, and organized economies. Agricultural centers included those in Greek, Roman, Mesoamerican, and Chinese civilizations where population densities and economic activity grew rapidly. Widespread sedimentation associated with these centers often resulted in theories that erosion and sedimentation were substantive factors leading to social decline, including in Mesopotamia, Greece, Macedonia, Rome, Roman provinces of northern Africa and the eastern Mediterranean, and Mesoamerica (Marsh 1864; Bennett 1926; Montgomery 2007). Substantial sedimentation has been documented around many of these centers, including a Mayan example outlined by a case study in this article, but we also point out periods of widespread conservation by indigenous people such as terrace proliferation (Beach et al. 2018). Stage IIb occurred where agrarian settlements intensified and adopted new technologies but population densities remained moderate (e.g., central Europe and Britain in the Middle Ages). For example, in nonurban Europe during the Middle Ages, accelerated upland erosion generated sediment pulses and valley-bottom sedimentation (Houben et al. 2013; Macklin, Lewin, and Jones 2014). This was complicated, however, by engineering works in channels (Brown et al. 2018), substantial lag times for sediment deliveries to large valley bottoms (Houben et al. 2013), and population declines during periods of epidemics.

Stage III of anthropogeomorphic development began with accelerated rates of geomorphic change in response to new technologies that facilitated land use intensification and to population growth land use expansion. Stage IIIa began when mechanical-age technologies became readily available, such as water-powered sawmills and a proliferation of other machines and metal tools. It was accompanied by increased population densities in rural areas and by settlement expansion in the Western Hemisphere. Steam and internal combustion engines and other mechanical innovations accelerated change as this stage evolved. Improved transportation and trade networks and extensive expansion of agriculture and resource extraction greatly encouraged land use changes during Stage IIIa. Geomorphic processes during this period accelerated rapidly in the Americas and Oceania, where previous land clearance had been limited. Stage IIIb, also known as the Great Acceleration, occurred after World War II, when rapid geographic spread of technologies such as heavily mechanized farming and logging was joined by the green revolution and medical and sanitary improvements that accelerated mid-twentieth-century global population and economic growth. Rates of geomorphic change associated with the second and third stages were often catastrophic from a geologic perspective, although widespread soil conservation measures had varying degrees of efficacy (Dotterweich 2013).

Stage IIIa began relatively rapidly in many regions of the western hemisphere in response to colonization and introductions of advanced technologies. Accelerated sedimentation and geomorphic change in these regions began with colonization, which brought advanced land clearance, farming, and resource extraction technologies including sawmills and plows with large draft animals, along with external economic markets for surplus or export products. Stratigraphic evidence of rapid sedimentation in Stage IIIa is particularly distinct in some landscapes where early intensive agriculture was limited until colonists cleared large tracts of land that had not previously been plowed. For example, thick postsettlement floodplain alluvium has been documented in North America (Happ 1945; Knox 1972, 1987, 2006; Trimble 1974, 1981), Australia (Nanson 1986; Cook 2019), and New Zealand (Griffiths 1979; Gomez et al. 1998). Stage IIIb includes a rapid global-scale intensification of erosion and land-moving activities that began after World War II, a time that corresponds to the Great Acceleration of the Anthropocene (Steffen et al. 2015). Substantial conservation efforts by indigenous and industrial societies also reduced erosion and sedimentation in complex ways during Stage III.

Not only did the anthropogeomorphic stages overlap in time between regions but they often involved long periods of transition between stages and distinctive accelerations in various regions. Rapid rates of alluviation in specific catchments tended to correspond to idiosyncratic settlement and technological histories. Given the complexities and regional differences, it follows that specific histories of geomorphic change must be unraveled separately for each region and cannot be deduced from general models. This, in turn, points out the importance of alluvial records for evidence in reconstructing environmental change.

Three Case Studies of Anthropic Sedimentation

The discussion thus far has focused on global and large regional perspectives, but the following sections present three regions in the Americas with widely diverse histories and processes of geomorphic change. They have in common the fact that they were all severely affected by extreme anthropogenically accelerated erosion and sedimentation. These examples of extreme regional sedimentation, as well as a large number of similar regional examples, occurred in Stages II and III, prior to the mid-twentieth century. The timing of change varied greatly between these three regions, which were selected for their diversity and as areas of the authors' long-term research.

The Maya Lowlands of Central America

Ancient Maya culture spanned from ~4,000 BP to ~1,000 BP and provides an example of the transition from Stage I to a densely centered Stage II of geomorphic change rates (Figure 1). Almost as early as ancient Maya research started, scientists started connecting the thin soils of the tropical karst Lowlands with ancient Maya severe soil erosion and collapse scenarios (Beach, Luzzadder-Beach, and Dunning 2006). For example, the famous soil scientist H. H. Bennett connected the Maya abandonment of ~1,000 BP to soil degradation (Bennett 1926). Bennett was writing about Guatemala's Petén, the ancient center of Maya culture, which in 1926 was largely mature tropical forest with a few temples

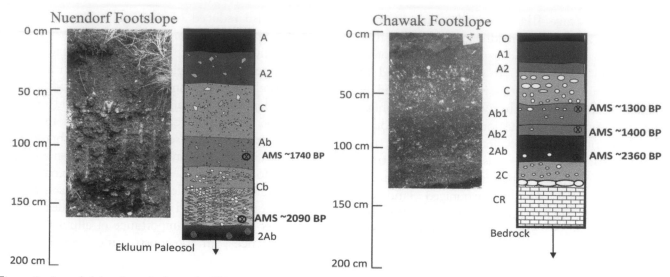

Figure 2. Aggraded footslopes built on the Eklu'um paleosol in Belize (Beach et al. 2018).

barely visible from above. Subsequent fieldwork has studied ancient Maya soil erosion and sedimentation, corroborating some examples of this intensive anthropogenic change.

The first empirical evidence of erosion and sedimentation came much later as excavations expanded from the Maya ruins into their hinterlands and focused on understanding how this populous culture procured enough food in a rainforest that many viewed as inimical to civilization. Multidisciplinary scientific teams started excavating sequences that exposed ancient soils buried by sediments that in turn had well-developed topsoils (Dunning et al. 2002; Beach et al. 2008). Some of the deposition was alluvium and colluvium in sinks and lower slopes from accelerated erosion and some was intentional burial by ancient farmers building up intensive farming systems (Beach et al. 2008; Luzzadder-Beach and Beach 2009). Beach et al. (2008) proposed that the buried soils were similar enough in age (often ~3,000–2,000 BP) to warrant the designation of a paleosol, which they called *Eklu'um* (Figure 2), Mayan for black earth, a buried paleosol (including mollisols, histosols, or vertisols). This late Holocene stable paleosol formed prior to the larger impacts of Maya landscape change, and the top of this soil and the sediments that buried it are often filled with artifacts and ecofacts (e.g., maize pollen) of ancient Maya use (Beach et al. 2019).

The first discussion of accelerated sedimentation came from lake coring that started in the 1960s that noted "Maya clays," or clayey sediments in lake cores sandwiched by pre- and post-Maya organic sediments

(>~4,000 BP and <~1,000 BP; Beach et al. 2009; Beach et al. 2015). These Maya clays often carry elemental, isotope, and ecofact evidence of Maya land use. The general lacustrine sequence begins with organic-rich "gyttja" sediment with little anthropogenic evidence covered by low organic clays with copious human evidence and followed again by gyttja with little anthropogenic evidence.

This pattern parallels the terrestrial buried soil sequences with their paleosols, sediment burial with little pedogenesis and choked with artifacts overlain by well developed, often Mollic, epipedons (Figure 3). A well-dated example of the lacustrine sequence from a lake surrounded by Maya ruins in Guatemala showed deposition rates and the presence of maize pollen in the region (Anselmetti et al. 2007). Agriculture surely started by 5000 BP, but soil erosion rates were only 16.3 t/km^2 yr^{-1}. But erosion rose to 134 t/km^2 yr^{-1} from 4000 BP to 2700 BP, rising again to 500 t/km^2 yr^{-1} (and peaking at 988 t/km^2 yr^{-1}) in the Late Preclassic until 1700 BP, then declining to 457 t/km^2 yr^{-1} in the Classic period and further declining to 49 t/km^2 yr^{-1} after ~1000 BP with widespread abandonment. Maize pollen and other ecofacts also rose and fell with sedimentation rates. Interestingly, a recent metastudy of world lake cores (Jenny et al. 2019) did not include Anselmetti et al. (2007), although the 2007 study findings largely parallel the 2019 study's global findings of erosion starting in many places ~4,000 BP.

Despite this clear Preclassic uptick in erosion and sedimentation, Anselmetti et al. (2007) produced the clearest example of decreased soil erosion and

Figure 3. Paleosol from the Rio Bravo floodplain, Belize, buried by ancient Maya wetland farming and fluvial aggradation (photo by T. Beach; after Beach et al. 2019).

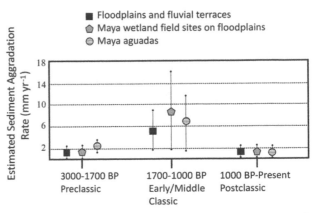

Figure 4. Sedimentation rates in sinks of northwestern Belize (after Beach et al. 2018).

sedimentation in the Classic period (after 1700 BP) when populations generally peaked. Several studies linked this with the proliferation of terraces and wetland fields over the landscape from the Late Preclassic to Classic periods (Beach and Dunning 1995; Beach et al. 2015; Beach et al. 2019). To test these ideas on slopes, the sources of the lake and depression sedimentation, Beach et al. (2018) studied soil catenas on a series of slopes mostly under tropical forest canopy and in locations near and distant from ancient ruins for erosion and sedimentation evidence. They found that most footslopes and sinks below backslopes had well-developed buried soils covered by Maya- period sedimentation, which in turn had well-developed topsoils generally developed over the last millennium of stability (Figure 4). The clearest such sequences occurred downslope from Maya ruins, which likely had high land use intensities. These studies indicate extensive Maya-induced erosion based on dated sedimentation in floodplains, karst sinks (*aguadas*), and lakes. These catena studies also indicated widespread anthropogenic terraces that provided support to the interpretation that recovery began before Maya land abandonment and soil and forest regeneration.

Summing these Maya lowlands studies, the three-stage model of anthropogenic sedimentation generally applies with gradualism and initial low, spotty erosion and sedimentation, followed by rapid, extensive, and high-magnitude erosion and sedimentation until the Maya abandonment. A millennium later, rapid deforestation and intensive erosion returned after the 1950s in some regions. Most studies focus on the ancient Maya period (Figure 4), but this last Great Acceleration of erosion is clear throughout this region (Beach and Dunning 1995). Two additions to this model from Maya examples were the decline of erosion in the Maya Classic period (~1700 to 1100 BP) when land use intensity and conservation (e.g., terraces) were high and recovery of soils for a millennium after the Maya abandonment until recent deforestation and its attendant sediment cascades.

Hydraulic Mining Sediment in the Sierra Nevada, California

The advent of hydraulic mining in 1853 and production of 1.1×10^9 m³ of sediment by 1884 when mining was suddenly halted by a federal court injunction led to an unparalleled episode of rapid floodplain sedimentation downstream of the mines (Gilbert 1917; James, Monohan, and Ertis 2019). This episode provides an extreme example of Stage IIIa geomorphic evolution, post-mechanical-age land use change, and sediment production prior to the Great Acceleration. The relatively abrupt onset and termination of sediment production over the thirty-one-year period is more a pulse than a time-transgressive forcing that often characterizes anthropogenic alluvial events at regional scales. Little evidence can be found in these watersheds of earlier (pre-1853) anthropogenic sedimentation that was presumably negligible because the local indigenous peoples were primarily hunter-gatherers and

Figure 5. Terraces of hydraulic mining sediment in Greenhorn Creek near Buckeye Ford. The terrace at this location (distant right) was 25 m above the channel, which flows on an additional ~15 m mining sediment remaining on the valley floor. Photo taken 23 October 2019.

practiced limited cultivation. Thus, little gradual accumulation of anthropogenic sediment preceded mining in this region and the influx of anthropic sediment was sudden in the late nineteenth century. The historical mine tailings and the underlying stratigraphic boundary are easily recognized by the highly distinctive lithology of the mining sediment (James 1991) and its high concentrations of gold and total mercury (Singer et al. 2013). Substantial deposits remain in the mountains (Figure 5) and in subsiding basins of the Sacramento Valley that will have long residence times.

Volumetric sediment budgets developed for alluvial deposits in two mountain catchments based on 1-m airborne LiDAR topographic data indicate local sediment deposits up to 60 m in depth. Sediment production by hydraulic mining totaled 23.5×10^6 m^3 in the upper Steephollow Creek catchment (54.6 km^2), which represents denudation (average lowering of the land surface) of 43.0 cm over the catchment or 1.39 cm yr^{-1} for thirty-one years (James, Monohan, and Ertis 2019). Almost twice as much mining sediment (41.3×10^6 m^3) was produced in the smaller upper Greenhorn Creek catchment (42.2 km^2; Figure 6), representing an average denudation of 97.9 cm or 3.16 cm yr^{-1} for thirty-one years (unpublished data). These denudation rates are orders of magnitude greater than natural geologic rates and represent a rapid anthropogenic shift of sediment from ridges to valley bottoms.

The Piedmont of the Southeastern United States

Extensive erosion and floodplain sedimentation occurred in the southeastern Piedmont throughout the nineteenth and early twentieth centuries in response to land clearance associated with agricultural settlement (Ireland, Sharpe, and Eargle 1939; Happ 1945; Trimble 1974). This provides another North American example of greatly elevated sediment production in Stage IIIa geomorphic evolution prior to the Great Acceleration. The region is known today as one of the most agriculturally eroded and gullied physiographic regions in North America. Over 90 percent of the area of the Piedmont are erosional summits and hillslopes, and less than 10 percent are floodplains and receiving areas; thus, the region's lower slopes, terraces, and floodplains are typically covered with thick strata (>0.5 m) of legacy sediment (James 2013; Wade et al. 2020). A number of mill ponds collected fine-textured sediments (Wegmann, Lewis, and Hunt 2012), much like well-documented deposits in the northern Piedmont (Walter and Merritts 2008; Merritts et al. 2011). For the most part, legacy sediments in the southeastern Piedmont are in natural levee, backswamp, and channel-fill deposits that were deposited by overbank flows.

The main source of accelerated erosion was cultivation-based cotton, corn, and wheat, but grazing and logging on steep slopes added to the upland

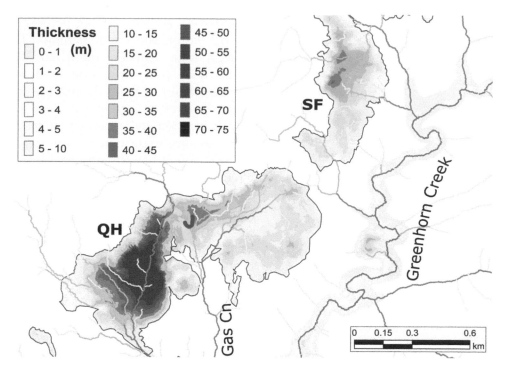

Figure 6. Production and storage of mining sediment in upper Greenhorn Creek. QH and SF are two of the largest of twenty-six mine pits in the upper Greenhorn Catchment. The maximum thickness of sediment removed from a mine in this catchment was 71.6 m in QH. Sediment storage in valley bottoms reached a maximum thickness of almost 27 m in this image, but reached a maximum of 47.6 m downstream. QH = Quaker Hill; SF = Sailor Flat.

erosion as well. Cultivation in the Piedmont decreased precipitously by the mid-twentieth century (Richter and Markewitz 2001) and was replaced by grasslands for grazing and hay, secondary forests, and suburban development. Valley morphology has been significantly transformed, with former valley-bottom wetlands and floodplains buried by 0.5 to 5 m of legacy sediments. Although these sediments are temporarily stabilized and subject to pedogenesis (Wade et al. 2020), they will be major sources of sediment in streams and rivers for many decades and centuries to come.

A detailed study of legacy sediment in the ~500-ha Holcombe's Branch watershed of South Carolina found thicknesses ranging from 0.35 to 1.25 m with an average of 1.01 m (Figure 7). Radiocarbon dating of wood indicates that these anthropogenic sediments began to be deposited by the mid-1800s. Fall-out ^{137}Cs and ^{210}Pb indicate that post-European settlement deposition rates had decreased substantially by the mid-1900s (Wade et al. 2020). Living trees currently growing in the legacy sediments have been cored with tree-ring dates prior to 1885. An analysis of soil organic carbon in the coarse-textured floodplains of Holcombe's Branch indicate that buried floodplain A horizons do not contain large concentrations of soil organic carbon (Wade et al. 2020). Tree stumps can occasionally be found buried under legacy sediments, indicating the rapidity with which sedimentation occurred (Figure 7).

Conclusion

This article synthesizes how records of long-term anthropogeomorphic change can be recorded in bottomland alluvium, and it frames phenomena within three stages of evolution of the Anthropocene. Three specific examples show how human-accelerated sedimentation is a time-transgressive forcing of geomorphology that enriches our understanding of the environment and of the stratigraphers' Anthropocene epoch that will soon be proposed to be set with a mid-twentieth-century date. Evidence for early global environmental change, even with pronounced spatial and temporal heterogeneity, suggests that the Anthropocene is best understood with knowledge of processes operating at fairly long timescales. Anthropogeomorphic change was not sudden and synchronous, and the change was not a recent phenomenon. Substantial anthropic environmental

Figure 7. Legacy sediment in South Carolina and North Carolina. (Left) Typical soil profile of a fluvent in Holcombe Branch, South Carolina, showing that the floodplain was raised by about 1.2 m with coarse-textured, well-drained historic sediments. The buried sediments were redox-active and frequently anaerobic. (Right) A 70-cm-diameter *Pinus* stump inundated by about 1 m of legacy sediment. The stump has been uncovered by an expanding tributary that drains contemporary road runoff. Carbon-14 dating of the pith and exterior rings of the pine indicated that the tree grew between approximately 1740 and 1850. The site is on the campus of Duke University along Sandy Creek, which drains an urbanized part of Durham, North Carolina. The legacy sediments precede the city's urban development and are thus agricultural in origin. The hillslopes of Sandy Creek catchment have numerous agricultural gullies.

changes that induced rapid erosion and sedimentation began more than 1,000 years ago at many locations globally. Although anthropogenic floodplain alluvium has limitations as a source of strata to be used to define the Anthropocene epoch, these deposits are rich repositories of evidence for terrestrial paleoenvironmental change that provide contextual information about the nature, timing, and magnitude of anthropic transformation of valley bottoms in the Anthropocene. In many locations, floodplain alluvium provides evidence of local discontinuities in the geological record caused by land use change. This record integrates changes throughout the watershed above by subsuming inputs from multiple smaller catchments.

Three general stages of human-induced geomorphic changes are recorded in valley-bottom strata at various sites ranging from millennial to recent time scales. In Stage I of the anthropogeomorphic evolution, early human societies began to clear land with hand tools and fire for the purpose of crop cultivation, pastoralism, and mining. The geomorphic effects were generally subtle and discontinuous in time and space, which resulted in gradual local change. Later in Stage II, with the growth of agricultural and urban centers in some regions and the intensification and expansion of agricultural land use and resource extraction in other regions, accelerated geomorphic change and large-scale sedimentation occurred. In the relatively recent postcolonial Stage III, machinery such as water-powered grist and saw-mills and mass-produced steel-tipped plows became available, and widespread land use change, mining, and construction generated episodes of rapid sedimentation. In all stages, evidence for indigenous and state-level conservation occurred widely, although often in fits and starts. At some rare sites, all three of these stages might appear in a single vertical section exposed in the strata of a streambank or floodplain core. More likely, an extensive stratigraphic sequence will only be preserved in a variety of locations including floodplains and terraces of various ages. The record at each site, however, can provide the context for environmental changes before and after the strata were deposited. This information on the nature, timing, and magnitude of complex anthropic environmental changes is entirely complementary to stratigraphers' needs to identify a stratum that divides the Holocene from the Anthropocene epochs. The three stages clearly point to the long, complex history of geomorphic change that is an important element of the present humanized Earth.

Acknowledgments

The authors thank David Leigh and an anonymous reviewer for constructive comments and suggestions on an earlier version of this article that led to substantial improvements.

ORCID

L. Allan James http://orcid.org/0000-0002-2623-1216
Timothy P. Beach http://orcid.org/0000-0003-0097-7973

References

Anselmetti, F. S., D. A. Hodell, D. Ariztegui, M. Brenner, and M. F. Rosenmeier. 2007. Quantification of soil erosion rates related to ancient Maya deforestation. *Geology* 35 (10):915–18. doi: 10.1130/G23834A.1.

Anthony, E. J., N. Marriner, and C. Morhange. 2014. Human influence and the changing geomorphology of Mediterranean deltas and coasts over the last 6000 years: From progradation to destruction phase? *Earth-Science Reviews* 139:336–61. doi: 10.1016/j.earscirev.2014.10.003.

Beach, T. 1994. The fate of eroded soil: Sediment sinks and sediment budgets of agrarian landscapes in southern Minnesota, 1851–1988. *Annals of the Association of American Geographers* 84 (1):5–28. doi: 10.1111/j.1467-8306.1994.tb01726.x.

Beach, T., and N. P. Dunning. 1995. Ancient Maya terracing and modern conservation in the Petén rain forest of Guatemala. *Journal of Soil and Water Conservation* 50:138–45.

Beach, T., S. Luzzadder-Beach, D. Cook, N. Dunning, D. J. Kennett, S. Krause, R. Terry, D. Trein, and F. Valdez. 2015. Ancient Maya impacts on the Earth's surface: An Early Anthropocene analog? *Quaternary Science Reviews* 124 (15):1–30. doi: 10.1016/j.quascirev.2015.05.028.

Beach, T., S. Luzzadder-Beach, D. Cook, S. Krause, C. Doyle, S. Eshleman, G. Wells, N. Dunning, M. Brennan, N. Brokaw, et al. 2018. Stability and instability on Maya Lowlands tropical hillslope soils. *Geomorphology* 305:185–208. doi: 10.1016/j.geomorph.2017.07.027.

Beach, T., S. Luzzadder-Beach, and N. Dunning. 2006. A soils history of Mesoamerica and the Caribbean Islands. In *Soils and societies: Perspectives from environmental history*, ed. J. R. McNeill and V. Winiwarter, 51–90. Cambridge, UK: White Horse Press.

Beach, T., S. Luzzadder-Beach, N. Dunning, and D. Cook. 2008. Human and natural impacts on fluvial and karst depressions of the Maya Lowlands. *Geomorphology* 101 (1–2):308–31. doi: 10.1016/j.geomorph.2008.05.019.

Beach, T., S. Luzzadder-Beach, N. Dunning, J. Jones, J. Lohse, T. Guderjan, S. Bozarth, S. Millspaugh, and T. Bhattacharya. 2009. A review of human and natural changes in Maya Lowlands wetlands over the Holocene. *Quaternary Science Reviews* 28 (17–18):1710–24. doi: 10.1016/j.quascirev.2009.02.004.

Beach, T., S. Luzzadder-Beach, S. Krause, T. Guderjan, F. Valdez, Jr., J. C. Fernandez-Diaz, S. Eshleman, and C. Doyle. 2019. Ancient Maya wetland fields revealed under tropical forest canopy from laser scanning and multiproxy evidence. *Proceedings of the National Academy of Sciences of the United States of America* 116 (43):21469–77. doi: 10.1073/pnas.1910553116.

Bennett, H. H. 1926. Agriculture in Central America. *Annals of the Association of American Geographers* 16 (2):63–84. doi: 10.1080/00045602609356957.

Bintliff, J. 2002. Time, process and catastrophism in the study of Mediterranean alluvial history: A review. *World Archaeology* 33 (3):417–35. doi: 10.1080/00438240120107459.

Braudel, F. 1949. *La Méditerranée et le Monde Méditerranéen a l'époque de Philippe II* [The Mediterranean and the world: The Mediterranean at the epoch of Philippe II], 3 vols. Paris: Editorial Armand Colin.

Braudel, F. 1958. Histoire et sciences sociales: La longue durée [History and social sciences: The long duration]. *Annales. Histoire, Sciences Sociales* 13 (4):725–53. doi: 10.3406/ahess.1958.2781.

Bravard, J.-P. 2019. *Sedimentary crisis at the global scale 1: Large rivers, from abundance to scarcity*. New York: Wiley.

Brown, A. G., L. Lespez, D. A. Sear, J.-J. Macaire, P. Houben, K. Klimek, R. E. Brazier, K. V. Oost, and B. Pears. 2018. Natural vs. anthropogenic streams in Europe: History, ecology and implications for restoration, river-rewilding and riverine ecosystem services. *Earth-Science Reviews* 180:185–205. doi: 10.1016/j.earscirev.2018.02.001.

Brown, A. G., P. Toms, C. Carey, and E. Rhodes. 2013. Geomorphology of the Anthropocene: Time-transgressive discontinuities of human-induced alluviation. *Anthropocene* 1:3–13. doi: 10.1016/j.ancene.2013.06.002.

Cook, D. 2019. Butzer "Down Under": Debates on anthropogenic erosion in early colonial Australia. *Geomorphology* 331:160–74. doi: 10.1016/j.geomorph.2018.12.011.

Crutzen, P. J. 2002. Geology of mankind—The Anthropocene. *Nature* 415 (6867):23. doi: 10.1038/415023a.

Dotterweich, M. 2008. The history of soil erosion and fluvial deposits in small catchments of central Europe: Deciphering the long-term interaction between humans and the environment—A review. *Geomorphology* 101 (1–2):192–208. doi: 10.1016/j.geomorph.2008.05.023.

Dotterweich, M. 2013. The history of human-induced soil erosion: Geomorphic legacies, early descriptions and research, and the development of soil conservation—A global synopsis. *Geomorphology* 201:1–34. doi: 10.1016/j.geomorph.2013.07.021.

Dunning, N., S. Luzzadder-Beach, T. Beach, J. Jones, V. Scarborough, and T. Culbert. 2002. Arising from the Bajos: The evolution of a neotropical landscape and the rise of Maya civilization. *Annals of the Association of American Geographers* 92 (2):267–83. doi: 10.1111/1467-8306.00290.

Edgeworth, M., D. D. Richter, C. N. Waters, P. Haff, C. Neal, and S. J. Price. 2015. Diachronous beginnings of the Anthropocene: The lower bounding surface of anthropogenic deposits. *The Anthropocene Review* 2 (1):33–58. doi: 10.1177/2053019614565394.

Ellis, E. C., J. O. Kaplan, D. Q. Fuller, S. Vavrus, K. K. Goldewijk, and P. H. Verburg. 2013. Used planet: A global history. *Proceedings of the National Academy of Science* 110 (20):7976–85. www.pnas.org/cgi/doi/10.1073/pnas.1217241110.

Foley, S. F., D. Gronenborn, M. O. Andreae, J. W. Kadereit, J. Esper, D. Scholz, U. Pöschl, D. E. Jacob, B. R. Schöne, R. Schreg, et al. 2013. The Palaeoanthropocene—The beginnings of anthropogenic environmental change. *Anthropocene* 3:83–88. doi: 10.1016/j.ancene.2013.11.002.

Gilbert, G. K. 1917. Hydraulic mining debris in the Sierra Nevada, California. U.S. Geological Survey Professional Paper 108, U.S. Government Printing Office, Washington, DC.

Gomez, B., D. N. Eden, D. H. Peacock, and E. J. Pinkney. 1998. Floodplain construction by recent, rapid vertical accretion: Waipaoa River, New Zealand. *Earth Surface Processes and Landforms* 23 (5):405–13. doi: 10.1002/(SICI)1096-9837(199805)23:5<405::AID-ESP854>3.0.CO;2-X.

Griffiths, G. A. 1979. Recent sedimentation history of the Waimakariri River, New Zealand. *Journal of Hydrology (New Zealand)* 18:6–28.

Haff, P. K. 2010. Hillslopes, rivers, plows, and trucks: Mass transport on Earth's surface by natural and technological processes. *Earth Surface Processes and Landforms* 35 (10):1157–66. doi: 10.1002/esp.1902.

Happ, S. C. 1945. Sedimentation in South Carolina valleys. *American Journal of Science* 243 (3):113–26. doi: 10.2475/ajs.243.3.113.

Happ, S. C., G. Rittenhouse, and G. C. Dobson. 1940. Some principles of accelerated stream and valley sedimentation. U.S. Department of Agriculture Technical Bulletin 965, Washington, DC.

Hooke, R. L. 2000. On the history of humans as geomorphic agents. *Geology* 28 (9):843–46. doi: 10.1130/0091-7613(2000)28<843:OTHOHA>2.0.CO;2.

Hooke, R. L., J. F. Martín-Duque, and J. Pedraza. 2012. Land transformation by humans: A review. *GSA Today* 12 (12):4–10. doi: 10.1130/GSAT151A.1.

Houben, P., M. Schmidt, B. Mauz, A. Stobbe, and A. Lang. 2013. Asynchronous Holocene colluvial and alluvial aggradation: A matter of hydrosedimentary connectivity. *The Holocene* 23 (4):544–55. doi: 10.1177/0959683612463105.

Hughes, A. O., J. C. Croke, T. J. Pietsch, and J. M. Olley. 2010. Changes in the rates of floodplain and in-channel bench accretion in response to catchment disturbance, central Queensland, Australia. *Geomorphology* 114 (3):338–47. doi: 10.1016/j.geomorph.2009.07.016.

Hutton, J. 1795. *Theory of the Earth; with proofs and illustrations*. 2 vols. Edinburgh, UK: Creech.

Ireland, H. A., C. F. S. Sharpe, and D. H. Eargle. 1939. Principles of gully erosion in the Piedmont of South Carolina. U.S. Department of Agriculture Technical Bulletin 633, Washington, DC.

Jacobson, R. B., and D. J. Coleman. 1986. Stratigraphy and recent evolution of Maryland Piedmont flood plains. *American Journal of Science* 286 (8):617–37. doi: 10.2475/ajs.286.8.617.

James, L. A. 1991. Quartz concentration as an index of alluvial mixing of hydraulic mine tailings with other sediment in the Bear River, California. *Geomorphology* 4:125–44.

James, L. A. 2013. Legacy sediment: Definitions and processes of episodically produced anthropogenic sediment. *Anthropocene* 2:16–26. doi: 10.1016/j.ancene.2013.04.001.

James, L. A. 2019. Impacts of pre- vs. postcolonial land use on floodplain sedimentation in temperate North America. *Geomorphology* 331:59–77. doi: 10.1016/j.geomorph.2018.09.025.

James, L. A., C. Monohan, and B. Ertis. 2019. Long-term hydraulic mining sediment budgets: Connectivity as a management tool. *Science of the Total Environment* 651 (2):2024–35. doi: 10.1016/j.scitotenv.2018.09.358.

James, L. A., J. D. Phillips, and S. A. Lecce. 2017. A centennial tribute to G.K. Gilbert's "Hydraulic-Mining Debris in the Sierra Nevada." *Geomorphology* 294:4–19. doi: 10.1016/j.geomorph.2017.04.004.

Jenny, J.-P., S. Koirala, I. Gregory-Eaves, P. Francus, C. Niemann, B. Ahrens, V. Brovkin, A. Baud, A. E. K. Ojala, A. Normandeau, et al. 2019. Human and climate global-scale imprint on sediment transfer during the Holocene. *Proceedings of the National Academy of Sciences of the United States of America* 116 (46):22972–76. doi: 10.1073/pnas.1908179116.

Knox, J. C. 1972. Valley alluviation in southwestern Wisconsin. *Annals of the Association of American Geographers* 62 (3):401–10. doi: 10.1111/j.1467-8306.1972.tb00872.x.

Knox, J. C. 1987. Historical valley floor sedimentation in the Upper Mississippi Valley. *Annals of the Association of American Geographers* 77 (2):224–44. doi: 10.1111/j.1467-8306.1987.tb00155.x.

Knox, J. C. 2006. Floodplain sedimentation in the Upper Mississippi Valley: Natural versus human accelerated. *Geomorphology* 79 (3–4):286–310. doi: 10.1016/j.geomorph.2006.06.031.

Leigh, D. S. 1997. Mercury tainted overbank sediment from past gold mining in north Georgia, USA. *Environmental Geology* 30:244–51.

Luzzadder-Beach, S., and T. Beach. 2009. Arising from the wetlands: Mechanisms and chronology of landscape aggradation in the northern coastal plain of Belize. *Annals of the Association of American Geographers* 99 (1):1–26. doi: 10.1080/00045600802458830.

Macklin, M. G., and J. Lewin. 2019. River stresses in anthropogenic times: Large-scale global patterns and extended environmental timelines. *Progress in Physical Geography: Earth and Environment* 43 (1):3–23. doi: 10.1177/0309133318803013.

Macklin, M. G., J. Lewin, and A. F. Jones. 2014. Anthropogenic alluvium: An evidence based meta-analysis for the UK Holocene. *Anthropocene* 6:26–38. doi: 10.1016/j.ancene.2014.03.003.

Marsh, G. P. 1864. *Man and nature; or, physical geography as modified by human action*. New York: Scribner.

Merritts, D., R. Walter, M. Rahnis, J. Hartranft, S. Cox, A. Gellis, N. Potter, W. Hilgartner, M. Langland, L. Manion, et al. 2011. Anthropocene streams and base-level controls from historic dams in the unglaciated mid-Atlantic region, USA. *Philosophical Transactions Series A: Mathematical, Physical, and Engineering Sciences* 369 (1938):976–1009. doi: 10.1098/rsta.2010.0335.

Montgomery, D. R. 2007. *Dirt: The erosion of civilizations*. Berkeley: University of California Press.

Mossa, J., and L. A. James. 2013. Impacts of mining on geomorphic systems. In *Geomorphology of human* disturbances, climate change, and natural hazards, ed. L. A. James, C. Harden, and J. Clague, vol. 13, 74–95. San Diego, CA: Academic Press.

Nanson, G. C. 1986. Episodes of vertical accretion and catastrophic stripping: A model of disequilibrium floodplain formation. Geological Society of America Bulletin 97 (12):1467–75. doi: 10.1130/0016-7606(1986)97<1467:EOVAAC>2.0.CO;2.

Notebaert, B., and G. Verstraeten. 2010. Sensitivity of west and central European river systems to environmental changes during the Holocene: A review. Earth-Science Reviews 103 (3–4):163–82. doi: 10.1016/j.earscirev.2010.09.009.

Richardson, J. M., I. C. Fuller, K. A. Holt, N. J. Litchfield, and M. G. Macklin. 2014. Rapid post-settlement floodplain accumulation in Northland, New Zealand. Catena 113:292–305. doi: 10.1016/j.catena.2013.08.013.

Richter, D. D. 2020. The Anthropocene in soil science and pedology. Journal of Plant Nutrition and Soil Science 183 (1):5–11. doi: 10.1002/jpln.201900320.

Richter, D. D., and D. Markewitz. 2001. Understanding soil change: Soil sustainability over millennia, centuries, and decades. Cambridge, UK: Cambridge University Press.

Ruddiman, W. F. 2003. The anthropogenic greenhouse era began thousands of years ago. Climatic Change 61 (3):261–93. doi: 10.1023/B:CLIM.0000004577.17928.fa.

Ruddiman, W. F., and E. C. Ellis. 2009. Effect of per-capita land use changes on Holocene forest clearance and CO_2 emissions. Quaternary Science Reviews 28 (27–28):3011–15. doi: 10.1016/j.quascirev.2009.05.022.

Rustomji, P., and T. Pietsch. 2007. Alluvial sedimentation rates from southeastern Australia indicate post-European settlement landscape recovery. Geomorphology 90 (1–2):73–90. doi: 10.1016/j.geomorph.2007.01.009.

Shi, C., Z. Dian, and L. You. 2002. Changes in sediment yield of the Yellow River basin of China during the Holocene. Geomorphology 46 (3–4):267–83. doi: 10.1016/S0169-555X(02)00080-6.

Singer, M. B., R. Aalto, L. A. James, N. E. Kilham, J. L. Higson, and S. Ghoshal. 2013. Enduring legacy of a toxic fan via episodic redistribution of California gold mining debris. Proceedings of the National Academy of Sciences of the United States of America 110 (46):18436–41. doi: 10.1073/pnas.1302295110.

Steffen, W., W. Broadgate, L. Deutsch, O. Gaffney, and C. Ludwig. 2015. The trajectory of the Anthropocene: The Great Acceleration. The Anthropocene Review 2 (1):81–98. doi: 10.1177/2053019614564785.

Stephens, L., D. Fuller, N. Boivin, T. Rick, N. Gauthier, A. Kay, B. Marwick, C. G. Armstrong, C. M. Barton, T. Denham, et al. 2019. Archaeological assessment reveals Earth's early transformation through land use. Science 365 (6456):897–902. doi: 10.1126/science.aax1192.

Trimble, S. W. 1974. Man-induced soil erosion on the southern Piedmont 1700–1970. Ankeny, IA: Soil Conservation Society of America.

Trimble, S. W. 1981. Changes in sediment storage in the Coon Creek basin, Driftless Area, Wisconsin, 1853 to 1975. Science 214 (4517):181–83. doi: 10.1126/science.214.4517.181.

van Andel, T. H., E. Zangger, and A. Demitrack. 1990. Land use and soil erosion in prehistoric and historical Greece. Journal of Field Archaeology 17 (4):379–96. doi: 10.2307/530002.

Vörösmarty, C. J., M. Meybeck, B. Fekete, K. Sharma, P. Green, and J. P. M. Syvitski. 2003. Anthropogenic sediment retention: Major global impact from registered river impoundments. Global and Planetary Change 39 (1–2):169–90. doi: 10.1016/S0921-8181(03)00023-7.

Wade, A. M., D. D. Richter, A. Cherkinsky, C. B. Craft, and P. R. Heine. 2020. Limited carbon contents of centuries old soils forming in legacy sediment. Geomorphology 354:107018. doi: 10.1016/j.geomorph.2019.107018.

Walling, D. E. 2006. Human impact on land–ocean sediment transfers by the world's rivers. Geomorphology 79 (3–4):192–216. doi: 10.1016/j.geomorph.2006.06.019.

Walling, D. E., and D. Fang. 2003. Recent trends in the suspended sediment loads of the world's rivers. Global and Planetary Change 39 (1–2):111–26. doi: 10.1016/S0921-8181(03)00020-1.

Walter, R. C., and D. J. Merritts. 2008. Natural streams and the legacy of water-powered mills. Science 319 (5861):299–304. doi: 10.1126/science.1151716.

Wasson, R., R. Mazari, B. Starr, and G. Clifton. 1998. The recent history of erosion and sedimentation on the Southern Tablelands of southeastern Australia: Sediment flux dominated by channel incision. Geomorphology 24 (4):291–308. doi: 10.1016/S0169-555X(98)00019-1.

Waters, C. N., J. Zalasiewicz, C. Summerhayes, A. D. Barnosky, C. Poirier, A. Gałuszka, A. Cearreta, M. Edgeworth, E. C. Ellis, M. Ellis, et al. 2016. The Anthropocene is functionally and stratigraphically distinct from the Holocene. Science 351 (6269):aad2622. doi: 10.1126/science.aad2622.

Wegmann, K. W., R. Q. Lewis, and M. C. Hunt. 2012. Historic mill ponds and piedmont stream water quality: Making the connection near Raleigh, North Carolina. In From the Blue Ridge to the coastal plain: Field excursions in the Southeastern United States. Geological Society of America field guide, ed. M. C. Eppes and M. J. Bartholomew, vol. 29, 93–121. Denver, CO: Geological Society of America. doi: 10.1130/2012.0029(03).

Wilkinson, B. H., and B. J. McElroy. 2007. The impact of humans on continental erosion and sedimentation. Geological Society of America Bulletin 119 (1–2):140–56. doi: 10.1130/B25899.1.

Williams, G. P., and M. G. Wolman. 1984. Downstream effects of dams on alluvial rivers. U.S. Geological Survey Professional Paper 1286. Washington, DC: U.S. Government Printing Office. https://doi.org/10.3133/pp1286

Xu, J. 1998. Naturally and anthropogenically accelerated sedimentation in the lower Yellow River, China, over the past 13,000 years. Geografiska Annaler: Series A

Physical Geography 80 (1):67–78. doi: 10.1111/j.0435-3676.1998.00027.x.

Zalasiewicz, J., C. N. Waters, M. Williams, A. D. Barnosky, A. Cearreta, P. Crutzen, E. Ellis, M. A. Ellis, I. J. Fairchild, J. Grinevald, et al. 2015. When did the Anthropocene begin? A mid-twentieth century boundary level is stratigraphically optimal. *Quaternary International* 383:196–207. doi: 10.1016/j.quaint.2014.11.045.

Zalasiewicz, J., C. N. Waters, M. Williams, and C. P. Summerhayes, eds. 2019. *The Anthropocene as a geological time unit: A guide to the scientific evidence and current debate.* Cambridge, UK: Cambridge University Press.

Zhuang, Y., and T. R. Kidder. 2014. Archaeology of the Anthropocene in the Yellow River region, China, 8000–2000 cal. BP. *Holocene* 24 (11):1602–23. doi: 10.1177/0959683614544058.

L. ALLAN JAMES is a Distinguished Professor Emeritus in the Department of Geography at the University of South Carolina, Columbia, SC 29208. E-mail: ajames@sc.edu. His research interests include fluvial responses to episodic sedimentation, interactions between anthropogenic sediment and flood risks, and applications of geospatial science in geomorphology.

Hotter Drought as a Disturbance at Upper Treeline in the Southern Rocky Mountains

Grant P. Elliott, Sydney N. Bailey, and Steven J. Cardinal

As we progress into the Anthropocene, rising temperatures have amplified evaporative demand and rendered heat-induced drought stress, or hotter drought, as the hallmark of climate change moving forward. It remains unknown, however, whether upper treeline environments have been affected. For this study, we grouped previously published and unpublished data from study sites within the southern Rocky Mountains by slope aspect to provide a possible baseline for what we expect for the Anthropocene. We returned to and resampled some of these study sites in the summer of 2019 after twelve years of sharply rising temperatures to measure patterns of seedling establishment. We also returned to a high-elevation site after seventeen years of warming to perform repeat photography in an attempt to capture visual evidence of threshold changes since 2002 in a location where little change had occurred during the twentieth century. Results from this research can be summarized into two main findings: (1) trends in recruitment over the past thirty years suggest that north-facing slopes are increasingly hospitable for successful seedling establishment compared to south-facing slopes at upper treeline and (2) spruce beetle–induced mortality is evident at upper treeline. Conceptually, this means that hotter drought could be progressively enveloping upper treeline along topoclimatic gradients. Unless ongoing trends in temperature deviate from expectations or unless precipitation increases considerably, there is little reason to justify the idea that upper treeline will continue to respond positively to hotter drought conditions during the Anthropocene, especially on south-facing slopes in the Northern Hemisphere. *Key Words: climate change, hotter drought, Rocky Mountains, slope aspect, treeline.*

Biogeographers studying climate–vegetation interactions during the Anthropocene face challenges from both top-down and bottom-up perspectives, as global climate change and human land use increasingly transform the biosphere (Wolkovich et al. 2014). Even within the remaining 20 percent of global terrestrial biomes still classified as remote or wild woodlands (Ellis et al. 2010), for example, we must still contend with the fact that human activities have modified the Earth's climate system and component biogeochemical cycles to such an extent that our collective influence now rivals nature in governing how they function and interact (Steffen et al. 2011). Based on the magnitude of human carbon emissions alone, continued warming is virtually guaranteed for centuries to come (Solomon et al. 2009). This is striking given that amplified rates of warming during recent decades produced the highest mean temperature for any century (1917–2016) of the past 11,000 years (Marsicek et al. 2018), surpassing the warm mid-Holocene (ca. 6,000 years BP) when summer insolation was >6 percent than present (Shuman 2012). Although a temperature regime distinct from the

Holocene supports the idea of the Anthropocene (Corlett 2015), it underscores the need to improve our predictive capabilities for what a "biogeography of the Anthropocene" might resemble given that traditional paradigms in plant geography might need to be reconceptualized as temperature thresholds are exceeded (Young 2014).

Elevational zonation in general and upper treeline in particular are interesting candidates for this because ongoing and projected rises in temperature are sharper across high-elevation mountain environments compared to adjacent lowlands, with unknown consequences for the structure and coverage of the mountain forest belt (Nogués-Bravo et al. 2007; Pepin et al. 2015). Elevational zonation is a fundamental concept in mountain biogeography and refers to the formation of different forest belts along an elevational gradient in response to changing temperature and precipitation regimes with cold temperatures eventually limiting the uppermost extent of mountain forest (von Humboldt and Bonpland 1805; Daubenmire 1943; Peet 2000; Fattorini, Biase, and Chiarucci 2019). Upper treeline (or alpine treeline) represents the uppermost extent of the mountain forest belt and forms along the variable-width transition zone between closed-canopy subalpine forest and alpine tundra (Holtmeier and Broll 2005; Malanson et al. 2007; Elliott 2017). Here, tree function and growth are limited by temperature, whereby upslope in tundra it is too cold for trees to survive (Körner 2012). Yet despite widespread warming since about 1950, only a slight majority of treelines have advanced worldwide (Harsch et al. 2009). This highlights the importance of considering other factors such as disturbance and moisture in modifying or overriding the influence of rising temperatures, particularly with respect to seedling establishment (Noble 1993; Lloyd and Graumlich 1997; Daniels and Veblen 2004; Stueve et al. 2009; Moyes, Germino, and Kueppers 2015).

Several types of biotic and abiotic disturbances impact seedling establishment at upper treeline across various spatial and temporal scales. For example, herbivory (Cairns et al. 2007; Munier et al. 2010), animals as zoogeomorphic agents (Butler 1995), invasive forest pathogens (Resler and Tomback 2008; Tomback et al. 2016), and geomorphic processes such as frost heaving (Butler et al. 2009) complicate climate–vegetation linkages at a local scale. At a coarser scale, mass movements including snow avalanches and debris flows can impede establishment (Walsh et al. 1994; Butler et al. 2009). Fire, with a rotation interval that exceeds 1,000 years at upper treeline (Baker 2009), can influence the rate and pattern of seedling establishment by altering soil properties and neighboring biotic interactions (Shankman and Daly 1988; Noble 1993; Stueve et al. 2009; Stine and Butler 2015; Wang et al. 2019). Collectively, disturbance regimes at upper treeline impact climate–vegetation interactions from a bottom-up perspective. This distinguishes treeline from subalpine forest below where the two main disturbances are large stand-replacing fires and bark beetle outbreaks, which are driven by the top-down influence of drought (Schoennagel et al. 2007; DeRose and Long 2012). Although multidecadal drought led to mortality-induced declines in the elevation of upper treeline in the Sierra Nevada and Wind River Range of the Central Rocky Mountains between 950 and 550 years BP (Lloyd and Graumlich 1997; Morgan, Losey, and Trout 2014), the influence of ongoing drought conditions at treeline remain less clear (Holtmeier and Broll 2019).

Across the Western United States, heat-induced drought stress, or hotter drought, has emerged as the hallmark of climate change during the Anthropocene (Allen, Breshears, and McDowell 2015). Because diminished snowpack retention during spring accompanies sharp increases in temperature and evaporative demand during the growing season (Barnett et al. 2008; Clow 2010; Cowell and Urban 2010; Allen, Breshears, and McDowell 2015; Mote et al. 2018), hotter drought has proven lethal for most of the mountain forest belt. Since approximately 2000, for example, drought-induced mortality has spread from low-elevation pinyon–juniper woodlands up into cool and wet subalpine forests of the Rocky Mountains (Breshears et al. 2005; van Mantgem et al. 2009; Hart et al. 2014; Smith et al. 2015). Mortality results from direct physiological stress or by weakened tree defense mechanisms against bark beetle attack, which has become so widespread that the collective influence of hotter drought has been termed a forest megadisturbance because of how it tests forest resilience (Millar and Stephenson 2015). Moving forward, intensifying drought stress is expected to reach forests not currently considered moisture limited (Allen et al. 2010; A. P. Williams et al. 2013; Allen, Breshears, and McDowell 2015), such as upper treeline. Ecotonal environments are indeed responsive to drought (Allen and Breshears 1998), yet it remains unknown whether hotter drought has engulfed upper treeline in the Rocky Mountains.

A central challenge is understanding how hotter drought would look if affecting upper treeline. This is particularly true on the timescale of decades because vegetation typically has a lagged response to changes in climate (Lloyd 2005; Alexander et al. 2018; Malanson et al. 2019). Post hoc explanations of climate thresholds are relatively common in time-series data spanning centuries to millennia (J. W. Williams, Blois, and Shuman 2011; Minckley, Shriver, and Shuman 2012), but we have a limited understanding of when and where climate extremes, including drought, will initially trigger biogeographic responses (Wolkovich et al. 2014; Martínez-Vilalta and Lloret 2016). Despite this, some spatial patterns have emerged with respect to tree mortality and tree growth that could lend insight into how hotter drought might influence tree recruitment, which overall, remains poorly understood (Clark et al. 2016; Martínez-Vilalta and Lloret 2016). Across subalpine forests of the Colorado Front Range, for example, drought-induced mortality was confined to south-facing slopes until a threshold was crossed in 2008 whereby even patterns developed between south- and north-facing slopes through 2013 (Smith et al. 2015). In the White Mountains of California, a threshold was passed in 1992 where trees on south-facing slopes that were previously limited by temperature, for millennia, are now more responsive to moisture stress (Bunn et al. 2018). It is therefore plausible that patterns of tree recruitment at upper treeline are being shaped by slope aspect in response to hotter drought, but research is lacking.

For this study, we grouped previously published and unpublished data from study sites within the southern Rocky Mountains by slope aspect to provide a longer term context or possible baseline for what we expect to see moving forward in the Anthropocene. We returned to and resampled some of these study sites in the summer of 2019 after twelve years of sharply rising temperatures to measure for the influence of hotter drought on patterns of seedling establishment. To expand our coverage of treeline environments in the region, we also returned to a high-elevation site after seventeen years of warming to perform repeat photography in an attempt to capture visual evidence of threshold changes since 2002 in a location where little change had occurred during the twentieth century. Taken together, the purpose of this study is to provide empirical evidence documenting that hotter drought has begun engulfing upper treeline ecotones in the southern Rocky Mountains by asking the following research questions. (1) Do spatio-temporal patterns of tree establishment differ by slope aspect? (2) Is there evidence of hotter drought as a megadisturbance? Our previous work has shown that top-down threshold changes in climate related to rising temperatures and a decrease in snowpack during the 1950s triggered abrupt increases in tree establishment at upper treeline in the U.S. Rocky Mountains (Elliott and Kipfmueller 2011; Elliott 2012). Further analyses of these data, however, showed that dry periods disproportionately favored tree establishment on north-facing slopes during the twentieth century (Elliott and Cowell 2015). A continuation of this trend provides regional-scale support for the bottom-up modification of hotter drought by slope aspect during the Anthropocene.

Study Area

Our study area extends across three mountain ranges within an approximate $100,000\,km^2$ area south of 40°N (ca. 35.5°N–39.5°N) in the southern Rocky Mountain region of northern New Mexico and Colorado (Figure 1). The southern Rocky Mountains contain the highest treeline ecotones along the spine of the Rocky Mountains and include the broadest expanse of climatically sensitive boundaries where subalpine forest borders alpine tundra capable of supporting treeline advance upslope (Butler et al. 2007; Elliott and Petruccelli 2018). We sampled at upper treeline on opposite north- and south-facing slopes of seven mountain peaks for a total of fourteen study sites in the Sangre de Cristo ($n = 6$), San Juan ($n = 4$), and Front Range ($n = 4$) mountains (Figure 1). The elevation of these sites ranged from 3,542 to 3,734 m. We systematically stratified our study sites in this manner because mountain topography partitions broader scale climate into distinct topoclimates based on slope angle and aspect, with opposing temperature–moisture regimes on north- versus south-facing slopes (Barry 2008).

The geology of these ranges varies, but the Sangre de Cristo and Front Range are oriented north–south and primarily composed of Precambrian granitic rock with coarse texture and high silica content, along with metamorphic portions of schist and gneiss (Peet 2000). The San Juan Mountains are an east–west-oriented range formed from Miocene to Quaternary volcanic deposits (Baker 2009). Soil properties including moisture and total N and P are highest on the lee side of existing vegetation at treeline (Bourgeron et al. 2015). Annual soil water deficits are low in this region and gradually increase moving north (Elliott and Cowell 2015).

Synoptic climate patterns are relatively uniform across the southern Rocky Mountains south of 40°N, with infrequent intrusions of Pacific air masses during the winter and an influx of monsoonal moisture from July through September (Mitchell 1976; Dettinger et al. 1998). From the monsoon, the Sangre de Cristos experience a summer maximum in precipitation (Elliott 2011). To examine temperature and precipitation variability, we calculated regional summaries for minimum temperature (T_{min}), maximum temperature (T_{max}), and precipitation during the warm season (June–September) and the cool season (October–May) by combining Parameter-elevation Regressions on Independent Slopes Model

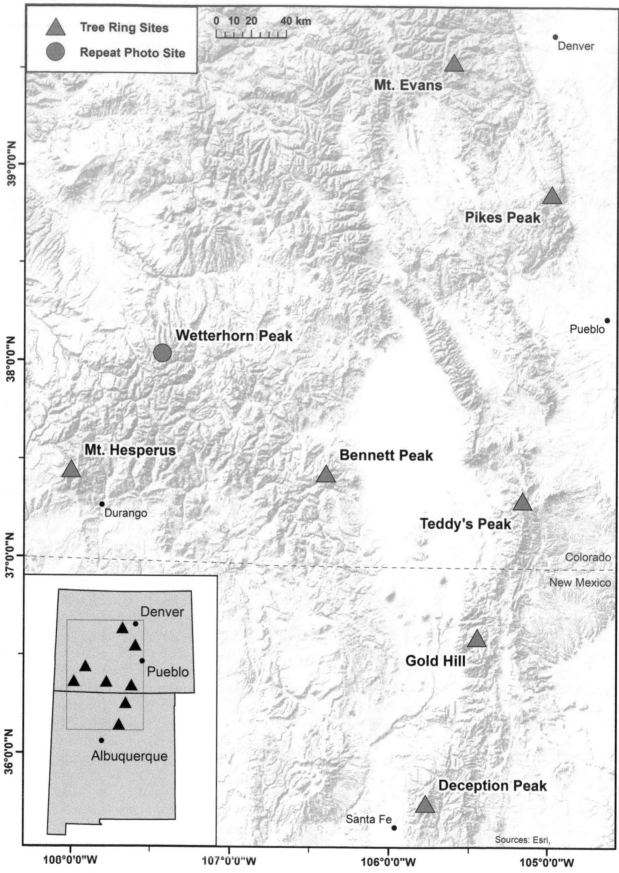

Figure 1. Study area map.

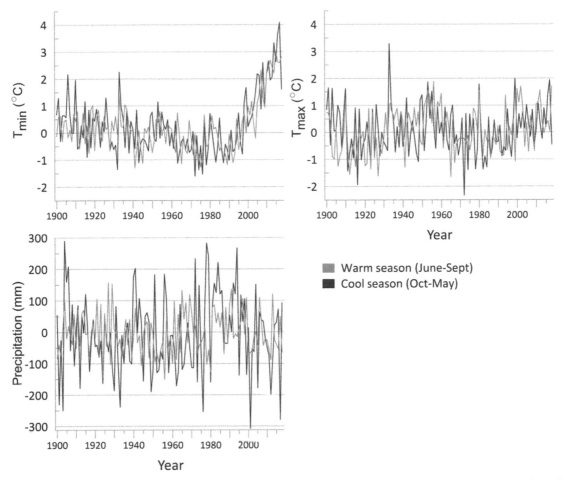

Figure 2. Annual climate data expressed as deviations from twentieth-century mean during the warm (1900–2018) and cool season (1901–2019). Data are regional average calculated from Parameter-elevation Regressions on Independent Slopes Model data grids (PRISM Group, Oregon State University, Corvallis, Oregon; Daly et al. 2008) used for each mountain peak in this study (n = 7).

(PRISM) climate data from each of the seven mountain peaks (PRISM Group, Oregon State University, www.prismclimate.org, created 10 November 2019). PRISM data are modeled to take into account changes in elevation and topography across mountain environments and are advantageous given the dearth of high-elevation weather stations across the region (Daly et al. 2008).

Tmin has been rising sharply since the early 1990s and has reached 3°C to 4°C above the average from the last century (1900–1999; Figure 2). T_{max} has been steadily rising since the 1980s and is near the consistently high levels seen during the 1950s drought (Figure 2). Warm season rain from the monsoon has been hovering around average and remains above the lowest levels recorded during the 1950s (Figure 2). Cool season snow has been decreasing, with 58 percent of years since 2000 (n = 11) falling below the twentieth-century average (Figure 2). Overall, temperature increases coupled with decreases in precipitation since

the early 1990s suggest that drought stress could be intensifying across upper treeline in the southern Rocky Mountains (Figure 2).

Vegetation at upper treeline is dominated by Englemann spruce (*Picea engelmannii* Parry ex Engelm.) and subalpine fir (*Abies lasiocarpa* [Hook] Nutt.), with Colorado bristlecone pine (*Pinus aristata* Engelm.) on south-facing slopes east of the Continental Divide. Spruce often form monospecific stands in the southern Sangre de Cristo range on north-facing slopes, and the relatively steep slopes throughout the range make krummholz less common (Peet 1978).

Methods

Field Data Collection

The field methods that informed results from this work have three distinct components and have been combined in this study to determine whether over a decade of intensifying warmth has brought hotter drought as a megadisturbance to upper treeline in the southern Rocky Mountains. Study sites from previous tree-ring studies were selected based on location south of 40°N and were originally selected because they contained a climatic treeline on opposite north- and south-facing slopes, meaning that possible treeline advance upslope was not limited by local topography or geomorphological constraints including steep and rocky slopes with no soil formed or avalanche tracks (Butler et al. 2007).

All sites shared a common sampling methodology, where tree-ring data were collected along a variable-length transect that began at the farthest upright tree (≥5 cm diameter at breast height) or sapling (<5 cm diameter at breast height, ≥1.2 cm diameter at ground level), which we termed the outpost tree (Paulsen, Weber, and Körner 2000), and extended downslope to the uppermost extent of closed-canopy subalpine forest, also called the timberline or forest line. Transect width was 20 m on each side for a total width of 40 m. These data were collected during the summer field season in 2007, 2008, and 2014, with the notable distinction that tree seedlings (<1.2 cm diameter at ground level) were not sampled and instead their presence was recorded by species.

In the summer of 2019 we returned to the six study sites in the Sangre de Cristo range (Figure 1) to destructively harvest tree seedlings in the same areas above timberline that were previously sampled in 2007. We repositioned the outpost tree, if regeneration extended farther upslope or if the previous one was a sapling harvested at ground level. We classified seedlings as <1 m in height and for each one, recorded tree height, species, x and y coordinates to the nearest 0.1 m, and presence of facilitative shelter, such as boulders, shrubs and grasses, and other larger trees. We excavated them with a garden shovel by removing the surrounding soil to retrieve a section of the stem that extended at least 10 cm above and below the root–shoot boundary (League and Veblen 2006). After uprooting them, we measured seedling diameter.

In the summer of 2019 we also returned to the San Juan Mountains to update a repeat photo pair originally retaken in 2002 (Elliott 2003) to test for visual evidence of hotter drought in a scene at upper treeline that had experienced relatively little change during the twentieth century. We added this photo pair to expand our coverage across upper treeline ecotones in the southern Rocky Mountains. Repeat photography is a trial-and-error process of locating the precise spot for the photograph and is limited in that it represents two qualitative snapshots, generally forcing inferences on possible causal mechanisms underlying observed changes (Zier and Baker 2006).

Laboratory Methods

Tree cores and sapling cross sections from our first round of data collection between 2007 and 2014 were prepared following standard dendrochronological methods and are detailed elsewhere (Elliott and Kipfmueller 2010; Elliott 2012). All samples were cross-dated and age to coring height corrections were calculated for each species by destructively sampling saplings at both the base and 30 cm above ground to add a given number of years to tree cores obtained from coring trees at the same height. Tree establishment dates were combined into a hybrid age–structure chronology, consisting of five-year age classes for the period 1900 to 2000 and ten-year age class before 1900 (Elliott and Kipfmueller 2010).

Data Analysis

To quantify differences in age–structure data between opposite slope aspects, we used regime-shift analysis to measure significant switches ($p \leq 0.01$) in the rate of tree establishment (Rodionov 2004) to compare temporal differences. This method uses a sequential t test procedure on time-series data to detect regime shifts. We used a ten-year cutoff because we are interested in decadal-scale trends (Elliott 2012). For differences in overall distribution, we used a two-sample Kolmogorov–Smirnov test. To quantify differences in seedling frequency between north- and south-facing slopes, we used chi-square goodness-of-fit tests to determine whether the proportional differences were greater than expected. Expected values were half of the total observed, which is what we would assume if climatic conditions did not favor a given topoclimatic setting. Seedling height and seedling diameter data were not normally distributed, so we compared them statistically with a Mann–Whitney U test. With open-grown seedlings at treeline, both height and diameter likely represent a suitable proxy for age.

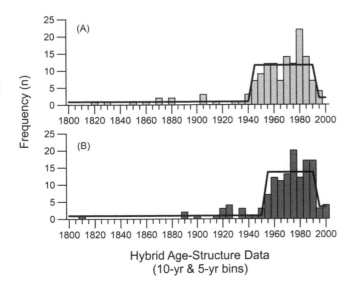

Figure 3. Hybrid age-structure data (1810–2000) for all trees sampled above timberline on (A) south-facing slopes ($n = 137$) and (B) north-facing slopes ($n = 136$). Trees that established before 1900 are in ten-year age classes, and five-year age classes are used for post-1900 establishment dates. Bold line denotes value of regime-shift analysis and the abrupt change point in time reflects a significant positive or negative shift ($p < 0.01$) toward a new regime.

Thus, we interpret a statistically significant difference ($p < 0.01$) in seedling height and diameter to mean that the tallest, largest, or both are older. Overall, these statistical analyses are intended to quantify differences in treeline structure that can be used to support inferences regarding how much hotter drought is driving observed patterns. Our repeat photo pair was visually assessed for qualitative evidence of spruce beetle–induced mortality.

Results

Regeneration Dynamics

A total of 273 trees were cross-dated from slope positions above timberline (ATL) in the southern Rocky Mountains to create an approximate 200-year record of establishment beginning in 1810 (data collected in 2007, 2008, and 2014). These trees were relatively young on both south- and north-facing slopes, with 87.6 percent and 88.2 percent established since 1945, respectively (Figure 3). As a result, no significant difference in the temporal patterns of establishment ($p > 0.05$) were measured using a two-sample Kolmogorov–Smirnov test. Regime-shift analysis detected a significant increase in the rate of

establishment ($p < 0.01$) in 1945 (1945–1949 age class) for trees on south-facing slopes and then a decade later in 1955 (1955–1959 age class) for trees on north-facing slopes (Figure 3). More recently, significant and synchronous declines in establishment occurred in 1995 (Figure 3). Despite this, more than twice as many trees established on north-facing slopes compared to south-facing slopes since 1990 (24 [17.6 percent] vs. 11 [8.0 percent]).

This discrepancy is further highlighted when comparing tree to seedling frequency by slope aspect (Figure 4). Of the 186 recorded seedlings, 133 (71.5 percent) were on north-facing slopes (Figure 4B), and based on a chi-square goodness-of-fit test, this difference was significant, $\chi^2(1) = 17.2$, $p < 0.01$. Recent tree seedling data collected in 2019 from the same sites at upper treeline in the Sangre de Cristo range ($n = 6$) perpetuate these trends after a decade of warming (Figures 2, 5, 6). Once again, significantly more seedlings ($p < 0.01$) were found on north-facing slopes (71.9 percent vs. 28.1 percent) compared to south-facing slopes, $\chi^2(1) = 45.2$, $p < 0.01$. Quantitative comparisons of seedling height and diameter showed that seedlings were significantly shorter ($p < 0.01$) and smaller in diameter ($p < 0.01$) on north facing slopes (Figure 6). Overall, results from data collected between 2007 and 2014 and again in 2019 show that younger seedlings disproportionately occupied north-facing slopes.

Disturbance

A spruce beetle (*Dendroctonus rufipennis* Kirby) outbreak was underway at a treeline study site during the summer of 2014 in the eastern San Juan Mountains, as evidenced by bore holes in the trunks of defoliated Engelmann spruce trees at treeline (Figure 7). Further outbreaks were evident in the eastern San Juans where we did not sample at treeline and, notably, trees across all size class were dead (Figure 7C). In addition, a spruce beetle outbreak has been evident at upper treeline along the shoulder of Wetterhorn Peak in the San Juan Mountains since 2002 (Figure 8).

Discussion

Results from this research can be summarized into two main findings, both of which provide evidence

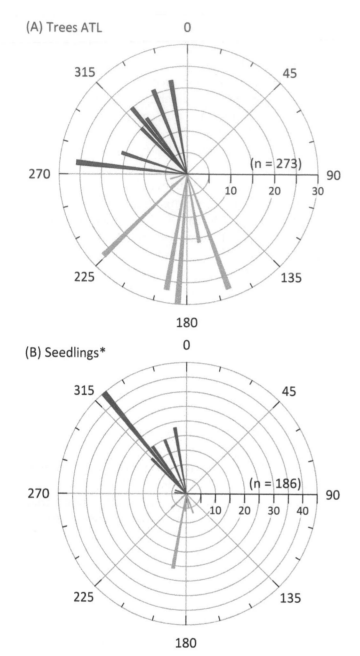

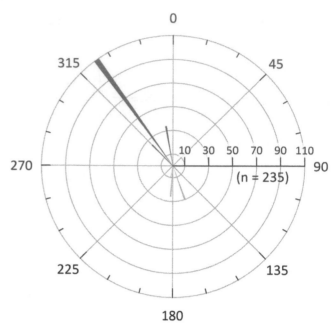

Figure 5. Seedlings harvested from upper treeline sites in Sangre de Cristo Mountains during summer 2019. Data from north-facing slopes plotted in dark gray and south-facing slopes in light gray. Radius angle shows slope aspect in degrees and radius length denotes frequency (n). Based on chi-square goodness-of-fit test, significantly more seedlings (p < 0.01) found on north-facing slopes.

Figure 4. Polar bar charts for (A) trees positioned ATL and (B) seedlings (<1.2 cm diameter at ground level) ATL at each site (n = 14). Data from north-facing slopes plotted in dark gray and south-facing slopes in light gray. Radius angle shows slope aspect in degrees and radius length denotes frequency (n). *Based on chi-square goodness-of-fit test, significantly more seedlings (p < 0.01) found on north-facing slopes. No significant difference measured for trees ATL. ATL = above timerline.

that hotter drought is likely affecting patterns of tree recruitment and mortality at upper treeline across the southern Rocky Mountain region. First, trends in recruitment over the past thirty years suggest that north-facing slopes are increasingly hospitable for successful seedling establishment compared to south-

facing slopes at upper treeline (Figures 3–6). Second, spruce beetle–induced mortality is evident at upper treeline in the San Juan Mountains (Figures 7 and 8). To our knowledge, this research represents the first documented case of spruce beetles affecting upper treeline in the Rocky Mountains.

Seedling establishment was concentrated on north-facing slopes compared to the spatial distribution of all trees ATL, which were split evenly between opposite aspects. Shifts in the spatial pattern of seedling establishment might indicate that a demographic response to climate is already underway (Bell, Bradford, and Lauenroth 2014). Indeed, seedling establishment and survival represent the most climatically sensitive phase of tree development and ultimately govern shifts in the structure and extent of upper treeline (Malanson et al. 2009). This is particularly true under the influence of drought stress, where shallow-rooted seedlings are more susceptible to death than mature trees (Smithers et al. 2018). South-facing slopes are more prone to drought and drying soil has been linked to high rates of seedling mortality at upper treeline (Dolanc et al. 2014; Moyes, Germino, and Kueppers 2015; Lazarus et al. 2018). In addition, higher solar insolation coupled

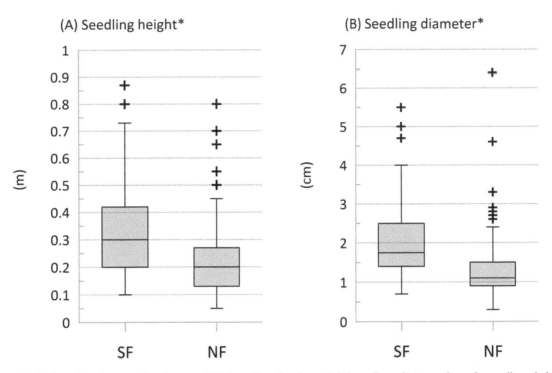

Figure 6. Box-and-whisker plots showing distribution of (A) seedling height and (B) seedling diameter from data collected during summer 2019 at upper treeline in the Sangre de Cristo Mountains (*n* = 235) on north-facing (*n* = 169) and south-facing slopes (*n* = 66). Crosses denote outliers and (*) indicates significant differences ($p < 0.01$) measured using a Mann–Whitney U test. NF = north-facing; SF = south-facing.

with increased climatic water deficits impede soil-forming processes, especially where vegetation cover is sparse (Schickoff 2005). On north-facing slopes, however, water has a longer residence time and is stored more effectively in the near surface of deeper, well-developed soils (Hinckley et al. 2014; Zapata-Rios et al. 2016). Under this scenario, increased soil water availability could be playing a key role in the observed slope aspect–mediated response of seedling establishment to increasingly warm and dry conditions over the past few decades.

Results from this research align with previous studies at upper treeline that examined the influence of moisture and drought stress on regeneration dynamics, including in the Himalayas, where increasing moisture-mediated responses of tree and shrub establishment have been noted along the highest treelines in the world (Sigdel et al. 2018; Lu et al. 2019). Across Western North America, seedling establishment at upper treeline has been more common on north-facing slopes compared to warm and dry south-facing slopes (Weisberg and Baker 1995; Germino, Smith, and Resor 2002; Gill, Campbell, and Karlinsey 2015; Millar et al. 2015; Elliott and Petruccelli 2018). At a local scale, recent experimental evidence from a southwest-facing

treeline site on Niwot Ridge in the Colorado Front Range found that growing-season drought stress limited seedling establishment (Moyes et al. 2013; Moyes, Germino, and Kueppers 2015; Kueppers et al. 2017). Significantly lower rates of successful establishment on south-facing slopes since about 2000 in this study suggest that these trends might persist at a regional scale across the southern Rocky Mountains and, thus, treeline advance could be more confined to north-facing slopes moving forward.

Spruce beetle–induced mortality is evidence that hotter drought as a megadisturbance has in fact reached upper treeline in the southern Rocky Mountains. Prior to this research, the only reported instance of spruce beetle at treeline was at arctic treeline along the eastern coast of Hudson Bay, Canada (Caccianiga, Payette, and Filion 2008). As a disturbance, spruce beetle outbreaks have the same potential as drought-induced mortality reconstructed from the Holocene to lower the elevational extent of upper treeline (Lloyd and Graumlich 1997; Morgan, Losey, and Trout 2014). Furthermore, this highlights the importance of understanding postdisturbance regeneration within the context of hotter drought (Martínez-Vilalta

Figure 7. Spruce beetle (*Dendroctonus rufipennis* Kirby) at upper treeline in the eastern San Juan Mountains. (A) Elevational extent of outbreak at 3,700 m. (B) Bore holes in Engelmann spruce at treeline. (C) Beetle outbreak is lethal across tree size classes at upper treeline in San Juans.

and Lloret 2016). Given that adequate moisture is needed for successful Engelmann spruce regeneration (Gill, Campbell, and Karlinsey 2015; Moyes, Germino, and Kueppers 2015; Kueppers et al. 2017), the ability to repopulate decimated stands will be contingent on local site conditions, including slope aspect. Considered together, this supports the idea that spruce beetle outbreaks are capable of lowering upper treeline, which could lead to an overall contraction in the range of spruce under the influence of hotter drought (Conlisk et al. 2017).

Conclusion

Predicting the response of upper treeline to climate in general and hotter drought in particular on the scale of decades is wrought with uncertainty (Lloyd 2005; Malanson et al. 2019), especially

considering the complex topography inherent to these environments (Holtmeier and Broll 2005; Mensing et al. 2012). Yet results from this research support the idea that slope aspect is modifying the influence of increasingly warm and dry conditions at upper treeline. Despite the fact that annual tree growth and seedling establishment can be governed by disparate climate conditions (Daniels and Veblen 2004; Holtmeier and Broll 2005; Fajardo and McIntire 2012), increased rates of seedling establishment on north-facing slopes support the idea that temperature-limited area for trees on south-facing slopes is contracting upslope as moisture-limited area fills in from below (Salzer et al. 2014). Conceptually, this means that hotter drought could be progressively enveloping upper treeline along topoclimatic gradients as reported for subalpine forests (Smith et al. 2015) and that drought-driven disturbances are beginning to affect

Figure 8. Wetterhorn Peak: (A) 2002 and (B) 2019. Photo B is purposefully offset to highlight spruce beetle outbreak at upper treeline along left side of photo. Repeat photo location is at 3,615 m and slope in scene is facing east (aspect = 90°). Photo (A) from Elliott (2003).

the highest upper treeline environments in the Rocky Mountains.

Moving forward, intensifying soil moisture deficits during the growing season are expected to reduce conifer forest cover across mountains of the western United States as more drought-tolerant grasses and shrubs expand (Jiang et al. 2013). Findings from this regional-scale study across the southern Rocky Mountains suggest that this could in part be driven by both a lack of successful establishment on dry south-facing slopes and spruce beetle outbreaks along the uppermost extent of the mountain forest belt. Unless ongoing trends in temperature deviate from expectations or unless precipitation increases considerably, there is little reason to justify the idea that upper treeline will continue to respond positively to hotter drought conditions during the Anthropocene, especially on south-facing slopes in the Northern Hemisphere.

References

Alexander, J. M., L. Chalmandrier, J. Lenoir, T. I. Burgess, F. Essl, S. Haider, C. Kueffer, K. McDougall, A. Milbau, M. A. Nuñez, et al. 2018. Lags in the response of mountain plant communities to climate change. *Global Change Biology* 24 (2):563–79. doi: 10.1111/gcb.13976.

Allen, C. D., and D. D. Breshears. 1998. Drought-induced shift of a forest–woodland ecotone: Rapid landscape response to climate variation. *Proceedings of the National Academy of Sciences* 95 (25):14839–42. doi: 10.1073/pnas.95.25.14839.

Allen, C. D., D. D. Breshears, and N. G. McDowell. 2015. On underestimation of global vulnerability to tree mortality and forest die-off from hotter drought in the Anthropocene. *Ecosphere* 6 (8):art129. doi: 10.1890/ES15-00203.1.

Allen, C. D., A. K. Macalady, H. Chenchouni, D. Bachelet, N. McDowell, M. Vennetier, T. Kitzberger, A. Rigling, D. D. Breshears, E. H. Hogg, et al. 2010. A global overview of drought and heat-induced tree mortality reveals emerging climate change risks for forests. *Forest Ecology and Management* 259 (4):660–84. doi: 10.1016/j.foreco.2009.09.001.

Baker, W. L. 2009. *Fire ecology in Rocky Mountain landscapes*. Washington, DC: Island.

Barnett, T. P., D. W. Pierce, H. G. Hidalgo, C. Bonfils, B. D. Santer, T. Das, G. Bala, A. W. Wood, T. Nozawa, A. A. Mirin, et al. 2008. Human-induced changes in the hydrology of the Western United States. *Science* 319 (5866):1080–83. doi: 10.1126/science.1152538.

Barry, R. G. 2008. *Mountain weather and climate*. 2nd ed. Cambridge, UK: Cambridge University Press.

Bell, D. M., J. B. Bradford, and W. K. Lauenroth. 2014. Mountain landscapes offer few opportunities for high-elevation tree species migration. *Global Change Biology* 20 (5):1441–51. doi: 10.1111/gcb.12504.

Bourgeron, P. S., H. C. Humphries, D. Liptzin, and T. R. Seastedt. 2015. The forest–alpine ecotone: A multiscale approach to spatial and temporal dynamics of treeline change at Niwot Ridge. *Plant Ecology & Diversity* 8 (5–6):763–79. doi: 10.1080/17550874.2015.1126368.

Breshears, D. D., N. S. Cobb, P. M. Rich, K. P. Price, C. D. Allen, R. G. Balice, W. H. Romme, J. H. Kastens, M. L. Floyd, J. Belnap, et al. 2005. Regional vegetation die-off in response to global-change-type drought. *Proceedings of the National Academy of Sciences of the United States of America* 102 (42):15144–48. doi: 10.1073/pnas.0505734102.

Bunn, A. G., M. W. Salzer, K. J. Anchukaitis, J. M. Bruening, and M. K. Hughes. 2018. Spatiotemporal variability in the climate growth response of high elevation bristlecone pine in the White Mountains of California. *Geophysical Research Letters* 45 (24):13312–21. doi: 10.1029/2018GL080981.

Butler, D. R. 1995. *Zoogeomorphology: Animals as geomorphic agents*. Cambridge, UK: Cambridge University Press.

Butler, D. R., G. P. Malanson, L. M. Resler, S. J. Walsh, F. D. Wilkerson, G. L. Schmid, and C. F. Sawyer. 2009. Geomorphic patterns and processes at alpine treeline. In *Developments in earth surface processes*, ed. D. R. Butler, G. P. Malanson, S. J. Walsh, and D. B. Fagre, 63–84. Elsevier.

Butler, D. R., G. P. Malanson, S. J. Walsh, and D. B. Fagre. 2007. Influences of geomorphology and geology on alpine treeline in the American West—More important than climatic influences? *Physical Geography* 28 (5):434–50. doi: 10.2747/0272-3646.28.5.434.

Caccianiga, M., S. Payette, and L. Filion. 2008. Biotic disturbance in expanding subarctic forests along the eastern coast of Hudson Bay. *The New Phytologist* 178 (4):823–34. doi: 10.1111/j.1469-8137.2008.02408.x.

Cairns, D. M., C. Lafon, J. Moen, and A. Young. 2007. Influences of animal activity on treeline position and pattern: Implications for treeline responses to climate change. *Physical Geography* 28 (5):419–33. doi: 10.2747/0272-3646.28.5.419.

Clark, J. S., L. Iverson, C. W. Woodall, C. D. Allen, D. M. Bell, D. C. Bragg, A. W. D'Amato, F. W. Davis, M. H. Hersh, I. Ibanez, et al. 2016. The impacts of increasing drought on forest dynamics, structure, and biodiversity in the United States. *Global Change Biology* 22 (7):2329–52. doi: 10.1111/gcb.13160.

Clow, D. W. 2010. Changes in the timing of snowmelt and streamflow in Colorado: A response to recent warming. *Journal of Climate* 23 (9):2293–2306. doi: 10.1175/2009JCLI2951.1.

Conlisk, E., C. Castanha, M. J. Germino, T. T. Veblen, J. M. Smith, and L. M. Kueppers. 2017. Declines in low-elevation subalpine tree populations outpace growth in high-elevation populations with warming. *Journal of Ecology* 105 (5):1347–57. doi: 10.1111/1365-2745.12750.

Corlett, R. T. 2015. The Anthropocene concept in ecology and conservation. *Trends in Ecology & Evolution* 30 (1):36–41. doi: 10.1016/j.tree.2014.10.007.

Cowell, C. M., and M. A. Urban. 2010. The changing geography of the U.S. water budget: Twentieth-century patterns and twenty-first-century projections. *Annals of the Association of American Geographers* 100 (4):740–54. doi: 10.1080/00045608.2010.497117.

Daly, C., M. Halbleib, J. I. Smith, W. P. Gibson, M. K. Doggett, G. H. Taylor, J. Curtis, and P. P. Pasteris. 2008. Physiographically sensitive mapping of climatological temperature and precipitation across the conterminous United States. *International Journal of Climatology* 28 (15):2031–64. doi: 10.1002/joc.1688.

Daniels, L. D., and T. T. Veblen. 2004. Spatiotemporal influences of climate on altitudinal treeline in northern Patagonia. *Ecology* 85 (5):1284–96. doi: 10.1890/03-0092.

Daubenmire, R. F. 1943. Vegetational zonation in the Rocky Mountains. *The Botanical Review* 9 (6):325–93. doi: 10.1007/BF02872481.

DeRose, R. J., and J. N. Long. 2012. Drought-driven disturbance history characterizes a southern Rocky Mountain subalpine forest. *Canadian Journal of Forest Research* 42 (9):1649–60. doi: 10.1139/x2012-102.

Dettinger, M. D., D. R. Cayan, H. F. Diaz, and D. M. Meko. 1998. North–south precipitation patterns in western North America on interannual-to-decadal timescales. *Journal of Climate* 11 (12):3095–111. doi: 10.1175/1520-0442(1998)011<3095:NSPPIW>2.0.CO;2.

Dolanc, C. R., H. D. Safford, J. H. Thorne, and S. Z. Dobrowski. 2014. Changing forest structure across the landscape of the Sierra Nevada, CA, USA, since the 1930s. *Ecosphere* 5 (8):art101. doi: 10.1890/ES14-00103.1.

Elliott, G. P. 2003. Quaking aspen (*Populus tremuloides* Michx.) and conifers at treeline: A century of change in the San Juan Mountains, Colorado, USA. Master's thesis, University of Wyoming.

Elliott, G. P. 2011. Influences of 20th-century warming at the upper tree line contingent on local-scale interactions: Evidence from a latitudinal gradient in the Rocky Mountains, USA. *Global Ecology and Biogeography* 20 (1):46–57. doi: 10.1111/j.1466-8238.2010.00588.x.

Elliott, G. P. 2012. Extrinsic regime shifts drive abrupt changes in regeneration dynamics at upper treeline in the Rocky Mountains, USA. *Ecology* 93 (7):1614–25. doi: 10.1890/11-1220.1.

Elliott, G. P. 2017. Treeline ecotones. In *International encyclopedia of geography*, ed. N. Castree, M. F. Goodchild, A. Kobayashi, W. Liu, and R. A. Marston, 1–10. Chichester, U.K.: Wiley. https://onlinelibrary.wiley.com/doi/abs/10.1002/9781118786352.wbieg0539.

Elliott, G. P., and C. M. Cowell. 2015. Slope aspect mediates fine-scale tree establishment patterns at upper treeline during wet and dry periods of the 20th century. *Arctic, Antarctic, and Alpine Research* 47 (4):681–92. doi: 10.1657/AAAR0014-025.

Elliott, G. P., and K. F. Kipfmueller. 2010. Multi-scale influences of slope aspect and spatial pattern on ecotonal dynamics at upper treeline in the southern Rocky Mountains. *U.S.A. Arctic, Antarctic, and Alpine Research* 42 (1):45–56. doi: 10.1657/1938-4246-42.1.45.

Elliott, G. P., and K. F. Kipfmueller. 2011. Multiscale influences of climate on upper treeline dynamics in the southern Rocky Mountains, USA: Evidence of intraregional variability and bioclimatic thresholds in response to twentieth-century warming. *Annals of the Association of American Geographers* 101 (6):1181–1203. doi: 10.1080/00045608.2011.584288.

Elliott, G. P., and C. A. Petruccelli. 2018. Tree recruitment at the treeline across the Continental Divide in the northern Rocky Mountains, USA: The role of spring snow and autumn climate. *Plant Ecology & Diversity* 11 (3):319–33. doi: 10.1080/17550874.2018.1487475.

Ellis, E. C., K. Klein Goldewijk, S. Siebert, D. Lightman, and N. Ramankutty. 2010. Anthropogenic transformation of the biomes, 1700 to 2000. *Global Ecology and Biogeography* 19 (5):589–606. doi: 10.1111/j.1466-8238.2010.00540.x.

Fajardo, A., and E. J. B. McIntire. 2012. Reversal of multicentury tree growth improvements and loss of synchrony at mountain tree lines point to changes in key drivers. *Journal of Ecology* 100 (3):782–94. doi: 10.1111/j.1365-2745.2012.01955.x.

Fattorini, S., L. D. Biase, and A. Chiarucci. 2019. Recognizing and interpreting vegetational belts: New wine in the old bottles of a von Humboldt's legacy. *Journal of Biogeography* 46 (8):1643–51. doi: 10.1111/jbi.13601.

Germino, M. J., W. K. Smith, and A. C. Resor. 2002. Conifer seedling distribution and survival in an alpine-treeline ecotone. *Plant Ecology* 162 (2):157–68. doi: 10.1023/A:1020385320738.

Gill, R. A., C. S. Campbell, and S. M. Karlinsey. 2015. Soil moisture controls Engelmann spruce (*Picea engelmannii*) seedling carbon balance and survivorship at timberline in Utah, USA. *Canadian Journal of Forest Research* 45 (12):1845–52. doi: 10.1139/cjfr-2015-0239.

Harsch, M. A., P. E. Hulme, M. S. McGlone, and R. P. Duncan. 2009. Are treelines advancing? A global meta-analysis of treeline response to climate warming. *Ecology Letters* 12 (10):1040–49. doi: 10.1111/j.1461-0248.2009.01355.x.

Hart, S. J., T. T. Veblen, K. S. Eisenhart, D. Jarvis, and D. Kulakowski. 2014. Drought induces spruce beetle (*Dendroctonus rufipennis*) outbreaks across northwestern Colorado. *Ecology* 95 (4):930–39. doi: 10.1890/13-0230.1.

Hinckley, E.-L. S., B. A. Ebel, R. T. Barnes, R. S. Anderson, M. W. Williams, and S. P. Anderson. 2014. Aspect control of water movement on hillslopes near the rain–snow transition of the Colorado Front Range. *Hydrological Processes* 28 (1):74–85. doi: 10.1002/hyp.9549.

Holtmeier, F.-K., and G. Broll. 2005. Sensitivity and response of northern hemisphere altitudinal and polar treelines to environmental change at landscape and local scales. *Global Ecology and Biogeography* 14 (5):395–410. doi: 10.1111/j.1466-822X.2005.00168.x.

Holtmeier, F.-K., and G. Broll. 2019. Treeline research—From the roots of the past to present time: A review. *Forests* 11 (1):1–38. doi: 10.3390/f11010038.

Jiang, X., S. A. Rauscher, T. D. Ringler, D. M. Lawrence, A. P. Williams, C. D. Allen, A. L. Steiner, D. M. Cai, and N. G. McDowell. 2013. Projected future changes in vegetation in western North America in the twenty-first century. *Journal of Climate* 26 (11):3671–87. doi: 10.1175/JCLI-D-12-00430.1.

Körner, C. 2012. *Alpine treelines: Functional ecology of the global high elevation tree limits.* Basel, Switzerland: Springer.

Kueppers, L. M., E. Conlisk, C. Castanha, A. B. Moyes, M. J. Germino, P. de Valpine, M. S. Torn, and J. B. Mitton. 2017. Warming and provenance limit tree recruitment across and beyond the elevation range of subalpine forest. *Global Change Biology* 23 (6):2383–95. doi: 10.1111/gcb.13561.

Lazarus, B. E., C. Castanha, M. J. Germino, L. M. Kueppers, and A. B. Moyes. 2018. Growth strategies and threshold responses to water deficit modulate effects of warming on tree seedlings from forest to alpine. *Journal of Ecology* 106 (2):571–85. doi: 10.1111/1365-2745.12837.

League, K., and T. Veblen. 2006. Climatic variability and episodic *Pinus ponderosa* establishment along the forest-grassland ecotones of Colorado. *Forest Ecology and Management* 228 (1–3):98–107. doi: 10.1016/j.foreco.2006.02.030.

Lloyd, A. H. 2005. Ecological histories from Alaskan tree lines provide insight into future change. *Ecology* 86 (7):1687–95. doi: 10.1890/03-0786.

Lloyd, A. H., and L. J. Graumlich. 1997. Holocene dynamics of treeline forests in the Sierra Nevada. *Ecology* 78 (4):1199–210. doi: 10.1890/0012-9658(1997)078[1199:HDOTFI]2.0.CO;2.

Lu, X., E. Liang, Y. Wang, F. Babst, S. W. Leavitt, and J. J. Camarero. 2019. Past the climate optimum: Recruitment is declining at the world's highest juniper shrublines on the Tibetan Plateau. *Ecology* 100 (2):e02557. doi: 10.1002/ecy.2557.

Malanson, G. P., D. G. Brown, D. R. Butler, D. M. Cairns, D. B. Fagre, and S. J. Walsh. 2009. Ecotone dynamics: Invasibility of alpine tundra by tree species from the subalpine forest. In *Developments in earth surface processes*, ed. D. R. Butler, G. P. Malanson, S. J. Walsh, and D. B. Fagre, 35–61. Elsevier.

Malanson, G. P., D. R. Butler, D. B. Fagre, S. J. Walsh, D. F. Tomback, L. D. Daniels, L. M. Resler, W. K. Smith, D. J. Weiss, D. L. Peterson, et al. 2007. Alpine treeline of western North America: Linking organism-to-landscape dynamics. *Physical Geography* 28 (5):378–96. doi: 10.2747/0272-3646.28.5.378.

Malanson, G. P., L. M. Resler, D. R. Butler, and D. B. Fagre. 2019. Mountain plant communities: Uncertain sentinels? *Progress in Physical Geography: Earth and Environment* 43 (4):521–43. doi: 10.1177/0309133319843873.

Marsicek, J., B. N. Shuman, P. J. Bartlein, S. L. Shafer, and S. Brewer. 2018. Reconciling divergent trends and millennial variations in Holocene temperatures. *Nature* 554 (7690):92–96. doi: 10.1038/nature25464.

Martínez-Vilalta, J., and F. Lloret. 2016. Drought-induced vegetation shifts in terrestrial ecosystems: The key role of regeneration dynamics. *Global and Planetary Change* 144:94–108. doi: 10.1016/j.gloplacha.2016.07.009.

Mensing, S., J. Korfmacher, T. Minckley, and R. Musselman. 2012. A 15,000 year record of vegetation and climate change from a treeline lake in the Rocky Mountains, Wyoming, USA. *The Holocene* 22 (7):739–48. doi: 10.1177/0959683611430339.

Millar, C. I., and N. L. Stephenson. 2015. Temperate forest health in an era of emerging megadisturbance. *Science* 349 (6250):823–26. doi: 10.1126/science.aaa9933.

Millar, C. I., R. D. Westfall, D. L. Delany, A. L. Flint, and L. E. Flint. 2015. Recruitment patterns and growth of high-elevation pines in response to climatic variability (1883–2013), in the western Great Basin, USA. *Canadian Journal of Forest Research* 45 (10):1299–1312. doi: 10.1139/cjfr-2015-0025.

Minckley, T. A., R. K. Shriver, and B. Shuman. 2012. Resilience and regime change in a southern Rocky Mountain ecosystem during the past 17 000 years.

Ecological Monographs 82 (1):49–68. doi: 10.1890/11-0283.1.

Mitchell, V. L. 1976. The regionalization of climate in the western United States. *Journal of Applied Meteorology* 15 (9):920–27. doi: 10.1175/1520-0450(1976)015<0920:TROCIT>2.0.CO;2.

Morgan, C., A. Losey, and L. Trout. 2014. Late-Holocene paleoclimate and treeline fluctuation in Wyoming's Wind River Range, USA. *The Holocene* 24 (2):209–19. doi: 10.1177/0959683613516817.

Mote, P. W., S. Li, D. P. Lettenmaier, M. Xiao, and R. Engel. 2018. Dramatic declines in snowpack in the western U.S. *NPJ Climate and Atmospheric Science* 1 (1):1–6. doi: 10.1038/s41612-018-0012-1.

Moyes, A. B., C. Castanha, M. J. Germino, and L. M. Kueppers. 2013. Warming and the dependence of limber pine (*Pinus flexilis*) establishment on summer soil moisture within and above its current elevation range. *Oecologia* 171 (1):271–82. doi: 10.1007/s00442-012-2410-0.

Moyes, A. B., M. J. Germino, and L. M. Kueppers. 2015. Moisture rivals temperature in limiting photosynthesis by trees establishing beyond their cold-edge range limit under ambient and warmed conditions. *New Phytologist* 207 (4):1005–14. doi: 10.1111/nph.13422.

Munier, A., L. Hermanutz, J. D. Jacobs, and K. Lewis. 2010. The interacting effects of temperature, ground disturbance, and herbivory on seedling establishment: Implications for treeline advance with climate warming. *Plant Ecology* 210 (1):19–30. doi: 10.1007/s11258-010-9724-y.

Noble, I. R. 1993. A model of the responses of ecotones to climate change. *Ecological Applications: A Publication of the Ecological Society of America* 3 (3):396–403. doi: 10.2307/1941908.

Nogués-Bravo, D., M. B. Araújo, M. P. Errea, and J. P. Martínez-Rica. 2007. Exposure of global mountain systems to climate warming during the 21st century. *Global Environmental Change* 17 (3–4):420–28. doi: 10.1016/j.gloenvcha.2006.11.007.

Paulsen, J., U. M. Weber, and C. Körner. 2000. Tree growth near treeline: Abrupt or gradual reduction with altitude? *Arctic, Antarctic, and Alpine Research* 32 (1):14–20. doi: 10.2307/1552405.

Peet, R. K. 1978. Latitudinal variation in southern Rocky Mountain forests. *Journal of Biogeography* 5 (3):275–89. doi: 10.2307/3038041.

Peet, R. K. 2000. Forests and meadows of the Rocky Mountains. In *North American terrestrial vegetation*, ed. M. G. Barbour and W. D. Billings, 79–121. Cambridge, UK: Cambridge University Press.

Pepin, N., R. S. Bradley, H. F. Diaz, M. Baraer, E. B. Caceres, N. Forsythe, H. Fowler, G. Greenwood, M. Z. Hashmi, X. D. Liu, et al. 2015. Elevation-dependent warming in mountain regions of the world. *Nature Climate Change* 5 (5):424–30.

PRISM Climate Group. 2019. PRISM Climate Data. Accessed 10 November 2019. https://prism.oregonstate.edu/.

Resler, L. M., and D. F. Tomback. 2008. Blister rust prevalence in Krummholz whitebark pine: Implications for treeline dynamics, northern Rocky Mountains,

Montana, U.S.A. *Arctic, Antarctic, and Alpine Research* 40 (1):161–70. doi: 10.1657/1523-0430(06-116)[RESLER]2.0.CO;2.

Rodionov, S. N. 2004. A sequential algorithm for testing climate regime shifts. *Geophysical Research Letters* 31 (9):L09204. doi: 10.1029/2004GL019448.

Salzer, M. W., E. R. Larson, A. G. Bunn, and M. K. Hughes. 2014. Changing climate response in near-treeline bristlecone pine with elevation and aspect. *Environmental Research Letters* 9 (11):114007. doi: 10.1088/1748-9326/9/11/114007.

Schickoff, U. 2005. The upper timberline in the Himalayas, Hindu Kush and Karakorum: A review of geographical and ecological aspects. In *Mountain ecosystems: Studies in treeline ecology*, ed. G. Broll and B. Keplin, 275–354. Berlin: Springer.

Schoennagel, T., T. T. Veblen, D. Kulakowski, and A. Holz. 2007. Multidecadal climate variability and climate interactions affect subalpine fire occurrence, western Colorado (USA). *Ecology* 88 (11):2891–902. doi: 10.1890/06-1860.1.

Shankman, D., and C. Daly. 1988. Forest regeneration above tree limit depressed by fire in the Colorado Front Range. *Bulletin of the Torrey Botanical Club* 115 (4):272–79. doi: 10.2307/2996159.

Shuman, B. 2012. Recent Wyoming temperature trends, their drivers, and impacts in a 14,000-year context. *Climatic Change* 112 (2):429–47. doi: 10.1007/s10584-011-0223-5.

Sigdel, S. R., Y. Wang, J. J. Camarero, H. Zhu, E. Liang, and J. Peñuelas. 2018. Moisture-mediated responsiveness of treeline shifts to global warming in the Himalayas. *Global Change Biology* 24 (11):5549–59. doi: 10.1111/gcb.14428.

Smith, J. M., J. Paritsis, T. T. Veblen, and T. B. Chapman. 2015. Permanent forest plots show accelerating tree mortality in subalpine forests of the Colorado Front Range from 1982 to 2013. *Forest Ecology and Management* 341:8–17. doi: 10.1016/j.foreco.2014.12.031.

Smithers, B. V., M. P. North, C. I. Millar, and A. M. Latimer. 2018. Leap frog in slow motion: Divergent responses of tree species and life stages to climatic warming in Great Basin subalpine forests. *Global Change Biology* 24 (2):e442. doi: 10.1111/gcb.13881.

Solomon, S., G.-K. Plattner, R. Knutti, and P. Friedlingstein. 2009. Irreversible climate change due to carbon dioxide emissions. *Proceedings of the National Academy of Sciences of the United States of America* 106 (6):1704–9. doi: 10.1073/pnas.0812721106.

Steffen, W., J. Grinevald, P. Crutzen, and J. McNeill. 2011. The Anthropocene: Conceptual and historical perspectives. *Philosophical Transactions. Series A, Mathematical, Physical, and Engineering Sciences* 369 (1938):842–67. doi: 10.1098/rsta.2010.0327.

Stine, M. B., and D. R. Butler. 2015. Effects of fire on geomorphic factors and seedling site conditions within the alpine treeline ecotone, Glacier National Park, MT. *Catena* 132:37–44. doi: 10.1016/j.catena.2015.04.006.

Stueve, K. M., D. L. Cerney, R. M. Rochefort, and L. L. Kurth. 2009. Post-fire tree establishment patterns at the alpine treeline ecotone: Mount Rainier National Park, Washington, USA. *Journal of Vegetation Science* 20 (1):107–20. doi: 10.1111/j.1654-1103.2009.05437.x.

Tomback, F. D., M. L. Resler, E. R. Keane, R. E. Pansing, J. A. Andrade, and C. A. Wagner. 2016. Community structure, biodiversity, and ecosystem services in tree-line whitebark pine communities: Potential impacts from a non-native pathogen. *Forests* 7 (12):1–21. doi: 10.3390/f7010021.

van Mantgem, P. J., N. L. Stephenson, J. C. Byrne, L. D. Daniels, J. F. Franklin, P. Z. Fulé, M. E. Harmon, A. J. Larson, J. M. Smith, A. H. Taylor, et al. 2009. Widespread increase of tree mortality rates in the Western United States. *Science (New York, N.Y.)* 323 (5913):521–24. doi: 10.1126/science.1165000.

von Humboldt, A., and A. Bonpland. 1805. *Essai sur la géographie des plantes* (Essay on the Geography of Plants). Paris Levrault, Schoell et Compagnie.

Walsh, S. J., D. R. Butler, T. R. Allen, and G. P. Malanson. 1994. Influence of snow patterns and snow avalanches on the alpine treeline ecotone. *Journal of Vegetation Science* 5 (5):657–72. doi: 10.2307/3235881.

Wang, Y., B. Case, X. Lu, A. M. Ellison, J. Peñuelas, H. Zhu, E. Liang, and J. J. Camarero. 2019. Fire facilitates warming-induced upward shifts of alpine tree-lines by altering interspecific interactions. *Trees* 33 (4):1051–61. doi: 10.1007/s00468-019-01841-6.

Weisberg, P. J., and W. L. Baker. 1995. Spatial variation in tree regeneration in the forest–tundra ecotone, Rocky Mountain National Park, Colorado. *Canadian Journal of Forest Research* 25 (8):1326–39. doi: 10.1139/x95-145.

Williams, A. P., C. D. Allen, A. K. Macalady, D. Griffin, C. A. Woodhouse, D. M. Meko, T. W. Swetnam, S. A. Rauscher, R. Seager, H. D. Grissino-Mayer, et al. 2013. Temperature as a potent driver of regional forest drought stress and tree mortality. *Nature Climate Change* 3 (3):292–97. doi: 10.1038/nclimate1693.

Williams, J. W., J. L. Blois, and B. N. Shuman. 2011. Extrinsic and intrinsic forcing of abrupt ecological change: Case studies from the late Quaternary. *Journal of Ecology* 99 (3):664–77. doi: 10.1111/j.1365-2745.2011.01810.x.

Wolkovich, E. M., B. I. Cook, K. K. McLauchlan, and T. J. Davies. 2014. Temporal ecology in the Anthropocene. *Ecology Letters* 17 (11):1365–79. doi: 10.1111/ele.12353.

Young, K. R. 2014. Biogeography of the Anthropocene: Novel species assemblages. *Progress in Physical Geography: Earth and Environment* 38 (5):664–73. doi: 10.1177/0309133314540930.

Zapata-Rios, X., P. D. Brooks, P. A. Troch, J. McIntosh, and Q. Guo. 2016. Influence of terrain aspect on water partitioning, vegetation structure and vegetation greening in high-elevation catchments in northern New Mexico. *Ecohydrology* 9 (5):782–95. doi: 10.1002/eco.1674.

Zier, J. L., and W. L. Baker. 2006. A century of vegetation change in the San Juan Mountains, Colorado: An analysis using repeat photography. *Forest Ecology and Management* 228 (1–3):251–62. doi: 10.1016/j.foreco.2006.02.049.

Onset of the Paleoanthropocene in the Lower Great Lakes Region of North America: An Archaeological and Paleoecological Synthesis

Albert E. Fulton II and Catherine H. Yansa

Beyond its centrality to debates on the definition of a global chronostratigraphic geologic unit, the Anthropocene concept has served as a useful theoretical construct with which to assess regional-scale anthropogenic impacts on prehistoric ecological systems through the recognition of earlier "Paleoanthropocene" events predating the onset of modern, industrial, global-scale effects. To this end, we present data derived from the archaeological and paleoecological records of the lower Great Lakes region of northeastern North America to evaluate the nature, magnitude, and timing of Native American land use impacts over the course of the Holocene. We identified three phases of emerging and progressively intensifying anthropogenic influence coinciding with initial human paleopopulation increase (5400–2500 BP), regional introduction of maize (*Zea mays*; 2500–1100 BP), and the regional adoption of maize-based agriculture (1100–300 BP). Each phase was accompanied by notable shifts in one or more proxy indicators of amplified fire regimes (increased soil charcoal deposition, higher lake sediment charcoal influx, greater percentages of fire-tolerant pollen taxa), decreased forest canopy density (increased herbaceous pollen taxa, enriched speleothem $\delta^{13}C$ values), paleopopulation growth (increased archaeological ^{14}C date frequencies), and dietary innovations (increased cultigen ^{14}C date frequencies, enriched pottery residue $\delta^{13}C$ values). Although prominent climate excursions also greatly influenced forest species composition, forest structure, and disturbance regimes, demographic and cultural factors impinging on Native American subsistence regimes and settlement patterns became increasingly important modulators of ecological processes over the course of the Holocene. *Key Words: agriculture, Anthropocene, maize, Native Americans, Paleoanthropocene.*

The Anthropocene is currently used as an informal chronostratigraphic term to denote the approximate onset of pervasive, global influence of human populations on Earth's environmental systems that is visible as one or more stratigraphic markers in the geological record (Carruthers 2019). Although 1950 CE is presently the most favored starting date for a formal Anthropocene epoch (Zalasiewicz et al. 2015; Waters et al. 2018), this date is contentious, because there is also evidence of human alteration of global environmental systems prior to the mid-twentieth century CE. For example, an Anthropocene commencing at 1800 CE coincides with the onset of the Industrial Revolution and atmospheric greenhouse gas emissions originating in fossil fuel combustion (Waters et al. 2016). Other researchers have proposed an onset at circa 3000 BP (years before present) based on archaeological evidence for pervasive global land use transformations beginning at this time (Stephens et al. 2019). More controversially, Ruddiman (2003) postulated an Early Anthropocene hypothesis in which the spread of agriculture in the Middle East and Europe from 10000 to 4000 BP was the primary transformative event leading to human domination of global ecosystem processes.

The difficulty of establishing a precise starting point for the global onset of the Anthropocene—and defining what the Anthropocene represents (Chin et al. 2017)—highlights the equally important problem of contextualizing evidence of subtler, regional- and landscape-scale environmental transformations produced by preindustrial societies. To this end, Foley et al. (2013) introduced the term *Paleoanthropocene* to denote all paleoenvironmental signals of anthropogenic land use practices predating the development of Western industrial economies. In North America, the scale and magnitude of environmental changes wrought by prehistoric human populations might not be directly comparable to those of contemporary societies, yet elucidating the spatiotemporal dynamics of prior centuries and millennia of anthropogenesis is critical to our understanding of how and why ecosystems have developed through time. Doing so also allows us to contextualize the primary drivers of such change, including human agency (settlement systems, subsistence economies, technological innovations), environmental variables (vegetation, soils, landforms), and natural climate variability. Additionally, developing more nuanced estimates of the nature, timing, and magnitude of Native American land use impacts has a direct bearing on contemporary ecological restoration paradigms, particularly calling into question the assumption of "baseline" ecological conditions (e.g., Oswald et al. 2020) derived from the immediate pre-European settlement period, which do not automatically represent "natural" pre-anthropogenic states (Guerrero-Gatica, Aliste, and Simonetti 2019). Knowledge of past anthropogenic impacts on the landscapes of North America can thus help to inform how present ecosystems might respond in the future to heightened climate variability and accelerated biodiversity loss (Johnson et al. 2017). In this study, we report the results of a synthetic, integrated analysis of selected archaeological and paleoecological data sets from the lower Great Lakes region of North America over the last 10,000 years to (1) assess the validity of the Paleoanthropocene concept, (2) establish its approximate commencement, (3) determine whether a regional Paleoanthropocene was uniform in character or possessed spatiotemporal variability, and (4) evaluate the utility of the Paleoanthropocene concept to regional archaeological and paleoenvironmental research paradigms.

Materials and Methods

The study area broadly encompasses the lower Great Lakes region of northeastern North America (Figure 1). The regional vegetation is dominated by broadleaf deciduous trees including beech (*Fagus grandifolia*), sugar maple (*Acer saccharum*), birch (*Betula* spp.), oak (*Quercus* spp.), and hickory (*Carya* spp.), with coniferous species such as eastern hemlock (*Tsuga canadensis*) and white pine (*Pinus strobus*) also common regionally (Braun 2001). At the time of initial European contact circa 1600 CE, the study area was occupied by numerous Native American groups having a diverse array of subsistence economies and settlement systems, with some groups—most notably Northern Iroquoians (Engelbrecht 2003)—practicing intensive forms of village-centered, maize-based agriculture; others engaging in a residentially mobile, hunter-gatherer pattern (Tanner 1987); and still others combining aspects of agricultural and nonagricultural subsistence patterns (Chilton 1999).

We analyzed long-term trends in regional human population through temporal analysis of archaeological radiocarbon date frequencies downloaded from the Canadian Archaeological Radiocarbon Database (CARD; Martindale et al. 2016). We obtained dates ($n = 2,934$) from a representative sample of archaeological sites ($n = 384$) within the study area. All archaeological radiocarbon dates were calibrated using Calib 7.0.4 software, which uses the Intcal13 calibration curve (Reimer et al. 2013). We used the median probability of the 2-sigma age range of each radiocarbon date's calibration curve and grouped the resulting age estimates into 100-year bins, which were subsequently tallied and plotted as a histogram. Direct radiocarbon dates ($n = 180$) on macro- and microbotanical cultigen remains including maize (*Zea mays*), cucurbit (*Cucurbita* spp.), and common bean (*Phaseolus vulgaris*) from regional archaeological sites ($n = 55$) were compiled from published sources (Crawford, Smith, and Bowyer 1997; Hart and Scarry 1999; Hart, Thompson, and Brumbach 2003; Chilton 2006; Hart, Brumbach, and Lusteck 2007; Stinchcomb et al. 2011; Hart 2012; Hart et al. 2012; Hart and Lovis 2013; Gates St-Pierre and Thompson 2015). Bulk $\delta^{13}C$ values of pottery cooking residues from central New York ($n = 34$ sites, $n = 106$ samples; Hart 2012) were plotted chronologically to track changes in maize consumption and, by extension, to identify potential changes in the proportion of C_3 (closed-canopy forest taxa) and C_4 (maize; warm-season grasses) plants within regional ecosystems. Trends were further compared to the 7,600-year-long $\delta^{13}C_{calcite}$ record of a speleothem from McFails Cave in eastern New York, which provides a continuous record of dripwater-mediated calcite deposition in a closed-canopy, humid-temperate forested environment characteristic of the lower Great Lakes region (van Beynen, Schwarcz, and Ford 2004). We used temporal variations in the McFails Cave $\delta^{13}C_{calcite}$ record to infer changes in both the

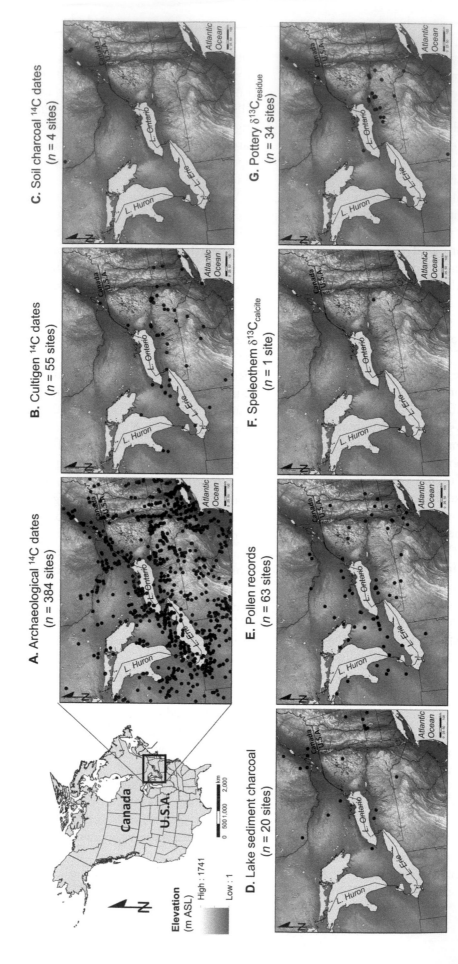

Figure 1. Location of the lower Great Lakes study area and proxy data sets analyzed in this study. (A) Archaeological ^{14}C dates ($n = 384$ sites). (B) Cultigen ^{14}C dates ($n = 55$ sites). (C) Soil charcoal ^{14}C dates ($n = 4$ sites). (D) Lake sediment charcoal ($n = 20$ sites). (E) Pollen records ($n = 63$ sites). (F) Cave speleothem $\delta^{13}C_{calcite}$ ($n = 1$ site). (G) Pottery $\delta^{13}C_{residue}$ ($n = 34$ sites).

proportion of C_3 versus C_4 plant biomass and the relative density and connectivity of the forest canopy. We hypothesized that enriched values of calcite $\delta^{13}C$ likely reflect forest contraction, reduced vegetation density, greater C_4 plant biomass, and expansion of open-canopy savanna, grassland, active cropland, and successional old-field communities, whereas depleted calcite $\delta^{13}C$ values suggest greater canopy density, forest expansion, and increased C_3 biomass (Ehleringer and Cerling 2002).

Holocene fire and vegetation dynamics were reconstructed using soil and lacustrine sediment charcoal and lacustrine pollen. We compiled direct radiocarbon dates ($n = 305$) on macroscopic soil charcoal fragments recovered from forest soils in the St. Lawrence River valley of southern Québec, Canada ($n = 4$ sites; Talon et al. 2005; Pilon and Payette 2015; Payette et al. 2016; Payette et al. 2017). Charcoal samples were derived from surface and mineral soil horizons primarily within fire-sensitive, old-growth sugar maple (*Acer saccharum*) stands and were selected for radiocarbon dating based on particle dry weight, botanical identification, spatial distribution of soil cores, and random selection. Lacustrine sediment charcoal data from selected coring sites were obtained from the National Oceanic and Atmospheric Administration's (2020) online paleoecology data archive. Because the charcoal data were summarized in multiple formats (e.g., raw count, volume, and influx), we selected sites ($n = 20$) in which the data were calculated as influx of charcoal particles ($mm^{-2}cm^{-2}yr^{-1}$). Pollen data from lake and wetland sites ($n = 63$) were downloaded from the Neotoma Paleoecology Database Williams et al. (2018). We computed pollen sums for major upland arboreal taxa ($n = 17$) including fir (*Abies*), maple (*Acer*), birch (*Betula*), hickory (*Carya*), chestnut (*Castanea*), beech (*Fagus*), ash (*Fraxinus*), butternut (*Juglans cinereal*), black walnut (*J. nigra*), hop-hornbeam/ironwood (*Ostrya/Carpinus*), spruce (*Picea*), pine (*Pinus*), aspen/poplar (*Populus*), oak (*Quercus*), basswood (*Tilia*), hemlock (*Tsuga*), and elm (*Ulmus*). Percentage of fire-tolerant pollen taxa was calculated by summing the pollen of upland arboreal taxa adapted to low-intensity ground fires (Thomas-Van Gundy and Nowacki 2013) including *Carya*, *Castanea*, *Juglans nigra*, *Pinus*, *Populus*, and *Quercus*; dividing by the total upland arboreal pollen sum; and multiplying by 100. Percentage herbaceous pollen was computed as the quotient of combined upland herbaceous pollen taxa (Amaranthaceae, *Ambrosia*, *Artemisia*, Asteroideae, *Plantago*, Poaceae, *Pteridium*, and *Rumex*) and the sum of combined upland arboreal and herbaceous pollen taxa, multiplied by 100. Age–depth models were created for each pollen profile using the Bacon v2.3.3 software package (Blaauw and Christen 2011), which uses the IntCal13 calibration curve (Reimer et al. 2013). We compared temporal trends in fire-tolerant and herbaceous pollen percentages averaged across all pollen sites ($n = 63$) at 400-year intervals to a subset of pollen records ($n = 22$) situated within a maximum 10 km linear distance of at least one archaeological site. Sites were selected using the Near tool of the ArcGIS 10.7 Spatial Analyst toolbox (Environmental Systems Research Institute 2019). Pollen percentage values were used to create percentage difference maps from individual pollen records using an inverse distance weighted interpolation algorithm in the Geostatistical Analyst toolbox of ArcGIS 10.7 (Environmental Systems Research Institute 2019).

Results

The archaeological and paleoecological proxy data suggest the existence of three phases of pre-European anthropogenic influence: (1) Paleoanthropocene Phase 1 (PA1; 5400–2500 BP); (2) Paleoanthropocene Phase 2 (PA2; 2500–1100 BP); and (3) Paleoanthropocene Phase 3 (PA3; 1100–200 BP). Temporal analysis of the CARD radiocarbon data indicates a general trend of increasing radiocarbon date frequency over the course of the Holocene (Figure 2A), with five stages evident: (1) a prolonged period from 10,000–5400 BP of very low (less than fifteen dates/century) radiocarbon date frequencies ("Pre-Paleoanthropocene") encompassing the Early Archaic through mid-Late Archaic periods; (2) a second phase of gradual increase following the mid-Late Archaic, plateauing at ~3900 BP (twenty-nine dates/century), with a subsequent decline terminating during the mid-Early Woodland period at 2600 BP (nine dates/century; Zone PA1); (3) a subsequent zone extending across the second half of the Early Woodland through the Middle Woodland periods, marked by a moderate increase in radiocarbon date frequencies by 1700 to 1600 BP (twenty-five dates/century), followed by a sudden rise to about forty dates per century near the

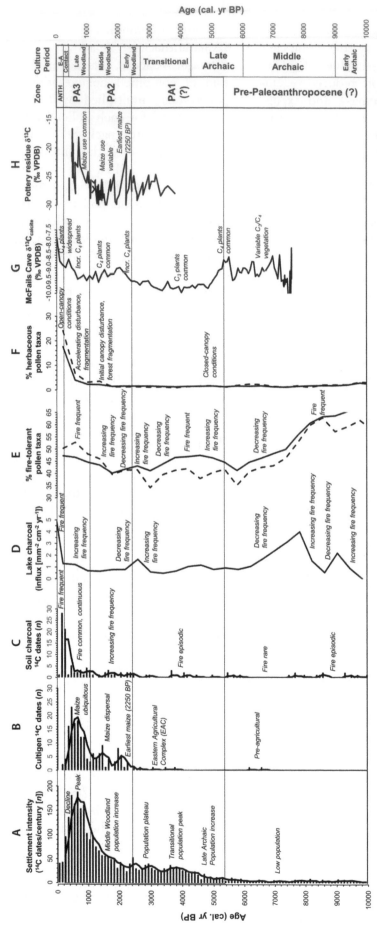

Figure 2. Chronology of proxy data sets. (A) Archaeological ^{14}C date frequency with 300-year moving average (solid black line). (B) Cultigen ^{14}C dates with 300-year moving average (solid black line). (C) Soil charcoal ^{14}C dates (400-year averaged values); solid black line = all pollen profiles; dashed black line = pollen profiles within 10 km of archaeological sites. (E) Percentage fire-tolerant upland arboreal pollen taxa (400-year averaged values); solid black line = all pollen profiles; dashed black line = pollen profiles within 10 km of archaeological sites. (F) Percentage herbaceous pollen taxa (400-year averaged values); solid black line = all pollen profiles; dashed black line = pollen profiles within 10 km of archaeological sites. (G) McFails Cave $\delta^{13}C_{calcite}$ values. (H) Pottery bulk $\delta^{13}C_{residue}$ values. Zone PA1 = Paleoanthropocene Phase 1 (5400–2500 BP); PA2 = Paleoanthropocene Phase 2 (2500–1100 BP); PA3 = Paleoanthropocene Phase 3 (1100–300 BP); ANTH = Anthropocene (300 BP–present). Major archaeological culture periods derived from Carr and Moeller (2015). E = European (220 BP–present).

top of the zone (Zone PA2); (4) a period centered on the Late Woodland and Contact Periods beginning ~1100 BP of very high radiocarbon date frequencies, peaking from 700 to 600 BP (ninety-one dates/century), followed by a sharp decline extending though the end of the period at 300 to 200 BP (forty-three dates/century; Zone PA3); and (5) a final zone of low radiocarbon date frequencies (about twenty dates/century) coinciding with European settlement and expansion (Anthropocene).

Aside from a small number ($n = 6$) of relatively early, temporally discontinuous cultigen radiocarbon dates from 6800 to 2700 BP, there is scant evidence of cultigens prior to circa 2300 BP ($M = 0.17$ dates/century; Figure 2B). From 2300 to 1600 BP, radiocarbon date frequencies on cultigens—primarily maize (*Zea mays*)—increase slightly ($M = 3.1$ dates/century), with a subsequent prominent rise around 900 BP, with overall high radiocarbon date frequencies persisting through the Late Woodland period until the Contact period circa 400 BP ($M = 16.3$ dates/century).

Low frequencies of macroscopic soil charcoal from 10,000 to 6000 BP ($M = 0.9$ dates/century; Figure 2C) suggest that infrequent, episodic, low-intensity fires were the norm. This prolonged period of limited charcoal deposition is followed by a subsequent phase (2500–1500 BP) of slightly higher date frequencies ($M = 2.5$ dates/century), culminating in a third period of very high charcoal deposition ($M = 14.2$ dates/century) from 1500 to 100 BP. A final period of very low charcoal deposition is evident in the last 100 years ($M = 1.0$ dates/century). Lacustrine sediment charcoal influx values for the Early Holocene indicate an initial period of variability but with an overall tendency toward increasing values from 9800 BP ($0.8\,\mathrm{mm}^{-2}\,\mathrm{cm}^{-2}\,\mathrm{yr}^{-1}$) to a premodern peak at 7800 BP ($3.6\,\mathrm{mm}^{-2}\,\mathrm{cm}^{-2}\,\mathrm{yr}^{-1}$; Figure 2D). The Middle Holocene is characterized by declining influx values ($<1.0\,\mathrm{mm}^{-2}\,\mathrm{cm}^{-2}\,\mathrm{yr}^{-1}$) extending through the Late Holocene, with slight rises centered at 2600 BP ($1.6\,\mathrm{mm}^{-2}\,\mathrm{cm}^{-2}\,\mathrm{yr}^{-1}$) and 600 BP ($1.7\,\mathrm{mm}^{-2}\,\mathrm{cm}^{-2}\,\mathrm{yr}^{-1}$). Influx values rise suddenly after 200 BP to attain values ($4.8\,\mathrm{mm}^{-2}\,\mathrm{cm}^{-2}\,\mathrm{yr}^{-1}$) not seen since the Early Holocene.

The frequency of fire-tolerant upland arboreal pollen taxa during the Early Holocene (10,000–8500 BP) is very high, with mean values ranging from 55.8 percent to 66.7 percent for all sites to 52.9 percent to 62.9 percent for settlement-proximal sites (Figure 2E).

Thereafter, a gradual decline in fire-tolerant pollen taxa ensues, reaching a low of 41.3 percent (combined sites) and 35.2 percent (settlement-proximal sites) by 5800 BP. A progressive, extended period of increase occurs after 5800 BP, culminating in peak values of 47.5 percent at 4600 BP (combined sites) and 42.1 percent at 4200 BP (settlement-proximal sites). This is followed by a gradual decline to the lowest levels of fire-tolerant taxa of the entire Holocene by 2800 BP at settlement-proximal sites (34.2 percent) and at combined sites by 1800 BP (40.1 percent). Percentages rise prominently at both pollen site types after 1800 BP, with values ranging from 43.4 percent (combined sites, 1400 BP) to 52.8 percent (settlement-proximal sites, 600 BP). Although a minor decline from 52.8 percent to 50.3 percent is noted from 600 to 200 BP at settlement-proximal sites, fire-tolerant pollen percentages continue to rise slightly at all sites during this time, from 46.6 percent to 47.2 percent. Herbaceous pollen percentages are characterized by both inertia and uniformly low percentages, with average values of 1.7 percent at both combined and settlement-proximal pollen sites from 10,000 to 1800 BP (Figure 2F). Values begin to rise at settlement-proximal sites after 1800 BP, from 1.4 percent to 3.6 percent by 1400 BP. This upward trend continues through the Late Woodland period, with a pre-Contact peak of 6.7 percent by 600 BP. A sharp rise in herbaceous pollen percentages is evident after 400 BP, with percentages attaining values of 17.6 percent (all sites) to 24.1 percent (settlement-proximal sites).

The McFails Cave $\delta^{13}C_{calcite}$ record (Figure 2G) indicates an initial period (7600–5300 BP) of relatively enriched $\delta^{13}C_{calcite}$ values (−9.0 to −8.5‰ Vienna Pee Dee Belemnite standard [VPDB]), followed by an extended period of relative depletion from 5000 to 3000 BP (−9.8 to −9.6‰ VPDB; Figure 2D). After 3000 BP, $\delta^{13}C_{calcite}$ values become progressively enriched in a step-like manner, with prominent increases at 600 BP, attaining peak Holocene values (−7.8‰ VPDB) at the present. The $\delta^{13}C$ pottery residue record shows a similar trend of progressive enrichment (Figure 2H). A short-term episode of higher $\delta^{13}C$ values is evident from 2500 to 2000 BP, peaking at −21.0‰ at 2250 BP, followed by a more prominent, unidirectional increase after 1500 BP, attaining a maximum value of −16.6‰ by 500 BP during the Late Woodland period.

The Late Holocene (post-2600 BP) distribution of fire-tolerant (Figure 3) and herbaceous (Figure 4) pollen taxa shows considerable spatiotemporal

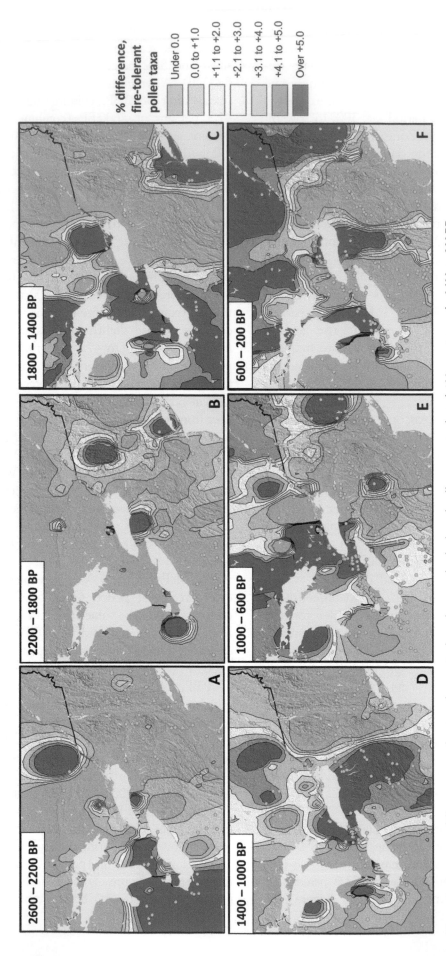

Figure 3. Percentage difference maps showing changes in fire-tolerant upland arboreal pollen taxa at selected 400-year intervals, 2600 to 200 BP.

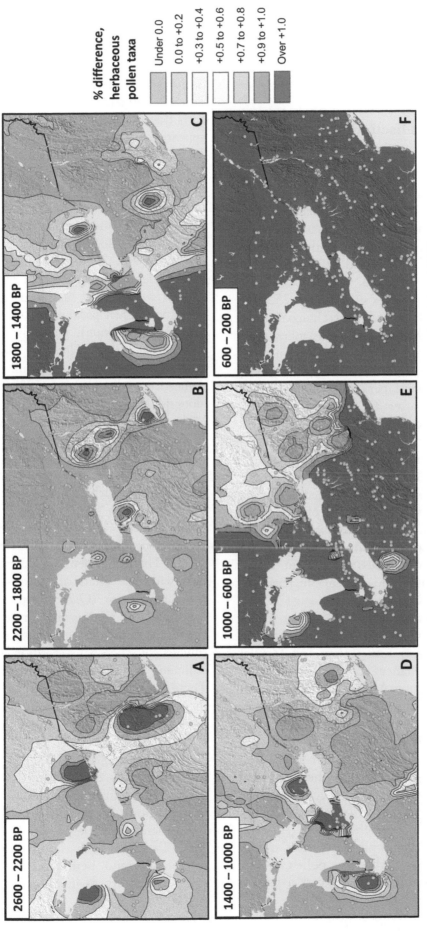

Figure 4. Percentage difference maps showing temporal changes in herbaceous pollen taxa at selected 400-year intervals, 2600 to 200 BP.

variability. A recurring pattern of expansion from the southwestern Great Lakes basin interrupted by periods of localization and decline is evident prior to 1400 BP, however, followed by northward and eastward shifts by 200 BP.

Interpretation and Discussion

Proxy data indicate important changes in prehistoric demography, subsistence patterns, and paleoenvironmental conditions within the lower Great Lakes region over the course of the Holocene. The most visible trends include (1) increasing human population densities after 5400 BP, with prominent episodes of growth at 2400 BP and 1200 BP; (2) the sudden appearance of cultigens after 2500 BP; (3) continuous soil charcoal deposition after 2500 BP; (4) a decrease in lacustrine sediment charcoal influx through the Early and Middle Holocene, followed by increases at 1000 BP and 200 BP; (5) declining frequency of fire-tolerant pollen taxa through the Early Holocene, with subsequent periods of increase from 5800 to 3500 BP and after 1800 BP; (6) very low abundances of herbaceous pollen taxa through most of the Holocene, with notable gains only after 1800 BP; and (7) increasingly enriched $\delta^{13}C_{calcite}$ values after 3000 BP. Although the number, timing, and amplitude of major excursions are different for the various proxies—likely a function of differences in proxy sensitivity to climatic, environmental, and anthropogenic forcing mechanisms—episodes of synchronous change across proxies are evident beginning at 5400 BP with the initial increase in archaeological radiocarbon dates. The cause of this acceleration is obscure, because there are no major cultural innovations evident in the regional archaeological record during this time (Carr and Moeller 2015). An increased reliance on nuts and seeds is indicated during the Late Archaic, however, with the routine processing of wild cereals and tree nuts suggested by the first appearance of pestles, mortars, and other milling equipment (Lavin 2013).

A second phase of inferred human population growth occurs after 2400 BP during the Early and Middle Woodland periods and is accompanied by coeval increases in cultigen and soil charcoal radiocarbon date frequencies, enriched $\delta^{13}C_{calcite}$ and $\delta^{13}C_{residue}$ values, and somewhat later rises in the pollen of fire-tolerant and herbaceous pollen taxa by 1800 BP. Together, these changes suggest a growth in the regional human population accompanied by increased frequency of fires, a progressive turnover in forest composition toward more fire-adapted species, and a transition to more prevalent open-canopy conditions. Our interpretation of an intensified anthropogenic influence postdating 2400 BP agrees well with the regional archaeological record. The regular appearance of cucurbit (*Cucurbita* spp.) after 3800 BP and, most important, maize (*Zea mays*) after 2250 BP in the regional archaeological record suggests that nonintensive forms of incipient horticulture might have been practiced during the Early and Middle Woodland periods, likely centered on river and stream floodplains (Crawford, Smith, and Bowyer 1997). Archaeobotanical remains of indigenous, ruderal, floodplain seed plants such as marshelder (*Iva annua*), sunflower (*Helianthus annuus*), goosefoot (*Chenopodium berlandieri*), erect knotweed (*Polygonum erectum*), maygrass (*Phalaris caroliniana*), and little barley (*Hordeum pusillum*)—collectively known as the Eastern Agricultural Complex—were domesticated beginning ~5000 BP and were grown alongside maize beginning ~2200 BP in parts of the Midcontinent and eastern United States (Mueller et al. 2017). For example, archaeobotanical remains from the Chenango River floodplain in south-central New York indicate the presence of goosefoot, marshelder, false buckwheat (*Polygonum scandens*), tick trefoil (*Desmodium* spp.), and giant ragweed (*Ambrosia trifida*) during Transitional and Early Woodland occupations from 2900 to 2150 BP (Asch Sidell 2008). These cultigens were stratigraphically associated with the charcoal of fire-tolerant nut-bearing taxa such as oak (*Quercus* spp.) and hickory (*Carya* spp.), and the timing of their collective presence in the region coincides with the period of low, but continuous soil charcoal deposition and accelerated lake sediment charcoal influx observed in our proxies. These trends lend support to our hypothesis of increased human influence on forested ecosystems at local, landscape, and regional scales by 2400 BP. Localized anthropogenic disturbance related to incipient horticulture after 3000 BP might have favored the expansion of open-canopy ecosystems (e.g., grassland, savanna, successional old fields) on some of the region's floodplains, a process suggested by the slight increase in the pollen of herbaceous taxa after 1800 BP. Changes in the geographic distribution of fire-tolerant and herbaceous pollen taxa suggest a recurrent pattern of incursion from the southwest at

2600 to 2200 BP and again at 1800 to 1400 BP. The relationship between this spatiotemporal pattern and the regional introduction of cultigens is unclear but might reflect the combined effects of human migration, cultigen adoption, and climate change. We note that the pattern observed in our maps is generally consistent with Crawford, Smith, and Bowyer's (1997) hypothesis of maize incursion into the Great Lakes region from the south over a period of several centuries during the Middle Woodland period.

After 1800 BP, a trend of continuous increase is evident across most proxies, suggesting major growth of human paleopopulation, the growing utilization of maize as a dietary component, the development of more active fire regimes, expansion of fire-tolerant forest taxa, and a greater prevalence of open-canopy habitats dominated by herbaceous C_4-pathway plants. By 1100 BP, a considerable portion of the region was likely characterized by novel, open-canopy, fire-prone, humanized landscapes inhabited by Native American groups practicing relatively intensive forms of maize-based agriculture. Such an ecosystem had no prior Holocene analogue. Land survey records of the eighteenth and nineteenth centuries CE indicate that this landscape contrasted discrete vegetation communities partitioned primarily on the basis of fire tolerance and secondarily according to soil productivity, both exhibiting strong, statistically significant spatial correlation with the distribution of Native American agricultural settlements (Fulton and Yansa 2019, 2020). Land use practices of Late Woodland agriculturists, including the habitual use of fire as a tool for preserving open conditions, clearing of extensive tracts of forest for village construction and establishment of cropland, thinning of stands from firewood collection and acquisition of construction materials, and the deliberate maintenance of economically favored forest taxa (Abrams and Nowacki 2008), likely drove the changes observed in the regional proxy records.

Interestingly, the spatiotemporal dynamics of terminal Native American anthropogenesis closely follow the emergence and decline of major Iroquoian groups documented in the ethnohistorical record. From 1000 to 600 BP, areas of the greatest net increase in fire-tolerant pollen taxa occur around the northern and western end of Lake Ontario where Northern Iroquoian cultural patterns are hypothesized to have originated and coalesced during the Late Woodland period (Birch 2015). This area was inhabited historically by agricultural Wendat, Petun, and Attiwandaron tribal groups whose land use practices might have contributed to the increase in fire-tolerant taxa observed in local pollen profiles. By 600 to 200 BP, this area had experienced a net loss in fire-tolerant pollen taxa. Although the main area of increase shifted to the north and east during this time—largely driven by increases in the pollen of fire-tolerant *Pinus* associated with Little Ice Age (~500–150 BP) cooling and drying (e.g., Muñoz and Gajewski 2010; Clifford and Booth 2015)—a secondary, isolated center of growth appears south of Lake Ontario in the Finger Lakes region of New York. Beginning circa 350 BP, Ontario Iroquoians were dispersed by competing groups from New York, particularly the Seneca Nation, whose ancestral territory was centered in the Finger Lakes (Engelbrecht 2003). To our knowledge, this study is the first of its kind to recognize a regional-scale spatiotemporal shift in pollen-inferred vegetation composition associated with indigenous sociopolitical developments.

Based on our findings, we believe that the Paleoanthropocene concept has obvious utility as a means to holistically conceptualize past environmental and cultural change in a relatively simple manner, particularly as an increasing number of paleoecological studies are focused on synthesizing rich compendia of archaeological, paleoecological, and paleoclimate data sets (Muñoz and Gajewski 2010; Bird et al. 2017; Abrams and Nowacki 2019; Gajewski et al. 2019; Oswald et al. 2020), each with its own classification scheme that is often difficult to reconcile with others. For example, archaeologists have developed complex culture-historic taxonomic classifications based on the remains of past societies' material culture (Hart and Brumbach 2003). Yet behaviors traditionally inferred from such classification systems have been challenged as a result of methodological advances (e.g., pottery residue analysis) that have redefined the pattern, process, and temporal sequence of important cultural transitions (e.g., the development of maize agriculture), rendering these categories largely irrelevant in a synthetic paleoecological research context. Adopting the Paleoanthropocene as standard terminology would allow disciplinary nonspecialists to evaluate interdisciplinary data sets for perceived anthropogenic influences, examine temporal trends, and refine existing subregional and regional chronologies using novel or high-resolution proxies without the need for overly complex or redundant terminology. We hope that the present analysis serves as a step toward this goal.

Conclusion

Paleoecological signals of a regional Paleoanthropocene event initiated by Native American land use impacts, and predating European settlement of the lower Great Lakes region, are ephemeral beginning ~5400 BP (PA1; 5400–2500 BP), with the first major increase in archaeological radiocarbon date frequencies, suggesting regional population growth. Proxy indicators of important changes in settlement intensity, land use, fire regimes, and forest composition and structure become increasingly visible after 2500 BP (PA2; 2500–1000 BP). These trends intensify and culminate by the Late Woodland period (PA3; 1100–300 BP). Spatiotemporal analysis of pollen indicators of prehistoric anthropogenesis indicates a time-transgressive pattern, generally from southwest to northeast, that was likely complicated by environmental, climatic, and cultural factors requiring further detailed study. Despite these confounding factors, an overall strengthening of response over three distinct temporal phases is suggested. We hypothesize that the regional introduction of maize between 2500 and 2000 BP—and the subsequent modification of human settlement patterns and subsistence regimes engendered by its prolonged incorporation into indigenous hunter-gatherer diets—was the primary catalyst for an emerging anthropogenic influence on the landscape beginning at this time. Notable human land use impacts on regional landscapes are clearly reflected in the paleoecological record of forest species composition, vegetation structure, and fire regimes, changes that cannot be explained by paleoclimate change alone. Agricultural intensification, sedentism, and settlement nucleation, all well documented in the regional archaeological record, are the most plausible factors accounting for the marked excursions seen in the region's proxy records.

The Paleoanthropocene encompassed temporally and processually distinct stages in the development of anthropogenic ecosystems that were largely modulated by cultural innovations that altered existing subsistence regimes and settlement systems and, ultimately, vegetation and fire regimes. At first, forest species composition was shifted toward greater frequencies of fire-tolerant forest taxa after the initial appearance of maize. Subsequently, forest structure was profoundly modified during a period of accelerating human population growth attendant on the region-wide adoption of maize-based agriculture. Our tripartite classification scheme is a necessary first attempt to delineate the broad outlines of the chronological development of the anthropogenic landscapes of the lower Great Lakes region. As such, this conceptual model is highly recommended for wider use in regional paleoecological studies but must remain open to substantial temporal revision pending the compilation and analysis of additional high-resolution archaeological and paleoecological data sets. The spatiotemporal dynamics of a multiphase Paleoanthropocene model should furthermore be tested using proxy data from other regions of eastern North America to refine its likely time-transgressive nature. This would provide a critical contextual perspective to the long-term trajectory of human–environment interactions over the course of the Holocene.

References

Abrams, M. D., and G. J. Nowacki. 2008. Native Americans as active and passive promoters of mast and fruit trees in the eastern U.S.A. *The Holocene* 18 (7):1123–37.

Abrams, M. D., and G. J. Nowacki. 2019. Global change impacts on forest and fire dynamics using paleoecology and tree census data for eastern North America. *Annals of Forest Science* 76 (1):1–23. doi: 10.1007/s13595-018-0790-y.

Asch Sidell, N. A. 2008. The impact of maize-based agriculture on prehistoric plant communities in the Northeast. In *Current Northeast paleoethnobotany II*, ed. J. P. Hart, 29–51. Albany: New York State Museum.

Birch, J. 2015. Current research on the historical development of Northern Iroquoian societies. *Journal of Archaeological Research* 23 (3):263–323.

Bird, B. W., J. J. Wilson, W. P. Gilhooly, B. A. Steinman, and L. Stamps. 2017. Midcontinental Native American population dynamics and late Holocene hydroclimate extremes. *Scientific Reports* 7:1–12. doi: 10.1038/srep41628.

Blaauw, M., and J. A. Christen. 2011. Flexible paleoclimate age-depth models using an autoregressive gamma process. *Bayesian Analysis* 6:457–74.

Braun, E. L. 2001. *Deciduous forests of eastern North America*. Caldwell, ID: The Blackburn Press.

Carr, K. W., and R. W. Moeller. 2015. *First Pennsylvanians: The archaeology of Native Americans in Pennsylvania*. Harrisburg: Pennsylvania Historical and Museum Commission.

Carruthers, J. 2019. The Anthropocene. *South African Journal of Science* 115 (7–8). doi: 10.17159/sajs.2019/6428.

Chilton, E. S. 1999. Mobile farmers of pre-Contact southern New England: The archaeological and ethnohistoric evidence. In *Current Northeast paleoethnobotany*, ed. J. Hart, 157–76. Albany: New York State Museum.

Chilton, E. S. 2006. The origin and spread of Maize (*Zea mays*) in New England. In *Histories of maize: Multidisciplinary approaches to the prehistory, linguistics, biogeography, domestication, and evolution of maize*, ed. B. F. Benz, R. M. Tykot, and J. E. Staller, 539–47. Cambridge, UK: Elsevier.

Chin, A., T. Beach, S. Luzzadder-Beach, and W. Solecki. 2017. Challenges of the "Anthropocene." *Anthropocene* 20:1–3.

Clifford, M. J., and R. K. Booth. 2015. Late-Holocene drought and fire drove a widespread change in forest community composition in eastern North America. *The Holocene* 25 (7):1102–10.

Crawford, G. W., D. G. Smith, and V. E. Bowyer. 1997. Dating the entry of corn (*Zea mays*) into the lower Great Lakes Region. *American Antiquity* 62 (1):112–19.

Ehleringer, J. R., and T. E. Cerling. 2002. C$_3$ and C$_4$ photosynthesis. In *Encyclopedia of global environmental change*, ed. H. A. Mooney and J. G. Canadell, 186–90. Chichester, UK: Wiley.

Engelbrecht, W. 2003. *Iroquoia: The development of a native world*. Syracuse, NY: Syracuse University Press.

Environmental Systems Research Institute. 2019. ArcGIS (version 10.7). Redlands, CA: ESRI.

Foley, S. F., D. Gronenborn, M. O. Andreae, J. W. Kadereit, J. Esper, D. Scholz, U. Pöschl, D. E. Jacob, B. R. Schöne, R. Schreg, et al. 2013. The Paleoanthropocene—The beginnings of anthropogenic environmental change. *Anthropocene* 3:83–88.

Fulton, A. E., II, and C. H. Yansa. 2019. Native American land use impacts on a temperate forested ecosystem, West Central New York State. *Annals of the American Association of Geographers* 109 (6):1706–28. doi: 10.1080/24694452.2019.1587281.

Fulton, A. E., II, and C. H. Yansa. 2020. Characterization of Native American vegetation disturbance in the forests of central New York State, USA during the late 18th century CE. *Vegetation History and Archaeobotany* 29 (2):259–75. doi: 10.1007/s00334-019-00741-6.

Gajewski, K., B. Kriesche, M. A. Chaput, R. Kulik, and V. Schmidt. 2019. Human–vegetation interactions during the Holocene in North America. *Vegetation History and Archaeobotany* 28 (6):635–47. doi: 10.1007/s00334-019-00721-w.

Gates St-Pierre, C., and R. G. Thompson. 2015. Phytolith evidence for the early presence of maize in southern Québec. *American Antiquity* 80 (2):408–15.

Guerrero-Gatica, M., E. Aliste, and J. A. Simonetti. 2019. Shifting gears for the use of the shifting baseline syndrome in ecological restoration. *Sustainability* 11:1–12.

Hart, J. P. 2012. Pottery wall thinning as a consequence of increased maize processing: A case study from central New York. *Journal of Archaeological Science* 39 (11):3470–74.

Hart, J. P., and H. J. Brumbach. 2003. The death of Owasco. *American Antiquity* 68 (4):737–52.

Hart, J. P., H. J. Brumbach, and R. Lusteck. 2007. Extending the Phytolith evidence for early maize (Zea mays ssp. mays) and squash (Cucurbita sp.) in Central New York. *American Antiquity* 72:563–583.

Hart, J. P., and W. A. Lovis. 2013. Reevaluating what we know about the histories of maize in northeastern North America: A review of current evidence. *Journal of Archaeological Research* 21 (2):175–216.

Hart, J. P., W. A. Lovis, R. J. Jeske, and J. D. Richards. 2012. The potential of bulk δ^{13}C on encrusted cooking residues as independent evidence for regional maize histories. *American Antiquity* 77 (2):315–25.

Hart, J. P., and C. M. Scarry. 1999. The age of common beans (*Phaseolus vulgaris*) in the northeastern United States. *American Antiquity* 64 (4):653–58.

Hart, J. P., R. G. Thompson, and H. J. Brumbach. 2003. Phytolith evidence for Early Maize (*Zea mays*) in the northern Finger Lakes region of New York. *American Antiquity* 68 (4):619–40.

Johnson, C. N., A. Balmford, B. W. Brook, J. C. Buettel, M. Galetti, L. Guangchun, and J. M. Wilmshurst. Biodiversity losses and conservation responses in the Anthropocene. *Science* 356:270–75.

Lavin, L. 2013. *Connecticut's Indigenous peoples: What archaeology, history, and oral traditions teach us about their communities and cultures*. New Haven, CT: Yale University Press.

Martindale, A., R. Morlan, M. Betts, M. Blake, K. Gajewski, M. Chaput, A. Mason, and P. Vermeersch. 2016. Canadian Archaeological Radiocarbon Database (CARD 2.1). https://www.canadianarchaeology.ca.

Mueller, N. G., G. J. Fritz, P. Patton, S. Carmody, and E. T. Horton. 2017. Growing the lost crops of eastern North America's original agricultural system. *Nature Plants* 3:1–5. doi: 10.1038/nplants.2017.92.

Muñoz, S. E., and K. Gajewski. 2010. Distinguishing prehistoric human influence on late-Holocene forests in southern Ontario, Canada. *The Holocene* 20:967–81.

National Oceanic and Atmospheric Administration (NOAA). 2020. National Centers for Environmental Information: Paleo data search. Accessed January 12, 2020. https://www.ncdc.noaa.gov/paleo-search/

Oswald, W. W., D. R. Foster, B. N. Shuman, E. S. Chilton, D. L. Doucette, and D. L. Duranleau. 2020. Conservation implications of limited Native American impacts in pre-contact New England. *Nature Sustainability* 3 (8):241–46. doi: 10.1038/s41893-019-0466-0.

Payette, S., M. Frégeau, P.-L. Couillard, V. Pilon, and J. Laflamme. 2016. Long-term fire history of maple (*Acer*) forest sites in the central St. Lawrence Lowland, Quebec. *Canadian Journal of Forest Research* 46 (6):822–31.

Payette, S., V. Pilon, P.-L. Couillard, and J. Laflamme. 2017. Fire history of Appalachian forests of the Lower St. Lawrence Region (southern Quebec). *Forests* 8 (4):120. doi: 10.3390/f8040120.

Pilon, V., and S. Payette. 2015. Sugar maple (*Acer saccharum*) forests at their northern distribution limit are recurrently impacted by fire. *Canadian Journal of Forest Research* 45 (4):452–62.

Reimer, P. J., E. Bard, A. Bayliss, J. W. Beck, P. G. Blackwell, C. B. Ramsey, C. E. Buck, H. Cheng, R. L. Edwards, M. Friedrich, et al. 2013. IntCal13 and Marine13 radiocarbon age calibration curves 0–50,000 cal BP. *Radiocarbon* 55 (4):1869–87. doi: 10.2458/azu_js_rc.55.16947.

Ruddiman, W. F. 2003. The Anthropogenic greenhouse era began thousands of years ago. *Climatic Change* 61 (3):261–93.

Stephens, L., D. Fuller, N. Boivin, T. Rick, N. Gauthier, A. Kay, B. Marwick, C. G. Armstrong, C. M. Barton, T. Denham, et al. 2019. Archaeological assessment reveals Earth's early transformation through land use. *Science* 365 (6456):897–902. doi: 10.1126/science.aax1192.

Stinchcomb, G. E., T. C. Messner, S. G. Driese, L. C. Nordt, and R. M. Stewart. 2011. Pre-colonial (A.D. 1100–1600) sedimentation related to prehistoric maize agriculture and climate change in eastern North America. *Geology* 39 (4):363–66.

Talon, B., S. Payette, L. Filion, and A. Delwaide. 2005. Reconstruction of the long-term fire history of an old-growth deciduous forest in Southern Québec, Canada, from charred wood in mineral soils. *Quaternary Research* 64 (1):36–43. doi: 10.1016/j.yqres.2005.03.003.

Tanner, H. H. 1987. *Atlas of Great Lakes Indian history*. Norman: University of Oklahoma Press.

Thomas-Van Gundy, M. A., and G. J. Nowacki. 2013. The use of witness trees as pyro-indicators for mapping past fire conditions. *Forest Ecology and Management* 304:333–44.

van Beynen, P. E., H. P. Schwarcz, and D. C. Ford. 2004. Holocene climatic variation recorded in a speleothem from McFail's Cave, New York. *Journal of Cave and Karst Studies* 66:20–27.

Waters, C. N., J. Zalasiewicz, C. Summerhayes, A. D. Barnosky, C. Poirier, A. Ga Uszka, A. Cearreta, M. Edgeworth, E. C. Ellis, M. Ellis, et al. 2016. The Anthropocene is functionally and stratigraphically distinct from the Holocene. *Science* 351 (6269). doi: 10.1126/science.aad2622.

Waters, C. N., J. Zalasiewicz, C. Summerhayes, I. J. Fairchild, N. L. Rose, N. J. Loader, W. Shotyk, A. Cearreta, M. J. Head, J. P. M. Syvitski, et al. 2018. Global boundary stratotype section and point (GSSP) for the Anthropocene series: Where and how to look for potential candidates. *Earth-Science Reviews* 178:379–429.

Williams, J. W., E. G. Grimm, J. Blois, D. F. Charles, E. Davis, S. J. Goring, R. Graham, A. J. Smith, M. Anderson, J. Arroyo-Cabrales, A. C. Ashworth, J. L. Betancourt, B. W. Bills, R. K. Booth, P. Buckland, B. Curry, T. Giesecke, S. Hausmann, S. T. Jackson, C. Latorre, J. Nichols, T. Purdum, R. E. Roth, M. Stryker, and H. Takahara. 2018. The Neotoma Paleoecology Database: A multi-proxy, international community-curated data resource. *Quaternary Research* 89:156–77.

Zalasiewicz, J. C., N., Waters, M. Williams, A. D. Barnosky, A. Cearreta, P. Crutzen, E. Ellis, M. A. Ellis, I. J. Fairchild, J. Grinevald, et al. 2015. When did the Anthropocene begin? A mid-twentieth century boundary level is straigraphically optimal. *Quaternary International* 383:196–203. doi: 10.1016/j.quaint.2014.11.045.

Identifying a Pre-Columbian Anthropocene in California

Anna Klimaszewski-Patterson, ⓘD Christopher T. Morgan, ⓘD and Scott Mensing ⓘD

The beginning of the Anthropocene, a proposed geological epoch denoting human-caused changes to Earth's systems, and what metrics signify its onset is currently under debate. Proposed initiation points range from the beginning of the Atomic Age to the Industrial Revolution to the adoption of agriculture in the early Holocene. Most of the debate centers on the effects of modern industrially oriented technological and economic development. The effects of preindustrial and preagricultural populations on Earth's systems are less commonly evaluated. Because the utility of the Anthropocene concept is to denote measurable impacts of human activity on Earth's systems, we argue that focusing on an exact date or single event ignores time-transgressive, spatially variable processes of anthropogenic ecosystem engineering. We argue instead for a flexible, anthropologically and ecologically informed conceptualization of the Anthropocene—one that recognizes spatial, temporal, and scalar variability in the effects of humans on Earth systems. We present evidence in support of an ecologically informed pre-Columbian Anthropocene in California using a meta-analysis of sedimentological, palynological, and archaeological data sets from California mountains. We argue that use of fire for resource management by pre-Columbian populations was sufficiently frequent and extensive enough to result in widescale anthropogenic modification of California's biota and that an Anthropocene therefore began in California by at least 650 years ago, centuries before the arrival of Europeans. Recognizing a pre-Columbian Anthropocene in California constructively conceptualizes a marker for human economic–ecological intensification processes that could be more meaningful for policy, resource management, and research than focusing on any single historical event. *Key Words: Anthropocene, California, fire, Native Americans, Paleoecology.*

The term Anthropocene refers to a proposed new geological epoch triggered by human-caused changes to the Earth's systems (Zalasiewicz et al. 2017). Defining the Anthropocene is contentious largely because there is no single agreed-on metric to mark its onset. Geoscientists, for example, use radionucelotides to argue that the Anthropocene began AD 1945 at the start of the Atomic Age (Zalasiewicz et al. 2015) or use evidence for the rapid accumulation of geochemicals such as carbon, nitrogen, and phosphorous in the lithosphere and atmosphere to argue that this new epoch began ca. AD 1800, at the start of the Industrial Revolution (Raupach and Canadell 2010; Steffen et al. 2011). Social scientists tend to argue for deeper time, pointing to long-term archaeological and ecological records for humans acting as land managers via agriculture and pastoralism (Ruddiman 2005; Ellis et al. 2016; L. Stephens et al. 2019). The Anthropocene's starting point consequently depends to a large degree on the characteristic used to identify it.

Because the utility of the Anthropocene concept is to denote human activity affecting Earth's systems to a measurable extent, we argue that focusing on the exact date for the beginning of the epoch discounts the profound effects of ecosystem engineering by nonindustrial or nonagrarian people (Ellis et al. 2016). Further, it ignores the fact that these events likely resulted from long-term processes of human economic and technological development in an ecological context, rather than any one event or set of events. Geographers recognize that space and place matter—that events do not unfold synchronously in all places and that impactful events can be time-transgressive. We consequently operate from the assumption that a flexible, anthropologically and ecologically informed conceptualization of the Anthropocene—one that recognizes spatial, temporal, and scalar variability in humankind's effect on biota—is a more useful approach than stricter geological conceptualizations (Ruddiman 2018). In this article we present evidence in support of an ecologically informed Anthropocene construct, demonstrating widespread human impacts via ecosystem engineering in California spanning at least 650 years. California represents an interesting challenge in identifying anthropogenic impacts because its pre-Columbian inhabitants were hunter-gatherers who used tools such as fire to manage natural resources.

Humans have potentially transformed their physical environments and affected ecosystem function and biodiversity patterns for thousands of years (Ruddiman 2005). Although fire occurs naturally, evidence suggests that it has also been used by humans for at least 400,000 years (Roebroeks and Vill 2011; Walker et al. 2016). By using fire, humans can alter fuels, creating large-scale successional shifts that affect ecosystem structure and buffer the effect of climate on fire size (Whitlock et al. 2010; Bliege Bird et al. 2012).

A recent global analysis found that hunter-gatherer groups living in wildfire-prone areas were more likely to use fire as a tool for ecological management precisely because the vegetation was adapted to frequent fires (Coughlan, Magi, and Derr 2018). The ethnographic record provides overwhelming evidence that low-intensity surface fires were ubiquitously set across California's predominately Mediterranean landscape to remove underbrush, clear travel corridors, facilitate hunting and drive game, promote seed germination, and generally increase yields of natural resources. Through historic photographs and written accounts, European explorers documented "park-like settings" free of underbrush with mixed-age and patchy tree cohorts (Muir 1894; Jepson 1923; Lewis 1973; Parker 2002). Fire scientists generally attribute such conditions to frequent low-intensity fires that consume shrubs and seedlings without damaging mature trees. These fire conditions are consistent with both frequent unsuppressed lightning-set fires and the pattern of Native Californian–set fires documented in ethnographies (M. K. Anderson and Moratto 1996; Codding and Bird 2013). The challenge in identifying a signal of anthropogenic fire becomes one of disentangling the effects of human ignitions from climatic ones. If evidence for anthropogenic fire can be consistently and unambiguously identified, then the argument can be made for the onset of an ecologically informed Anthropocene.

We use meta-analyses of available paleoecological and archaeological data sets to identify when a widespread signal of fire use affiliated with rapidly growing populations occurs within the ecological context of the mountains of California. We identify climate- versus human-driven environmental change with unambiguous pollen taxa that exhibit contrasting life history traits. Paleoecological studies based on pollen and sedimentary charcoal provide a biosphere record stored in a geologic context of sedimentary records. Radiocarbon dating of macrobotanical remains from these sedimentary records provides stratigraphically constrained timing of ecological change. Archaeological data sets provide insights into prehistoric population densities, traditional resource management practices, and natural resource use. Intensification-caused vegetation changes can be identified through reconstructions of vegetation history and timing of land use and occupancy (Munoz and Gajewski 2010; Crawford et al. 2015; Fulton and Yansa 2019).

We use the persistence of fire-adapted taxa and their associated systems during cool, wet periods when fire-sensitive taxa should dominate, coupled with evidence for increased human population densities and resource use, as an identifiable signal of a human-modified landscape. We challenge the assumption that the mountainous forests encountered by the first Europeans to reach California were created primarily by climatic conditions. We consequently argue that the process of large-scale anthropogenic modification of California's biota, and hence an Anthropocene, is identifiable in California at least 650 years ago, centuries before the arrival of Europeans.

Climate–Fire–Vegetation Dynamics

A warmer, drier climate is associated with more frequent fires due to less soil moisture and an increase in combustible fuels. These conditions lead to more open forest canopies and a greater prevalence of fire-tolerant and shade-intolerant taxa (Mohr, Whitlock, and Skinner 2000; Bond and Keeley 2005; Fites-Kaufman et al. 2007; Swetnam et al. 2009) such as oaks (Quercus spp.), pines (Pinus spp.), grasses (Poaceae), and bracken fern (Pteridium aquilinum). Fire-tolerant and open-canopy taxa provide important food sources for aboriginal Californians (acorns, pine nuts, seeds, shoots). Less frequent fire is typically associated with cooler, wetter climate, although large, high-severity fires might still occur. Greater availability of soil moisture and decreased fire frequency allow for the succession of

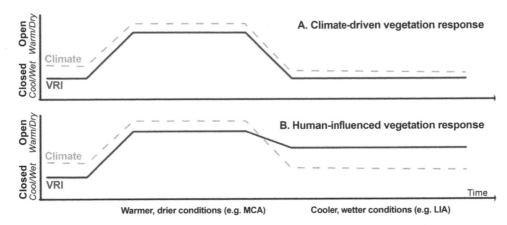

Figure 1. Conceptual model of forest canopy (open/closed) in relation to climate–fire–vegetation dynamics over time. (A) Expected forest canopy (solid dark gray line) driven by climate-only factors. (B) Expected forest canopy influenced by human behavior. Gray dashed line represents climate. VRI = vegetation response index; MCA = Medieval Climate Anomaly; LIA = Little Ice Age.

shade-tolerant and fire-sensitive taxa such as fir (*Abies*), incense cedar (*Calocedrus decurrens*), sedges (Cyperaceae), Douglas fir (*Pseudotsuga menziesii*), mountain hemlock (*Tsuga mertensiana*), and tanoak (*Notholithocarpus*; Swetnam 1993; Dale et al. 2001; Lenihan et al. 2003)

In the last 1,300 years, California has experienced multicentury periods of both warmer and drier and cooler and wetter conditions against which to test drivers of vegetation change. The Medieval Climate Anomaly (MCA), from ca. AD 900 to 1200, is regionally identified as a predominantly warmer and drier period with increased occurrences of regional fires. The Little Ice Age (LIA), from ca. AD 1200 to 1850, is identified broadly as a cooler and wetter period (Bowerman and Clark 2011), with a decrease in regional fires and a corresponding decrease in background charcoal accumulation (Swetnam 1993; Mohr, Whitlock, and Skinner 2000; Swetnam et al. 2009).

If climate and lightning-caused fires are the driving factors of vegetation response we expect more open-canopy and fire-tolerant taxa during warmer and drier climatic periods and more closed-canopy and fire-sensitive taxa during cooler and wetter climatic periods (Figure 1A). If human behavior and resource intensification are the driving factor, then we expect the persistence of open-canopy and fire-tolerant taxa even during suboptimal cooler and wetter periods (Figure 1B).

Paleoecological Evidence

Fires were ubiquitous throughout pre-Columbian California, with an estimated 6 to 16 percent (2–5 million ha) of nondesert land burned annually (Martin and Sapsis 1992). Pre-Columbian median fire return intervals are estimated in the range of five to ten years in oak woodlands and mixed conifer forests, twenty to thirty years in shrublands, and three years in grasslands, potentially a result of both lightning and human ignitions (S. L. Stephens, Martin, and Clinton 2007). Over the last century, vast portions of California have experienced no fire activity. Although large, severe fires occurred in the past, the paleoenvironmental record demonstrates a decrease in fire intensity over the last several hundred years, especially during the LIA (Swetnam 1993; Marlon et al. 2012).

Much paleoecological work in California has focused on vegetation change as a proxy for climate change (Davis et al. 1985; Edlund 1996; Mensing 2001; Barnosky et al. 2016). Only a handful of studies have focused on pre-Columbian anthropogenic influences with an emphasis on identifying human impact rather than climate proxies (e.g., Gassaway 2009; R. S. Anderson and Stillick 2013; Lightfoot et al. 2013; Ejarque et al. 2015). Even fewer data sets have been made publicly available.

We performed a meta-analysis of pollen-based paleoecological studies at five sites (R. S. Anderson and Carpenter 1991; Crawford et al. 2015; Klimaszewski-Patterson 2016; Klimaszewski-Patterson and Mensing 2016) from the Sierra Nevada and Klamath Ranges in California (Figure 2) to investigate an identifiable Anthropocene. Of the forty-five identified late- Holocene paleoecological sites reported throughout California's mountainous regions since AD 1980 (e.g., West 1982; Davis and

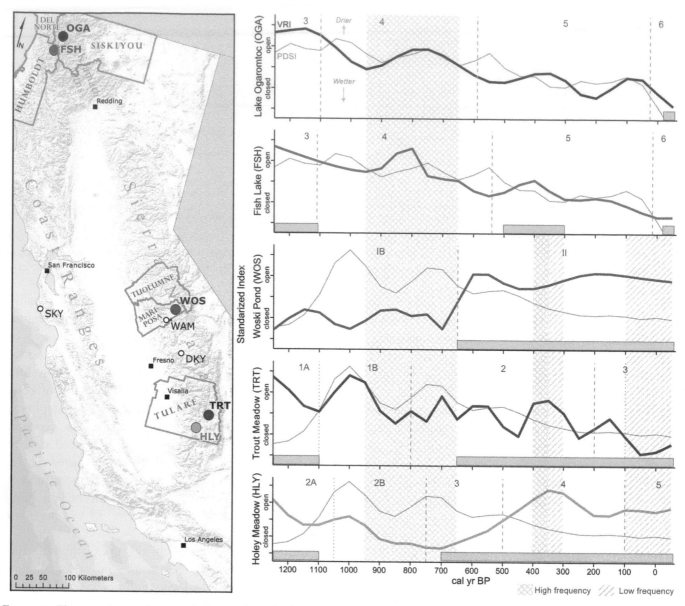

Figure 2. Thirteen thousand years of climate (gray line), vegetation (colored lines), and anthropogenic interpretations (by original authors; orange bars) from subcentennial sites in California mountains. Colored lines (matching colored circles, left) represent standardized vegetation response index of shade-tolerant and fire-sensitive (closed-canopy) versus shade-intolerant and fire-adapted (open-canopy) taxa. Gray line represents inferred climate from the North American Drought Atlas (E. R. Cook et al. 1999; E. R. Cook et al. 2004; E. R. Cook et al. 2008; Herweijer et al. 2007). Higher values indicate drier conditions; lower values indicate wetter conditions. Hashed shading represents regional fire scar studies (Swetnam, Touchan, and Baisan 1991; Swetnam and Anderson 2008). Double hash indicates high frequency of regional fires; single hash represents a low frequency. White dots represent other paleoecological sites discussed but not used in analysis. SKY = Skylark Pond; WAM = Wawona Meadow; DKY = Dinkey Meadow.

Moratto 1988; Edlund 1996; Wahl 2002; Wanket 2002), only these five published palynological studies meet all of the following criteria: (1) conducted at a subcentennial resolution, (2) for at least the last 1,300 years, (3) where identifiable life-history distinctions in vegetation changes could be identified, and (4) the data set was available for reanalysis.

At each of the five sites we calculated a vegetation response index (VRI; term coined in Klimaszewski-Patterson and Mensing 2016) between fire-sensitive or shade-tolerant (FSST) and fire-adapted or shade-intolerant (FASI) taxa. We used each study's identified nonambiguous FSST (e.g., *Abies, Pseudotsuga*) and FASI (e.g., *Quercus*, Poaceae) taxa to calculate VRI as (FSST − FASI)/(FSST + FASI). A negative VRI indicates more fire-adapted or shade-intolerant taxa and a positive VRI more fire-sensitive or shade-tolerant taxa. We then compared VRI changes

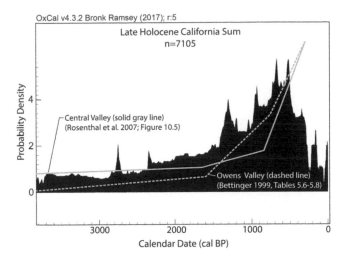

Figure 3. Population trajectories in Late Holocene aboriginal California. Radiocarbon summed probability distributions generated with normalized data from the CARD 2.1 data set (Martindale et al. 2016), using OxCal 4.3 (Bronk Ramsey 2009) and the Intcal 13 calibration curve (Reimer et al. 2013).

against a fifty-year smooth-spline climate reconstruction based on local, annually resolved tree-ring data (E. R. Cook et al. 1999; E. R. Cook et al. 2004; E. R. Cook et al. 2008; Herweijer et al. 2007). Our results demonstrate signals of anthropogenically altered landscapes by increases in fire-adapted or shade-intolerant taxa (e.g., *Quercus*) during cool, wet periods (LIA). Each site indicates periods of anthropogenic influence on the landscape, with Sierran sites Woski Pond (WOS; R. S. Anderson and Carpenter 1991) and Holey Meadow (HLY; Klimaszewski-Patterson and Mensing 2016) showing the strongest signals starting ca. 700 cal BP (Figure 2). Lake Ogaromtoc (OGA) and Fish Lake (FSH; Crawford et al. 2015) in the Klamath Mountains show the weakest signals, likely due to wetter conditions associated with their proximity to the Pacific Ocean. All five sites show anomalous changes in VRI compared to climatic expectations starting no later than 650 cal BP (AD 1300), 250 years prior to European contact.

Archaeological Evidence

Although there is little direct archaeological evidence for landscape-scale burning and other land management practices in California (see Cuthrell 2013; Lightfoot et al. 2013; M. K. Anderson and Rosenthal 2015), the ethnographic literature is replete with descriptions of managing wild plants through coppicing, pruning, or whipping (Fowler

2008); deliberately planting and tending tobacco (Todt 2007); irrigating wild plants to increase seed yield (Lawton et al. 1976); and especially deliberately setting fires (Jordan 2003). The benefits of the latter include marking territorial boundaries, increasing forage for game, clearing land for travel, producing better basketry material, and increasing annual yields of wild grasses, berries, and acorns (Lewis 1973; M. K. Anderson 2005). Fire was so important to aboriginal Californians that its use has been linked to the development of the complex sociopolitical organization (Bean and Lawton 1973).

When these deliberate fire practices began is unclear, but it seems reasonable to assume that their frequency increased in proportion to human population growth. Estimates for the number of people living in pre-Columbian California vary from 133,000 to 1.52 million (Merriam 1905; Kroeber 1925; Powers 1976), with most estimates hovering around 300,000 (Baumhoff 1963; S. F. Cook 1976). Although several researchers argue that these estimates are too low, because they are typically based on postcontact data (Preston 2002), many agree that population density in aboriginal California, a land of hunter-gatherers, rivaled or exceeded that of contemporaneous agriculturalists living in adjacent regions (Baumhoff 1963).

Several lines of archaeological evidence point to Late Holocene population growth in California. In Owens Valley, Bettinger (1999) identified exponential population increase after 1700 cal BP by tracking the frequency of time-sensitive projectile points. Rosenthal, White, and Sutton (2007) identified a similar pattern in their evaluation of the frequency of dated archaeological components in California's Central Valley. Arguably the most powerful indicator of population growth is radiocarbon data, the idea being that the probability of finding and dating cultural carbon increases with the frequency by which people generated cultural carbon in the past (Williams 2012). A summed probability distribution (SPD) generated from 7,105 radiocarbon dates (Martindale et al. 2016) from Late Holocene California (i.e., <3500 BP) generates a curve similar to those of Bettinger (1999) and Rosenthal, White, and Sutton (2007), with a marked increase in radiocarbon frequencies after 1500 cal BP and a peak from 800 to 500 cal BP (Figure 3).

In northwestern California, near OGA and FSH, ethnographic groups are the Yurok, Karok, and

Table 1. Demographic data for California ethnolinguistic groups at time of European contact

Ethnolinguistic group	Population[a]	Population density[a] (people per 100 km[b])	Associated paleoecological records
Yurok	2,500	131.00	Lake Ogaromtoc, Fish Lake
Karok	1,500	46.90	
Shasta	2,925	25.00	
Tubatulabal	1,000	17.20	Holey Meadow, Trout Meadow
Miwok	1,212	24.54	Woski Pond
California	91,364[b]	58.8 (avg.)	

[a]Data from Binford (2001, table 5.01).
[b]This number does not include all California ethnolinguistic groups.

Shasta, all of whom subsisted chiefly on anadromous fish (Tushingham and Bettinger 2013). Ethnographic sources indicate that these were among the largest and most densely packed pre-Columbian populations in California (Table 1). The radiocarbon SPD for counties containing or immediately adjacent to OGA and FSH suggest pronounced population growth after 1500 cal BP (and decline after 500 cal BP), but the VRI for both sites more closely correlates with climate than demography (Figures 2 and 4), possibly because as fishing-oriented groups they had far less incentive to manage terrestrial resources than groups in central and southern California. Lighting and managing the low-intensity fires required by indigenous traditional resource and environmental management (TREM) practices was also likely far less tenable in the moist coniferous forests of California's Pacific Northwest.

Like all Sierra Nevada groups, the Miwok, affiliated with the Yosemite region and the WOS site, were accomplished hunters and prodigious acorn storers, subsisting through the winter largely on acorns stored in granaries (Barrett and Gifford 1933; Bates 1983). Both hunting and an acorn-centered diet would have benefited from landscape management by fire. The Miwok also had some of the highest population densities in the Sierra Nevada at the time of European contact (Table 1), with SPD-derived population levels in the area peaking along with VRI ca. 500 cal BP. Population declines thereafter, inverse to the more open-canopy biota (likely the result of anthropogenic burning).

The Tubatulabal and Foothill Yokuts, of the HLY and Trout Meadow sites, had a subsistence economy similar to that of the Miwok (Harvey 2019). At the time of European contact, the Tubatulabal had among the lowest population density in California outside the Mojave Desert and the lowest in the Sierra Nevada (Binford 2001). Their SPD, however,

suggests rapid population growth after 1000 cal BP and a peak in population ca. 600 cal BP (Figure 4). HLY indicates the most pronounced VRI signal, suggesting strong anthropogenic effects and the greatest deviation from climate from 350 to 200 cal BP. Trout Meadow demonstrates a more equivocal signal, with punctuated periods of inferred anthropogenic effects at ca. 650 to 550 cal BP, 450 to 300 cal BP, and 250 to 150 cal BP.

Discussion

The process of large-scale anthropogenic modification of California's biota, and hence an Anthropocene, is identifiable in the Sierra Nevada of California at least 650 years ago. In the Sierra Nevada, paleoecological records indicate a shift toward more open-canopy forests after 700 cal BP, when climate is modeled to have encouraged proliferation of closed-canopy coniferous forests. Although populations appear to decline after peaking ca. 500 cal BP, VRI data from at least two Sierran sites suggest that aboriginal burning continued unabated or even increased after this period of time. We speculate that growing populations between 1500 and 500 cal BP (roughly contemporaneous with the MCA) faced a new challenge during the LIA with the onset of pronounced climatic change and increased precipitation. Climate during the LIA ought to have favored the spread of *Abies* over *Quercus*. Nowhere would this have been more pronounced than in the Sierra Nevada. Given this predicament, perhaps aboriginal burning increased after 700 cal BP as a deliberate attempt to maintain MCA-type environments with abundant oak woodlands. We suggest that aboriginal inhabitants of the Sierra Nevada not only maintained their habitat through burning at scales recognizable in the

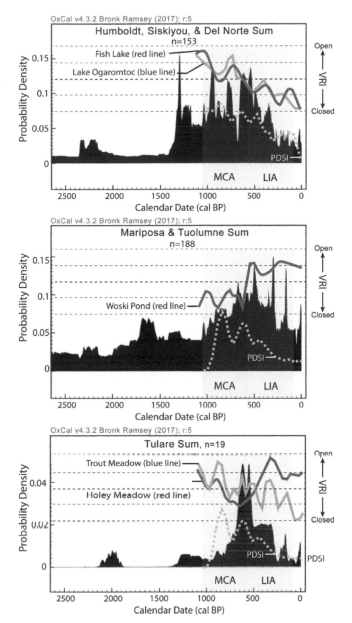

Figure 4. Radiocarbon (SPDs) and affiliated paleoclimatic and paleoenvironmental data for reviewed paleoecological and archaeological data sets. SPDs generated using normalized data from the CARD 2.1 data set (Martindale et al. 2016), using OxCal 4.3 (Bronk Ramsey 2009) and the Intcal 13 calibration curve (Reimer et al. 2013). MCA = Medieval Climate Anomaly; LIA = Little Ice Age; VRI = vegetation response index; PDSI = Palmer Drought Severity Index; SPD = summed probability distribution.

paleoecological record but they actively did so in response to altered paleoclimatic circumstances.

The argument for aboriginal burning maintaining oak woodlands is further supported by paleolandscape models reconstructing the last 1,100 years of forest succession at Holey Meadow (Klimaszewski-Patterson et al. 2018) and Trout Meadow (Klimaszewski-Patterson and Mensing 2020).

Both studies analyzed various scenarios of climate-driven and human-influenced fire regimes to explore which models of forest succession best approximated the observed palynological record. Both studies indicate that the most likely fire regime to explain the empirical record is through the addition of TREM-like low-intensity surface fires. In short, climate alone could not reproduce the observed pollen record with statistical relevance, whereas the addition of TREM fires best approximated the observed record, with statistical significance. This ecosystem engineering is most noticeable at both sites during the LIA.

We recognize that of our five sites, the southern and central Sierra Nevada sites by far show the strongest support for a pre-Columbian ecological–Anthropocene signal, but other studies from the Sierra Nevada and coastal California also support identifiable pre-Columbian human impacts. At Wawona Meadow (R. S. Anderson and Stillick 2013) in Yosemite National Park (Figure 2), the authors indicate evidence for frequent surface fires starting ca. 650 cal BP (AD 1300). At Dinkey Meadow (Davis and Moratto 1988) in Sierra National Forest the published reconstructed pollen diagram shows a sharp decline in *Abies* throughout the LIA while *Quercus* remains steady. At Skylark Pond (Cowart and Byrne 2013) near Point Año Nuevo (Figure 2), the authors interpret evidence for nonclimatic fires starting no later than ca. 550 cal BP, possibly sooner. Additional well-dated, high-resolution palynological data sets are necessary throughout California to further refine onset of an Anthropocene. Given the spatial distribution of reanalyzed paleoecological sites from central-east to northwest California, we think a minimum date of 650 cal BP is reasonable, especially given that mountainous regions in the interior are thought to have had lower populations and less intense land use.

Aside from the Anthropocene, epochal boundaries have been retrospectively assigned based on extinction events identified from an incomplete fossil record. It remains to be seen whether the amalgamation of modern-day extinctions and radionucleotide markers is sufficiently concentrated in the geologic record to represent a distinguishable signal at the scale of geologic time. What is more important to the current debate is that the Anthropocene concept has greater value than a strictly defined point in time because the idea can help change thinking about the depth and extent of human impacts in the present day. Humans have clearly altered the biota and chemistry of Earth, and traces of these changes over thousands of years are evident in the geologic record. Traditional geologic methods never attempted to consider human impacts, and those impacts are time-transgressive and cumulative. People had an impact on Earth's systems prior to the Industrial Revolution or the Atomic Age. Adopting a flexible, ecologically informed approach to the question of an Anthropocene, as we have in pre-Columbian California, can have important consequences for how we conceptualize modern ecology and the long-term role of human manipulation of Earth systems. In the case of California, this way of thinking can potentially transform modern land use management and fire policy at a time when new approaches to fire ecology are desperately needed.

Conclusions

We propose that an Anthropocene began in present-day California no later than AD 1300 (650 cal BP), well before Europeans entered the western Americas. We base this conclusion on identifiable and quantifiable signals from independent paleoecological and archaeological evidence. We offer that intentionally set fires resulting in low-intensity burning by pre-Columbian Native Californian populations were responsible for altering forest structure in California such that open oak/mixed conifer forests persisted into and through the LIA instead of climatically expected closed coniferous forests. As per the Anthropocene concept, human modification of the environment is observable in the geologic sedimentary record and was spread throughout California by this date.

We argue that the term Anthropocene is more constructively conceptualized as a marker of the intensification of human economic–ecological processes rather than any one historical event. Identifying a signal of preindustrial, or even preagricultural, human behavior in the geologic record can be difficult, but the effort can be accomplished by employing a multidisciplinary and multiproxy approach focused on human behavior.

The onset of an Anthropocene will consequently not be contemporaneous in all locations, much in the same way that there are variations in timing or effect of climatic events (e.g., LIA, MCA); however, this does not make the construct any less valuable.

Thinking of the term Anthropocene as a flexible concept expressed in the multitude of geophysical changes that people have caused across the globe, through the lenses of society, politics, and economics, is particularly useful in informing policy and management of natural resources. The debate will continue as to defining a formal geologic boundary. Given that such temporal boundaries are typically defined from an incomplete fossil record spanning millions of years, we argue that the consideration of the full record of human activity is an integral part of this debate.

Funding

This study was supported by the National Science Foundation (NSF GSS 0964261, "Did Native Americans Significantly Alter Forest Structure in California? A Paleoecological Reconstruction of Vegetation and Fire History from Two Different Ecosystems," and NSF GSS 1740918, "Fire, Vegetation Change, and Human Settlement").

ORCID

Anna Klimaszewski-Patterson (iD) http://orcid.org/0000-0001-7765-8802
Christopher T. Morgan (iD) http://orcid.org/0000-0002-1219-7993
Scott Mensing (iD) http://orcid.org/0000-0003-4302-112X

References

Anderson, M. K. 2005. *Tending the wild: Native American knowledge and the management of California's natural resources*. Berkeley: University of California Press.

Anderson, M. K., and M. J. Moratto. 1996. Native American land-use practices and ecological impacts. In *Sierra Nevada Ecosystem Project: Final report to Congress*, 187–206. Davis: University of California, Centers for Water and Wildland Resources. https://pubs.usgs.gov/dds/dds-43/VOL_II/VII_TOC.PDF

Anderson, M. K., and J. Rosenthal. 2015. An ethnobiological approach to reconstructing indigenous fire regimes in the foothill chaparral of the western Sierra Nevada. *Journal of Ethnobiology* 35 (1):4–36.

Anderson, R. S., and S. L. Carpenter. 1991. Vegetation change in Yosemite Valley, Yosemite National Park, California, during the Protohistoric period. *Madro* 38 (1):1–13.

Anderson, R. S., and R. D. Stillick. 2013. 800 years of vegetation change, fire and human settlement in the Sierra Nevada of California, USA. *The Holocene* 23 (6):823–32.

Barnosky, A. D., E. L. Lindsey, N. A. Villavicencio, E. Bostelmann, E. A. Hadly, J. Wanket, and C. R. Marshall. 2016. Variable impact of late-Quaternary megafaunal extinction in causing ecological state shifts in North and South America. *Proceedings of the National Academy of Sciences of the United States of America* 113 (4):856–61. doi: 10.1073/pnas.1505295112.

Barrett, S. A., and E. W. Gifford. 1933. Miwok material culture: Indian life of the Yosemite region. *Bulletin of the Public Museum of the City of Milwaukee* 2 (4):117–376.

Bates, C. D. 1983. Acorn storehouses of the Yosemite Miwok. *Masterkey* 57 (1):19–27.

Baumhoff, M. A. 1963. Ecological determinants of Aboriginal California populations. *University of California Publications in American Archaeology and Ethnology* 49 (2):155–236.

Bean, L. J., and H. W. Lawton. 1973. Some explanations for the rise of cultural complexity in Native California with comments on proto-agriculture and agriculture. In *Patterns of Indian burning in California: Ecology and ethnohistory*, ed. L. J. Bean, v–xlvii. Ramona, CA: Ballena Press.

Bettinger, R. L. 1999. What happened in the Medithermal. In *Models for the millennium: Great Basin anthropology today*, ed. C. Beck, 62–74. Salt Lake City: The University of Utah Press.

Binford, L. R. 2001. *Constructing frames of reference: An analytical method for archaeological theory building using hunter-gatherer and environmental data sets*. Berkeley: University of California Press.

Bliege Bird, R., B. F. Codding, P. G. Kauhanen, and D. W. Bird. 2012. Aboriginal hunting buffers climate-driven fire-size variability in Australia's *Spinifex* grasslands. *Proceedings of the National Academy of Sciences of the United States of America* 109 (26):10287–92. doi: 10.1073/pnas.1204585109.

Bond, W. J., and J. E. Keeley. 2005. Fire as a global "herbivore": The ecology and evolution of flammable ecosystems. *Trends in Ecology & Evolution* 20 (7):387–94. doi: 10.1016/j.tree.2005.04.025.

Bowerman, N. D., and D. H. Clark. 2011. Holocene glaciation of the central Sierra Nevada, California. *Quaternary Science Reviews* 30 (9–10):1067–85.

Bronk Ramsey, C. 2009. Development of the radiocarbon calibration program OxCal. *Radiocarbon* 43 (2A):355–63.

Codding, B. F., and D. W. Bird. 2013. Forward: A global perspective on traditional burning in California. *California Archaeology* 5 (2):199–208.

Cook, E. R., D. M. Meko, D. W. Stahle, and M. K. Cleaveland. 1999. Drought reconstructions for the continental United States. *Journal of Climate* 12 (4):1145–63.

Cook, E. R., C. A. Woodhouse, C. M. Eakin, D. M. Meko, and D. W. Stahle. 2004. Long-term aridity changes in the Western United States. *Science* 306 (5698):1015–18. doi: 10.1126/science.1102586.

Cook, E. R., C. A. Woodhouse, C. M. Eakin, D. M. Meko, and D. W. Stahle. 2008. *North American summer PDSI reconstructions, Version 2a*. Boulder, CO: IGBP PAGES/World Data Center for Paleoclimatology.

Cook, S. F. 1976. *The population of the California Indians, 1769–1970*. Berkeley: University of California Press.

Coughlan, M. R., B. Magi, and K. Derr. 2018. A global analysis of hunter-gatherers, broadcast fire use, and lightning-fire-prone landscapes. *Fire* 1 (3):41. https://doi.org/10.3390/fire1030041

Cowart, A., and R. Byrne. 2013. A paleolimnological record of Late Holocene vegetation change from the Central California coast. *California Archaeology* 5 (2):337–52.

Crawford, J. N., S. A. Mensing, F. K. Lake, and S. R. Zimmerman. 2015. Late Holocene fire and vegetation reconstruction from the western Klamath Mountains, California, USA: A multi-disciplinary approach for examining potential human land-use impacts. *The Holocene* 25 (8):1341–57.

Cuthrell, R. Q. 2013. Archaeobotanical evidence for indigenous burning practices and foodways at CA-SMA-113. *California Archaeology* 5 (2):265–90.

Dale, V. H., L. A. Joyce, S. McNulty, R. P. Neilson, M. P. Ayres, M. D. Flannigan, P. J. Hanson, L. C. Irland, A. E. Lugo, C. J. Peterson, et al. 2001. Climate change and forest disturbances. *BioScience* 51 (9):723–34. doi: 10.1641/0006-3568(2001)051[0723:CCAFD]2.0.CO;2.

Davis, O. K., R. S. Anderson, P. L. Fall, M. K. O'Rourke, and R. S. Thompson. 1985. Palynological evidence for early Holocene aridity in the Southern Sierra Nevada, California. *Quaternary Research* 24 (3):322–32.

Davis, O. K., and M. J. Moratto. 1988. Evidence for a warm dry early Holocene in the western Sierra Nevada of California: Pollen and plant macrofossil analysis of Dinkey and Exchequer Meadows. *Madroño* 35 (2):132–49.

Edlund, E. G. 1996. Late Quaternary environmental history of montane forests of the Sierra Nevada, California. PhD dissertation, University of California, Berkeley.

Ejarque, A., R. S. Anderson, A. R. Simms, and B. J. Gentry. 2015. Prehistoric fires and the shaping of colonial transported landscapes in Southern California: A paleoenvironmental study at Dune Pond, Santa Barbara County. *Quaternary Science Reviews* 112:181–96.

Ellis, E., M. Maslin, N. Boivin, and A. Bauer. 2016. Involve social scientists in defining the Anthropocene. *Nature* 540 (7632):192–93.

Fites-Kaufman, J. A., P. Rundel, N. L. Stephenson, and D. A. Weixelman. 2007. Montane and subalpine vegetation of the Sierra Nevada and Cascade Ranges. In *Terrestrial vegetation of California*, ed. M. G. Barbour, T. Keeler-Wolf, and A. A. Schoenherr, 456–501. Berkeley: University of California Press.

Fowler, C. S. 2008. Historical perspectives on Timbisha Shoshone land management practices, Death Valley, California. In *Case studies in environmental archaeology*, ed. E. J. Reitz, C. M. Scarry, and S. J. Scudder, 43–57. New York: Springer Science + Business Media.

Fulton, A. E., and C. H. Yansa. 2019. Native American land-use impacts on a temperate forested ecosystem, West Central New York State. *Annals of the American Association of Geographers* 109 (6):1706–28.

Gassaway, L. 2009. Native American fire patterns In Yosemite Valley: Archaeology, dendrochronology, subsistence, and culture change in the Sierra Nevada. *SCA Proceedings* 22:1–19.

Harvey, D. C. 2019. Habitat distribution, settlement systems, and territorial maintenance in the southern Sierra Nevada, California. PhD dissertation, University of Nevada, Reno.

Herweijer, C., R. Seager, E. R. Cook, and J. Emile-Geay. 2007. North American droughts of the last millennium from a gridded network of tree-ring data. *Journal of Climate* 20 (7):1353–76.

Jepson, W. 1923. *The trees of California*. 2nd ed. Berkeley, CA: Sather Gate Bookshop.

Jordan, T. A. 2003. Ecological and cultural contributions of controlled fire use by Native Californians: A survey of literature. *American Indian Culture and Research Journal* 27 (1):77–90.

Klimaszewski-Patterson, A. 2016. Climate, fire, and Native Americans: Identifying forces of paleoenvironmental change in the southern Sierra Nevada, California. PhD dissertation, University of Nevada, Reno.

Klimaszewski-Patterson, A., and S. A. Mensing. 2016. Multi-disciplinary approach to identifying Native American impacts on Late Holocene forest dynamics in the southern Sierra Nevada range, California, USA. *Anthropocene* 15:37–48.

Klimaszewski-Patterson, A., and S. A. Mensing. 2020. Paleoecological and paleolandscape modeling support for pre-Columbian burning by Native Americans in the Golden Trout Wilderness Area, California, USA. *Landscape Ecology* 35:2659–78. doi: 10.1007/s10980-020-01081-x.

Klimaszewski-Patterson, A., P. J. Weisberg, S. A. Mensing, and R. M. Scheller. 2018. Using paleolandscape modeling to investigate the impact of Native American–set fires on pre-Columbian forests in the southern Sierra Nevada, California, USA. *Annals of the American Association of Geographers* 108 (6):1635–20.

Kroeber, A. L. 1925. *Handbook of the Indians of California*. Washington, DC: Bureau of American Ethnology.

Lawton, H. W., P. J. Wilke, M. Dedecker, and W. M. Mason. 1976. Agriculture among the Paiute of Owens Valley. *Journal of California Anthropology* 3 (1):13–50.

Lenihan, J. M., R. Drapek, D. Bachelet, and R. P. Neilson. 2003. Climate change effects on vegetation distribution, carbon, and fire in California. *Ecological Applications* 13 (6):1667–81.

Lewis, H. T. 1973. *Patterns of Indian burning in California: Ecology and ethnohistory*. Ramona, CA: Ballena Press.

Lightfoot, K. G., R. Q. Cuthrell, C. M. Boone, R. Byrne, A. S. Chavez, L. Collins, A. Cowart, R. R. Evett, P. V. A. Fine, D. Gifford-Gonzalez, et al. 2013. Anthropogenic burning on the Central California coast in Late Holocene and early historical times: Findings, implications, and future directions. *California Archaeology* 5 (2):371–90.

Marlon, J. R., P. J. Bartlein, D. G. Gavin, C. J. Long, R. S. Anderson, C. E. Briles, K. J. Brown, D. Colombaroli, D. J. Hallett, M. J. Power, et al. 2012. Long-term perspective on wildfires in the western USA. *Proceedings of the National Academy of Sciences of the United States of America* 109 (9):E535–43. doi: 10.1073/pnas.1112839109.

Martin, R. E., and D. B. Sapsis. 1992. Fires as agents of biodiversity: Pyrodiversity promotes biodiversity. In *Proceedings of the Symposium on Biodiversity of Northwestern California*, ed. H. M. Kerner, 150–57. Berkeley, CA: Wildland Resources Centre, University of California.

Martindale, A., R. Morlan, M. Betts, M. Blake, K. Gajewski, M. Chaput, A. Mason, and P. Vermeersch. 2016. Canadian Archaeological Radiocarbon Database (CARD 2.1). Accessed November 4, 2019. https://www.canadianarchaeology.ca/

Mensing, S. A. 2001. Late-glacial and early Holocene vegetation and climate change near Owens Lake, eastern California. *Quaternary Research* 55 (1):57–65.

Merriam, C. H. 1905. The Indian population of California. *American Anthropologist* 7 (4):594–606.

Mohr, J. A., C. Whitlock, and C. N. Skinner. 2000. Postglacial vegetation and fire history, eastern Klamath Mountains, California, USA. *The Holocene* 10 (5):587–601.

Muir, J. 1894. *The mountains of California*. Berkeley, CA: Ten Speed Press.

Munoz, S. E., and K. Gajewski. 2010. Distinguishing prehistoric human influence on late-Holocene forests in southern Ontario, Canada. *Holocene* 20 (6):967–81.

Parker, A. J. 2002. Fire in Sierra Nevada forests: Evaluating the ecological impact of burning by Native Americans. In *Fire, native peoples, and the natural landscape*, ed. T. R. Vale, 233–67. Washington, DC: Island.

Powers, S. 1976. *Tribes of California*. Berkeley, CA: University of California Press.

Preston, W. L. 2002. Portents of plague from California's protohistoric period. *Ethnohistory* 49 (1):69–121.

Raupach, M. R., and J. G. Canadell. 2010. Carbon and the Anthropocene. *Current Opinion in Environmental Sustainability* 2 (4):210–18.

Reimer, P. J., E. Bard, A. Bayliss, J. W. Beck, P. G. Blackwell, M. Bronk, P. M. Grootes, T. P. Guilderson, H. Haflidason, I. Hajdas, et al. 2013. IntCal 13 and Marine 13 radiocarbon age calibration curves 0–50,000 years cal BP. *Radiocarbon* 55:1869–87.

Roebroeks, W., and P. Vill. 2011. On the earliest evidence for habitual use of fire in Europe. *Proceedings of the National Academy of Sciences of the United States of America* 108 (13):5209–14. doi: 10.1073/pnas.1018116108.

Rosenthal, J. S., G. G. White, and M. Q. Sutton. 2007. The Central Valley: A view from the catbird's seat. In *California prehistory: Colonization, culture and complexity*, ed. T. L. Jones and K. A. Klar, 147–64. Lanham, MD: AltaMira Press.

Ruddiman, W. F. 2005. *Plows, plagues, and petroleum: How humans took control of climate*. Princeton, NJ: Princeton University Press.

Ruddiman, W. F. 2018. Three flaws in defining a formal "Anthropocene." *Progress in Physical Geography* 42 (4):451–61.

Steffen, W., J. Grinevald, P. Crutzen, and J. McNeill. 2011. The Anthropocene: Conceptual and historical perspectives. *Philosophical Transactions of the Royal Society A* 369:842–67. https://doi.org/10.1098/rsta.2010.0327

Stephens, L., D. Fuller, N. Boivin, T. Rick, N. Gauthier, A. Kay, B. Marwick, C. Geralda, D. Armstrong, C. M. Barton, et al. 2019. Archaeological assessment reveals Earth's early transformation through land use. *Science* 365 (6456):897–902. doi: 10.1126/science.aax1192.

Stephens, S. L., R. E. Martin, and N. E. Clinton. 2007. Prehistoric fire area and emissions from California's forests, woodlands, shrublands, and grasslands. *Forest Ecology and Management* 251 (3):205–16.

Swetnam, T. W. 1993. Fire history and climate change in giant sequoia groves. *Science* 262 (5135):885–89. doi: 10.1126/science.262.5135.885.

Swetnam, T. W., and R. S. Anderson. 2008. Fire climatology in the western United States: Introduction to special issue. *International Journal of Wildland Fire* 17:1–7.

Swetnam, T. W., C. H. Baisan, A. C. Caprio, P. M. Brown, R. Touchan, R. S. Anderson, and D. J. Hallett. 2009. Multi-millennial fire history of the Giant Forest, Sequoia National Park, California, USA. *Fire Ecology* 5 (3):120–50. doi: 10.4996/fireecology.0503120.

Swetnam, T. W., R. Touchan, and C. H. Baisan. 1991. Giant sequoia fire history in Mariposa Grove, Yosemite National Park. *Yosemite Centennial Symposium Proceeding-Natural Areas and Yosemite: Prospects for the Future.*

Todt, D. L. 2007. Upriver and downriver: A gradient of tobacco intensification along the Klamath River, California and Oregon. *Journal of California and Great Basin Anthropology* 27 (1):1–14.

Tushingham, S., and R. L. Bettinger. 2013. Why foragers choose acorns before salmon: Storage, mobility, and risk in aboriginal California. *Journal of Anthropological Archaeology* 32 (4):527–37.

Wahl, E. R. 2002. Paleoecology and testing of paleoclimate hypotheses in Southern California during the Holocene. PhD dissertation, University of California, Berkeley.

Walker, M. J., D. Anesin, D. E. Angelucci, A. Avilés-Fernández, F. Berna, A. T. Buitrago-López, Y. Fernández-Jalvo, M. Haber-Uriarte, A. López-Jiménez, M. López-Martínez, et al. 2016. Combustion at the late Early Pleistocene site of Cueva Negra del Estrecho del Río Quípar (Murcia, Spain). *Antiquity* 90 (351):571–89.

Wanket, J. A. 2002. Late Quaternary vegetation and climate of the Klamath Mountains. PhD dissertation, University of California, Berkeley.

West, G. J. 1982. Pollen analysis of sediments from Tule Lake: A record of Holocene vegetation/climatic changes in the Mendocino National Forest, California Proceedings, Symposium of Holocene Climate and Archeology of California's Coast and Desert San Diego, California, Feb. 1982 Special Publication, Anthropology Department, San Diego State University.

Whitlock, C., P. E. Higuera, D. B. McWethy, and C. E. Briles. 2010. Paleoecological perspectives on fire ecology: Revisiting the fire-regime concept. *The Open Ecology Journal* 3 (2):6–23.

Williams, A. N. 2012. The use of summed radiocarbon probability distributions in archaeology: A review of methods. *Journal of Archaeological Science* 39 (3):578–89.

Zalasiewicz, J., C. N. Waters, C. P. Summerhayes, A. P. Wolfe, A. D. Barnosky, A. Cearreta, P. Crutzen, E. Ellis, I. J. Fairchild, A. Gałuszka, et al. 2017. The Working Group on the Anthropocene: Summary of evidence and interim recommendations. *Anthropocene* 19 (January):55–60.

Zalasiewicz, J., C. N. Waters, M. Williams, A. D. Barnosky, A. Cearreta, P. Crutzen, E. Ellis, M. A. Ellis, I. J. Fairchild, J. Grinevald, P. K. Haff, I. Hajdas, R. Leinfelder, J. McNeill, E. O. Odada, C. Poirier, D. Richter, W. Steffen, C. Summerhayes, J. P. M. Syvitski, D. Vidas, M. Wagreich, S. L. Wing, A. P. Wolfe, Z. An, and N. Oreskes. 2015. When did the Anthropocene begin? A mid-twentieth century boundary level is stratigraphically optimal. *Quaternary International* 383:196–203.

Wetland Farming and the Early Anthropocene: Globally Upscaling from the Maya Lowlands with LiDAR and Multiproxy Verification

Sheryl Luzzadder-Beach,[iD] Timothy P. Beach,[iD] and Nicholas P. Dunning[iD]

Of multiple ways to assess the geography of the early Anthropocene, three ongoing efforts are establishing the extent, intensity, and chronology of human impacts on landscapes and connecting impacts to global change through greenhouse gas (GHG) fluxes. Landscapes interact with GHGs, and these have global climate implications. LiDAR, capable of precisely mapping through forest gaps, has revolutionized our ability to characterize and quantify humanized landscapes. In many cases, though, LiDAR is only as good as its accompanying ground verification. This article forges these together to compare a mature literature on wetland contributions to the early Anthropocene in Asia through methane from paddy rice agriculture with the growing literature on a large area of wetland agriculture in the Americas, focusing on the newest discoveries in Central America. Several studies have linked the ~20 ppm rise in atmospheric CO_2 from ~7000 to 1000 BP with deforestation for global farming; the 100 ppb rise in CH_4 from ~5000 to 1000 BP with wetland farming; and the 7 to 10 ppm decline in CO_2 in the sixteenth century CE with reforestation and population collapses of the Americas after the European Conquest. We synthesize the evidence for the onset, duration, and impacts of wetland agriculture in the Maya Lowlands of Mesoamerica to compare their impacts on GHGs and, thus, their contributions to global impacts on climate. This article builds from three decades of studying neotropical humanized landscapes and wetland agroecosystems and more recent quantification from ground-verified LiDAR imagery and synthesizes this growing research and the challenges ahead to gauge the early Anthropocene. *Key Words: early Anthropocene, LiDAR, Maya Lowlands, Mesoamerica, wetland farming.*

Denevan (1992) shook the human environmental interaction literature in 1992 when he and others, especially Butzer (1992), cast considerable doubt on the Pristine Myth, proposing that Prehispanic human impacts were more apparent on the landscape than those after 1492 CE. These pioneering ideas from that special *Annals* issue on the 500th anniversary of the Columbian Encounter showed that indigenous peoples of the Americas were perfectly capable of long and intensive land uses. That issue now nearly thirty years ago set the stage for an early or paleo Anthropocene hypothesis, emerging in works by Ruddiman (2003) and Ruddiman, Crucifix, and Oldfield (2011), Alberts (2011), Syvitski and Kettner (2011), Abrams and Nowacki (2015), Beach et al. (Beach, Luzzadder-Beach, Cook, et al. 2015; Beach, Luzzadder-Beach, Krause, et al. 2015; Beach, Luzzadder-Beach, Krause, et al. 2019), Kunnas (2017), and the ArchaeoGLOBE Project (Stephens et al. 2019). Ruddiman, Crucifix, and Oldfield (2011), in a special issue of *Holocene*, developed this concept after years of debate (Ruddiman 2003). The *Holocene* special issue included authors exploring natural and anthropogenic hypotheses to explain increasing atmospheric CO_2 and CH_4 concentrations over the last 5,000 to 7,000 years. Another compendium grew out of the American Association of Geographers (AAG) Annual Meeting sessions in 2017 (Chin et al. 2017), which explored the challenges of the Anthropocene. In this *Anthropocene* issue, Ruddiman (2017) presented the geographic and agricultural evidence for an early Anthropocene, using ground-based paleoecological, archaeobotanical, and archaeological data to focus on early forest clearance in Europe and Asia emitting CO_2 and on early agricultural and livestock husbandry producing methane in Asia (Ruddiman 2017).

Until recently, there has been a gap in ground-verified evidence for Mesoamerican wetland agriculture extensive enough to contribute to an early

Anthropocene, paralleling the wetlands of Asia (Ruddiman 2017). Here we piece together evidence to fill this gap by synthesizing existing and newly emerging field-based evidence and airborne LiDAR imaging for an early Anthropocene in tropical Central America, driven in part by early Mesoamerican wetland agriculture in the Maya Lowlands (Canuto et al. 2018; Beach, Luzzadder-Beach, Krause, et al. 2019; Dunning et al. 2019). The first piece of the puzzle to gauge global impacts is a comprehensive assessment of the early Anthropocene. For example, Beach, Luzzadder-Beach, Cook, et al. (2015) identified and synthesized six stratigraphic markers of the early Anthropocene in the Maya world. These markers include Maya clays accumulating in lakes, sinks, and depressions; paleosol and anthrosol sequences related to Maya land use; ^{13}C isotopic ratios enriched by C_4 species that date to Maya land use; material remains of buildings, building materials, and other landesque capital; sediments chemically enriched in such elements as phosphorous and mercury that date to Maya periods (Lentz et al. 2020); and evidence for human-induced climate change, as modeled in several studies of forest cover change in the Maya world (Turner and Sabloff 2012; Beach, Luzzadder-Beach, Cook, et al. 2015). We add a seventh sedimentary marker for verifying wetland agriculture: the package of multiproxy evidence (remains of economic or agricultural species; i.e., pollen, phytoliths, and seed remains) for intensive wetland use that dates to the Maya period (~2000–1000 BP; Beach, Luzzadder-Beach, Krause, et al. 2019; Krause, Beach, Luzzadder-Beach, Cook, et al. 2019). To scale these up to a global impact, we need to know their extent, connections to climate, and chronology.

The growing extent of LiDAR imaging studies in Mesoamerica—the "geospatial revolution in archaeology" (Chase et al. 2012)—has been the key to understanding the greater spatial extent and possible global significance of wetland farming. Therefore, we cross the threshold of a new era of fusing deep temporal, ecological, and economic information gained from excavations and field exploration, with high-resolution but geographically expansive LiDAR mapping, allowing unprecedented exploration in tropical regions difficult to access and underserved in the scientific literature. The result is a powerful set of integrated tools to expand our diachronic and geographical understanding of early human impacts (Dunning et al. 2015). This article chronicles the advancements that brought us to this stage and proposes a protocol for the next steps of exploration and understanding. We use these frameworks as markers to evaluate the state of our knowledge of the extent and chronology of early wetland agriculture in the Maya Lowlands, its impacts on atmospheric greenhouse gases (GHGs), and contribution to an early Anthropocene.

Asian Paddy Polycultures and the Early Anthropocene

Known archaeological sites and their accompanying land clearance and cultivation in the Yangtze and Yellow River Valleys increased exponentially between 8000 to 7000 BP and 5000 to 4000 BP (Ruddiman 2017). Ruddiman (2017) noted that estimates for nonpasture cultivation increased from $300\,km^2$ to $16,500\,km^2$ during that time. Before the advent of widespread wetland farming and agricultural deforestation between 10000 and 5000 BP, CO_2 and CH_4 declined (as during other interglacials) from ~266 to 258 ppm (CO_2) and ~700 ppb to 550 ppb (CH_4). After this from ~5000 BP to 1800 CE (unlike other interglacials), CO_2 began to rise to 280 ppm and CH_4 to 710 ppb (Ruddiman 2017). Ruddiman (2017) attributed the reversal of these downward trends in GHGs from 5000 BP forward to the rise of agriculture in Europe and Southwest Asia, and paddy rice diffusion in East and South Asia. If we place the advent and growth of ancient Maya forest clearing and farming, including terrace agriculture, kitchen gardens, milpas, and wetland agriculture, into the timeline of this global increase, it coincides with the time when the GHG concentrations of CO_2 (from forest clearing) and CH_4 (from wetland farming) are continuing to increase, from about 2100 to 900 BP, the Maya Late Preclassic to the Early Postclassic periods (Table 1; Ruddiman 2017; Beach, Luzzadder-Beach, Krause, et al. 2019). Recent discoveries of greater extents of Maya wetland agriculture, including the first efforts to couple multiproxy ground evidence with multiscalar LiDAR mapping (Beach, Luzzadder-Beach, Krause, et al. 2019), point to substantially greater Mesoamerican wetland impacts and potential contributions to GHGs. What makes quantification difficult is that there is little field verification or dating of Maya wetland fields (Beach, Luzzadder-Beach, Krause, et al. 2019); therefore, we need much more

Table 1. Ancient Maya timeline

Maya periods	Gregorian date range BCE/CE[a,c]	Years before present (BP)[b,c]
Late Postclassic	1250–1450 CE	700–500 BP
Early Postclassic	950–1250 CE	1000–700 BP
Terminal Classic	850–950 CE	1120–1000 BP
Late Classic	550–850 CE	1400–1120 BP
Early Classic	250–550 CE	1700–1400 BP
Late Preclassic	350 BCE–250 CE	2300–1700 BP
Middle Preclassic	850–350 BCE	2800–2300 BP
Early Preclassic	2500–850 BCE	4450–2800 BP
Archaic	7200–2500 BCE	9150–4450 BP

[a]BCE/CE (Before the Common Era and Common Era) are the secular equivalents of years BC and AD.
[b]Conversion from Gregorian dates to years BP is relevant to the standard date of 1950 CE as present.
[c]Data from Leyden et al. (1996) and Beach, Luzzadder-Beach, and Dunning (2019).

multiproxy evidence leveraged with the growing corpus of LiDAR remote sensing to quantify the spatial and chronological extent of ancient Maya wetland farming. This fits with the larger challenges of synthesizing evidence for pre–Industrial Revolution human–nature interactions (e.g., Koch et al. 2019) and acknowledges a fairer assessment of the contributions of the indigenous peoples of the Americas to the adaptation capital (Beach, Luzzadder-Beach, and Dunning 2019). Although much Anthropocene research focuses on the Great Acceleration as the advent for the Anthropocene (Steffen et al. 2011; Zalasiewicz et al. 2015), we concentrate here on these deeper diachronic connections to global change.

Sources of Greenhouse Gases from Ancient Land Use Change

Anthropogenic Middle Holocene Greenhouse Gases

Human sources of middle Holocene GHGs include a complex of land disturbance activities related to urbanization, agriculture, and water management (Stephens et al. 2019). For example, land clearance (devegetation) leads to oxidation of C into CO_2 through cutting and burning vegetation, which leads to decreased transpiration and water table rise in some conditions. Intentional wetland draining can also lead to oxidation of C from soil organic matter and peat releasing CO_2. Other activities lead to CH_4 releases. For example, the creation of wetlands and their accompanying anaerobic

conditions, whether intentionally for agriculture and water management or by sea level rise, lead to CH_4 release. Raising livestock above prefarming levels of ruminants also contributes CH_4 (Fuller et al. 2011). Seasonal water storage and drainage of agricultural lands, wetlands and rice paddies, and reservoirs lead to alternating releases of CO_2 and CH_4. Finally, other GHGs like nitrous oxide (N_2O) can derive from permafrost and agricultural fertilization of soils using waste (e.g., night soil or animal manure) and also from animal waste treatment and wastewater management (U.S. Energy Information Administration 2011). The equifinalities to consider that contribute natural GHGs include the top three sources of CH_4—natural wetlands, termites, and oceans (Heilig 1994)—or CO_2 from volcanic eruptions, miniscule compared with modern anthropogenic inputs (Gerlach 2011), or N_2O from microbial degradation of soil and oceanic nitrogen (Thomson et al. 2012). Background concentrations determined from ice cores (Heilig 1994; Ruddiman 2003) of these three GHGs for the past 100,000 years, however, have been lower than after the advent of early agricultural activities (Ruddiman 2003, 2017). Moreover, research (by Koch et al. [2019] and others before) demonstrates that the post-1492 CE decline of pre-Columbian Indigenous populations from disease and associated decreasing levels of land uses, abandonment, and forest regrowth correlates with measurable declines in atmospheric CO_2. Koch et al. (2019) also provided pre-Columbian population and land use estimates for several regions including Central America; the latter is useful for a point of comparison in this work but focuses on extensive swidden systems (i.e., *milpas*) rather than intensive systems (i.e., wetlands, terraces, reservoirs) for Mexico and Central America. To understand potential early Anthropocene human impacts on global climate, however, we need to upscale these land use estimates with areal and temporal estimates from research employing LiDAR and multiple proxy field evidence (Beach, Luzzadder-Beach, Krause, et al. 2019).

General Factors for Quantifying Ancient GHGs from Wetland Agricultural Systems

How do we begin to quantify the impacts of particular human activities on global GHGs? Ruddiman (2003) and Koch et al. (2019) demonstrated that if we break down the individual parts, by extent, intensity, and time, we can begin to make reasonable estimates of anthropogenic input. Spanning complementary scales of time and space, ice cores from Greenland provide global chronologies and measures of atmospheric concentration of GHGs, and geoarchaeological proxies and remote sensing technologies provide better spatiotemporal insight into the land use changes of ancient human activities to contribute GHGs. Comparing these data to modern systems inputs can form the basis for estimating ancient activities' GHG production. For ancient Central America, we need to quantify devegetation, gas cycles in a range of conditions, burning, draining, field surface building, and water logging. Obtaining a representative sample, however, has been difficult because of the inaccessibility of sites and the equifinalities of separating human and natural features and events (e.g., natural and anthropogenic burning, canal building vs. patterned ground; natural sea level rise and water table rise, and wetland formation processes). Distinguishing the natural from the anthropogenic has been an enduring source of tension in geoarchaeology and paleoenvironmental change, and particularly in the Maya Lowlands (Beach, Luzzadder-Beach, Krause, et al. 2019). LiDAR evidence verified with multiple proxy approaches, however, provides additional evidence of human agency by increasing spatial contexts (Canuto et al. 2018; Beach, Luzzadder-Beach, Krause, et al. 2019; Dunning et al. 2019).

Pre-Columbian Wetland Agroecosystems and the Maya Lowlands

A brief review of the long history of pre-Columbian wetland agricultural research is in order based on the rich geographic research (Denevan 1970, 2001; Turner and Harrison 1981, 1983; Siemens 1983; Sluyter 1994, 2006). Denevan (1970) estimated that there were around 170,000 ha (1,700 km^2) of abandoned, drained, or ridged agricultural fields in pre-Columbian Amazonia. Additionally, Sluyter (1994) estimated about 36,000 ha (360 km^2) of wetlands coming under cultivation in Mesoamerica between 4,500 years ago (the onset of the Mesoamerican Preclassic period) and 520 years ago (the end of the Postclassic), with some continuity to the present. These coincide with Ruddiman's (2003, 2017) estimated CO_2 and CH_4 acceleration time

frame. For the Maya Lowlands portion of Mesoamerica, Sluyter (1994) also quantified a total area of clearance and agriculture ranging from about 9,200 to 22,900 ha. Earlier key research on ancient Maya wetland agriculture (e.g., Turner and Harrison 1981, 1983) estimated a larger areal extent of wetland agriculture in the Maya Lowlands, discussed in more detail later. In the early stages of this research, the assumption of human-made canals and wetland fields formed an enduring tension in ancient Maya research from 1970 through to the early 1990s. This tension was between the seemingly obvious point of canals everywhere and the reality of very little field evidence and the need for more hypothesis testing of natural processes that could mimic wetland complexes (Jacob 1995; Pope, Pohl, and Jacob 1996; Luzzadder-Beach and Beach 2009). Moreover, there was an intellectual roadblock that the predominant mode of agriculture was slash and burn, which required less capital labor and infrastructure (Dunning et al. 1998; Dunning and Beach 2004). Nonetheless, much research (Armillas 1971; Sluyter 1994, 2006) showed strong evidence of historical use of wetland fields from Mesoamerica in the *chinampas* around the Basin of Mexico, which covered ~10,000 to 20,000 ha in precolonial times and perhaps fed one half to two thirds of Tenochtitlan's ~400,000 people in the fifteenth century (Sluyter 2006). Even by 1803, after three centuries of European conquest, von Humboldt called the *chinampas* very productive (Sluyter 2006). We have no historical information about *chinampas* use southeast in the Maya Lowlands, however, where we must depend on hard-won field and laboratory proxy evidence gathered over decades (Dunning et al. 2015). Even around the Basin of Mexico with written and historical records, there has been precious little multiproxy excavation, with a few exceptions (Morehart and Frederick 2014; Frederick and Cordova 2019).

In the Maya world, Turner (1974) and Harrison (Harrison and Turner 1978; Turner and Harrison 1981, 1983) built from their own fieldwork, following that of Denevan (1970), Siemens and Puleston (1972), and others, which grew into a *new orthodoxy* of Maya intensive agriculture (Turner 1993; Dunning et al. 1998). Using mapping, aerial photography, excavating, and multiple proxies at the site of Pulltrouser Swamp, Turner and Harrison (1981, 1983) showed that the ancient Maya built raised and ditched agricultural fields, dating between 2,200

Table 2. Criteria for demonstrating ancient wetland agriculture

Equifinalities, other explanations, falsification
Archaeological control
Soils geomorphological control
Chronological control
Ecofacts: Evidence for use in time
Large-scale criteria:
Mapping at multiple scales
Evidence for management within the system: diversion, damming, sourcing

Note: Data from Pope and Dahlin (1989) and Pope et al. (1993).

and 1,170 years ago (late Preclassic 150–200 BCE to late Classic 850 CE). The aerial photography conducted by these early projects inspired research into other airborne technologies, including experimentation with airborne radar (synthetic aperture radar) mapping in the early 1980s to detect subtle (and not so monumental) linear canal and field patterns, hidden in heavily vegetated tropical environments (Adams, Brown, and Culbert 1981; Pope and Dahlin 1989; Pope et al. 1993).

The early experimentation with radar, however, produced two steps forward and one step backward, providing a continuing lesson on the need for field verification in science (Beach and Luzzadder-Beach 2019; Beach, Luzzadder-Beach, Krause, et al. 2019). Adams, Brown, and Culbert's (1981) article in *Science* misidentified large areas with radar as wetland canals, but the work showed innovative thinking to attempt to resolve identification, location, and measurement of large areas of remote archaeological features in forested tropical environments. Pope and Dahlin (1989; Pope et al. 1993) derived lessons from this remote sensing setback to propose a rubric to determine the presence of wetland agriculture we still use (Table 2).

Following the early 1990s controversies, lack of remote sensing verification, and sparse field evidence, a new initiative examined wetland field agriculture from 2000 to the present in the Belize, Mexico, and Guatemala transboundary area (Beach et al. 2008; Luzzadder-Beach and Beach 2009; Dunning, Beach, and Luzzadder-Beach 2012; Luzzadder-Beach, Beach, and Dunning 2012; Beach, Luzzadder-Beach, Cook, et al. 2015; Beach, Luzzadder-Beach, Krause, et al. 2015; Luzzadder-Beach et al. 2016; Beach, Luzzadder-Beach, Krause, et al. 2019; Krause, Beach, Luzzadder-Beach, Cook, et al. 2019; Krause, Beach, Luzzadder-Beach,

Guderjan, et al. 2019). This new region provided exposures, albeit disturbed from clearing for modern Anthropocene wetland (rice) farming and pastures, and preserved areas under the canopy of a forest and archaeology reserves. These projects brought together aerial photography, satellite imagery, water and soil chemistry, and scores of field excavations to establish the chronology, formation, use, and stratigraphic sequences of multiple wetland agricultural fields. This work started with the earlier, healthy scientific debate about whether the Maya created ditched fields and intensive agriculture and branched into new questions: Why did the Maya invest in this complex agricultural infrastructure and what were the geographical and chronological extents? Were the ditched and raised fields a response to major environmental or social shifts (Beach et al. 2009; Dunning, Beach, and Luzzadder-Beach 2012; Luzzadder-Beach, Beach, and Dunning 2012), were these (and other land uses) related to the Maya droughts (Hodell, Brenner, and Curtis 1995; Luzzadder-Beach et al. 2016), and how can we detect a more representative sample of these structures in inaccessible and forest-covered regions to gauge their landscape and global environmental reach (Beach, Luzzadder-Beach, Cook, et al. 2015; Guderjan et al. 2016)? Beach, Luzzadder-Beach, Krause, et al. (2019) extended the inquiry to verify that LiDAR could detect known wetland fields and canals and also detect unknown fields and canals under the forest canopy. We note that most of this Mesoamerican wetlands work occurs as salvage under the urgency of ongoing modern land use changes that can erase the indigenous past.

LiDAR in the Maya Tropical Forest

Experiments using LiDAR for detecting Mesoamerican and Maya archaeological features are increasing (Chase et al. 2011; Chase et al. 2012; Fernandez-Diaz, Carter, Shrestha, and Glennie 2014). Chase et al. (2011) reported on many terraces and reservoirs at the ancient Maya site of Caracol in Belize. Other tropical sites using LiDAR to detect ancient architecture include sites in Honduras (Fernandez-Diaz, Carter, Shrestha, Leisz, et al. 2014; Fisher et al. 2016), Tabasco, Mexico (Inomata et al. 2020), and Angkor Wat in Cambodia (Evans et al. 2013). Many projects document copious ancient infrastructure, spread in a palimpsest across the landscape, unsorted in time, but covering an impressive amount of land. Fisher et al. (2016), working in Michoacán, Mexico, identified 300 linear kilometers of potential canals (among other hydrologic features) but were unable to field check them. These footprints point to a heavy and early human presence and alteration of the landscape both by urbanization and by agricultural infrastructure. All of these land uses including ancient wetland agroecosystems affected GHGs, distributed across space and time, which conditioned the early Anthropocene and the ranges for human adaptations and sustainability (Dunning, Beach, and Luzzadder-Beach 2012; Luzzadder-Beach et al. 2016; Beach, Luzzadder-Beach, Krause, et al. 2019).

A recent case of field verification of LiDAR imagery provides a roadmap for future studies. At present, only a handful of LiDAR studies have detected linear features with a high potential to be ancient Maya wetland fields; even fewer have actually field verified these with excavation, geoarchaeological proxies, and rigorous dating (Beach, Luzzadder-Beach, Krause, et al. 2015; Beach, Luzzadder-Beach, Krause, et al. 2019; Krause, Beach, Luzzadder-Beach, Cook, et al. 2019); and some have reported field visits and surface (nonexcavation) surveys (Canuto et al. 2018; Garrison, Houston, and Firpi 2019). These together provide a baseline model for detection, site survey, and field verification. This central theme of testing the early Anthropocene hypothesis also presents an opportunity for greater project-to-project collaboration, to begin to systematically understand and quantify the greater potential spatial extent of ancient Maya wetland fields and estimate their combined contribution. We can add or subtract new potential fields to the lexicon of already identified and verified fields (Sluyter 1994; Luzzadder-Beach and Beach 2009; Beach, Luzzadder-Beach, Cook, et al. 2015; Beach, Luzzadder-Beach, Krause, et al. 2015; Beach, Luzzadder-Beach, Krause, et al. 2019; Krause, Beach, Luzzadder-Beach, Cook, et al. 2019; Krause, Beach, Luzzadder-Beach, Guderjan, et al. 2019). Although there remain few such verified LiDAR-detected wetland features, the new multidisciplinary methodology for spatial and chronological verification marks a new opportunity for understanding early human impacts in the Neotropics and on the Earth. Multiple projects owe their data quality, consistency, and comparability to the National Center for Airborne Laser Mapping

(NCALM; Fernandez-Diaz, Carter, Shrestha, Leisz, et al. 2014) because NCALM flew scores of missions and processed data in support of Maya and Mesoamerican projects.

Promising recent results come from four projects: in northern Guatemala; southern Campeche, Mexico; Campeche, Mexico; and northwestern Belize, representing different hydrogeomorphic zones of the Maya Lowlands. Canuto et al. (2018) reported in *Science* on using LiDAR to identify 67 km^2 (6,659 ha) of unverified wetland fields across the Petén District of Guatemala and classified them as zones of intensive cultivation. They used these classified areal estimates to calculate food production based on swidden studies. The field, however, needs to develop new estimates, computed according to wetland agroecosystem production parameters. These parameters will become more accurate only with more paleoecological research and integrated with more modeling of field functions (Arco and Abrams 2006).

One of the subprojects of Guatemala's Fundación Patrimonio Cultural y Natural Maya Consortium (Canuto et al. 2018) is at the site of El Zotz in the Petén about 15 km west of the great Maya center of Tikal (Luzzadder-Beach et al. 2017; Beach, Luzzadder-Beach, Doyle, and Delgado 2019; Garrison, Houston, and Firpi 2019). The Zotz-Palmar LiDAR shows new possibilities for previously unexplored wetland fields, and Garrison, Houston, and Firpi (2019) estimated about 95 ha of unverified areas that could represent canalized fields. In June 2018, the authors ground-located and visited the site for a brief surface reconnaissance (Garrison, Houston, and Firpi 2019). For future field campaigns, the seven markers of the early Anthropocene (Beach, Luzzadder-Beach, Cook, et al. 2015) as adapted to ancient Maya wetland agriculture will serve as guiding principles for verification (Beach, Luzzadder-Beach, Krause, et al. 2019).

LiDAR survey also indicates areas of potential ancient wetland fields in seasonal, upland wetlands called *bajos* of the elevated interior region (EIR) of southern Campeche and Quintana Roo (Mexico), typically with canalization concentrated along the edges of these depressions. Dunning et al. (2019) reported potential wetland field and canal sites detected with the LiDAR in bajos near Chactun and Tamchen, Campeche, and in multiple bajos in Quintana Roo (visible in swaths of the National Aeronautic and Space Administration's G-LiHT

LiDAR survey). These areas are not yet measured. Like the fields reported in Canuto et al. (2018), these sites present an important opportunity to investigate their origins, extent, and chronology along the edges of the EIR, constructed in a zone with a deep groundwater table far from the coastal margin. This area is immune from the influence of sea level rise on groundwater tables but is highly sensitive to drought (Luzzadder-Beach and Beach 2009; Beach, Luzzadder-Beach, Krause, et al. 2015). Field verifying the existence of wetland fields in the EIR will further underscore the human agency in their origins and their connections to climate change (Luzzadder-Beach and Beach 2009; Dunning, Beach, and Luzzadder-Beach 2012; Luzzadder-Beach, Beach, and Dunning 2012). The features' occurrence in varying hydrologic, geomorphic, and ecological zones across the Maya Lowlands obviates a single explanation for their origin, natural or anthropogenic (Dunning, Beach, and Luzzadder-Beach 2012; Luzzadder-Beach, Beach, and Dunning 2012).

What appears to be the largest contiguous area of wetland fields in the Maya Lowlands, at least $119 \, km^2$ (11,900 ha), lies near the northeastern edge of Laguna de Terminos, Campeche (Dunning et al. 2020). This extensive array of fields and canals includes a pattern of larger, amorphous fields grading westward into dense clusters of smaller, rectangular fields and might span an age range as long as 500 BCE to 1500 CE, contingent on verification.

In 2016, Beach and Luzzadder-Beach initiated the Northwestern Belize LiDAR Consortium project with their archaeological collaborators and NCALM to collect $275 \, km^2$ of LiDAR data over northwestern Belize. This project tests the hypothesis that LiDAR will detect known, verified wetland fields and will work at a sufficient scale to detect undiscovered sites under forest cover (Beach, Luzzadder-Beach, Krause, et al. 2019). Thus far, we have identified $14.08 \, km^2$ (1,408 ha) of verified wetland agricultural fields in northwest Belize (Beach, Luzzadder-Beach, Krause, et al. 2019), including a previously unknown and the largest verified complex yet (Beach, Luzzadder-Beach, Krause, et al. 2019; Figure 1). The serendipity of this project was that NCALM collected these LiDAR data in 2016 just before it collected the Fundación Patrimonio Cultural y Natural Maya Consortium's LiDAR over the Petén of Guatemala. In addition to our remote sensing hypotheses, we used the LiDAR findings to guide more extensive

studies of soils, ecology, geology, and ancient land use. Our prior work aided by aerial photography, commercial imagery, and exploration and excavation had verified about $1 \, km^2$ of wetland fields, but the LiDAR mapping expanded our estimate of verified fields by at least five times in the Birds of Paradise site alone and uncovered a previously unknown $7.7 \, km^2$ of patterned wetland features in the central Rio Bravo, where field verification is in progress. Together these increase our new findings to more than ten times the area of previously known wetland fields at this site in northwestern Belize.

In summation of the main studies of wetland complexes, Sluyter (1994) synthesized all known (pre-LiDAR), mapped wetland fields for Mesoamerica at about $36,400 \, ha$ $(364 \, km^2)$ and estimated known fields (all time periods) in the Maya Lowlands to be about 9,200 to $22,900 \, ha$ $(92–229 \, km^2$; the range depended on the confidence at the time in the site of Bajo Acatuch). These represent conservative totals because at the time they excluded known wetland areas like Cobweb Swamp in Belize and the Usumacinta River of Mexico, and the northern Belize numbers presented far underestimated the extent that Turner and Harrison (1983) reported. Sluyter (1994) noted that these Maya wetland fields represented the largest proportion of Mesoamerica fields known even then. The four new Maya Lowlands LiDAR projects we review here have added at least $19,967 \, ha$ from new sites $(199.67 \, km^2)$ to Sluyter's estimate, for a new estimated range of 32,317 to $40,317 \, ha$ $(32.3–40.3 \, km^2)$, although not all are ground verified. This exercise produces a substantial areal increase. Beach, Luzzadder-Beach, Krause, et al. (2019) demonstrated and verified through multiple lines of evidence that areal extent identified through conventional means without LiDAR is ~10 percent of what is potentially hidden under forest cover. If we apply the Birds of Paradise findings alone, an increase of five times, we can begin to understand the potential upscaling of the extent of other ancient wetland agricultural sites. For example, we can further project a range of estimated areal coverage of Mesoamerican fields by scaling up Sluyter's Mesoamerican estimate of $36,400 \, ha$ $(364 \, km^2)$ fivefold to about $182,000 \, ha$ $(1,820 \, km^2)$. Finally, by adding in the approximately $20,000 \, ha$ of new LiDAR-detected Maya Lowlands wetland fields (which still require ground verification), the potential Mesoamerican total becomes about $202,000 \, ha$

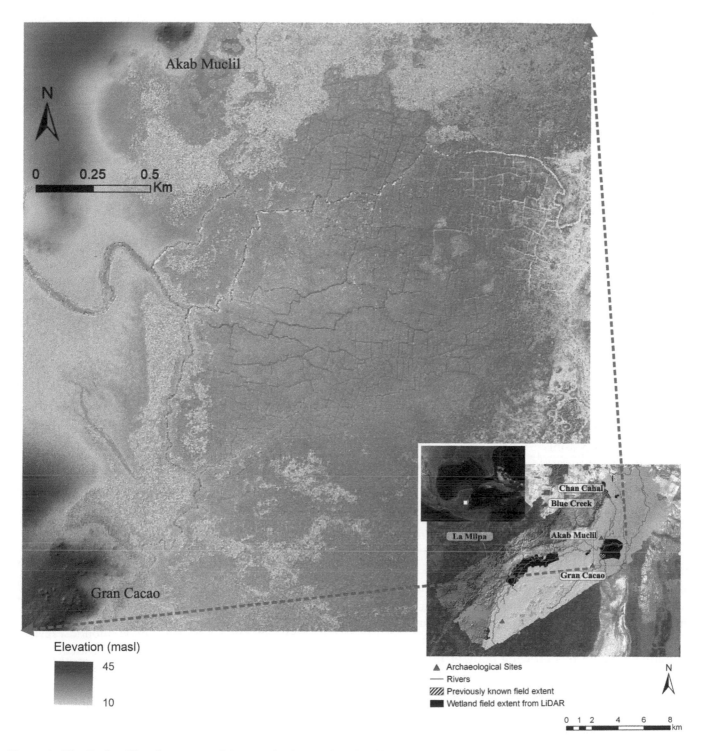

Figure 1. The Birds of Paradise ancient Maya wetland agricultural field system and parts of the nearby Maya sites of Gran Cacao (southwest corner) and Akab Muclil (northwest corner) in northwestern Belize. The authors created this 2.5 km × 2.67 km image from a LiDAR-derived DEM; a semitransparent hill-shaded DEM overlays a colorized DEM with a color stretch that emphasizes anthropogenic features. This image contains 5.13 km^2 of wetland fields and 71 linear kilometers of canals (Beach, Luzzadder-Beach, Krause, et al. 2019). DEM = digital elevation model. Image courtesy of T. Beach, S. Luzzadder-Beach, S. Eshleman, the Beach/Butzer Labs at The University of Texas at Austin, and the Northwestern Belize LiDAR Consortium. Cartography by A. Zimmer.

(2,020 km^2). Returning to the contributions to an early Anthropocene, these Mesoamerican wetland agriculture estimates of ~2,020 km^2 are about 12 percent of the peak estimate of 16,500 km^2 for Asia alone by Ruddiman. Future modeling will need to consider GHG inputs of ancient wetland fields from the growing corpus of LiDAR for Central America, and elsewhere, such as the vast complexes of South America (Denevan 1970, 2001), sites mentioned (but not quantified) by Ruddiman (2003). We reiterate that these areal estimates are only as good as their reliable chronologies (Sluyter 1994) and multiproxy field substantiation (e.g., Beach, Luzzadder-Beach, Krause, et al. 2019).

Discussion and Conclusions

LiDAR provides a better spatial footprint to potentially calculate GHG inputs from early agriculture than do earlier attempts to map on the ground, by airplane in inaccessible tropical sites, and using conventional imagery. We can bridge the gap of slow fieldwork and fast LiDAR by a concerted focus on machine learning and other big data analyses—possibly with citizen science—because these can provide rapid upscaling of feature identification and integrate the knowledge base of local groups by volunteers participating in examining digital LiDAR data images and field checking potential features. This participation therefore increases stakeholders' shares and voices in the knowledge capital and their investment in protecting geoheritage (Lambers, Verschoof-van der Vaart, and Bourgeouis 2019). We return, though, to the limiting factors for all remote sensing—ground verification and geospatial, ecological, and chronological control. There is a growing body of LiDAR data from multiple projects and consortia to estimate areas and some ground control (geoarchaeology) of wetlands and reservoirs. To move beyond rough estimates, we need to add more elements to our exploration model as introduced at the beginning of the article, such as excavation, multiple proxies of chronologically verified environmental data, GHG monitoring, and wetland agroecosystem food production rates, and recognize the error brackets around these data.

GHG monitoring in tropical wetlands needs to occur, similar to programs monitoring temperate wetlands for GHG flux. To understand the equifinalities and the spatial and temporal consequences of global upscaling, Morrison et al. (2018) are developing a shared terminology to better standardize past land use and land cover categories for modeling global change. This will enhance collaborative and reproducible geoarchaeological and remote sensing research that combines and field tests these model elements. Excavation is an imperative to show stratigraphy, soil evidence of reduction and oxidation, and chronology, and we need to emplace GHG monitoring and modeling and alternative remote sensing data sets and modeling where LiDAR sensing fails.

We also need to return to the fundamentals of testing multiple working hypotheses and lines of evidence (Dunning et al. 2015), including testing the equifinalities of canal-like rectilinear features. Although fieldwork in the Neotropics is challenging, remote sensing has also been challenging, not only because of deep forest cover and perhaps more confounding deep savannas but also due to the scale of the features: Canals or ditches could be 1 to 4 m wide and many meters long. From a chronological viewpoint, LiDAR shows a cumulative landscape without chronological control, and some of it is erased over time like a palimpsest, and other infrastructure is built directly atop earlier works. Raised or drained field complexes can be built piecemeal over time or all at once as single planned public works. Alternative hypotheses to consider are whether there are yet possible natural explanations (equifinalities) for patterned ground in this complex landscape: Linear patterns and mounded topography can emerge from global processes of geological jointing, faults, karst weathering, tree islands, clay cracking, polygons, hog wallows, mima mounds, gypsic laccoliths, pathways, walls, or, later, historical period fields for logwood or tropical fruits. Even with excavation, the nature and origins of linear features in wetlands can remain problematic (Dunning et al. 2017). The challenges of testing the early Anthropocene hypothesis are legion but not unsurmountable, like all long-term environmental change research endeavors.

Acknowledgments

The authors thank the Programme for Belize Rio Bravo Conservation Management Area, the Institute of Archaeology, and the San Felipe and Blue Creek

Communities, Belize; the National Center for Airborne Laser Mapping; The Northwestern Belize LiDAR Consortium; J.-C. Fernandez-Diaz, F. Valdez, Jr., T. Guderjan, N. Brokaw, S. Ward, T. Garrison, S. Houston, S. Krause, S. Eshleman, C. Doyle, A. Zimmer, G. Wells, J. Dale, and L. Donn; the anonymous reviewers; and our many collaborators in Mesoamerica.

Funding

This work was supported in part by the National Science Foundation (0924501, 0924510, 1114947, 1550204); the National Geographic Society Committee for Research and Exploration (7861-05, 7506-03); The C.B. Smith Sr. Centennial Chair, The R. C. Dickson Centennial Professorship, College of Liberal Arts Faculty Study Leave, Planet Texas 2050, and Programme for Belize Archaeological Project, University of Texas at Austin; the Maya Research Program, University of Texas at Tyler; and the Cinco Hermanos Chair, Georgetown University.

ORCID

Sheryl Luzzadder-Beach http://orcid.org/0000-0002-9184-2427
Timothy P. Beach http://orcid.org/0000-0003-0097-7973
Nicholas P. Dunning http://orcid.org/0000-0002-1843-3088

References

Abrams, M. D., and G. J. Nowacki. 2015. Exploring the early Anthropocene burning hypothesis and climate-fire anomalies for the eastern U.S. *Journal of Sustainable Forestry* 34 (1–2):30–48. doi: 10.1080/10549811.2014.973605.

Adams, R. E. W., W. E. Brown, Jr., and T. P. Culbert. 1981. Radar mapping, archeology, and ancient Maya land use. *Science* 213 (4515):1457–63. doi: 10.1126/science.213.4515.1457.

Alberts, P. 2011. Responsibility towards life in the Early Anthropocene. *Angelaki* 16 (4):5–17. doi: 10.1080/0969725X.2011.641341.

Arco, L., and E. Abrams. 2006. An essay on energetics: The construction of the Aztec chinampa system. *Antiquity* 80 (310):906–18. doi: 10.1017/S0003598X00094503.

Armillas, P. 1971. Gardens on swamps. *Science* 174 (4010):653–61. doi: 10.1126/science.174.4010.653.

Beach, T., and S. Luzzadder-Beach. 2019. The perpetuation of error in the science of wetlands. Paper presented at the American Association of Geographers Annual Meeting, Washington, DC, April 5.

Beach, T., S. Luzzadder-Beach, D. Cook, N. Dunning, D. Kennett, S. Krause, R. Terry, D. Trein, and F. Valdez. 2015. Ancient Maya impacts on the Earth's surface: An early Anthropocene analog? *Quaternary Science Reviews* 124:1–30. doi: 10.1016/j.quascirev.2015.05.028.

Beach, T., S. Luzzadder-Beach, C. Doyle, and W. Delgado. 2019. Environments of El Zotz: Water and soil chemistry, the El Zotz Dam, and long-term environmental change. In *An inconstant landscape: The archaeology of El Zotz, Guatemala*, ed. T. G. Garrison and S. Houston, 163–88. Boulder: The University Press of Colorado.

Beach, T., S. Luzzadder-Beach, and N. P. Dunning. 2019. Out of the soil: Soil (dark matter biodiversity) and societal "collapses" from Mesoamerica to Mesopotamia and beyond. In *Biological extinction: New perspectives*, ed. P. Dasgupta, P. H. Raven, and A. L. McIvor, 138–74. Cambridge, UK: Cambridge University Press.

Beach, T., S. Luzzadder-Beach, N. Dunning, and D. Cook. 2008. Human and natural impacts on fluvial and karst depressions of the Maya Lowlands. *Geomorphology* 101(1–2): 301–31. doi: 10.1016/j.geomorph.2008.05.019.

Beach, T., S. Luzzadder-Beach, N. Dunning, J. Jones, J. Lohse, T. Guderjan, S. Bozarth, S. Millspaugh, and T. Bhattacharya. 2009. A Review of Human and Natural Changes in Maya Lowlands Wetlands over the Holocene. *Quaternary Science Reviews* 28:1710–24.

Beach, T., S. Luzzadder-Beach, S. Krause, T. Guderjan, F. Valdez, Jr., J.-C. Fernandez-Diaz, S. Eshleman, and C. Doyle. 2019. Ancient Maya wetland fields revealed under tropical forest canopy from laser scanning and multiproxy evidence. *Proceedings of the National Academy of Sciences of the United States of America* 116 (43):21469–77. doi: 10.1073/pnas.1910553116.

Beach, T., S. Luzzadder-Beach, S. Krause, S. Walling, N. Dunning, J. Flood, T. Guderjan, and F. Valdez. 2015. "Mayacene" floodplain and wetland formation in the Rio Bravo watershed of northwestern Belize. *The Holocene* 25 (10):1612–26. doi: 10.1177/0959683615591713.

Butzer, K. 1992. The Americas before and after 1492: An introduction to current geographical research. *Annals of the Association of American Geographers* 82 (3):345–68.

Canuto, M. A., F. Estrada-Belli, T. G. Garrison, S. D. Houston, M. J. Acuña, M. Kováč, D. Marken, P. Nondédéo, L. Auld-Thomas, C. Castanet, et al. 2018. Ancient lowland Maya complexity as revealed by airborne laser scanning of northern Guatemala. *Science* 361 (6409):eaau0137. doi: 10.1126/science.aau0137.

Chase, A. F., D. Z. Chase, C. T. Fisher, S. J. Leisz, and J. F. Weishampel. 2012. Geospatial revolution and remote sensing LiDAR in Mesoamerican archaeology. *Proceedings of the National Academy of Sciences of the*

United States of America 109 (32):12916–21. doi: 10.1073/pnas.1205198109.

Chase, A. F., D. Z. Chase, J. F. Weishampel, J. B. Drake, R. L. Shrestha, K. C. Slatton, J. J. Awe, and W. E. Carter. 2011. Airborne LiDAR, archaeology, and the ancient Maya landscape at Caracol, Belize. *Journal of Archaeological Science* 38 (2):387–98. doi: 10.1016/j.jas.2010.09.018.

Chin, A., T. Beach, S. Luzzadder-Beach, and W. Solecki. 2017. Challenges of the Anthropocene. *Anthropocene* 20:1–90. doi: 10.1016/j.ancene.2017.12.001.

Denevan, W. M. 1970. Aboriginal drained-field cultivation in the Americas: Pre-Columbian reclamation of wet lands was widespread in the savannas and highlands of Latin America. *Science* 169 (3946):647–54. doi: 10.1126/science.169.3946.647.

Denevan, W. M. 1992. The pristine myth: The landscape of the Americas in 1492. *Annals of the Association of American Geographers* 82 (3):369–85. doi: 10.1111/j.1467-8306.1992.tb01965.x.

Denevan, W. M. 2001. *Cultivated landscapes of native Amazonia and the Andes.* New York: Oxford University Press.

Dunning, N. P., A. Anaya Hernández, T. Beach, C. Carr, R. Griffin, J. G. Jones, D. L. Lentz, S. Luzzadder-Beach, K. Reese-Taylor, and I. Šprajc. 2019. Margin for error: Anthropogenic geomorphology of *Bajo* edges in the Maya lowlands. *Geomorphology* 331:127–45. doi: 10.1016/j.geomorph.2018.09.002.

Dunning, N. P., and T. Beach. 2004. Noxious or nurturing nature? Maya civilization in environmental context. In *Continuities and change in Maya archaeology: A millennial perspective*, ed. C. Golden and G. Borgstede, 124–41. London and New York: Routledge.

Dunning, N. P., T. Beach, P. Farrell, and S. Luzzadder-Beach. 1998. Prehispanic agrosystems and adaptive regions in the Maya Lowlands. *Culture & Agriculture* 20 (2–3):87–101. doi: 10.1525/cag.1998.20.2-3.87.

Dunning, N. P., T. P. Beach, and S. Luzzadder-Beach. 2012. Kax and kol: Collapse and resilience in lowland Maya civilization. *Proceedings of the National Academy of Sciences of the United States of America* 109 (10):3652–57. doi: 10.1073/pnas.1114838109.

Dunning, N. P., R. Griffin, T. Sever, W. Saturno, and J. G. Jones. 2017. The nature and origin of linear features in the Bajo de Azúcar, Guatemala: Implications for ancient Maya adaptations to a changing environment. *Geoarchaeology* 32 (1):107–29. doi: 10.1002/gea.21568.

Dunning, N. P., C. McCane, T. Swinney, M. Purtill, J. Sparks, A. Mann, J.-P. McCool, and C. Ivenso. 2015. Geoarchaeological investigations in Mesoamerica move into the 21st century: A review. *Geoarchaeology* 30 (3):167–99. doi: 10.1002/gea.21507.

Dunning, N. P., T. Ruhl, C. Carr, T. Beach, C. Brown, and S. Luzzadder-Beach. 2020. The ancient Maya wetland fields of Acalán. *Mexicon* 42 (4): 91–105.

Evans, D. H., R. J. Fletcher, C. Pottier, J.-B. Chevance, D. Soutif, B. S. Tan, S. Im, D. Ea, T. Tin, S. Kim, et al. 2013. Uncovering archaeological landscapes at Angkor using lidar. *Proceedings of the National*

Academy of Sciences of the United States of America 110 (31):12595–600. doi: 10.1073/pnas.1306539110.

Fernandez-Diaz, J.-C., W. Carter, R. Shrestha, S. Leisz, C. Fisher, A. Gonzalez, D. Thompsan, and S. Elkins. 2014. Archaeological prospection of north Eastern Honduras with airborne mapping LiDAR. In *2014 Geoscience and Remote Sensing Symposium (IGARSS), 2014 IEEE International*, 902–5. Quebec City: IEEE. https://ieeexplore.ieee.org/xpl/conhome/6919813/proceeding doi: 10.1109/IGARSS.2014.6946571.

Fernandez-Diaz, J.-C., W. E. Carter, R. L. Shrestha, and C. L. Glennie. 2014. Now you see it … now you don't: Understanding airborne mapping LiDAR collection and data product generation for archaeological research in Mesoamerica. *Remote Sensing* 6 (10):9951–10001. doi: 10.3390/rs6109951.

Fisher, C. T., J. C. Fernández-Diaz, A. S. Cohen, O. N. Cruz, A. M. Gonzáles, S. J. Leisz, F. Pezzutti, R. Shrestha, and W. Carter. 2016. Identifying ancient settlement patterns through LiDAR in the Mosquitia Region of Honduras. *PLoS ONE* 11 (8):e0159890. doi: 10.1371/journal.pone.0159890.

Frederick, C. D., and C. E. Cordova. 2019. Prehispanic and colonial landscape change and fluvial dynamics in the Chalco Region, Mexico. *Geomorphology* 331:107–26. doi: 10.1016/j.geomorph.2018.10.009.

Fuller, D. Q., J. van Etten, K. Manning, C. Castillo, E. Kingwell-Banham, A. Weisskopf, Q. Ling, Y.-I. Sato, and R. J. Hijmans. 2011. The contribution of rice agriculture and livestock pastoralism to prehistoric methane levels. *The Holocene* 21 (5):743–59. doi: 10.1177/0959683611398052.

Garrison, T. G., S. Houston, and O. A. Firpi. 2019. Recentering the rural: Lidar and articulated landscapes among the Maya. *Journal of Anthropological Archaeology* 53:133–46. doi: 10.1016/j.jaa.2018.11.005.

Gerlach, T. 2011. Volcanic versus anthropogenic carbon dioxide. *Eos: Transactions of the American Geophysical Union* 92 (24):201–2. doi: 10.1029/2011EO240001.

Guderjan, T. H., S. Krause, S. Luzzadder-Beach, T. Beach, and C. Brown. 2016. Visualizing Maya agriculture along the Rio Hondo: A remote sensing approach. In *Perspectives on the ancient Maya of Chetumal Bay*, ed. D. Walker, 92–106. Gainesville: University Press of Florida.

Harrison, P. D., and B. L. Turner, II. 1978. *Pre-Hispanic Maya agriculture*. Albuquerque: University of New Mexico Press.

Heilig, G. K. 1994. The greenhouse gas methane (CH_4): Sources and sinks, the impact of population growth, possible interventions. *Population and Environment* 16 (2):109–37. doi: 10.1007/BF02208779.

Hodell, D. A., M. Brenner, and J. Curtis. 1995. Possible role of climate in the collapse of classic Maya civilization. *Nature* 375 (6530):391–94. doi: 10.1038/375391a0.

Inomata, T., D. Triadan, V. A. Vázquez López, J.-C. Fernandez-Diaz, T. Omori, M. Belén Méndez Bauer, M. García Hernández, T. Beach, C. Cagnato, K. Aoyama, et al. 2020. Monumental architecture at Aguada Fénix and the rise of Maya civilization. *Nature* 582 (7813):530–33. doi: 10.1038/s41586-020-2343-4.

Jacob, J. S. 1995. Archaeological pedology in the Maya Lowlands. In *Pedological perspectives in archaeological research*, Special Publication 44, ed. M. E. Collins, B. J. Carter, B. G. Gladfelter, and R. J. Southard, 51–80. Madison, WI: Soil Science Society of America.

Koch, A., C. Brierley, M. M. Maslin, and S. L. Lewis. 2019. Earth system impacts of the European arrival and Great Dying in the Americas after 1492. *Quaternary Science Reviews* 207:13–36. doi: 10.1016/j.quascirev.2018.12.004.

Krause, S., T. Beach, S. Luzzadder-Beach, D. Cook, G. Islebe, M. R. Palacios-Fest, S. Eshleman, C. Doyle, and T. H. Guderjan. 2019. Wetland geomorphology and paleoecology near Akab Muclil, Rio Bravo floodplain of the Belize coastal plain. *Geomorphology* 331:146–59. doi: 10.1016/j.geomorph.2018.10.015.

Krause, S., T. Beach, S. Luzzadder-Beach, T. Guderjan, F. Valdez, S. Eshleman, C. Doyle, and S. Bozarth. 2019. Ancient Maya wetland management in two watersheds in Belize: Soils, water, and paleoenvironmental change. *Quaternary International* 502 (Part B):280–95. doi: 10.1016/j.quaint.2018.10.029.

Kunnas, J. 2017. Storytelling: From the early Anthropocene to the good or the bad Anthropocene. *The Anthropocene Review* 4 (2):136–50. doi: 10.1177/2053019617725538.

Lambers, K., W. B. Verschoof-van der Vaart, and Q. P. J. Bourgeois. 2019. Integrating remote sensing, machine learning, and citizen science in Dutch archaeological prospection. *Remote Sensing* 11 (7):794–820. doi: 10.3390/rs11070794.

Lentz, D. L., T. L. Hamilton, N. P. Dunning, V. L. Scarborough, T. P. Luxton, A. Vonderheide, E. J. Tepe, C. J. Perfetta, J. Brunemann, L. Grazioso, et al. 2020. Molecular genetic and geochemical assays reveal severe contamination of drinking water reservoirs at the ancient Maya city of Tikal. *Scientific Reports* 10 (1):10316. doi: 10.1038/s41598-020-67044-z.

Leyden, B., M. Brenner, T. Whitmore, J. Curtis, D. Piperno, and B. Dahlin. 1996. A record of long- and short-term climatic variation from northwest Yucatan: Cenote San Jose Chulchaca. In *The managed mosaic: Ancient Maya agriculture and resource use*, ed. S. Fedick, 30–50. Salt Lake City: University of Utah Press.

Luzzadder-Beach, S., and T. Beach. 2009. Arising from the wetlands: Mechanisms and chronology of landscape aggradation in the northern coastal plain of Belize. *Annals of the Association of American Geographers* 99 (1):1–26. doi: 10.1080/00045600802458830.

Luzzadder-Beach, S., T. Beach, and N. Dunning. 2012. Wetland fields as mirrors of drought and the Maya abandonment. *Proceedings of the National Academy of Sciences of the United States of America* 109 (10):3646–51. doi: 10.1073/pnas.1114919109.

Luzzadder-Beach, S., T. Beach, T. Garrison, S. Houston, J. Doyle, E. Roman, S. Bozarth, R. Terry, S. Krause, and J. Flood. 2017. Paleoecology and geoarchaeology at El Palmar and the El Zotz region, Guatemala. *Geoarchaeology* 32 (1):90–106. doi: 10.1002/gea.21587.

Luzzadder-Beach, S., T. Beach, S. Hutson, and S. Krause. 2016. Sky-earth, lake-sea: Climate and water in Maya history and landscape. *Antiquity* 90 (350):426–42. doi: 10.15184/aqy.2016.38.

Morehart, C. T., and C. Frederick. 2014. The chronology and collapse of pre-Aztec raised field (*chinampa*) agriculture in the northern Basin of Mexico. *Antiquity* 88 (340):531–48. doi: 10.1017/S0003598X00101164.

Morrison, K. D., E. Hammer, L. Popova, M. Madella, N. Whitehouse, M.-J. Gaillard, and LandCover6k Land-Use Group Members. 2018. Global-scale comparisons of human land use: Developing shared terminology for land-use practices for global change. *Past Global Change Magazine* 26 (1):8–9. doi: 10.22498/pages.26.1.8.

Pope, K. O., and B. Dahlin. 1989. Ancient Maya wetland agriculture: New insights from ecological and remote sensing research. *Journal of Field Archaeology* 16 (1):87–106. doi: 10.1179/jfa.1989.16.1.87.

Pope, K. O., B. H. Dahlin, and R. E. W. Adams. 1993. Radar detection and ecology of ancient Maya canal systems—Reply to Adams et al. *Journal of Field Archaeology* 20 (3):379–83. doi: 10.1179/jfa.1993.20.3.379.

Pope, K. O., M. D. Pohl, and J. S. Jacob. 1996. Formation of ancient Maya wetland fields: Natural and anthropogenic processes. In *The managed mosaic: Ancient Maya agriculture and resource use*, ed. S. L. Fedick, 165–76. Salt Lake City: University of Utah Press.

Ruddiman, W. F. 2003. The anthropogenic greenhouse era began thousands of years ago. *Climatic Change* 61 (3):261–93. doi: 10.1023/B:CLIM.0000004577.17928.fa.

Ruddiman, W. F. 2017. Geographic evidence of the early anthropogenic hypothesis. *Anthropocene* 20:4–14. doi: 10.1016/j.ancene.2017.11.003.

Ruddiman, W. F., M. C. Crucifix, and F. A. Oldfield. 2011. Introduction to the early-Anthropocene special issue. *The Holocene* 21 (5):713. doi: 10.1177/0959683611398053.

Siemens, A. H. 1983. Wetland agriculture in pre-Hispanic Mesoamerica. *Geographical Review* 73 (2):166–81. doi: 10.2307/214642.

Siemens, A. H., and D. E. Puleston. 1972. Ridged fields and associated features in Southern Campeche: New perspectives on the lowland Maya. *American Antiquity* 37 (2):228–39. doi: 10.2307/278209.

Sluyter, A. 1994. Intensive wetland agriculture in Mesoamerica: Space, time and form. *Annals of the Association of American Geographers* 84 (4):557–84. doi: 10.1111/j.1467-8306.1994.tb01877.x.

Sluyter, A. 2006. Humboldt's Mexican texts and landscapes. *Geographical Review* 96 (3):361–81. doi: 10.1111/j.1931-0846.2006.tb00256.x.

Steffen, W., J. Grinevald, P. Crutzen, and J. McNeill. 2011. The Anthropocene: Conceptual and historical perspectives. *Philosophical Transactions A: Mathematical, Physical, and Engineering Sciences* 369 (1938):842–67. doi: 10.1098/rsta.2010.0327.

Stephens, L., D. Fuller, N. Boivin, T. Rick, N. Gauthier, A. Kay, B. Marwick, C. G. Armstrong, C. M. Barton, T. Denham, et al. 2019. Archaeological assessment reveals Earth's early transformation through land use. *Science* 365 (6456):897–902. doi: 10.1126/science.aax1192.

Syvitski, J. P. M., and A. Kettner. 2011. Sediment flux and the Anthropocene. *Philosophical Transactions A: Mathematical, Physical, and Engineering Sciences* 369 (1938):957–75. doi: 10.1098/rsta.2010.0329.

Thomson, A. J., G. Giannopoulos, J. Pretty, E. M. Baggs, and D. J. Richardson. 2012. Biological sources and sinks of nitrous oxide and strategies to mitigate emissions. *Philosophical Transactions of the Royal Society of London. Series B: Biological Sciences* 367 (1593):1157–68. doi: 10.1098/rstb.2011.0415.

Turner, B. L., II. 1974. Prehistoric intensive agriculture in the Mayan lowlands. *Science* 185 (4146):118–24. doi: 10.1126/science.185.4146.118.

Turner, B. L., II. 1993. Rethinking the "New Orthodoxy": Interpreting ancient Maya agriculture and environment. In *Culture, form, and place: Essays in cultural and historical Geography*, ed. K. Mathewson, 57–88. Baton Rouge, LA: Geoscience.

Turner, B. L., II, and P. D. Harrison. 1981. Prehistoric raised-field agriculture in the Maya Lowlands. *Science* 213 (4506):399–405. doi: 10.1126/science.213.4506.399.

Turner, B. L., II, and P. Harrison, eds. 1983. *Pulltrouser Swamp: Ancient Maya habitat, agriculture, and settlement in northern Belize*. Austin: University of Texas Press.

Turner, B. L., II, and J. A. Sabloff. 2012. Classic Period collapse of the central Maya Lowlands: Insights about human–environment relationships for sustainability. *Proceedings of the National Academy of Sciences of the United States of America* 109 (35):13908–14. doi: 10.1073/pnas.1210106109.

U.S. Energy Information Administration. 2011. Emissions of greenhouse gases in the United States 2009. U.S. Department of Energy DOE/EIA, 0573(2009), Washington, DC.

Zalasiewicz, J., C. N. Waters, M. Williams, A. D. Barnosky, A. Cearreta, P. Crutzen, E. Ellis, M. A. Ellis, I. J. Fairchild, J. Grinevald, et al. 2015. When did the Anthropocene begin? A mid-twentieth century boundary level is stratigraphically optimal. *Quaternary International* 383:196–203. doi: 10.1016/j.quaint.2014.11.045.

Putting the Anthropocene into Practice: Methodological Implications

Christine Biermann, Lisa C. Kelley, and Rebecca Lave

The foundational premise of the Anthropocene, constant across the range of proposed definitions, is that the biophysical world is now profoundly social. This carries substantive methodological implications: If the environment is ecosocial, surely the way it is studied must be, too. Yet, as our bibliometric analysis demonstrates, the bulk of academic articles on the Anthropocene published between 2002 and 2019 focus on its conceptual implications rather than embracing its analytical consequences. Further, of the subset of articles that engage the Anthropocene empirically, fewer than a quarter employ interdisciplinary methods at even a cursory level. In response, we outline an alternative approach, critical physical geography (CPG), which enables researchers to pick up the methodological and conceptual gauntlet thrown down by the Anthropocene. Work in CPG begins from the premise that all biophysical questions are also social, but it goes beyond a simple mixed methods approach to emphasize the politics both of how knowledge is produced and of how it is taken up outside academia. We illustrate the utility of a CPG approach for analysis of the Anthropocene via succinct examples of research in critical dendrochronology, geomorphology, and remote sensing. *Key Words: Anthropocene, critical physical geography, interdisciplinarity, mixed methods.*

Since its popularization in the early 2000s, the Anthropocene concept has sparked a great deal of debate over its stratigraphic start date, primary causes, and intellectual implications. The methodological consequences of the Anthropocene have received much less attention (cf. Harden et al. 2014; F. Biermann et al. 2016; Verburg et al. 2016) but are equally consequential. If we accept the Anthropocene's foundational premise, constant across the range of proposed definitions, that the biophysical world is now profoundly social, surely our methods must be, too. What now justifies the separate physical and social science research traditions that have guided methodological decisions within physical and human geography for decades? Very little, we believe.

Taking the Anthropocene seriously requires a deeply integrative approach to the ecosocial dynamics that shape our field sites, but that is not what most of the rapidly growing body of literature that employs the Anthropocene concept has done. Our bibliometric analysis of literature on the Anthropocene reveals several striking characteristics. First, although there is widespread recognition that empirical work is needed to document and understand the Anthropocene, the existing literature is predominantly conceptual. Second, despite tremendous growth in this literature (Figure 1) across the humanities, social sciences, and biophysical sciences, the integration of biophysical and social methods is surprisingly rare. Third, even in the relatively few cases where this methodological integration has been achieved, mixed methods are often used to address objects of inquiry that are delimited as primarily biophysical or social, rather than acknowledged as fundamentally ecosocial.

Bluntly, existing work on the Anthropocene does not yet live up to the integrative potential of its core concept. Realizing that potential will require researchers to pick up both the methodological and the conceptual gauntlet thrown down by the Anthropocene. As one example of existing work that does so, we offer critical physical geography (CPG), illustrating its utility via succinct examples drawn from our research in critical dendrochronology, geomorphology, and remote sensing.

Methods

To build the bibliometric data set we analyze in this article, we began by searching for academic

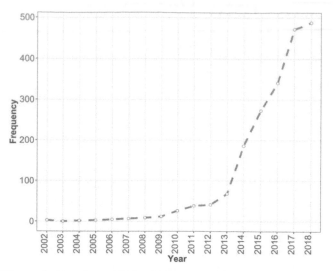

Figure 1 Growth in Web of Science publications mentioning the "Anthropocene" in title, abstract, journal name, or keywords, 2002–2018 (N = 1,974), excluding 17 articles with no publication year and 201 articles from 2019.

articles associated with the term "Anthropocene" in the Thomson Reuters' Web of Science database, which returned a list of 2,548 articles published between 2002 and 2019. Because we were primarily interested in articles that closely engaged the Anthropocene concept, we removed those that did not have the term "Anthropocene" in their title, abstract, journal name, or keywords, resulting in a data set of 2,192 articles in 960 journals (referred to as the full data set throughout).

These were the questions we hoped to answer: Just how methodologically and conceptually integrative is the literature on the Anthropocene? Within this literature, do certain kinds of disciplines, empirical objects, and methods dominate? Bibliometric analysis of such a large data set is a blunt instrument, which cannot do justice to the subtleties of any single article within it. We thus approached our analysis with humility and with awareness that our attempts to sort through the data set would involve subjective judgment calls and thus would likely reveal trends rather than precise delineations.

We began with a broad-brush analysis of the full data set to determine the growth in articles on the Anthropocene between 2002 and 2019 and the most common journals, author keywords, Web of Science keywords, and Web of Science subject categories, which roughly correspond to academic disciplines.[1] To enable finer grained analysis of the methodological approaches and empirical foci of the articles, we sorted the full data set by citation rate

and began to analyze the 100 most cited papers. It quickly became apparent, however, that approximately two thirds of the most cited articles were not empirical (i.e., engaging in primary data collection or new analysis of existing secondary or tertiary data) but instead conceptual (i.e., relying on existing analyses or developing new theoretical frameworks from the existing literature). When we analyzed 100 articles randomly selected from the full data set, we found the same pattern, suggesting that the high volume of conceptual work was not an artifact of citation practices but instead typical of work on the Anthropocene more generally.[2]

Because we were primarily interested in the implications of the Anthropocene for research methods, conceptual papers were not useful for our analysis. Thus, we began again, classifying papers within the full data set, from most to least cited, as either empirical or conceptual until we reached a data set of the 100 most highly cited empirical papers on the Anthropocene. We then read and analyzed each of the 100 most cited empirical papers. To avoid the need for calibration, each of the analyses described here was performed by a single researcher, with ambiguous cases flagged for discussion by all three authors. First, we coded each paper for its empirical focus and then sorted those foci into thirteen overarching categories (e.g., pollution, paleoclimate) using an inductive approach. Second, we coded each paper for the three or fewer primary methods used and again used an inductive approach to categorize specific methods into broader groups (e.g., computational or statistical analysis, remote sensing, and ethnography and interviews). Because research articles are complex and often deal with multiple interrelated phenomena and methods, we do not view these classifications as definitive but used them instead to give a rough indication of frequency.

Finally, we classified the methodological integration (or lack thereof) of each of the 100 most cited empirical papers based on the methods employed and data analyzed, again producing results that we view as illustrative rather than definitive. We classified papers as weakly integrative if they added an existing social or biophysical data set to a more comprehensive analysis, such as adding census data to a paper otherwise entirely focused on biophysical data on invasive species. We classified papers as strongly integrative if they collected and analyzed both social and biophysical data. Within the subset of methodologically integrative papers, we classified methods as social or biophysical, and quantitative or qualitative, resulting in a designation for each method as biophysical qualitative (e.g., soil classification or species distribution mapping), biophysical quantitative (e.g., hydraulic modeling or genomic analysis), social qualitative (e.g., archival or discourse analysis), or social quantitative (e.g., statistical analysis of census data or gross domestic product). This classification presented an analytical challenge, because these categories are not entirely separable. Are data on land use or fish catches social or biophysical? Clearly, they are both. For the purposes of this analysis, however, we chose to sort conservatively, so that even minimal attempts to cross the biophysical–social divide were registered. Thus, if there was any human component to the data analyzed in a predominantly biophysical paper, we grouped the method as social, and vice versa.

Although cognizant of the epistemological limitations of our analysis called out earlier, we nonetheless believe that bibliometric analysis is a useful tool for parsing the now enormous literature on the Anthropocene to identify dominant methodological, disciplinary, and topical orientations. To supplement and work around the limitations of conventional bibliometric analysis, we also engaged in a close reading of work we classified as strongly methodologically integrative. We used this more qualitative analysis to extend our understanding of the strengths and limitations of existing integrative research practice.

Bibliometric Results

As noted by other scholars, geography, particularly physical geography, is well represented in literature on the Anthropocene (see Knitter et al. 2019). Perhaps more surprising, our analyses of Web of Science subject categories (Table 1) and of journals (Figure 2) show that, despite its origin in the biophysical sciences, engagement with the Anthropocene concept is now common across the social sciences and humanities as well. Author keywords suggest the same pattern (Figure 3); although climate change is far and away the most common focus, terms such as "ethics," "political ecology," "posthumanism," and "capitalism" are well represented in the full data set.

Across disciplines, the Anthropocene has had a powerful conceptual impact. Based on our analysis, though, it seems to be primarily in the biophysical sciences that it has spurred large quantities of empirical work (Table 1, Figures 2B and 3B). As mentioned earlier, only roughly one third of the articles we analyzed—both in a random sample of 100 papers and among the most cited papers—engaged in primary data collection or new analysis of existing secondary or tertiary data. This suggests that scholars have predominantly engaged with the Anthropocene concept to propose new research agendas, make conceptual arguments, or reflect on past research. In the humanities and social sciences, the Anthropocene primarily has been engaged conceptually, provoking a wide range of responses: from critique for its undifferentiated assignment of responsibility for anthropogenic environmental change, to praise as an approach for overcoming nature–society binaries. In the biophysical sciences, the Anthropocene has been used to retheorize human–environment relations and to justify synthetic analysis and research that engages human impacts in some way.

We are not arguing that the wealth of conceptual research engaging the Anthropocene is a problem; this research is pushing scholars across academia to think more deeply about human–environment interactions. The relative lack of empirical work, however, suggests that scholars have not yet prioritized designing and carrying out research that reflects the Anthropocene's foundational premise that the biophysical world is now profoundly social as well.

Although empirical scholarship on the Anthropocene seems to have lagged behind conceptual work, there are certain topics where the Anthropocene concept has had a more powerful impact, as demonstrated in our analysis of empirical foci and author keywords. In both the full data set and in the 100 most cited empirical papers, climate change was the most common author keyword (Figures 3A and 3B). Our analysis of topics among the most cited empirical papers diverged sharply from the author keywords, however (Figure 4A). Although climate change provided key context for many of the articles in the smaller data set (and hence was commonly listed by authors as a keyword), it was not the primary empirical focus. This might seem like splitting hairs, but it is notable that the majority of papers listing climate change as a keyword analyzed how climate change affects other

Table 1. Most common Web of Science subject categories

All Web of Science articles on the Anthropocene (N = 2,192) (freq ≥ 20)		Most cited empirical articles on the Anthropocene (N = 100) (freq ≥ 2)	
Web of Science subject category	Frequency	Web of Science subject category	Frequency
Multidisciplinary geosciences/ physical geography	107	Multidisciplinary sciences	34
Multidisciplinary sciences	96	Multidisciplinary geosciences/ physical geography	12
Geography	88	Ecology	6
Multidisciplinary humanities	76	Geography	5
Literature	71	Environmental engineering/ environmental sciences	4
Environmental sciences/environmental studies/multidisciplinary geosciences	63	Environmental sciences/ multidisciplinary geosciences/limnology	4
Philosophy	58	Ecology/physical geography	3
Environmental studies	50	Environmental sciences/meterorology and atmospheric sciences/ multidisciplinary geosciences	3
Anthropology	48	Geochemistry and geophysics	3
Ecology	42	Biochemistry and molecular biology/biology	2
Education and educational research	37	Ecology/environmental sciences	2
Cultural studies	31	Environmental sciences/limnology/ water resources	2
Environmental sciences	31	Multidisciplinary geosciences	2
Multidisciplinary geosciences	31		
Environmental sciences/ multidisciplinary geosciences/ physical geography	30		
Geology	29		
Biology	27		
Religion	27		
Environmental studies/law	25		
Sociology	24		
Environmental sciences/meterorology and atmospheric sciences	23		
Architecture	22		
History	20		

systems rather than focusing on climate per se. Instead, biodiversity change was the most common primary empirical focus (20 percent of papers), with papers addressing nonnative species introductions, biotic homogenization, and other novel patterns of biodiversity produced through human activities and anthropogenic warming. Rivers, streams, and deltas were the second most common topic category among the most cited papers (11 percent). Again, although climate change served as a broad context for many articles in this category, the primary research topics and questions were not directly about climate but instead centered on damming and dam removal, channel dynamics, and water chemistry.[3]

Notably less common were empirical papers that focused on social dynamics, reinforcing our finding that the Anthropocene has been taken up differently in the biophysical sciences versus the social sciences and the humanities. Only 12 percent of the most cited empirical papers focused on social systems (social movements [5 percent]; energy, resources, and economy [4 percent]; and governance [3 percent]), but they shared a common thread: the new social, political, and economic dynamics arising in response to the ecological and climatic changes associated with the Anthropocene. In other words, although the majority of empirical work on the Anthropocene has focused on biophysical systems

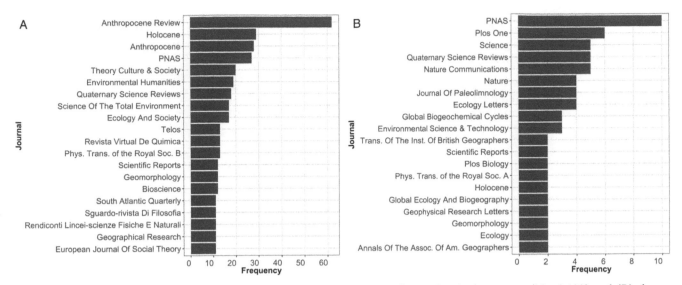

Figure 2 Top twenty most common journals for (A) all Web of Science articles on the Anthropocene ($N = 2,192$) and (B) the most cited empirical articles on the Anthropocene ($N = 100$).

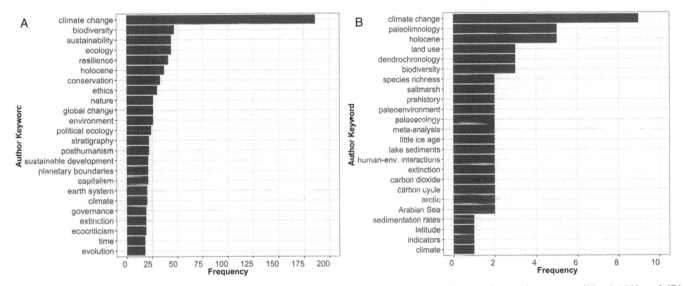

Figure 3 Top twenty-five most common author keywords for (A) all Web of Science articles on the Anthropocene ($N = 2,192$) and (B) the most cited empirical articles on the Anthropocene ($N = 100$).

with humans as catalysts of change, a small subset of empirical work (primarily from the humanities and social sciences) flips the switch to examine social systems, with biophysical systems as catalysts of change.

Our analysis of research methods in the most cited empirical articles highlights a much higher frequency of methods that we classified as quantitative and biophysical than of those that we classified as qualitative or social (Figure 4B), echoing the trends in empirical foci (Figure 4A). The preponderance of computational and statistical methods as well as modeling and simulation approaches is particularly

striking. Unsurprising, papers that focused on biophysical systems generally used methods from the biophysical sciences, whether those methods were primarily quantitative (e.g., isotope geochemistry) or qualitative (e.g., field observations of ecosystems). Conversely, papers that mainly addressed social systems tended to rely on methods associated with the social sciences, whether quantitative (e.g., demographic statistics) or qualitative (e.g., discourse analysis). Further, as shown in more detail later, qualitative social science methods were very rarely used to understand material landscapes or biophysical systems. Similarly, quantitative biophysical

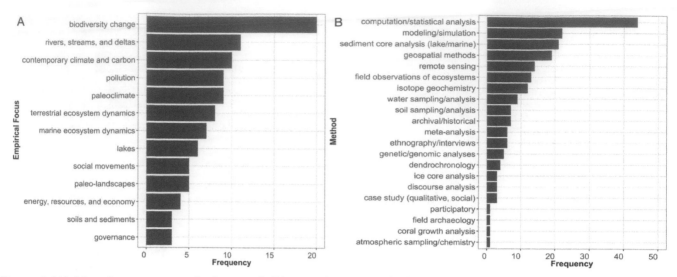

Figure 4 (A) Most frequent empirical object and (B) most frequent methods used in the most cited empirical articles on the Anthropocene (*N* = 100).

science methods were rarely used to investigate primarily social topics (e.g., governance or social movements).

This business-as-usual correspondence between methods and disciplines is perhaps to be expected. If we accept the core premise of the Anthropocene, however—that our world is now ecosocial, shaped by social power relations as much as biophysical processes—then social science methods are directly relevant to research on biophysical phenomena and the limitations and material consequences of that research. We contend that the opposite is true as well: If our social worlds are fundamentally ecological, biophysical science methods can help us understand them better.

Given the relative rarity of biophysical science methods in papers with social empirical foci and vice versa, it is not surprising that fewer than a quarter of the articles in our data set of the 100 most cited empirical papers combine biophysical and social analyses. We classified ten of the 100 articles in the most cited empirical papers data set as weakly integrative methodologically, meaning that they plugged a predominantly social data set into a predominantly biophysical analysis (there were no examples of primarily social analyses adding an existing biophysical data set), and twelve of the articles as strongly integrative, meaning that they collected and analyzed both biophysical and social data. There were no clear patterns among the specific empirical foci and methods used in the twenty-two methodologically integrative articles, but there were some striking absences. For example, none of these papers

focused on obviously ecosocial topics common in the data set of most cited empirical papers, such as pollution or contemporary climate and carbon. Qualitative methods were also far less common than quantitative approaches, with relatively few papers integrating qualitative social science analysis with quantitative analysis of biophysical data (Table 2, Figure 5). By far the most common combination in both weakly and strongly integrative papers was quantitative biophysical with quantitative social methods (eleven papers), and all twenty-two papers performed quantitative analysis of biophysical data.

A closer read of the methodologically integrative articles surprised us, because it demonstrated that combining biophysical and social methods does not in and of itself result in work that engages the intellectual challenge of the Anthropocene concept; it is clearly possible to dip just a toe into Anthropocene waters rather than diving into the deep end of the pool. Roughly half of the integrative articles in the 100 most cited empirical papers data set use mixed methods without disrupting assumed boundaries between biophysical and social objects of inquiry. The two papers that used all four of our methods categories illustrate this contrast well. To analyze the global distribution of invasive species and how that has changed over time, Dyer et al. (2017) expanded traditional biogeographical approaches by including data on gross domestic product (social quantitative, SQN) and the former colonial status of each country (social qualitative, SQL) but did so within an analysis that was otherwise biophysical,

Table 2. Methods used in weakly and strongly integrative papers (N = 22)

Method	Weak	Strong
BQN, SQN	5	6
BQN, SQL	1	1
BQL, BQN, SQL	0	2
BQL, BQN, SQN	3	1
BQN, SQL, SQN	0	1
BQL, BQN, SQL, SQN	1	1

Note: BQN = biophysical quantitative; SQN = social quantitative; SQL = social qualitative; BQL = biophysical qualitative.

adding social context to help answer an otherwise biophysical question about a biophysical topic. By contrast, Dotterweich et al. (2013) analyzed the interactions of land use change and historical soil erosion in Slovakia through a combination of soil mapping (biophysical qualitative, BQL), chemical analysis (biophysical quantitative, BQN), archival research (SQL), and anthropogenic charcoal dating (SQN). Their object of study—human-induced soil erosion—and the questions they asked about it were fundamentally ecosocial.

Even when research questions are framed as ecosocial, there are significant challenges to integrating biophysical and social methods and data in a way that does not oversimplify complex dynamics or reaffirm tired binaries. One subset of methodologically integrative work has used the Anthropocene concept to refashion ecological theories developed without reference to human influence (e.g., island biogeography theory). Yet modeling ecosocial processes requires additional reflexivity around the selection of proxies for social dynamics (e.g., population density) and the biases that certain data or categories inherit and reproduce. There are politics involved in even the most seemingly neutral knowledge production practices, and those politics have ecosocial consequences. To pick just two examples, Sayre's (2008) exploration of the history of the term "carrying capacity" shows both its colonial and deeply racist roots and the mass slaughters of livestock and devastating economic losses that the carrying capacity concept has justified. Similarly, Davis's (2007) ecological and political history of the colonization of North Africa shows how French ecologists' training in humid environments led them to view semiarid environments as degraded, justifying land seizures and other policies that destroyed native livelihoods.

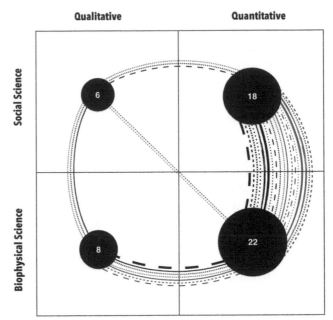

Figure 5. Classification of methods used in methodologically integrative papers by biophysical qualitative, biophysical quantitative, social qualitative, and social quantitative. Each of the twenty-two papers in this data set is represented by a distinctly colored and patterned line connecting the methods used, with the black circles indicating how many of the twenty-two studies had used each of the four categories of methods. The lines are shaded black or red to indicate strong or weak integration, respectively.

Our larger point is this: To understand the Anthropocene, we must draw on both biophysical and social methods. It is equally important, though, that the object of study and the questions guiding the research are themselves defined in ecosocial and reflexive ways. As the preceding examples suggest, and as we argue next, it is also critical to do so in ways attentive to the power relations shaping the uneven distribution of environmental benefits and burdens.

Putting the Anthropocene into Practice: Critical Physical Geography

Attempts to integrate social and biophysical science are increasingly common, encompassing approaches that range from paleoecology to political ecology. This diversity in scholarship adds to the vitality of existing conversations on methodologically and analytically integrative approaches. Yet we were struck by the fact that although geographical scholarship is very well represented in our subset of strongly integrative work, certain subfields predominated (e.g., historical geomorphology), whereas others were absent (e.g., critical human geography). We believe that geographers can more broadly engage the analytical challenge posed by the Anthropocene. CPG (Lave et al. 2014) is one example of how we can do so.

CPG bridges concepts and methods from critical human and physical geography and shares the foundational premise of the Anthropocene concept: The biophysical world that surrounds us is now profoundly social as well. Taking this a step further, CPG contends that to understand ecosocial systems, we must reach across conventional disciplinary boundaries to develop deeply integrative empirical research, investigating biophysical processes, knowledge politics, and structural power relations together (Lave, Biermann, and Lane 2018). As seen in our bibliometric analysis, this runs counter to the majority of existing work on the Anthropocene, which remains siloed in the biophysical sciences, social sciences, or humanities.

Before discussing three examples of CPG research, we outline its core tenets. First, in keeping with the Anthropocene concept, CPG begins from the premise that most material landscapes and biophysical systems are now deeply shaped by human actions. Further, CPG emphasizes the role of structural power relations and inequalities around race, gender, and class in shaping ecosocial systems (Urban 2018). A second core tenet of CPG is that the same social dynamics and power relations at work in the systems and landscapes we study also shape the production of knowledge about these systems. Reflecting on the scholarship we have surveyed here, this means that the Anthropocene framework, research questions, empirical foci, methods, and even findings are all imbricated in social, cultural, and political–economic relations (King and Tadaki 2018). Finally, the third tenet of CPG recognizes that academic scholarship has real material and political consequences for the people and landscapes studied (Law 2018).

Bringing together these three core tenets with an expansive methodological toolkit, CPG allows us to pursue empirical research that addresses the interconnectedness of material landscapes and systems, social power relations, and knowledge politics. In doing so, CPG allows for an analysis of empirical objects that are conceived as themselves both biophysical and social. For example, Engel-Di Mauro

(2014), Turner (2018), and Malone (Malone and Polyakov (2020) used critical pedological approaches to explain the class and gender implications of soil chemical profiles, soil fertility in agropastoral systems, and sediment runoff in agricultural watersheds. Wilcock (Wilcock, Brierley, and Howitt 2013), Arce-Nazario (2018), and Brierley (Brierley et al. 2018) developed a shared platform among indigenous groups and geomorphologists for understanding cultural–biophysical hydroscapes, demonstrating the environmental injustice of Environmental Protection Agency monitoring intended to produce more just water quality outcomes, and engaging geomorphologically with recent moves to grant legal personhood to rivers.

As we have come to understand from our own research, framing research questions as ecosocial from the outset through a CPG approach also informs subsequent methodological and conceptual possibilities. A first example of this is Biermann's work on tree rings and the researchers who study them. The annual growth rings of trees provide an archive of data on past ecosystems, with tree-ring-based climate reconstructions typically predicated on the idea that it is possible to identify a stable relationship between tree growth rates and prevailing climatic conditions. Yet this core uniformitarian assumption has been increasingly called into question by dendrochronological science (Wilmking et al. 2020). Beginning from a fundamentally ecosocial appreciation of this uncertain state of affairs informed Biermann's decision to combine biophysical analysis of mid-elevation pine–oak woodlands in the U.S. Great Smoky Mountains National Park with social analysis of how dendrochronologists are responding to this major epistemological shift in interpreting tree-ring data (C. Biermann and Grissino-Mayer 2018). In turn, bringing these data together enabled Biermann to conceptualize the linkages between tree-ring growth rates, climate science, and global-scale climate agendas more fully than is commonly done within either dendrochronology or a social analysis of climate science.

A second example comes from Lave's work, which integrates methods from fluvial geomorphology with political economy and science and technology studies to study a strongly ecosocial object: stream mitigation banking, a form of market-based environmental management under the U.S. Clean Water Act. Stream mitigation banking allows developers to offset damage to inconveniently located streams by purchasing mitigation credits produced by for-profit companies that speculatively restore damaged streams. This approach of "selling nature to save it" (McAfee 1999) has sparked a great deal of debate but remarkably little attention to whether it has better environmental outcomes than nonmarket approaches to conservation. Using geomorphological surveys of channel form, interviews, and policy analysis (among other methods), Lave and her colleagues investigated both how a market for streams has been constructed and the hydroscape that market produces, revealing that the contrast between the dynamism and complexity of ecosystems and the stability and simplicity required for functional markets radically limits the conservation potential of market-based approaches (Lave, Biermann, and Lane 2018).

A third example comes from Kelley's analysis of environmental change on the island of Sulawesi, Indonesia, which integrates computational approaches for mapping landscape change with ethnographic and survey analyses of cacao expansion. Kelley adopted this approach because ethnographic findings revealed that smallholder cacao plantings were facilitating the revegetation of grasslands and swidden fallows, complicating more essentializing narratives about the connection between cacao and deforestation on the basis of yield imperatives. Drawing on satellite data revealed the significance of this trend while highlighting how dominant remote sensing practices were often obscuring an analysis of variability in landscape change (Kelley, Evans, and Potts 2017). Ethnographic and survey data in turn helped to explain this variability in relation to uneven histories of state violence and displacement in Sulawesi. Together, these findings reveal the diverse ecosocial changes associated with cacao while also challenging the implicit politics of attributing most regional deforestation to smallholder cacao producers (Kelley 2018).

These examples demonstrate how work in CPG can address the methodological and analytical challenges inherent in the Anthropocene concept and apparent in our bibliometric analysis. Work in CPG is helping to retheorize the objects we study and engage, and to make visible the ways in which social dynamics (knowledge, power, identity) are implicated in environmental change processes and vice versa.

Further, these examples demonstrate how the core premise of the Anthropocene—that the biophysical world is also deeply social—extends beyond our field sites and into our research processes and findings. Whereas our bibliometric analysis reflects the need for greater methodological integration and more eco-social research objects, the preceding CPG examples take this integrative impulse a step further to consider the politics of knowledge production and the material consequences of these politics. For Biermann, this means reflecting on how tree ring researchers are moving toward more humble and relational analyses that give weight to local, more contextual environmental management approaches. For Lave, this involves investigating how homogenization of stream form follows from redefining something as messy and interconnected as a stream in the standardized terms of a marketable commodity. For Kelley, attention to knowledge politics led to methods that are not only integrative but multiscalar and that add visibility to the local historical, social, and spatial relations often silenced in analyses of deforestation and deforestation policy in Indonesia.

Thus, we believe that CPG is well suited to developing the empirics of the Anthropocene because it involves not only integrative methods, questions, and research objects but also a theoretical framework that insists on reflexivity in research and attention to the ecosocial consequences of knowledge production.

Conclusion

The Anthropocene concept undermines the pursuit of purely social or biophysical scientific explanations by focusing on how intensifying human activities have shaped new ecologies, new types of human–environment interactions, and new challenges for resource management and conservation. Despite the potential of this idea, however, we found more talk about integrative research than action. Not only is empirical work on the Anthropocene relatively rare vis-à-vis synthetic or conceptual pieces, but the empirical work that does exist appears to be dominated by methodological approaches from the biophysical sciences. How might geographers and other human–environment scholars better meet the challenge of the Anthropocene—particularly when, as we have suggested here, this challenge cannot be solved simply by adding a social variable to an otherwise biophysical analysis (or vice versa)?

Answers to this question are being developed in a range of fields. Although most political ecologists are not trying to integrate biophysical and social methods (notable exceptions include Nightingale, Robbins, and Turner), political ecology provides a strong platform for conceptualizing the environment in more ecosocial ways. As scholars such as Munroe (Munroe et al. 2014), Roy Chowdhury (Roy Chowdhury et al. 2011), and Napoletano (Napoletano, Paneque-Galvez, and Vieyra 2015) have demonstrated, there is work within land change science that is clearly ecosocial. It also seems that some objects of inquiry might lend themselves more to integrative approaches: In our analysis, most of the work on paleolandscapes, for example, was strongly integrative.

This article highlights CPG as one promising and proven way forward. Work in CPG begins from the premise that all biophysical questions are also social, thus orienting analytical attention to ecosocial interactions and relations from the get-go. Through its emphasis on power and the politics of knowledge production, CPG also provides a means of reflexively situating associated research approaches. CPG thus offers a platform for mixing methods while also going beyond the assumption that mixing methods (on its own) will be enough to adequately capture and explain human–environment relations and change.

Notes

1. We also examined Web of Science's categorization of journals by Research Areas but found that these larger buckets (e.g., technology, social sciences, arts and humanities) were less helpful than the more detailed categories, in part because the journal classifications were confusing (e.g., *PNAS* was listed as technology).
2. Bibliometric scholarship suggests other potential characteristics of highly cited papers that might distinguish them from the data set as a whole, such as the gender of lead authors (Larivière et al. 2013).
3. It is possible that climate change is used in a similar way in other areas of scholarship: as context rather than direct empirical focus. Our data set does not allow us to explore this question.

References

Arce-Nazario, J. 2018. The science and politics of water quality. In *The Palgrave handbook of critical physical geography*, ed. R. Lave, C. Biermann, and S. N. Lane, 465–84. London: Palgrave.

Biermann, C., and H. Grissino-Mayer. 2018. Shifting climate sensitivities, shifting paradigms: Tree-ring science in a dynamic world. In *The Palgrave handbook of critical physical geography*, ed. R. Lave, C. Biermann, and S. N. Lane, 201–22. London: Palgrave.

Biermann, F., X. Bai, N. Bondre, W. Broadgate, C.-T. Arthur Chen, O. P. Dube, J. W. Erisman, M. Glaser, S. van der Hel, M. C. Lemos, et al. 2016. Down to earth: Contextualizing the Anthropocene. *Global Environmental Change* 39:341–50. doi: 10.1016/j.gloenvcha.2015.11.004.

Brierley, G., M. Tadaki, D. Hikuroa, B. Blue, C. Sunde, J. Tunnicliffe, and J. Salmond. 2018. A geomorphic perspective on the rights of the river in Aotearoa New Zealand. *River Restoration Applications* 35 (10):1–12. doi: 10.1002/rra.3343.

Davis, D. 2007. *Resurrecting the granary of Rome: Environmental history and French colonial expansion in North Africa*. Athens: Ohio University Press.

Dotterweich, M., M. Stankoviansky, J. Minar, S. Koco, and P. Papco. 2013. Human induced soil erosion and gully system development in the Late Holocene and future perspectives on landscape evolution: The Myjava Hill Land, Slovakia. *Geomorphology* 201:227–45. doi: 10.1016/j.geomorph.2013.06.023.

Dyer, E. E., P. Cassey, D. W. Redding, B. Collen, V. Franks, K. J. Gaston, K. E. Jones, S. Kark, C. D. L. Orme, and T. M. Blackburn. 2017. The global distribution and drivers of alien bird species richness. *PLoS Biology* 15 (1):e2000942. doi: 10.1371/journal.pbio.2000942.

Engel-Di Mauro, S. 2014. *Ecology, soils, and the Left*. New York: Palgrave.

Harden, C. P., A. Chin, M. R. English, R. Fu, K. A. Galvin, A. K. Gerlak, P. F. McDowell, D. E. McNamara, J. M. Peterson, N. L. Poff, et al. 2014. Understanding human–landscape interactions in the "Anthropocene." *Environmental Management* 53 (1):4–13. doi: 10.1007/s00267-013-0082-0.

Kelley, L. 2018. The politics of uneven smallholder cacao expansion: A critical physical geography of agricultural transformation in Southeast Sulawesi, Indonesia. *Geoforum* 97:22–34. doi: 10.1016/j.geoforum.2018.10.006.

Kelley, L., S. Evans, and M. Potts. 2017. Richer histories for more relevant policies: 42 years of tree cover loss and gain in Southeast Sulawesi, Indonesia. *Global Change Biology* 23 (2):830–39. doi: 10.1111/gcb.13434.

King, L., and M. Tadaki. 2018. A framework for understanding the politics of science (core tenet #2). In *The Palgrave handbook of critical physical geography*, ed. R. Lave, C. Biermann, and S. N. Lane, 67–88. London: Palgrave.

Knitter, D., K. Augustin, E. Biniyaz, W. Hamer, M. Kuhwald, M. Schwanebeck, and R. Duttmann. 2019. Geography and the Anthropocene: Critical approaches needed. *Progress in Physical Geography: Earth and Environment* 43 (3):451–61. doi: 10.1177/0309133319829395.

Larivière, V., C. Ni, Y. Gingras, B. Cronin, and C. Sugimoto. 2013. Bibliometrics: Global gender disparities in science. *Nature News* 504 (7479):211–13. doi: 10.1038/504211a.

Lave, R., C. Biermann, and S. N. Lane. 2018. *The Palgrave handbook of critical physical geography*. London: Palgrave.

Lave, R., M. Doyle, M. Robertson, and J. Singh. 2018. Commodifying streams: A CPG approach to stream mitigation banking in the U.S. In *The Palgrave handbook of critical physical geography*, ed. R. Lave, C. Biermann, and S. N. Lane, 443–64. London: Palgrave.

Lave, R., M. Wilson, E. Barron, C. Biermann, M. Carey, C. Duvall, L. Johnson, M. Lane, N. McClintock, D. Munroe, et al. 2014. Critical physical geography. *The Canadian Geographer / Le Géographe Canadien* 58 (1):1–10. doi: 10.1111/cag.12061.

Law, J. 2018. The impacts of doing environmental research (core tenet #3). In *The Palgrave handbook of critical physical geography*, ed. R. Lave, C. Biermann, and S. N. Lane, 89–103. London: Palgrave.

Malone, M., and V. Polyakov. 2020. A physical and social analysis of how variations in no-till conservation practices lead to inaccurate sediment runoff estimations in agricultural watersheds. *Progress in Physical Geography: Earth and Environment* 44 (2):151–67. doi: 10.1177/0309133319873115.

McAfee, K. 1999. Selling nature to save it? Biodiversity and green developmentalism. *Environment and Planning D: Society and Space* 17 (2):133–54. doi: 10.1068/d170133.

Munroe, D., K. McSweeney, J. Olson, and B. Mansfield. 2014. Using economic geography to reinvigorate land-change science. *Geoforum* 52 (1):12–21. doi: 10.1016/j.geoforum.2013.12.005.

Napoletano, B., J. Paneque-Galvez, and A. Vieyra. 2015. Spatial fix and metabolic rift as conceptual tools in land-change science. *Capitalism Nature Socialism* 26 (4):198–214. doi: 10.1080/10455752.2015.1104706.

Roy Chowdhury, R., K. Larson, M. Grove, C. Polsky, E. Cook, J. Onsted, and L. Ogden. 2011. A multi-scalar approach to theorizing socio-ecological dynamics of urban residential landscapes. *Cities and the Environment* 4 (1):1–21. doi: 10.15365/cate.4162011.

Sayre, N. 2008. The genesis, history, and limits of carrying capacity. *Annals of the Association of American Geographers* 98 (1):120–34. doi: 10.1080/00045600701734356.

Turner, M. 2018. Questions of imbalance: Agronomic science and sustainability assessment in dryland West Africa. In *The Palgrave handbook of critical physical geography*, ed. R. Lave, C. Biermann, and S. N. Lane, 421–41. London: Palgrave.

Urban, M. 2018. In defense of crappy landscapes (core tenet #1). In *The Palgrave handbook of critical physical geography*, ed. R. Lave, C. Biermann, and S. N. Lane, 49–66. London: Palgrave.

Verburg, P., J. Dearing, J. Dyke, S. Van Der Leeuw, S. Seitzinger, W. Steffen, and J. Syvitski. 2016. Methods and approaches to modelling the Anthropocene. *Global Environmental Change* 39:328–40. doi: 10.1016/j.gloenvcha.2015.08.007.

Wilcock, D., G. Brierley, and R. Howitt. 2013. Ethnogeomorphology. *Progress in Physical Geography: Earth and Environment* 37 (5):573–600. doi: 10.1177/0309133313483164.

Wilmking, M. M., van der Maaten-Theunissen, E. van der Maaten, T. Scharnweber, A. Buras, C. Biermann, M. Gurskaya, M. Hallinger, J. Lange, R. Shetti, et al. 2020. Global assessment of relationships between climate and tree growth. *Global Change Biology* 26 (6):3212–20. doi: 10.1111/gcb.15057.

Part 4

Natural Hazards, Disasters, and the Anthropocene

The Changing Nature of Hazard and Disaster Risk in the Anthropocene

Susan L. Cutter (iD)

The concept of the Anthropocene provides the reflexive rubric for examining human alteration of the Earth's basic natural systems and, in turn, how society responds to and adapts to such changes in its life support systems. Nowhere is this more evident than in natural hazards and disaster risk. This article examines the changing nature of hazard and disaster risk from local to global scales highlighting three thematic areas. The first is the redefinition of what constitutes extremes (lower probability, higher consequence events) and the movement away from characterizing hazards as extreme events to a focus on the chronic or everyday events, the cumulative impacts compounding to produce impacts often far greater than the periodic extreme event. The second theme examines the confluence and complexity in the production and reproduction of hazards and disaster risk. The intersection of natural systems, human systems, and technology produces a tightly coupled and complex array of potential failure modes and large-scale impacts, so much so that when a natural hazard occurs in one part of the world, the consequences extend well beyond the affected region, which in turn affects the global economic system. The third theme is the increasing social, procedural, and spatial inequalities in disaster risk. These three themes are explored using historic and recent examples from North America. *Key Words: disaster risk, extreme events, hazards, social vulnerability.*

As noted elsewhere in this special issue, the exact starting markers of the Anthropocene are a subject of great interest and contested debates in contemporary society. Lewis and Maslin (2015) suggested two different potential "start dates." The first is the Orbis spike (~1610) representing hemispheric connectivity, the rise of globalization, and the development of the world economic system. The second, called the C^{14} spike (~1964), marks the so-called Great Acceleration of human population, capacities to alter fundamental biogeochemical systems and processes, discoveries and industrialization of new chemicals and materials, and technologies capable of delivering chemical elements (e.g., plutonium) into weapons of mass destruction. As noted by Dominey-Howes (2018), hazards and disasters research has much to contribute to the contemporary understanding of the Anthropocene given the crisis-oriented narratives in some of these debates on origins, evidence, and meanings. He suggests that scholarship in hazards and disasters offers useful guidance conceptually and methodologically to many of these conversations.

This article examines the changing nature of hazards and disaster risk and local to global consequences using the C^{14} spike as the benchmark. Three short arguments highlight the social construction of disaster risk from a geospatial perspective. First, there is increasing evidence that a singularly large extreme event, although impactful in proximate space and time, is far less consequential in the longer term than less extreme but chronic (or everyday) events on places. Second, cascading disasters are becoming the norm rather than the exception, with consequences well beyond the immediate area or time frame. Third, inequalities in disaster risk are increasing and in some instances accelerating beyond the capacity of places to cope with disasters. Confirmatory evidence based on a synthesis of the extant literature supports these arguments.

Hazard Paradigms Past and Present

The postwar paradigm of hazards and disasters focused on hazards (potential threats to people and the things they valued) and disasters (discrete events, spatially and temporally concentrated that disrupted a community, thus affecting its ability to function). The linear thinking employed in hazards and disasters scholarship assumed that the physical force (hazard or disaster agent) interacted with exposure and vulnerability (people, assets) to create risk (the probability of losses or the probability of positive or negative outcomes), which in turn produced the adverse outcome. The outcome labeled emergency, disaster, or catastrophe depends on the magnitude frequency of the initiating hazard event or disaster agent and the associated outcomes (Alexander and Pescaroli 2019).

The traditional hazards paradigm, best illustrated by White and his students, focused on a number of related themes including the following:

1. Range of choice of adjustments taken by decision makers to align human activities within natural constraints of the local environment.
2. Role of hazard perception and decision processes in the choice of adjustments.
3. Spatial delineation of the human occupancy of hazard zones (Cutter et al. 2019).

Critiques of the hazards orthodoxy began in the early 1980s with the rise of the vulnerability paradigm and the proclamation that without people and their associated vulnerability, there was no such thing as a natural disaster (O'Keefe, Westgate, and Wisner 1976; Hewitt 1983). Vulnerability became the root cause of disasters, partially explaining the degree to which different social groups in society were differentially at risk from the same natural forces or hazards. The new paradigm also helped to conceptualize how vulnerability accentuated by historical social, economic, political, and social processes affected local places, which in turn led to the unsafe conditions, thereby increasing vulnerability (Wisner et al. 2004).

The multidisciplinary vulnerability paradigm (Wisner 2016) continues and is coevolving along with companion concepts in hazards and disaster research such as resilience and transformation (Matyas and Pelling 2014). This new hazard and disaster science paradigm focuses on integrated evidentiary-based, action-oriented research coproduced with multiple stakeholders employing a variety of qualitative, quantitative, or mixed-method approaches (Gaillard and Mercer 2012; Ismail-Zadeh et al. 2017). It also reemphasizes the anthropocentric nature of hazards and disasters. As Tierney (2019) wrote, "The severity of a disaster is measured not by the magnitude of the physical forces involved, but rather by the magnitude of its societal impacts" (4).

Extreme Events and Extreme Consequences

Defining extreme events is an emergent area of multidisciplinary scholarship in hazards (McPhillips et al. 2018), with available outlets in publications such as the *Journal of Extreme Events* and *Earth's Future*. Yet, there are many definitions of what constitutes an extreme. What might be defined as rare (an oft-used synonym for extreme) in one context or discipline might be considered normal by another authoring group. Extreme climate or weather events are based on statistically defined thresholds; for example, the "occurrence of a value of the weather or climate variable above (or below) a threshold value near the upper (or lower) ends of the range of observed values of the variable" (Intergovernmental Panel on Climate Change 2012, 557). The most recent Intergovernmental Panel on Climate Change report, AR5, defines an extreme weather event as "an event that is rare at a particular place and time of year. Definitions of rare vary, but an extreme weather event would normally be as rare as or rarer than the 10th or 90th percentile of a probability density function estimated from observations" (Intergovernmental Panel on Climate Change 2014, 123). Advancements in extreme value theory and application have advanced understanding of the statistical distributions of solid earth extreme events such as earthquakes and volcanic eruptions, as well as the economic impacts of hazards (Katz 2015).

There are other metrics used to define extremes not in statistical or probability terms but rather based on human consequences. These include financial losses exceeding a monetary threshold such as the National Oceanic and Atmospheric Administration's (NOAA) "Billion-Dollar Weather and Climate Disasters" product (NOAA 2020) or human-centered consequences such as direct deaths or excess mortality (Hammer 2018; Arnold 2019). There is ongoing debate on the attribution of extreme events and impacts to the climate signal (National Research Council 2016), catalyzing a new multidisciplinary field of attribution science examining not only the statistical likelihood of change but also the relative contributions of multiple causal factors to such changes and their statistical significance.

Extreme Physical Forces Producing Extreme Consequences

The question of whether or not extreme events always produce extreme consequences has been addressed elsewhere (Cutter 2016). For example, an extreme precipitation event in 2015 in South Carolina (popularly labeled as a 1,000-year event) produced historical rainfall exceeding twenty inches (51 cm) over a three- to five-day period throughout the central and eastern portions of the state, with locally heavy downpours of two inches (5 cm) or more of rainfall per hour. This led to catastrophic flooding throughout the central and eastern portions of the state (National Weather Service 2016). Two years later, Hurricane Harvey, another extreme precipitation event, dumped a record-breaking fifty-plus inches (127 cm) of rainfall in the Houston metropolitan area, leading to disastrous flooding throughout the region (National Weather Service 2018).

Since 2017, a series of extreme Category 5 hurricane events have occurred in the Atlantic Basin. Hurricanes Irma and Maria, initially Category 5 storms, caused billions of dollars in property damage in the U.S. Virgin Islands, Puerto Rico, and Florida, as well as other island nations in the Caribbean, but were only Category 4 storms at landfall. A year later, Hurricane Michael (another Category 5 hurricane) struck the Florida panhandle, causing billions of dollars in damages. Hurricane Michael was the third Category 5 storm at landfall in the contiguous United States during the Anthropocene behind Hurricane Camille (1969) and Hurricane Andrew (1992). In 2019, Hurricane Dorian (another Category 5 hurricane) devastated the Bahamas, creating the greatest disaster in that island nation's history, with billions of dollars in damages and countless lives lost.

We assume that statistically defined extreme events produce disastrous consequences to the affected areas. What about events not defined as statistical extremes based on the magnitude or frequency distributions of the physical parameters of the hazard agent itself but that result in catastrophic impacts? Are these extreme events as well?

Extreme Consequences Produced by Nonextreme Physical Forces

At its peak, Hurricane Katrina was a Category 5 extreme wind event. The winds diminished to Category 3 at landfall, however. Was it a meteorological extreme? Or, given its location, was it an extreme consequence event produced by nonextreme physical forces based on antecedent conditions as well as the failure of the protective levee system for New Orleans? The locally experienced storm surge and weeks of flooding from levee breaches and failures resulted in catastrophic consequences for vulnerable populations in New Orleans and Mississippi. At the time, Hurricane Katrina was the most deadly (1,883 recorded deaths) and costliest (estimated $125 billion in damages) storm in U.S. history (National Hurricane Center 2018). Hurricane Sandy was not an extreme hurricane based on wind speeds (Category 1 winds at landfall near Atlantic City) but was extreme in its consequences (Finn 2020). Again, due to storm surge, the physical size of the storm, and landfall location (New York metropolitan region), significant flooding in New York City and the metropolitan area produced catastrophic damage to the major infrastructure in the region.

The recent experience in California with wildfires is equally illustrative. California's fire-adapted ecosystems have a long historical legacy, but anthropogenic climate change has played a significant role in enhancing the weather conditions that increase the frequency and intensity of the wildfire risk. When coupled with increased population density and expansion of suburban housing into the wildland–urban interface (WUI), the severity of wildfire impacts increased through more exposed residential property and more potential human ignition sources (either intentional or unintentional; Calkin, Short, and Traci 2020). Twenty of the most destructive wildfires in California occurred from 1999 to 2018. Once ignited, wildfires in WUI areas moved quickly and spread rapidly because of topography, weather (high winds during the dry season), and availability of fuel from neighboring burning structures in the WUI. By most accounts, wildfire risk in California is increasing, with the likelihood of continued catastrophic losses despite wildfire mitigation and management, enhanced building codes, and increased insurance rates in the WUI to cover the losses (Calkin, Short, and Traci 2020).

The bottom line is that you do not need an extreme (based on probability distributions) initiating event to produce extreme consequences from a disaster, based on damages, deaths, or both. Weather-related disasters in the Anthropocene no longer fit the frequency and magnitude estimations of low-probability, high-consequence events. Instead, higher frequencies of heavy precipitation events (Myhre et al. 2019) and high-loss wildfire events in California constitute the new normal of larger loss events occurring with greater frequency.

Moving from Extremes to the Everyday

Changes in the Earth's climate have transformed the production of extreme events into annual, seasonal, monthly, or everyday occurrences. A recent study (Corringham et al. 2019) found that storms associated with the most extreme atmospheric rivers caused about nearly half of all of the flood damage in the west from 1978 to 2017 and 99 percent of the flood damage in the coastal regions of California and Oregon during the same time period. Furthermore, the study emphasized that atmospheric rivers (or Ark Storms) will increase in size and intensity and affect more communities throughout the west as global warming continues. Atmospheric rivers now are predictable seasonal events with weather reports announcing their arrival along with local alert and preparedness information for residents.

Elsewhere in the country, heavy precipitation events show a steady increase in the so-called extreme single-day precipitation events, especially since the 1990s. A warmer climate means more frequent and intense heavy downpours from frontal systems. These heavy precipitation events (ranking in the 99th percentile of rainfall and snowfall daily events) have increased in all regions according to the fourth National Climate Assessment (Easterling et al. 2017), with predictions of increases by 40 percent or more of these events in the future, especially in the Northeast (Scott 2019).

Heavy precipitation downpours are a major contributor to the increases in pluvial flooding, especially in urban areas, with actual monetary losses from urban flooding exceeding the recorded losses by billions of dollars annually due to the inadequacy of the existing data (National Academies of Sciences, Engineering, and Medicine 2019). Notable heavy precipitation inland flooding events include the 2013 Boulder flood; the 2016 Ellicott City, Maryland, flash flood; and the 2016 West Virginia multiple flash flood events.

Nuisance or sunny-day flooding in coastal areas used to be unusual with the annual king tide but has now transformed into a routine occurrence, producing chronic flooding conditions that disrupt daily activities, affect public and private infrastructural systems such as roads and sewers, damage property, and most significantly threaten public safety (Moftakhari et al. 2018). Sea-level rise exacerbates the coastal nuisance flooding along the U.S. Atlantic and Gulf Coasts. One study found a 90 percent average increase in Atlantic Coast nuisance floods over the last twenty years, periodically inundating more than 7,500 miles (12,070 km) of roadways including major interstates (Jacobs et al. 2018). Such transportation disruptions and delays will increase at all locations up and down the East Coast; NOAA tidal gauge data show a linear increase in frequency of high tidal flooding in twenty-five locations since 2000 and accelerating frequencies of flood days at forty-two locations during the same time frame (Sweet et al. 2019).

Costly and deadly wildfires in California are now expected annual occurrences, with insurance companies refusing to renew homeowners' policies in many areas. More than half of California's top twenty most destructive wildfires occurred since 2010, destroying 34,320 structures and causing 144 fatalities, nearly all (80 percent) caused by electrical grids and powerlines (Cal Fire 2019). The most destructive was the November 2018 Camp Fire in Butte County, destroying Paradise and surrounding WUI communities.

Multiple sources of hazards, from heavy precipitation, to rising seas, to wildfires, are now frequent events producing nearly continuous impacts on the communities they affect. Managing such risks is one of the key challenges facing communities and emergency managers today.

Cascading Hazards Increase Human Insecurity

The concept of cascading hazards (also called compound hazards, secondary hazards, or complex disasters) is not new. Yet there has been a resurgence in interest in these types of large-scale events (Cutter 2018) during the last few decades. Some cascading events (e.g., the Tohoku earthquake, tsunami, and Fukushima reactor meltdown) reflect their potential to disrupt regional, national, and global economies. The global connectedness of systems of production and consumption, in addition to the extensive role of humans as geosocial actors (Dalby 2017), increases human insecurity in places subject to many of these cascading events. Other forms of cascading disasters (Pescaroli and Alexander 2015, 2016) reflect the complexity of modern society and the interdependent sociotechnical systems that support it, which expand the impact of the precipitating event well beyond the original location. Indeed, society itself is responsible for constructing such cascading risks either overtly or covertly, as Beck (1999) warned us decades ago. The reflexive nature of risk necessitates societal reconfigurations to manage it. Such changes are derivative from political processes infused with systemic power and wealth inequalities that contribute to the differential vulnerabilities, thus positioning most risk accumulation at the bottom of the power structure. Finally, the unintended consequences of cascades reflect an inability of society to anticipate the unexpected, further precipitating failures in critical infrastructure or accelerating preexisting conditions of vulnerability.

An example of a commonly known cascade is the wildfire–torrential rains–mudslides sequence especially prevalent in California. The occurrence of the postfire debris flows mechanism is well known (Oakley et al. 2018), but the timing and location of the cascade is quite variable from one wildfire event to another. The Thomas wildfire in December 2017 burned nearly 282,000 acres (114,000 hectares) in Ventura and Santa Barbara counties, destroying more than 1,000 structures and causing two fatalities (Cal Fire 2019). As the fire moved westward, pushed by the strong Santa Ana winds, it threatened hillside homes in the wealthy coastal enclave of Montecito, damaging or destroying dozens of homes. In early January 2018, heavy rains fell over the scarred landscape and triggered a debris flow that destroyed hundreds more homes in the community as it made its way to the Pacific Ocean, resulting in twenty-three fatalities in Montecito (Hamilton and Serna 2018). The early morning timing of the debris flow and less residential compliance with mudslide warnings and evacuations, unlike wildfire evacuations, contributed to the higher death toll.

Another dominant cascade type is a by-product of the risk society and is especially prevalent in regions where natural hazards intersect with industrial facilities and processes to create toxic and potentially deadly releases of chemicals into the air or water. These types of cascades or natech accidents (Showalter and Myers 1994; Cruz and Suarez-Paba 2019) have increased, necessitating different risk reduction strategies that are often regional in scope, regulatory in their approach, and in need of different governance and legal infrastructures to manage the risk (Krausmann, Girgin, and Necci 2019). Recent examples of natech events in North America include industrial toxic releases in the petrochemical corridor in Houston after Hurricane Harvey, which included both acute releases (industrial explosion and fire of the Arkema facility) and releases into the floodwaters from inundated facilities and damage done to Superfund sites from the floodwaters and the resultant soil contamination (Bajak and Olsen 2019).

Natech cascades also occur in rural areas. During Hurricane Florence (2018), floodwaters caused breaches in lagoons in the industrialized hog farms (or concentrated animal feeding operations) in North Carolina, as happened during Hurricane Floyd two decades earlier (Pierre-Louis 2018). Another rural example is flooding-induced breaches in coal ash containment ponds, which release the toxic residue from coal-fired power plants such as the case with Hurricane Florence flooding in North Carolina (Thrush and Pierre-Louis 2018). According to a recent study, more than 100 coal ash ponds and landfalls in the United States are located in Federal Emergency Management Administration–designated high-flood hazard areas (1 percent chance flood zone; Colman 2019). Power companies are aware of their liability for damages from such breaches but rarely disclose the risk to downstream communities. With likely increases in heavy precipitation events due to climate change, such flood-induced breaches could happen more often at both active and inactive sites, including surface flooding and groundwater contamination from rising water tables.

A newly emergent cascade is triggered by forecasts of strong Santa Ana winds in Southern California or Diablo winds in Northern California, leading the power company to shut off power as a precautionary measure to avoid the liability of igniting a wildfire from powerlines. The preemptive mitigation measure initially deployed by PG&E, the state's largest utility, on 9 October 2019 affected a half-million people in northern California, who had no power for days (Fuller 2019). Not only did the power shutdown severely disrupt daily lives (school and business closures, transportation safety, government services), but it also hastened a major health emergency. State and federal law requires hospitals to have backup diesel generators, but outpatient clinics and doctors' offices are not required to do so. There is no requirement for backup power for households who depend on the use of electricity-dependent medical equipment such as ventilators at home. The rise in electricity-dependent populations, especially among the chronically ill, disabled, aging and populations and other Medicare beneficiaries, prompted the U.S. Department of Health and Human Services to strengthen local disaster preparedness and the continuity of care for these at-risk and special needs populations (HHS emPower Program 2019).

Perhaps the poster child, however, for cascading impacts increasing human insecurity is the historical socially produced vulnerability of Puerto Rico and the catastrophic outcomes ensuing from repeated encounters with tropical storms and hurricanes—the most recent being Hurricanes Irma and Maria in 2017—and the series of earthquakes in January 2020 including a magnitude 6.4 earthquake. The trajectory of recovery is less than optimistic because of the preexisting conditions of financial insolvency, lack of transparent governance structures, and damage to critical infrastructures such as power, water, and health care. Some of the damage to critical infrastructures was due to the hurricanes, but some of it was also self-inflicted based on poor construction and maintenance, resulting in large disparities between urban centers and outlying rural areas that accelerated the exodus of Puerto Ricans to the mainland. The sociostructural impacts of mass outmigrations, increasing percentages of elderly, persons in poverty, and children raised by grandparents because their parents work off the island exacerbates the existing sociospatial vulnerability (Santos-Hernández, Méndez-Heavilin, and Álvarez-Rosario 2020).

Accelerating Disparities Create Unequal Impacts

The pace of social, political, economic, and technological change has quickened during the Anthropocene, producing even more disparities in the modern risk society, especially in affluent societies like the United States. Racial, economic, political, gender, and technological divides derived from long-standing historical social, economic, and political processes segment U.S. society into haves and have-nots. Households or communities that have are able to better prepare for, respond to, and recover from a disaster event than those who do not. As the nation's demographics change, the structural racism, classicism, and sexism that permeate U.S. society enhance and enlarge the existing social vulnerability in communities, further reducing capacities to adequately respond to unexpected disaster events and, even more significant, to respond to chronic events.

Wealth does not exempt everyone from hazards, although wealthier individuals and communities can reduce the impact through enhanced mitigation and resilience given their resources. Wealth (total assets minus total liabilities) inequality has steadily increased since 1995, so that 75 percent of the wealth is now concentrated in 20 percent of U.S. households (Sawhill and Pulliam 2019). More telling for the future is the rise in generational inequality in wealth accumulation, with younger households having less net worth than two decades ago.

Social inequality based on the intersection of race, class, gender, and age influences the impacts of disasters and the creation of vulnerability (Tierney 2014). It is the poor who disproportionately bear the burdens of disasters as a consequence of their preexisting vulnerability (Bolin and Kurtz 2018) and exposure (housing locations in hazard zones or high-risk areas, housing tenure as renters not owners (Lee and van Zandt 2019). Poorer households can rely on societal safety nets from government relief programs and private philanthropic services organizations to assist in postdisaster recovery. There are increasing procedural inequities in federal responses to disasters (Willison et al. 2019), however, as well as the allocation of individual assistance resources (Domingue and Emrich 2019).

Increasingly the middle class often endures the most impacts, because they are too rich to qualify for many of the governmental relief programs and too poor to have the accumulated assets to repair homes and livelihoods after an event, as was the case in New Orleans after Hurricane Katrina (Finch, Emrich, and Cutter 2010). Many middle-class families live paycheck to paycheck, with little available cash to cover an emergency, highlighting the economic fragility of U.S. households even in good economic times (Federal Reserve Board 2019). According to a Federal Reserve Board (2019) survey, 27 percent of adults would not be able cover an unexpected $400 expense (e.g., car repair, a medical expense, or temporary relocation due to flooding) and cover all of the current monthly bills.

Concluding Thoughts

This article began with the premise that hazard and disaster risk has accelerated with the beginning of the Anthropocene, using the C^{14} spike as the benchmark. The magnitude and frequency of disasters have changed in the Anthropocene as illustrated here. Events once thought extreme in their magnitude and frequency have transformed into annual or seasonal frequency. It no longer takes an extreme event to produce catastrophic damage at local to regional to national scales. The increasing societal exposure to hazards as a combination of vulnerability and society's own production of risks through development in high-hazard areas and alterations of the Earth's physical systems is increasing human insecurity from local to global scales. Industrialization, globalization, and technology have connected us in untoward ways, making societal systems interdependent, often producing unintended consequences from everyday events such as rain, snow, heat, or cold and the cascading impacts initiated by them. Demographic changes and widening disparities result in unequal impacts, with disaster risk concentrated in places and households at the bottom rung of the wealth ladder. Many of the processes and impacts described here apply globally, but in some places they are even more detrimental, especially the disproportionate effects of disasters on women and children (Cutter 2017).

ORCID

Susan L. Cutter ⓘ http://orcid.org/0000-0002-7005-8596

References

Alexander, D., and G. Pescaroli. 2019. What are cascading disasters? *UCL Open Environment* 1:03. doi: 10.14324/111.444/ucloe.000003.

Arnold, C. 2019. Death, statistics and a disaster zone: The struggle to count the dead after Hurricane Maria. *Nature News Feature*, February 5. Accessed January 8, 2020. https://www.nature.com/articles/d41586-019-00442-0.

Bajak, F., and L. Olsen. 2019. Silent spills: Environmental damage from Hurricane Harvey is just beginning to emerge. *Houston Chronicle*, March 5. Accessed January 4, 2020. https://www.chron.com/news/houston-weather/hurricaneharvey/article/Silent-Spills-Environmental-damage-from-12768677.php.

Beck, U. 1999. *World risk society.* Malden, MA: Polity Press.

Bolin, B., and L. C. Kurtz. 2018. Race, class, ethnicity, and disaster vulnerability. In *Handbook of disaster research*, ed. H. Rodríguez, W. Donner, and J. E. Trainer, 181–203. Dordrecht, The Netherlands: Springer

Cal Fire. 2019. Top 20 most destructive California wildfires. Accessed January 2, 2020. https://www.fire.ca.gov/media/5511/top20_destruction.pdf.

Calkin, D., K. Short, and M. Traci. 2020. California wildfires. In *Emergency management in the 21st century: From disaster to catastrophe*, ed. C. B. Rubin and S. L. Cutter, 155–82. London and New York: Routledge.

Colman, Z. 2019. The toxic waste threat that climate change is making worse. *Politico*, August 26. Accessed January 4, 2020. https://www.politico.com/story/2019/08/26/toxic-waste-climate-change-worse-1672998.

Corringham, T. W., F. M. Ralph, A. Gershunov, D. R. Dayan, and C. A. Talbot. 2019. Atmospheric rivers drive flood damages in the western United States. *Science Advances* 5 (12):eaax4631. doi: 10.1126/sciadv.aax4631.

Cruz, A. M., and M. C. Suarez-Paba. 2019. Advances in natech research: An overview. *Progress in Disaster Science* 1 (May):100013. doi: 10.1016/j.pdisas.2019.100013.

Cutter, S. L. 2016. The changing context of hazard extremes: Events, impacts, and consequences. *Journal of Extreme Events* 3 (2):1671005. doi: 10.1142/S2345737616710056.

Cutter, S. L. 2017. The forgotten casualties redux: Women, children, and disaster risk. *Global Environmental Change* 42:117–21. doi: 10.1016/j.gloenvcha.2016.12.010.

Cutter, S. L. 2018. Compounding, cascading, or complex disasters: What's in a name? *Environment: Science and Policy for Sustainable Development* 60 (6):16–25. doi: 10.1080/00139157.2018.1517518.

Cutter, S. L., R. H. Platt, I. Burton, J. K. Mitchell, M. Reuss, C. B. Rubin, J. L. Wescoat, Jr., and B. T. Richman. 2019. Reflections on Gilbert F. White: Scholar, advocate, friend. *Environment: Science and Policy for Sustainable Development* 61 (5):4–21. doi: 10.1080/00139157.2019.1637671.

Dalby, S. 2017. Anthropocene formations: Environmental security, geopolitics and disaster. *Theory, Culture & Society* 34 (2–3):233–52. doi: 10.1177/0263276415598629.

Dominey Howes, D. 2018. Hazards and disasters in the Anthropocene: Some critical reflections for the future. *Geoscience Letters* 5:7. doi: 10.1186/s40562-018-0107-x.

Domingue, S. J., and C. T. Emrich. 2019. Social vulnerability and procedural equity: Exploring the distribution of disaster aid across counties in the United States. *American Review of Public Administration* 49 (8):897–913. doi: 10.1177/0275074019856122.

Easterling, D. R., K. E. Kunkel, J. R. Arnold, T. Knutson, A. N. LeGrande, L. R. Leung, R. S. Vose, D. E. Waliser, and M. F. Wehner. 2017. Precipitation change in the United States. In *Climate science special report: Fourth national climate assessment*, ed. D. J. Wuebbles, D. W. Fahey, K. A. Hibbard, D. J. Dokken, B. C. Stewart, and T. K. Maycock, Vol. I, 207–30. Washington, DC: U.S. Global Change Research Program. doi: 10.7930/J0H993CC.

Federal Reserve Board. 2019. Report on the economic well-being of U.S. households in 2019. Board of Governors of the Federal Reserve System, Washington, DC. Accessed January 5, 2020. https://www.federalreserve.gov/consumerscommunities/files/2018-report-economic-well-being-us-households-201905.pdf.

Finch, C., C. T. Emrich, and S. L. Cutter. 2010. Disaster disparities and differential recovery in New Orleans. *Population and Environment* 31 (4):179–202. doi: 10.1007/s11111-009-0099-8.

Finn, D. 2020. Hurricane Sandy: The New York City experience. In *Emergency management in the 21st century: From disaster to catastrophe*, ed. C. B. Rubin and S. L. Cutter, 61–90. London and New York: Routledge.

Fuller, T. 2019. PG&E outage darkens northern California amid wildfire threat. *The New York Times*, October 9. Accessed January 4, 2020. https://www.nytimes.com/2019/10/09/us/california-power-outage-PGE.html.

Gaillard, J. C., and J. Mercer. 2012. From knowledge to action: Bridging gaps in disaster risk reduction. *Progress in Human Geography* 37 (1):93–114. doi: 10.1177/0309132512446717.

Hamilton, M., and J. Serna. 2018. Montecito braced for fire, but mud was a more stealthy, deadly threat. *Los Angeles Times*, January 12. https://www.latimes.com/local/lanow/la-me-montecito-mudslide-main-20180112-story.html.

Hammer, C. C. 2018. Understanding excess mortality from not-so-natural disasters. *The Lancet Planetary Health* 2:e471–e472. doi: 10.1016/S2542-5196(18)30222-5.

Hewitt, K. 1983. *Interpretations of calamity*. Winchester, MA: Allen and Unwin.

HHS emPower Program. 2019. *Executive summary: Shaping decisions to protect health in an emergency*. Washington, DC: U.S. Health and Human Services. Accessed January 4, 2020. https://empowermap.hhs.gov/emPOWER_Executive%20Summary_FINAL_508.pdf.

Intergovernmental Panel on Climate Change. 2012. *Managing the risks of extreme events and disasters to advance climate change adaptation. A special report of Working Groups I and II of the Intergovernmental Panel on Climate Change*, ed. C. B. Field, V. Barros, T. F. Stocker, D. Qin, D. J. Dokken, K. L. Ebi, M. D. Mastrandrea, K. J. Mach, G.-K. Plattner, S. K. Allen, M. Tignor and P. M. Midgley. Cambridge, UK: Cambridge University Press.

Intergovernmental Panel on Climate Change. 2014. Annex II: Glossary. In *Climate change 2014: Synthesis report. Contribution of Working Groups I, II and III to the fifth assessment report of the Intergovernmental Panel on Climate Change*, ed. Core Writing Team, R. K. Pachauri, and L. A. Meyer, 117–30. Geneva, Switzerland: Intergovernmental Panel on Climate Change.

Ismail-Zadeh, A. T., S. L. Cutter, K. Takeuchi, and D. Paton. 2017. Forging a paradigm shift in disaster science. *Natural Hazards* 86:969–88. doi: 10.1007/s11069-016-2726-x.

Jacobs, J. M., L. R. Cattaneo, W. Sweet, and T. Mansfield. 2018. Recent and future outlooks for nuisance flooding impacts on roadways on the U.S. East Coast. *Transportation Research Record* 2672 (2):1–10. doi: 10.1177/0361198118756366.

Katz, R. W. 2015. Economic impact of extreme events: An approach based on extreme value theory. In *Extreme events: Observations, modeling, and economics,*

ed. M. Chavez, M. Ghil, and J. Urrutia-Fucugauchi, 205–17. Washington, DC: American Geophysical Union.

Krausmann, E., S. Girgin, and A. Necci. 2019. Natural hazard impacts on industry and critical infrastructure: Natech risk drivers and risk management performance indicators. *International Journal of Disaster Risk Reduction* 40:101163. doi: 10.1016/j.ijdrr.2019.101163.

Lee, J. Y., and S. van Zandt. 2019. Housing tenure and social vulnerability to disasters: A review of the evidence. *Journal of Planning Literature* 34 (2):156–70. doi: 10.1177/0885412218812080.

Lewis, S. L., and M. A. Maslin. 2015. Defining the Anthropocene. *Nature* 519:171–80. doi: 10.1038/nature14258.

Matyas, D., and M. Pelling. 2014. Positioning resilience for 2015: The role of resistance, incremental adjustment and transformation in disaster risk management policy. *Disasters* 39 (S1):S1–S18. doi: 10.1111/disa.12107.

McPhillips, L. E., H. Chang, M. V. Chester, Y. Depietri, E. Friedman, N. B. Grimm, J. S. Kominoski, et al. 2018. Defining extreme events: A cross-disciplinary review. *Earth's Future* 6:441–55. doi: 10.1002/2017EF000686.

Moftakhari, H. R., A. AghaKouchak, B. F. Sanders, M. Allaire, and R. A. Matthew. 2018. What is nuisance flooding? Defining and monitoring an emerging challenge. *Water Resources Research* 54:4218–27. doi: 10.1029/2018WR022828.

Myhre, G., K. Alterskjaer, C. W. Stjern, Ø. Hodnebrog, L. Marelle, B. H. Samset, J. Sillmann, et al. 2019. Frequency of extreme precipitation increases extensively with event rareness under global warming. *Nature Scientific Reports* 9:16063. doi: 10.1038/s41598-019-52277-4.

National Academies of Sciences, Engineering, and Medicine. 2019. *Framing the challenge of urban flooding in the United States*. Washington, DC: National Academies Press. doi: 10.17226/25381.

National Hurricane Center. 2018. *Costliest U.S. tropical cyclones tables updated*. Miami, FL: National Oceanic and Atmospheric Administration, National Hurricane Center. Accessed December 19, 2019. https://www.nhc.noaa.gov/news/UpdatedCostliest.pdf.

National Oceanic and Atmospheric Administration, National Centers for Environmental Information. 2020. Billion-dollar weather and climate disasters. Accessed January 11, 2020. https://www.ncdc.noaa.gov/billions/.

National Research Council (NRC). 2016. *Attribution of extreme weather events in the context of climate change*. Washington, DC: National Academies Press.

National Weather Service. 2016. *The historic South Carolina floods of October 1–5, 2015*. Silver Spring, MD: U.S. Department of Commerce, National Oceanic and Atmospheric Administration, National Weather Service, Service Assessment. Accessed December 19, 2019. https://www.weather.gov/media/publications/assessments/SCFlooding_072216_Signed_Final.pdf.

National Weather Service. 2018. *August/September 2017 Hurricane Harvey*. Silver Spring, MD: U.S.

Department of Commerce, NOAA National Weather Service. Accessed January 8, 2020. https://www.weather.gov/media/publications/assessments/harvey6-18.pdf.

Oakley, N. S., F. Cannon, R. Munroe, J. T. Lancaster, D. Gomberg, and F. M. Ralph. 2018. Brief communication: Meteorological and climatological conditions associated with the 9 January 2018 post-fire debris flows in Montecito and Carpinteria, California, USA. *Natural Hazards Earth Systems Science* 18:3037–43. doi: 10.5194/nhess-18-3037-2018.

O'Keefe, P., K. Westgate, and B. Wisner. 1976. Taking the naturalness out of natural hazards. *Nature* 250:566–67. doi: 10.1038/260566a0.

Pescaroli, G., and D. Alexander. 2015. A definition of cascading disasters and cascading effects: Going beyond the "toppling dominos" metaphor. *GRF Davos Planet@Risk* 3 (1):58–67. Accessed April 9, 2020. https://pdfs.semanticscholar.org/0607/4f128dfc90ea9ad b5d54b629bcc586199089.pdf.

Pescaroli, G., and D. Alexander. 2016. Critical infrastructure, panarchies and the vulnerability paths of cascading disasters. *Natural Hazards* 82:175–92. doi: 10.1007/s11069-016-2186-3.

Pierre-Louis, K. 2018. Lagoons of pig waste are overflowing after Florence. Yes, that's as nasty as it sounds. *The New York Times*, September 19, 2018. Accessed January 4, 2020. https://www.nytimes.com/2018/09/19/climate/florence-hog-farms.html.

Santos-Hernández, J. M., A. J. Méndez-Heavilin, and G. Alvarez-Rosario. 2020. Hurricane María in Puerto Rico: Pre-existing vulnerabilities and catastrophic outcomes. In *Emergency management in the 21st century: From disaster to catastrophe*, ed. C. B. Rubin and S. L. Cutter, 183–208. London and New York: Routledge.

Sawhill, I. V., and C. Pulliam. 2019. Six facts about wealth in the United States. *Brookings Blog*, June 25. Accessed January 5, 2020. https://www.brookings.edu/blog/up-front/2019/06/25/six-facts-about-wealth-in-the-united-states/.

Scott, M. 2019. Prepare for more downpours: Heavy rain has increased across most the United States and is likely to increase further. Accessed December 30, 2019. https://www.climate.gov/news-features/featured-images/prepare-more-downpours-heavy-rain-has-increased-across-most-united-0.

Showalter, P. S., and M. F. Myers. 1994. Natural disasters in the United States as release agents of oil, chemicals, or radiological materials between 1980–9: Analysis and recommendations. *Risk Analysis* 14 (2):169–81. doi: 10.1111/j.1539-6924.1994.tb00042.x.

Sweet, W., G. Dusek, D. Marcy, G. Carbin, and J. Marra. 2019. 2018 State of U.S. high tide flooding with a 2019 outlook. NOAA Technical Report NOS-CO-OPS 090, National Oceanic and Atmospheric Administration, Washington, DC. Accessed January 2, 2020. https://tidesandcurrents.noaa.gov/publications/Techrpt_090_2018_State_of_US_HighTideFlooding_with_a_2019_Outlook_Final.pdf.

Thrush, G., and K. Pierre-Louis. 2018. Energy plan, sending toxic coal ash into river. *The New York Times*, September 21. Accessed January 4, 2020. https://www.nytimes.com/2018/09/21/climate/florences-floodwaters-breach-defenses-at-power-plant-prompting-shutdown.html.

Tierney, K. 2014. *The social roots of risk: Producing disasters, promoting resilience.* Palo Alto, CA: Stanford University Press.

Tierney, K. 2019. *Disasters: A sociological approach.* Medford, MA: Polity Press.

Willison, C. E., P. M. Singer, M. S. Creary, and S. L. Greer. 2019. Quantifying inequities in U.S. federal response to hurricane disaster in Texas and Florida compared with Puerto Rico. *BMJ Global Health* 4:e001191. doi: 10.1136/bmjgh-2018-001191.

Wisner, B. 2016. Vulnerability as concept, model, metric, and tool. In *Natural hazard science: Oxford research encyclopedias.* D. Benouar, Editor in chief, 1-51, Accessed January 4, 2020. New York: Oxford University Press. doi: 10.1093/acrefore/9780199389407.013.25.

Wisner, B., P. Blaikie, T. Cannon, and I. Davis. 2004. *At risk: Natural hazards, people's vulnerability and disasters.* 2nd ed. London and New York: Routledge.

Seismic Shifts: Recentering Geology and Politics in the Anthropocene

Ben A. Gerlofs ⓘD

A strident focus on atmospheric carbons and on climate change as its distinguishing feature has seen much debate and research surrounding the Anthropocene stray from its conceptual grounding in geology. Yet new research argues that hallmarks of the Anthropocene such as sea-level rise, melting ice sheets, and environmental engineering projects designed to mitigate chronic shortages of potable water all increase the potential of seismic activity, placing geological forces back at the center of conversations on planetary futures. Following calls to consider the political dimensions of environmental change in the Anthropocene, this article examines the multiple layerings of natural hazards, political crises and transformations, and mega-urbanization through the lens of catastrophic geological events, drawing evidence from an ongoing longitudinal study of the devastating Mexico City earthquakes of 1985 and 2017. This case suggests that such events intersect with contemporary political economy in a variety of ways, from the disruption of urban investment patterns and the initiation of processes of mass abandonment or of creative destruction and gentrification to the exacerbation of political crises at scales from the submunicipal to the supranational, the facilitation of social movement formation, and the acceleration of political revolutions such as Mexico's democratic transition of 2000. These insights offer productive avenues for engaging with the political economies of geological events and processes and the prescriptive powers of geology as a field of knowledge production and political power that increasingly demand attention across an urbanizing planet. *Key Words: Anthropocene, earthquakes, gentrification, geology, Mexico City.*

The recent focus on the myriad and multiplying catastrophes rooted in the systematized rapacity of the anthropogenic ravaging of the planet, not least in this journal and its discipline, is in the main a healthy and necessary move toward correction. The frames through which this attention is filtered and power organized, however, as Wainwright and Mann (2018) forcefully reminded, remain political questions despite diffuse postpolitical impulses to occlude, preempt, ignore, elide, and deny this reality across the globe (Swyngedouw 2010, 2013). Emergent political formations and those of the future will undoubtedly find much of their legitimacy in such questions, likely couched, as they often are at present, in languages of emergency. On the geophysical side of such catastrophic proofs loom encroachments of the deep from sea-level rise to superstorms of increasing ferocity, atmospheric apparitions from megalopolitan urban smog to the record-breaking temperatures of each new season, and the lachrymose knells and early dirges of vast forests and hordes of plant and animal species. The social forecasts of such narratives appear equally grim: typically colored with teeming masses of "climate refugees" (Bettini 2013) spun from the maelstrom by famine, flood, and resource conflicts and infested with the microbial and metabolic by-products of our epoch leeched or intentionally disposed into earth, sea, sky, and flesh. Casting the devastations of present and future planetary ecologies specifically as catastrophes discursively enrolls this sociopolitical dimension. As opposed to the language of crisis or disaster, Maldonado-Torres (2019) explained, catastrophe names a process of profoundly social rupture, a necessary decolonial extension of N. Smith's (2006) one-time polemic, "There's no such thing as a natural disaster." This era of human intervention into geological time has been named, controversially, the Anthropocene. Although this naming makes a strong attempt to give the lie to the immutable potency of nonhuman nature, the Anthropocene's empirical reach and political utility are limited by its very novelty, as well as by its polemical emergence and its oft-cited ambiguity. These issues, along with other factors, have produced an understandable proclivity toward certain analytically charismatic systems and processes, such as those described earlier, to the detriment of other geographies also being drastically transformed at the nature–culture nexus. As new research suggests, some of the Anthropocene's most powerful forces may be both scientifically and materially buried below the surface. In particular, I argue in this article that attention should be refocused on geology and geological events, especially earthquakes, in seeking to understand and respond to the politics of this unstable epoch.

A number of anthropogenic or anthropogenically induced processes have been scientifically linked to increased seismic activity, including groundwater depletion (González et al. 2012; Amos et al. 2014), disturbance from extractive industry such as wastewater and other waste fluid disposal wells (P. Smith 1978; see also Galchen 2015), and changing tectonic pressures associated with ice sheet depletion (McGuire 2016), the influence of storms of increasing severity (Liu, Linde, and Sacks 2009; Lovett 2013),[1] and glacial isostatic adjustment and associated sea-level change (see Chen 2016). Perhaps predictably, such connections have not received significant attention from policymakers. To take just one prominent example, despite an overwhelming and still growing scientific consensus that points the causative arrow emanating from the exponential explosion of earthquakes (induced seismicity, in a telling characterization) in Oklahoma firmly at the disposal wells of the oil and gas industry, the official statements and behaviors of industry representatives and, more significant, local officials remain obstinately mum at best and duplicitously obfuscatory at worst (Galchen 2015). Far more surprising is that Anthropocene seismic events have failed to garner much attention from the scientific community—particularly in academic geography (Melo Zurita, Munro, and Houston 2018),[2] a discipline distinctly suited to Anthropocene research (Johnson et al. 2014; Ziegler 2019). This lack of attention to the subterranean is what Bebbington and Bury (2013) referred to as a "surface bias" in geographical research.

Simply asserting the centrality of geological events and processes to Anthropocene political formations in a context defined increasingly by global exigency, however, obscures more profound and unsettling imbrications of geology with the histories and impulses of settler colonialism, patriarchy, racism, and innumerable other ills in the production of what Clark and Yusoff (2017) called "geosocial formations." Most fundamental, the academic and practical field of geology organizes the ontological power to separate life from nonlife, which Povinelli (2016) called "geontopower." This authority has been used to such ends, as Yusoff (2018) argued, as the drawing of scientifically gilded "geologic color line[s]" that produce naturalized "forced intimac[ies] with the inhuman" (xii) themselves "predicated on the presumed absorbent qualities of black and brown bodies to take up the body burdens of exposure to toxicities and to buffer the violence of the earth" (3). That geology presents itself as the naive and natural language for explaining and responding to events like earthquakes belies this ability to rupture and reconfigure political economies and social geographies across material and conceptual terrain alike. Quite distinct from and far preceding the depoliticizing effects of global climate discourses cited earlier, and in even more fundamental ways than those captured by Braun's (2000) explication of the modern discipline's way of seeing, the capacity to emplace entities (bodies, energies, matter) in particular relationship to one another (e.g., the human and the inhuman, the lively and the inert), or to displace them from such relationships, thereby affords geology a rare privilege among the sciences.

These developments suggest, on the one hand, that Anthropocene research and debate within and beyond geography must contend with the reality that the geophysical processes that geology seeks to understand are themselves changing due to and in concert with anthropogenic activity, provoking the forces that Grosz collectively called "geopower" in largely unforeseen and as yet only nascently understood ways (Grosz, Yusoff, and Clark 2017). On the other hand, such a recentering of geology should be done critically, with an eye toward its power, as a field of knowledge production, to define reality, regulate discourse, and dictate the path of change. In what remains of this article, I use evidence gathered from a longitudinal research project[3] based around the two most destructive earthquakes in Mexico City's recent history to demonstrate the significance of earthquakes and geology more broadly to Anthropocene political futures. These two events intersect in complex and catastrophic ways with geographical, ecological, economic, political, and sociocultural life in Mexico, seeping and surging into generational transformations decades on from the mere seconds of magnitude they are often afforded in political analyses. In the following section I present the catastrophic nature of Mexico City's 1985 and 2017 earthquakes from several angles, collectively offering a view toward the relevance of these and similar events to Anthropocene geographical inquiry. In the concluding section I draw on Maldonado-Torres's (2019) prescriptions for counter-catastrophic thinking and behavior to demonstrate the dramatic and multifaceted potential of earthquakes as social and political–economic actants. Refocusing inquiry on the political questions that unmistakably comprise the core of the Anthropocene, I argue, requires a reevaluation of the politics of geology in the most vulgar and extraordinary senses. I also offer some notes toward understanding earthquakes in their catastrophic magnitude from a geographical perspective, from the accumulating tensions of the prolonged and deceptive calm to the breakneck hysteria of seismic eruption and the machines[4] and solutions forged, fashioned, and deployed in the desperate social furnace of their aftermath.

Earthquake as Catastrophe in Mexico City

The devastating earthquakes of September 1985 are generally considered to have killed at least some 10,000 of the Mexican capital's residents, although estimates vary considerably into the tens of thousands. The first of these occurred at 7:19 a.m. on 19 September at a magnitude of 8.1, with a nearly equal successor following the next evening, destroying or damaging structures across the central sections of the city, burying countless residents in rubble, and leaving several hundred thousand displaced from their homes. Although the epicenter of the earthquake[5] was located several hundred kilometers away on the Pacific coast, a tragic convergence of geography, geology, and centuries of environmental engineering conspired to transport the earthquake's harshest effects across this distance with unfathomable ease. Mexico City is a sprawling megalopolis constructed on the floor of a vast intermontane valley, much of it atop the silty beds of a great system of interconnected lakes drained over the last several centuries. As a result of this underlying geology, a city official explained to a group of my students in spring 2019, seismic activity shakes Mexico City as if it sat upon a great bowl of gelatin. As Simon (2002) explained of the 1985 earthquake, "The seismic waves were trapped in the spongy soils under the ancient lakes;

they bounced around wildly, hurling themselves against the denser basaltic rock that once marked the lake-shores, and then vibrating back through the soft soil until they hit something solid" (534). The *damnificados* (victims, affected persons), who numbered in the hundreds of thousands, dug through rubble for loved ones and strangers, worked cooperatively to ensure the provision of medical care and basic needs, and otherwise banded together for survival in the relative absence of government support.[6]

Decidedly a destructive environmental event, the 1985 earthquake was also profoundly political. Many contemporary residents and some academic analyses place the earthquake squarely in a catalytic chain of democratization (e.g., Tavera-Fenollosa 1998; Gutmann 2002; Olsen and Gawronski 2003) leading to the country's democratic transition of 2000, some even characterizing it as the linchpin. From 1928 until 2000, Mexico was governed by the ruling party, the Partido Revolucionario Institucional (Institutional Revolutionary Party, hereafter PRI), with peculiar consequences for residents of the capital. The lack of local democracy and the party's increasingly violent authoritarian practices that coincided with massive population growth and unprecedented rates and extensions of urban development conjointly elicited and nurtured opposition militancy in the city, especially after the Tlatelolco massacre of 1968.[7] The earthquake struck at a particularly inopportune moment for the PRI, as the city and party struggled through the torturous throes of economic crisis, caustically prodigious neoliberal reforms, and the disquieting emergence of a schism between technocratic and traditional factions tearing at the core of the party (Ward 1990; Davis 1994; Bruhn 1997). An emergency response many consider grossly inadequate only fed the growing grassroots opposition to the PRI, as desperate residents turned toward each other, the church, and nongovernmental organizations for support (Ai Camp 2014). Banding together in solidarity and in opposition, *damnificados* and their allies soon coalesced into or joined organized political groups and social movements, many swelling the ranks of the increasingly prominent Movimiento Urbano Popular (Urban Popular Movement, hereafter MUP).[8] Davis (1994) made this claim most forcefully, carefully articulating the intrusion of the earthquake into several growing challenges to the hegemony of the PRI in precisely those areas of the city most affected by the 1985 earthquake:

Not only did these conditions breed even greater support for grassroots democratic control, they directly challenged the PRI's legitimate claim to authority, at least in Mexico City, which in previous administrations had been based in large part on the party's ability to deliver the goods. The party's failure to respond adequately to the circumstances motivated citizens to question past political practices and to take control of the urban situation themselves. Indeed, the earthquake spurred Mexico City residents to struggle for greater participation and control over urban servicing and their own political destinies, which challenged the PRI as never before. (282)

Although the precise magnitude of the 1985 earthquake's bearing on Mexico's electoral transition to democracy in 2000 remains a contentious subject, its role in the protracted concatenation of democratic advance in Mexico City (typically parceled under the aegis of political reform) is undeniable. Even the somewhat dubious Kandell (1988) argued that political unrest after the earthquake led to calls for the direct election of the city's mayoralty,[9] an important step in democratization eventually accomplished in 1997 in the person of Cuauhtémoc Cárdenas and a benchmark victory for opposition militants. Further reforms accomplished by legislation, jurisprudence, and constitutional amendment have subsequently transformed Mexico City's political geography episodically since the late 1980s, including most notably the recent abolition of the capital's status as a Federal District and (re-)ascension to equal status among the states of the Mexican Union and the provisioning of its first political constitution. The most dramatic of these changes were strategically timed to coincide with the centennial of Mexico's 1917 constitution, in a layering of auspicious gestures toward a glorious official narrative of the Mexican Revolution.

As if issuing from the cruelest autodidactic malice, a highly destructive magnitude 7.1 earthquake also marked an anniversary in 2017, striking the city on the same date as the earthquake of 1985 (19 September). Notable differences in epicenter and tectonic initiative marked the 1985 and 2017 earthquakes (see Melgar and Pérez-Campos 2018), although the geographies affected had considerable overlap in the capital. In both cases, neighborhoods near the city's historic center were hit particularly hard, including the colonias (officially designated neighborhoods) of Roma, Juárez, Centro Histórico, Doctores, Guerrero, Morelos, and Hipódromo/Condesa.[10] The toll was not as severe in 2017,

although dozens of structures collapsed and hundreds lost their lives. Although the city witnessed widespread spontaneous civilian responses as it had in 1985, there has by comparison been precious little in the way of antigovernment sentiment expressed in the aftermath. Aside from the national presidential election of former Mexico City Mayor Andrés Manuel López Obrador and a general municipal electoral shift toward his new Morena party—which, given long-standing predictions can hardly be lain at the feet of the earthquake—it remains to be seen what, if any, political ramifications might result. Rumors and accusations now increasingly abound in many of the affected areas, however, of opportunistic state interventions into the built environment and redevelopment processes said to favor and indeed ensure the rapidly accelerating alteration of a number of the city's neighborhoods. Here, too, the specter of 1985 haunts both residential mistrust and verifiable patterns of transformation.

Holding aside the lack of adequate emergency response, in the months and years that followed the 1985 earthquake the Mexican government did engage in a robust effort to rehouse many (although not all) of those displaced, most especially through a new program jointly funded by the World Bank and the Mexican state known as the Renovación Habitacional Popular (Popular Housing Renovation, hereafter RHP). According to the World Bank (1993), the RHP rehoused some 78,000 families in the years immediately after the disaster.[11] Although many of those rehoused had been renters living in tenement conditions in the vecindades of the city's historic center, the RHP focused on the construction of owner-occupied dwellings for resettlement, creating a massively subsidized windfall for some of the urban poor and working classes of the city center, and initiating a rift in these classes between those chosen for rehousing and those damnificados left on the outside of its benefits (Ward 1990).[12] Despite these and other diffuse efforts to repair and repopulate the city's central neighborhoods, many areas saw significant abandonment after the earthquake, and many structures and properties went without formal attention or improvement for years. Thus, notwithstanding the government's rapid class promotion of some of those displaced,[13] many of the affected neighborhoods suffered several decades of decline and developed local reputations as places to be avoided by the gente decente (decent people). In recent years, however, most of these same neighborhoods have seen an influx of new residents, accompanied by massive waves of property redevelopment, a flood of foreign and domestic capital pouring into commercial and residential real estate, and startling rises in prices, exploiting a "rent gap" (N. Smith 1979) produced by a peculiar alchemy of interscalar politics and the persuasive powers of subsurface geology and its prophets.

Although such a process is typically christened "gentrification" in Anglophone academic parlance, many local residents object to the use of its Hispanicized transposition, *gentrificación*, opting instead for an emergent endogenous concept, *blanqueamiento por despojo* (whitening by dispossession).[14] Building in part on Harvey's (2003) notion of "accumulation by dispossession," this conceptualization enrolls a constellation of forces and effects into a much broader understanding of neighborhood change, including especially the locally specific dynamics of race and property that often elude global languages like gentrification, as well as the geopolitical tensions produced and experienced around the sudden rush of foreign finance capital and the imposition of more international architecture and aesthetics. In addition to a steadily growing and auspicious series of efforts by public and private forces to clean up many of these neighborhoods in just such fashions in recent years (Crossa 2013; Díaz 2015), some residents have begun to report more insidious modes of transformation at work, especially after the 2017 earthquake. From subterranean power and sewer upgrades replacing "damaged" infrastructures and suspicious official designations for evacuation, demolition, and repair to brazen work-arounds in historic preservation law, residents increasingly point an accusatory finger at various levels and agencies of the state for complicity or collusion with predatory real estate interests and foreign capital exploiting disaster to reave yet further depths of the rapidly crowding rent gap in colonias like Roma, Juárez, Doctores, and Centro. In this telling, the earthquake plays the role Klein (2007) outlined for "shock" in several registers, from inducing pliable political conditions for intervention to violently clearing the slate of the built environment.

This brief sketch has introduced only some of the many ways in which the 1985 and 2017 earthquakes are entangled and collide with social and political–economic life in Mexico City. These events have produced tense intersections of state power across scales, drastically altered local and national political trajectories, thrust Mexico City's residents and institutions into relations with global actors and processes in new configurations, loosed the bonds of the "secondary circuit" (Harvey [1982] 2006) by destroying whole elements of the built environment in an instant, and thrown previously underappreciated choreographies of urban geographical change into startling relief. If geography and geographical research are to take politics seriously in the Anthropocene, the experience of Mexico City insists that earthquakes and geology find a prominent place in this agenda.

Conclusion: Toward a Broader Approach to Anthropocene Political Futures

"Certainly," Simon (2002) argued in a disciplinarily familiar Blaikiean refrain, "the [1985] earthquake was a 'natural disaster'" (533). "The impetus was a cataclysmic event," Simon continued, "that could not have been controlled or predicted. But a great deal of the tragedy was also manmade, a result of centuries of environmental abuse in the Valley of Mexico" (534). Such recognition of social precursor[15] is not enough, though. We must also take stock of the tumult produced in the aftermath of such events—the sociopolitical reverberations and ruptures wrought by an aggravated physical geography—and the role that scientific authority, from prediction to response, might be made to play. Distinctively Anthropocene urban geomorphologies (Dixon, Viles, and Garrett 2018), as earthquakes dramatically illustrate, are both product and productive of socionatural processes (Dalby 2017). Although they arrive in an instant, millennia of pressure and centuries of engineering shaking and pulverizing cities into rubble in seconds, their analysis as socionatural processes ironically requires liberally casting the investigative time scale both forward and back, as the case of Mexico City suggests. This case also demonstrates a geographical irony of earthquakes, in that highly localized events (affecting relatively small geographical regions and differentially experienced from one city block to the next, or even between different rooms of a single house) nevertheless pull together political and economic geographies that span the globe, and increasingly so. Like Schwartz's (2019) attempt to "liberate" commonplace but powerful Anthropocene notions like vulnerability from the shackles of neoliberal quantification that anemically cast them into a "perpetually hypothetical future," I argue that the politics of geological events such as earthquakes require significant analytical extension to make plain their centrality to Anthropocene futures.

The concept of catastrophe offers a productive frame for repositioning the politics of geology along the lines I have suggested here. Building on an etymological distinction between crisis, disaster, and catastrophe, Maldonado-Torres (2019) insisted on a reconsideration of events surrounding Puerto Rico's Hurricane María under catastrophic terms (see also Sierra-Rivera 2018). "Unlike disaster," Maldonado-Torres (2019) wrote, "which makes one wonder about fate, but like crisis, which calls for a diagnosis, catastrophe calls for thinking; however, catastrophe challenges all existing cognitive frameworks and 'induces new problematization and modes of questioning,' to use Aradau's and Van Munster's words, that are irreducible to critique" (336). Such a rupture is what the 1985 earthquake produced in Mexico City, completely upending the political dynamics of the city and country in an instant and setting Mexican democracy on new footing as alliances changed, patterns were interrupted, and the feelings and behaviors anchored to bodies and ideas were violently unmoored. Whether the 2017 earthquake will have similarly dramatic effects remains to be seen, although its destructive and productive intersection with the processes and politics of redevelopment and neighborhood change even now provides partial vantages on a sea change at the scale of several neighborhoods, which many real estate interests and policymakers hope and many longtime residents fear could swell to engulf much of the central city in the years to come. Undeniably, this future hinges in no small part on geology, in its incarnations as both the unstable substrata of concrete and steel and the terra firma of scientific assurance on which are erected the infrastructure allocations, eviction notices, mitigation contracts, parcel allotments, demolition orders, and broader policy adjustments that govern the pace and substance of change in the contemporary city.

Finally, these insights are crucial to Anthropocene research and thought because of their urban dimension. Whatever is to be made of the specific theses proffered on the global urban condition in the contemporary epoch (e.g., Ruddick 2015; Brenner 2018),

the urbanization of the planet at an increasing pace is hardly a matter of debate any longer. Although atmospheric debris, sea-level rise, tropical storms, and record temperatures will without question affect many of the world's urban residents, the Anthropocene's prophesied seismicity will also affect countless more, not least in megalopolitan urban regions like Mexico City, Los Angeles, Jakarta, and Tokyo, already at high risk of such events. An Anthropocene research agenda attentive to present and future political formations can hardly afford to overlook such a central axis of vulnerability and catastrophic potential.

Acknowledgments

My deepest thanks to David Butler and Jennifer Cassidento for editorial support throughout the writing and review processes for this article. Thanks also to the anonymous reviewers, whose critiques, comments, corrections, and questions were invaluable. I hope the article shows at least some of the improvement they inspired. All errors, omissions, and questionable aesthetic elections remain my own.

ORCID

Ben A. Gerlofs (iD) http://orcid.org/0000-0003-1118-5111

Notes

1. Although the U.S. Geological Survey cautions that this influence is not statistically significant (Johnston 2019).
2. This is in no way to suggest that no academic research has been conducted in this and other closely related areas (e.g., Bobbette 2016; Clark and Yusoff 2017; Squire and Dodds 2020) but, rather, to say that this is an emergent literature and that its relative incipiency is illustrative of a tendential focus on more politically or analytically charismatic themes. To provide an anecdotal snapshot of this topical privileging, on 19 December 2019 I ran a simple series of associative keyword searches on topics popularly related to the Anthropocene. I selected four prominent Anglophone geographical journals (*Annals of the American Association of Geographers*, *Transactions of the Institute of British Geographers*, *Progress in Human Geography*, and *Progress in Physical Geography*), three prominent academic databases (Web of Science, Scopus, and JSTOR), my current university's library home page and that of a nearby partner institution, and three search engines (Google, Google Scholar, and Bing). In all, I ran searches for twenty-seven terms or

groups of terms, beginning with Anthropocene and then adding a series of additional terms (e.g., Anthropocene earthquake). The search terms were selected strategically to test the hypothesis that certain phenomena, such as floods and sea-level rise, would predominate over others, such as earthquakes and volcanic eruptions, with reference to the Anthropocene. A clear difference emerged in support of this hypothesis. Three pairs of these results (Anthropocene earthquake and Anthropocene hurricane, Anthropocene earthquake and Anthropocene flood, and Anthropocene earthquake and Anthropocene sea level) illustrate this difference. On average across all of these sources, Anthropocene hurricane produced a 59 percent increase in search results over Anthropocene earthquake, and Anthropocene flood and Anthropocene sea level produced increases of 107 percent and 1,144 percent, respectively. The difference was less pronounced in the geography journals, with Anthropocene earthquake actually yielding 8 percent more results than Anthropocene hurricane, although still lagging behind Anthropocene flood and Anthropocene sea level, which returned search results 323 percent and 900 percent greater, respectively (*Transactions*, which produced no results for Anthropocene earthquake, was not included in these averages).

3. The majority of the evidence presented in or relied on for this article was collected as part of an ongoing research project begun in 2016–2017 that investigated the spatial, sociocultural, and political-economic contexts and impacts of the 1985 and 2017 earthquakes in Mexico City and to a lesser degree from a doctoral research project on grassroots politics in Mexico City conducted between 2013 and 2016 and comprising roughly twelve months of ethnographic and archival fieldwork. The lone piece of evidence referenced here gathered outside of these projects is the metaphoric explanation of the challenges that Mexico City's geology presents to planners and other officials, which was offered during a roundtable discussion with several city officials during the ten-day field experience portion of an undergraduate seminar I taught on contemporary Mexico City in the spring of 2019.

4. Following Gidwani (2008). See also Deleuze and Guattari ([1987] 2003).

5. Local convention favors discussion of these two events in the singular (i.e., "the earthquake"), a practice I follow here.

6. See Monsiváis (1987), Eckstein ([1977] 1988), Walker (2009), Joseph and Buchenau (2013), Aguilera Crivelli (2015), and Poniatowska (2015). A prominent explanation for the government's response revolves around executive hesitance to implement ready disaster management plans that relied on the military (see Ai Camp 2000).

7. See Poniatowska ([1971] 2012).

8. Bautista González (2015) argued that the increasing presence and concerns of the *damnificados* significantly reconfigured the geography of the MUP, which had previously focused largely on peripheral neighborhoods and informal settlements.

9. This position, previously titled *Regente* (Regent), had been acceded by direct appointment of the president of the Republic since 1928 and has often been considered a political stepping stone in the PRI state.

10. See Mier, Rocha, and Rabell Romero (1988) for a thorough official accounting of damaged residential structures and information about the *damnificados*.

11. This figure is complicated by varying estimates. Pentileć (1991) claimed that the program constructed "50,000 housing units." Adler (2015) reported the construction of "over 45,000 new housing units." Inam (1999) stated, "RHP was responsible for the construction, rehabilitation, and repair of 48 749 housing units within two years" (393). Ward (1990) provided the most thorough explanation of the program's own estimates: "Some 28000 previous *vecindad* tenant households were housed almost always on their previous sites, now in owner-occupied two-bedroom accommodation following one of four design prototypes. A further 11650 dwellings were rehabilitated and 4500 homes were subject to minor repairs" (195).

12. The divide between those included in and excluded from the program is one among several such rifts created by PRI policies and practices in the aftermath of the earthquake, a reality that squares easily with well-established patterns of clientelism and political nullification by incorporation under the PRI.

13. This assertion is difficult to precisely qualify, owing to the long-standing problem of reliably identifying analytically robust "middle class(es)" in Mexico. See Walker (2009).

14. Often shortened to simply *blanqueamiento* (whitening).

15. See also Candiani (2014) and Vitz (2018).

References

Adler, D. 2015. The Mexico City earthquake, 30 years on: Have the lessons been forgotten? *The Guardian*, September 18. Accessed December 24, 2019. https://www.theguardian.com/cities/2015/sep/18/mexico-city-earthquake-30-years-lessons.

Aguilera Crivelli, A. 2015. Natural disasters, state response capacity and political legitimacy: Assessing the 1985 earthquake in Mexico City. MA thesis, Leiden University.

Ai Camp, R. 2000. The time of the technocrats and deconstruction of the revolution. In *The Oxford history of Mexico*, ed. M. C. Meyer and W. H. Beezley, 609–36. New York: Oxford University Press.

Ai Camp, R. 2014. *Politics in Mexico: Democratic consolidation or decline?* 6th ed. New York: Oxford University Press.

Amos, C. B., P. Audet, W. C. Hammond, R. Bürgmann, I. A. Johanson, and G. Blewitt. 2014. Uplift and seismicity driven by groundwater depletion in central California. *Nature* 509 (7501):483–86. doi: 10.1038/nature13275.

Bautista González, R. 2015. *Movimiento urbano popular: Bitácora de la lucha, 1968-2011* [Urban popular movement: Diary of the struggle]. Mexico City: Casa y Ciudad.

Bebbington, A., and J. Bury. 2013. Political ecologies of the subsoil. In *Subterranean struggles: New dynamics of mining, oil, and gas in Latin America*, ed. A. Bebbington and J. Bury, 1–26. Austin: University of Texas Press.

Bettini, G. 2013. Climate barbarians at the gate? A critique of apocalyptic narratives on "climate refugees." *Geoforum* 45:63–72. doi: 10.1016/j.geoforum.2012.09.009.

Bobbette, A. 2016. Contortions of the unconsolidated: Hong Kong, landslides, and the production of urban grounds. *City* 20 (4):523–38. doi: 10.1080/13604813. 2016.1192417.

Braun, B. 2000. Producing vertical territory: Geology and governmentality in late Victorian Canada. *Ecumene* 7 (1):7–46. doi: 10.1177/096746080000700102.

Brenner, N. 2018. Debating planetary urbanization: For an engaged pluralism. *Environment and Planning D: Society and Space* 36 (3):570–90. doi: 10.1177/ 0263775818757510.

Bruhn, K. 1997. *Taking on Goliath: The emergence of a new left party and the struggle for democracy in Mexico*. State College: The Pennsylvania State University Press.

Candiani, V. S. 2014. *Dreaming of dry land: Environmental transformation in colonial Mexico City*. Stanford, CA: Stanford University Press.

Chen, J. 2016. Melting glaciers are wreaking havoc on Earth's crust. *Smithsonian Magazine*, September 1. Accessed December 24, 2019. https://www.smithsonianmag.com/science-nature/melting-glaciers-are-wreaking-havoc-earths-crust-180960226/.

Clark, N., and K. Yusoff. 2017. Geosocial formations and the Anthropocene. *Theory, Culture & Society* 34 (2–3):3–23. doi: 10.1177/0263276416688946.

Crossa, V. 2013. Play for protest, protest for play: Artisan and vendors' resistance to displacement in Mexico City. *Antipode* 45 (4):826–43. doi: 10.1111/j.1467-8330.2012.01043.x.

Dalby, S. 2017. Anthropocene formations: Environmental security, geopolitics, and disaster. *Theory, Culture & Society* 34 (2–3):233–52. doi: 10.1177/0263276415598629.

Davis, D. 1994. *Urban leviathan: Mexico City in the twentieth century*. Philadelphia: Temple University Press.

Deleuze, G., and F. Guattari. [1987] 2003. *A thousand plateaus: Capitalism and schizophrenia*. Minneapolis: The University of Minnesota Press.

Díaz, J. 2015. Gentrificación por la red: Nuevos actores de clase en el Centro Histórico de la Ciudad de México [Gentrification through the internet: New class actors in Mexico City's Historic Center]. In *Perspectivas del estudio de la gentrificación en México y América Latina*, ed. V. Delgadillo, I. Díaz, and L. Salinas, 303–22. Mexico City: Instituto de Geografía, Universidad Nacional Autónoma de México.

Dixon, S. J., H. A. Viles, and B. L. Garrett. 2018. Ozymandias in the Anthropocene: The city as an emerging landform. *Area* 50 (1):117–25. doi: 10.1111/ area.12358.

Eckstein, S. [1977] 1988. *The poverty of revolution: The state and the urban poor in Mexico*. Princeton, NJ: Princeton University Press.

Galchen, R. 2015. Weather underground: The arrival of man-made earthquakes. *The New Yorker* 91 (8):34–40.

Gidwani, V. 2008. *Capital interrupted: Agrarian development and the politics of work in India*. Minneapolis: The University of Minnesota Press.

González, P. J., K. F. Tiampo, M. Palano, F. Cannavó, and J. Fernández. 2012. The 2011 Lorca earthquake slip distribution controlled by groundwater crustal unloading. *Nature Geoscience* 5 (11):821–25. doi: 10. 1038/ngeo1610.

Grosz, E., K. Yusoff, and N. Clark. 2017. An interview with Elizabeth Grosz: Geopower, inhumanism, and the biopolitical. *Theory, Culture & Society* 34 (2–3):129–46. doi: 10.1177/0263276417689899.

Gutmann, M. 2002. *The romance of democracy: Compliant defiance in contemporary Mexico*. Berkeley: The University of California Press.

Harvey, D. 2003. *The new imperialism*. New York: Oxford University Press.

Harvey, D. [1982] 2006. *The limits to capital*. Brooklyn: Verso.

Inam, A. 1999. Institutions, routines, and crises: Post-earthquake housing recovery in Mexico City and Los Angeles. *Cities* 16 (6):391–407. doi: 10.1016/S0264-2751(99)00038-4.

Johnson, E., H. Morehouse, S. Dalby, J. Lehman, S. Nelson, R. Rowan, S. Wakefield, and K. Yusoff. 2014. After the Anthropocene: Politics and geographic inquiry for a new epoch. *Progress in Human Geography* 38 (3):439–56. doi: 10.1177/0309132513517065.

Johnston, M. 2019. Is there earthquake weather? Accessed December 24, 2019. https://www.usgs.gov/faqs/there-earthquake-weather?qt-news_science_products=0#qt-news_science_products.

Joseph, G. M., and J. Buchenau. 2013. *Mexico's once and future revolution: Social upheaval and the challenge of rule since the late nineteenth century*. Durham, NC: Duke University Press.

Kandell, J. 1988. *La capital: The biography of Mexico City*. New York: Random House.

Klein, N. 2007. *The shock doctrine: The rise of disaster capitalism*. New York: Picador.

Liu, C. C., A. T. Linde, and I. S. Sacks. 2009. Slow earthquakes triggered by typhoons. *Nature* 459 (7248):833–36. doi: 10.1038/nature08042.

Lovett, R. A. 2013. Hurricane may have triggered earthquake aftershocks. *Nature*. Accessed December 24, 2019. https://www.nature.com/articles/nature.2013.12839.

Maldonado-Torres, N. 2019. Afterword: Critique and decoloniality in the face of crisis, disaster, and catastrophe. In *Aftershocks of disaster: Puerto Rico before and after the storm*, ed. Y. Bonilla and M. Lebrón, 332–42. Chicago: Haymarket Books.

McGuire, B. 2016. How climate change triggers earthquakes, tsunamis and volcanoes. *The Guardian*, October 16. Accessed December 24, 2019. https:// www.theguardian.com/world/2016/oct/16/climate-change-triggers-earthquakes-tsunamis-volcanoes.

Melgar, D., and X. Pérez-Campos. 2018. Mexico City's potent 2017 earthquake was a rare "bending" quake—and it could happen again. *The Conversation*, March 12. Accessed December 24, 2019. https://theconversation.com/mexico-citys-potent-2017-earthquake-was-a-rare-bending-quake-and-it-could-happen-again-92994.

Melo Zurita, M. L., P. G. Munro, and D. Houston. 2018. Un-earthing the subterranean Anthropocene. *Area* 50 (3):298–305. doi: 10.1111/area.12369.

Mier, M. M., T. Rocha, and C. A. Rabell Romero. 1988. Ciudad de México: Características socioeconómicas de los damnificados de los sismos de septiembre [Socioeconomic characteristics of those displaced by the September earthquakes]. In *Atlas de la Ciudad de México*, ed. G. Garza, vol. 5, 162–66. Mexico City: Departamento del Distrito Federal and El Colegio de México.

Monsiváis, C. 1987. *Entrada libre: Crónicas de la sociedad que se organiza* [Free entry: Chronicles of the society in the process of organizing]. Mexico City: Ediciones Era.

Olsen, R. S., and V. T. Gawronski. 2003. Disasters as critical junctures? Managua, Nicaragua 1972 and Mexico City, 1985. *International Journal of Mass Emergencies and Disasters* 21 (1):5–35.

Pentilec, J. 1991. The link between reconstruction and development. *Land Use Policy* 8 (October):343–47.

Poniatowska, E. [1971] 2012. *La Noche de Tlatelolco: Testimonios de historia oral* [The night of Tlatelolco: Oral history testimonies]. Mexico City: Biblioteca Era.

Poniatowska, E. 2015. La gran solidaridad de los Mexicanos [The great solidarity of Mexicans]. In *07:19: A treinta años del terremoto en la Ciudad de México (1985–2015)*, ed. U. Castellanos and M. L. Passarge, 133–36. Mexico City: Gobierno de la Ciudad de México.

Povinelli, E. A. 2016. *Geontologies: A requiem to late liberalism*. Durham, NC: Duke University Press.

Ruddick, S. 2015. Situating the Anthropocene: Planetary urbanization and the anthropological machine. *Urban Geography* 36 (8):1113–30. doi: 10.1080/02723638.2015.1071993.

Schwartz, S. 2019. Measuring vulnerability and deferring responsibility: Quantifying the Anthropocene. *Theory, Culture & Society* 36 (4):73–93. doi: 10.1177/0263276418820961.

Sierra-Rivera, J. 2018. *Affective intellectuals and the space of catastrophe in the Americas*. Columbus: The Ohio State University Press.

Simon, J. 2002. The sinking city. In *The Mexico reader: History, culture, politics*, ed. G. M. Joseph and T. J. Henderson, 520–35. Durham, NC: Duke University Press.

Smith, N. 1979. Toward a theory of gentrification: A back to the city movement by capital, not people. *Journal of the American Planning Association* 45 (4):538–48. doi: 10.1080/01944367908977002.

Smith, N. 2006. There's no such thing as a natural disaster. Accessed December 24, 2019. https://items.ssrc.org/understanding-katrina/theres-no-such-thing-as-a-natural-disaster/.

Smith, P. 1978. Mining triggers earthquakes. *Nature* 271 (5642):207–8. doi: 10.1038/271207a0.

Squire, R., and K. Dodds. 2020. Introduction to the special issue: Subterranean geopolitics. *Geopolitics* 25 (1):4–16. doi: 10.1080/14650045.2019.1609453.

Swyngedouw, E. 2010. Apocalypse forever? Post-political populism and the spectre of climate change. *Theory, Culture & Society* 27 (2–3):213–32. doi: 10.1177/0263276409358728.

Swyngedouw, E. 2013. The non-political politics of climate change. *ACME* 12 (1):1–8.

Tavera-Fenollosa, L. 1998. Social movements and civil society: The Mexico City 1985 earthquake victim movement. PhD dissertation, Yale University.

Vitz, M. 2018. *A city on a lake: Urban political ecology and the growth of Mexico City*. Durham, NC: Duke University Press.

Wainwright, J., and G. Mann. 2018. *Climate leviathan: A political theory of our planetary future*. Brooklyn, NY: Verso.

Walker, L. E. 2009. Economic fault lines and middle-class fears: Tlatelolco, Mexico City, 1985. In *Aftershocks: Earthquakes and popular politics in Latin America*, ed. J. Buchenau and L. L. Johnson, 184–221. Albuquerque: University of New Mexico Press.

Ward, P. 1990. *Mexico City: The production and reproduction of an urban environment*. Boston: G.K. Hall.

World Bank. 1993. Performance audit report: Mexico earthquake rehabilitation and reconstruction project (Loan 1665-ME). Report no. 12149, World Bank, Washington, DC.

Yusoff, K. 2018. *A billion black Anthropocenes or none*. Minneapolis: University of Minnesota Press.

Ziegler, S. S. 2019. The Anthropocene in geography. *Geographical Review* 109 (2):271–80. doi: 10.1111/gere.12343.

Understanding Urban Flood Resilience in the Anthropocene: A Social–Ecological–Technological Systems (SETS) Learning Framework

Heejun Chang, [iD] David J. Yu, [iD] Samuel A. Markolf, [iD] Chang-yu Hong, [iD] Sunyong Eom, [iD] Wonsuh Song, [iD] and Deghyo Bae [iD]

Urban flooding is a major concern in many cities around the world. Together with continuous urbanization, extreme weather events are likely to increase the magnitude and frequency of flood hazards and exposure in populated regions. This article examines the changing pathways of flood risk management (FRM) in Portland, Oregon; Seoul, South Korea; and Tokyo, Japan, which have different histories of land development and flood severity. We used city governance documents to identify how FRM strategies have changed in the study cities. Using a combined framework of social learning with an integrated social–ecological–technological systems (SETS) lens, we show what components of SETS have been emphasized and how FRM strategies have diversified over time. In response to historical flood events, these cities built hard infrastructure such as levees to reduce flood risks. The recent paradigm shift in urban FRM, such as the adoption of socioecological elements in SETS, including floodplain restoration, green infrastructure, and public education, is a response to making cities more resilient or transformative to the anticipated future extreme floods. The pathways that cities have taken and the main emphasis across SETS elements differ by city, however, suggesting that opportunities exist for learning from each city's experience collectively to tackle global flooding issues. *Key Words: climate change, resilience, social learning, urban floods, urbanization.*

Flooding is one of the most destructive natural disasters on Earth, causing more than $40 billion in damage annually (Organization for Economic Co-operation and Development 2016). Together with the increasing concentration of people in cities, rising temperature and precipitation variability induced by anthropogenic forces have elevated flood risk in many urban areas (Chang and Franczyk 2008; National Academies of Sciences, Engineering, and Medicine 2019). Thus, urban flood risk management (FRM) and the broader strategies that underpin it have become of great interest worldwide (Luo et al. 2015; Allen et al. 2019). Many government efforts are underway to protect individuals and communities from flooding (Aerts and Botzen 2014). These measures not only affect people's exposure and vulnerability to flooding but

also modify floodplain hydrology and landscapes in the long run through changes to built structures and socioeconomic activities (Merz et al. 2015). Because these processes affect and are affected by both engineered and self-organizing components, it is useful to view FRM systems as integrated social–ecological–technological systems (SETS; Markolf et al. 2018); namely, understanding complex systems such as FRM requires consideration of complex social, ecological, and technological components and the interdependencies among them (McPhearson et al. 2016; Morrison, Westbrook, and Noble 2018).

Governments take actions and introduce changes that shape resistance, resilience, or adaptability of urban systems in relation to floods (Liao 2012). These changes are often the result of collective choices of the policy actors and broader stakeholders who learn as a group from past flooding experiences to develop some collective preferences for future FRM. This type of learning is known as *social learning*, and it goes beyond individuals and develops within groups (which can range from small workgroups to large-scale societies) through common experience (Reed et al. 2010). In this article, we argue that, depending on social learning, different levels or types of change can occur in the SETS elements of an FRM system. Currently, the prevailing strategy for FRM used by modern societies is the resistance paradigm, which relies on technocentric solutions (e.g., higher levees) to resist floods. With the recognition that such technocentric approaches can lead to even more catastrophic floods in the long run (Di Baldassarre et al. 2013; Yu et al. 2017), some societies have begun to experiment with more resilience-based approaches (e.g., Morrison, Noble, and Westbrook 2018; Zevenbergen, Gersonius, and Radhakrishan 2020). For example, if a city experiences a catastrophic flood even with highly robust infrastructures, its network of policy actors and broader stakeholders might undergo significant social learning that leads to revisions in its underlying assumptions and goals of FRM strategies and shift its focus to more nonstructural solutions (Solecki, Pelling, and Garschagen 2017). In other words, social learning can occur at different levels, subject to the severity and frequency of floods (e.g., Yu et al. 2016), and these different levels of learning, in turn, can lead to a different type, composition, and degree of change in the SETS elements of FRM. Some changes can be merely fine-tuning or intensification of existing measures in specific SETS elements, whereas others might involve more radical or transformative changes in goals and strategies and lead to fundamental shifts in focus in SETS elements. Little investigation has been done to date, however, to examine how transitions happen in each of the SETS elements of FRM in connection with different levels of social learning in response to the impacts of floods on the urban system (den Boer, Dieperink, and Mukhtarov 2019; Johannessen et al. 2019). Such knowledge would be useful in understanding the trajectory of FRM systems in terms of diversity of strategies and stakeholders' capacity for learning, two fundamental conditions of resilience (Biggs et al. 2012).

In this article, we provide an approach that can be used to assess and compare the composition and transition trajectory of FRM systems. We then apply the approach to real cases to explore its potential and to develop comparative insights. Specifically, we provide a common set of variables (SETS elements) that can be used to characterize the composition of FRM systems and a conceptual lens (social learning) to interpret patterns in the transition trajectory of FRM systems. We used the approach to examine SETS changes and social learning that likely happened in three test-bed cities (Portland, Oregon; Seoul, South Korea; and Tokyo, Japan) in the last few decades. Document coding and analysis were conducted to understand changes in flood policy and planning (social), floodplain landscape (ecological), and flood protection infrastructure (technological). By investigating and comparing the experiences of these three cities, we develop insights about whether and how those strategies lead to resistant, resilient, or transformative FRM. Although there have been a few studies comparing more than one city or country's FRM (e.g., Solecki, Pelling, and Garschagen 2017; Chan et al. 2018; Goodchild, Sharpe, and Hanson 2018), the SETS approach and social learning theory have rarely been used together to conduct such comparative analyses. This article thus contributes to cross-comparative cases in FRM by addressing this gap.

Approach: SETS Learning Framework

The aim of this article is to provide a framework that can be used to describe and diagnose the transition trajectory of FRM systems in terms of changes over time in SETS elements through social learning. On the theoretical and methodological levels, this framework represents a convergence of three related fields of study: resilience of complex self-organizing systems (Folke 2006), Bloomington School of Political Economy (Cole and McGinnis 2017), and resilience trajectories by geographers (Pelling, O'Brien, and Matyas 2015; Solecki, Pelling, and Garschagen 2017). Building on the perspective of ecological and social–ecological resilience, scholars who research resilience in the context of an urban environment increasingly call for a complete integration of the social, ecological, and technological systems into a fully integrated SETS approach (McPhearson et al. 2016; Markolf et al. 2018; Yu et al. 2020). We embrace this direction by viewing FRM systems as SETS and analyze how the constituent SETS elements might have evolved over time through social learning in the three testbed cities. The Bloomington School of Political Economy focuses on the analysis of how local governance systems remain robust or evolve over time in the face of a complex and ever-changing social and physical world (Ostrom 2005). We applied this political economic lens to FRM. Namely, we coded a set of FRM-related public documents produced by the local governments of the three testbed cities to extract various SETS elements and analyzed how these SETS elements might have changed over time. We then used this information to interpret the transition trajectory of FRM and what types of social learning have been experienced by the cities' networks of policy actors and broader stakeholders.

Common SETS Variables and Data

A first step toward our goal is to determine a set of common variables that can be used to characterize FRM systems in terms of SETS. Selection of these variables, in turn, requires identification of a principal data source from which SETS-related information on FRM can be readily extracted. Formal government documents containing important policies or strategies related to FRM are fitting in this regard. These documents include national flood policy and local river management plans, as well as published peer-reviewed articles. In this study, we focus on comprehensive plans and flood hazard plans or hazard mitigation plans for each of our three case cities (see Appendix). Comprehensive plans are documents that specify all zoning and land use regulations of a community. Hazard mitigation plans define rules and procedures that guide community stakeholder actions during the predisaster and disaster response phases. For each city, we collected and coded two versions of each document covering multiple periods and attempted to identify transitions in SETS elements and infer associated social learning. To the extent that digital files were available, we looked to include documents that covered as long of a time span as possible to facilitate the assessment of social learning.

Specifically, we categorized the specific flood-related actions, policies, or guidelines outlined in each document to gain a better sense of whether they tended to be socially oriented, ecologically oriented, technologically oriented, or some

combination of the three. To accomplish this, we first reviewed each document and identified specific sections related to flooding. Each identified item was then qualitatively coded according to its SETS emphasis based on key words and phrasing. Those key words and phrases were then grouped into five social, three ecological, and four technological elements based on the literature and discussions among the authors. To adequately capture the main points of each quote, the coding was updated several times by merging similar variables into a single category. Note that the selected elements do not constitute an exhaustive list of the components of FRM and that our effort necessarily involves some subjective judgment. The selected categories, however, largely cover important aspects discussed in the FRM literature and are consistent with the SETS perspective employed here.

Table 1 summarizes the final variables to which each action, policy, or guideline were categorized and provides an example quote associated with each SETS variable. We counted the frequency with which each specific action, policy, or guideline related to each variable, and in many cases a policy was associated with more than one SETS variable. Because each document has a different number of relevant policies, we calculated the relative proportion of S, E, T, SE, ST, ET, and SET for each document. Given the importance of diversity for enhancing resilience (Biggs et al. 2012; Yu et al. 2020), we hypothesize that if a given document contains a more diverse set of SETS elements, the system might be more resilient. At least two researchers coded each document independently for intercoder reliability. A third reviewer resolved any discrepancies between the two coders, following the method described by Syed and Nelson (2015). To ensure consistency in coding over time, the research team re-reviewed a random set of coding in April and May 2020.

Social Learning and Resilience Trajectory

We posit that transitions in FRM involve institutional learning or, more generally, social learning by a network of policy actors and broader stakeholders (den Boer, Dieperink, and Mukhtarov 2019; Johannessen et al. 2019). To understand why, it is important to recognize that FRM (and its SETS elements) generally involves public or common goods

Table 1. Description and example of each social (S), ecological (E), and technological (T) subcategory used in governance document coding analysis

	Category	Coding sample
Social systems	Emergency planning/preparation/safety/management, response	"Facilitate an identification and prioritization process for the purpose of defining a candidate list of localized inundation scenarios related to levee failures that result from different hazard events."
	Laws, regulations, standards	"Multnomah County has a 2010 Stormwater Management Plan (updated in 2011). The plan includes several urban pocket areas; the unincorporated area of Interlachen; and the roadways in Fairview, Troutdale and Wood Village (approximately 28 miles)."
	Promotion of participation and collaboration	"Concurrent Risk MAP 'Resilience' efforts will take place within the Planning Area between 2017 and 2018. Resilience meetings include all jurisdictions in the Planning Area working alongside FEMA and the State Risk MAP Coordinator to identify mitigation and risk reduction strategies, and discuss possible action and implementation opportunities."
	Knowledge transfer and communication	"Install high-water-mark signs to educate the public about flooding potential in targeted locations along or within the leveed areas."
	Economic mechanisms (e.g., insurance, land purchase, etc.)	"Identify target areas for flood mitigation projects, such as high-risk/repetitive risk problem areas. Identify specific mitigation projects and grants. Consider areas at risk to multiple hazards for increased cost benefit."
Ecological systems	Conservation, preservation, and restoration	"Johnson Creek floods with some regularity. Plans are in place to set aside the floodplain as open space."
	Green infrastructure and ecological engineering	"Protect and restore streams, wetlands and floodplains, reduce paved surfaces, utilize green infrastructure, update stormwater plans, manuals and drainage rules and prepare to manage increased stormwater runoff."
	Ecological services (e.g., benefits obtained from natural floodplains or improvement of floodplains)	"Conserve significant wetlands, riparian areas, and water bodies which have significant functions and values related to flood protection, sediment and erosion control, water quality, groundwater recharge and discharge, education, vegetation, and fish and wildlife habitat."
Technological systems	Design standards and codes (e.g., design storm criteria, buildings codes, etc.)	"The County's current regulations for new storm water drainage systems require control of the 10-year 24-hour storm; however, many older drainage systems are built to lower standards."
	Construction of engineered infrastructure (e.g., dams, levees, pumps)	"Replace the flow control structure regulating water levels on the TRIP wetland mitigation site within the next year. ... A new flow control structure with an adjustable concrete weir structure and larger diameter culvert with gate valve is needed to properly control the flow of stormwater with greater flexibility to adjust flow in support of flood control in the upstream segment of Salmon Creek and environmental protection."
	Operation and maintenance of existing engineered infrastructure	"Flood-proof wastewater manholes and pipelines within the 100-year floodplain."
	Development and implementation of data-driven solutions (e.g., hazard mappings, Web-based platforms, sensing, simulation)	"Complete an inventory and GIS mapping of structures, critical facilities and important transportation or utility system components within mapped floodplains and/or within areas subject to flood in the event of levee or dam failures, including elevation data."

Note: Some of the coding samples listed could apply to multiple categories (e.g., "Conserve significant wetlands, riparian areas, and water bodies which have significant functions and values related to flood protection, sediment and erosion control, water quality, groundwater recharge and discharge, education, vegetation, and fish and wildlife habitat" relates to both ecological services and conservation, preservation, and restoration). Data from Multnomah County (2009, 2017) and City of Portland (2011, 2018).

that affect and are affected by many stakeholders; for example, physical infrastructure, wetlands and reservoirs, flood response–related rules and procedures, and so on (Ostrom 2009). In democratic systems, agencies' management of such shared goods is likely to be heavily influenced by the collective choices of such a network of policy and domain actors. Thus, we assume that transitions happen in the SETS elements of FRM only when there are corresponding shifts in the collective choice. Such change in collective choice, in turn, likely occurs when the network of policy actors and broader stakeholders collectively learns and develops knowledge through common experience. The public documents we analyzed were not created in a vacuum; they reflect the collective preferences of the policy and broader organizations as a result of public hearings and workshops during the formation of the documents.

Resilience studies highlight that social learning is crucial for building community resilience because of its positive effect on decision making under uncertainty (Biggs, Schlüter, and Schoon 2015). The basic idea is that social groups develop knowledge about complex systems through looped learning composed of three steps: monitoring of system states, evaluation of the acquired information leading to revised knowledge, and adjustment of subsequent management trials (Pahl-Wostl 2009). More specifically, social learning can occur at multiple levels of such loops and lead to varying degrees of outcomes (Argyris and Schön 1978). Single-loop learning involves fine-tuning of specific practices or actions to better meet existing goals or assumptions. Double-loop learning entails changes in shared goals or assumptions. Triple-loop learning involves changes in deep-seated beliefs and mental models (Pahl-Wostl 2009). As social learning moves from single- to triple-loop, more fundamental changes are likely to happen in FRM systems: management response for short-term resistance (single-loop) and for longer range adaptation (double-loop) or transformation for resilience (triple-loop).

Based on this foundation, we provide a framework for evaluating social learning and categorizing FRM systems into resistant, resilient, or transformative types based on changes to the SETS elements of FRM across time and space. In particular, by characterizing the specific S, E, and T elements of specific flood management strategies outlined in the coded documents (and observing how those SETS elements change over time across different versions of the documents), we examine the breadth and diversity of FRM strategies that are implemented. A shift in emphasis (e.g., from technologically oriented FRM strategies to socially oriented FRM strategies) over time or an increase in the diversity of strategies (e.g., an increasingly proportional mix of different types of S, E, T, and combination of SET FRM strategies) allows for inferences about the type of social learning present, as well as the resilience trajectory (i.e., resistance vs. resilience vs. transformation) of a particular city. A key notion here is that (1) diversity in system components and (2) system capacity for learning are fundamental to building resilience (Biggs et al. 2012).

The top panel of Figure 1 depicts single-loop learning in a reactive or resistant society, which "primarily focuses on improving the efficiency of action" (Intergovernmental Panel on Climate Change [IPCC] 2012, 54; Pelling et al. 2008), and refinements primarily happen in technological elements. Flooding is suppressed through increasing dike heights or widths, and there are few opportunities for people to learn because there are minimal changes in the process or actions; that is, technocentric command-and-control approach. With double-loop learning (shown in the center panel of Figure 1), changes happen across one or more of the SETS components. This process reframes the current management goals and objectives in the context of anticipated climate change and assesses whether new FRM decisions are needed (Pelling 2011; IPCC 2012). Such new decisions might require intersectoral collaboration such as managing flow and water quality simultaneously during flood events. Fundamental characteristics, identities, and relationships among SETS components remain mostly the same, however. With triple-loop learning (bottom panel of Figure 1), new governance and institutional arrangements are adopted (Pahl-Wostl 2009), and fundamental shifts happen in one or more of the SETS components—particularly in terms of the use of an increasingly diverse set of strategies. Further, transformation in one SETS component can propagate and induce significant changes in other SETS components. Following Pelling (2011), Pelling, O'Brien, and Matyas (2015), and Solecki, Pelling, and Garschagen (2017), we describe some main characteristics of resistance, resilience, and transformation across SETS, including some possible examples (see Table 2).

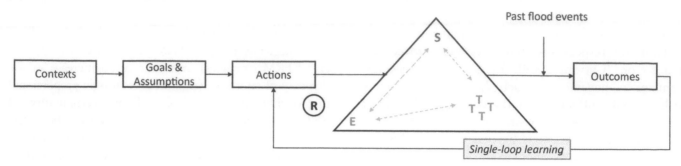

R: Refinements happen (e.g., further adjustment, optimization, etc.)

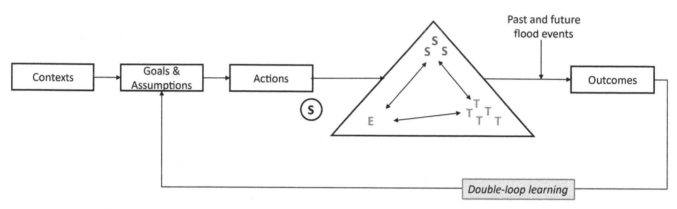

S: Significant changes happen but fundamental characteristics, identities, and relationships remain the same

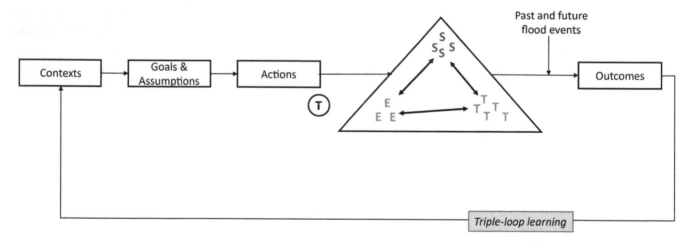

T: Transformations (fundamental changes in characteristics) occur in one or more of SETS components. Further, transformation in one of SETS components can propagate and induce significant changes in other SETS components

Figure 1. The combined framework of social learning and the SETS framework. The number of S, E, T strategies and the thickness of arrows indicate the relative emphasis on each SETS component and the degree of interactions among the SET components. SETS = social–ecological–technological systems.

Case Study Cities

The three study cities—Portland, Seoul, and Tokyo—are representative cases of growing urban regions with increasing exposure to floods, thus serving as testbed cities for understanding the evolving FRM strategies (Table 3; Chang et al. 2010;

Table 2. Characteristics of social–ecological–technological systems social learning framework

	Resistant (single-loop)	Resilient (double-loop)	Transformative (triple-loop)
Social	Offer socioeconomic aid after disaster (e.g., disaster relief fund)	Change in practice or organization (e.g., collaborate among different departments)	Change in norms, principles, and value systems (e.g., radical change in sociopolitical system)
Ecological	Maintain the existing environment (e.g., waterway, vegetation maintenance)	Modify the current environment to reduce flood (e.g., small-scale restoration, installation of green infrastructure)	Change in the whole landscape for different uses (e.g., major floodplain restoration, designation of new protected area)
Technological	Focus on repairing damaged infrastructure (e.g., increase levee heights, construct pump stations)	Improve or change existing technologies (e.g., introduce automatic flood warning system)	Reform technological paradigm (e.g., large-scale system-wide technical reconfiguration)

Nakamura and Oki 2018; Bae and Chang 2019; Fahy et al. 2019). As major rivers pass through the densely populated center of each city, they all have a long history of chronic flooding and flood damage, with increasing trends in the number of heavy precipitation events since the 1980s (Figure 2). These heavy precipitation events caused major flooding in the study cities (e.g., Portland in 1996 and 2009, Seoul in 1998 and 2011, and Tokyo in 2015 and 2019), which in turn prompted the creation of new laws and policies in the respective cities. The manner in which each city responded to these events, however, varied due to institutional and cultural differences.

Concomitant with economic development, both Seoul and Tokyo's population increased rapidly during the latter half of the twentieth century (Cho 2011), and Portland has been one of the fastest growing cities in the Western United States since the 1980s (Figure 3A). Such rapid growth has inevitably resulted in the loss of natural areas such as wetlands or forests (Figure 3B) and reduced the natural capacity of soils to absorb rainwater. Many low-lying areas, once used for rice paddies or crop fields in Seoul and Tokyo, have been converted to urban residential, commercial, or industrial areas. As a result of intensive land development in Seoul and Tokyo, some communities along the river are particularly susceptible to floods. The dense urban infill development along the Willamette River in Portland has also created some challenges for FRM, particularly under projected sea-level rise and river flow increase scenarios (Helaire et al. 2020). Climate change is also likely to increase the probability of intense rain events in these locations (IPCC 2012).

Flood Prevention Strategies: Historical Perspective

Taking a historical perspective of FRM in each study city, we seek to better understand the context of which SETS elements have been emphasized over time. Portland's FRM strategies have been a mix of structural and nonstructural measures. After a great flood in 1894 that inundated downtown areas, the city built 10-m walls to protect it from flooding. During the Great Depression era in the 1930s, the Works Progress Administration constructed dams in the upstream areas of both the Columbia River and the Willamette River. Despite the dam building, the 1964 Christmas flood inundated the Columbia, with river levels ultimately peaking at 10.6 m. Although major flood events were relatively rare in the large rivers throughout the latter half of the twentieth century, local flooding persisted in the Johnson Creek Basin, where the Works Progress Administration created rock walls along the creek. In response to this persistent flooding, as well as increased challenges of stormwater management associated with urban development, the City of Portland enacted the 1991 Johnson Creek Basin Protection Plan, with the objective to "protect and preserve the scenic, recreation, fishery, wildlife, flood control, water quality, and other natural resource values of the Johnson creek basin" (City of Portland 1991, 2011). Concerns about flood risks were reignited after a large snowpack, rapid temperature increases, and record-setting rainfall led to severe flooding throughout Oregon (including the Portland area) during February 1996. Partly in response to this event and addressing climate change concerns,

Table 3. Population, urbanization, precipitation, and flood characteristics of Portland, Seoul, and Tokyo

Characteristics	Portland	Seoul	Tokyo
Population density	1,830/km^2 (2017)	16,204/km^2 (2018)	6,519/km^2 (2018)
% Urban areas	77.5 (2016)	57.8 (2016)	68.2 (2017)
Annual precipitation	1,092.2 mm	1,312.3 mm	1,435 mm
Major flood years	2015, 1996, 1964, 1894	2018, 2013, 2011, 2010, 2006, 2001, 1998, 1990, 1984, 1925	2019, 2015, 1958, 1947, 1910
Major cause of floods	Heavy precipitation, rain on snow events	Heavy precipitation, typhoon	Typhoon, heavy precipitation, tidal surge

Note: % Urban areas are calculated from land cover classification in each city. *Source:* Tokyo Metropolitan Government (2018, 2019).

the City of Portland developed the Johnson Creek Willing Seller Land Acquisition Program in 1997. This program was intended to facilitate and encourage the movement of people and property out of flood-prone areas and restore the acquired land back to its original state to restore wetlands and increase flood storage capacity (City of Portland 2019b). More recently, the State of Oregon established the Lents Stabilization and Job Creation Collaborative in 2016 to "protect floodplain owners, prevent displacement ... and increase local jobs by reducing Johnson Creek flood impacts" (City of Portland 2019a). In other words, a flood mitigation project was employed as a way to revitalize the local economy.

Seoul's early FRM efforts primarily focused on technological elements such as improving sewers and enhancing pipes in the 1980s. The city widened the main stem of the Han River while creating some recreational facilities within floodplains. After the big flood of September 1990 (Figure 4), the Seoul City government constructed twenty-eight pump stations while strengthening levees and dredging 30 km to avoid chronic flooding in low-lying residential areas along the Han River. These gray infrastructure systems were not sufficient to handle subsequent big floods that occurred in 1998 and 2001, however. Immediately after these two flood events, the Seoul government constructed nineteen additional pump stations with automatic sensors, spending over US$6 billion. In 2007, under the green city slogan and anticipation of climate change–induced extreme precipitation events, the city enhanced the capacity of storm design from 75 mm per hour to 95 mm per hour. Similar to the previous era, investment in gray infrastructure continued. Two consecutive urban flood events occurred in 2010 and 2011, inundated city hall, and initiated more aggressive climate

change adaptation strategies. The strategies established a ten-year, US$8 billion investment in flood prevention structures and designated thirty-four major flood-vulnerable regions. In 2016, a comprehensive flood damage reduction plan was established, which includes both structural and nonstructural measures. This plan identifies 240 flood hazard regions within the city. At the national level, since the establishment of the Ministry of Environment in 1990 in South Korea, the River Act was revised in 1993, allowing more ecologically oriented restoration in many urban streams. A good example is the Cheonggye Stream restoration in downtown Seoul. Following the Cheonggye Stream restoration, local districts opened covered streams or removed parking lots in floodplains along several other tributaries of the Han River, which had been frequently flooded by heavy summer precipitation. River restoration projects were promoted to prevent flooding as well as to improve water quality.

Tokyo's modern flood prevention efforts trace back to the late nineteenth century (Nakamura and Oki 2018). Although the first River Act was established in 1896, it was not until 1910, when the massive flooding of the Arakawa River happened in central Tokyo, that the government initiated the first national flood prevention master plan. The government embarked on a project to divert part of the river into a human-made channel between 1911 and 1924 that is called the 22 km Arakawa floodway. After the 1947 severe east Tokyo flood by Typhoon Kathleen, which caused 132,991 deaths and damage to 30,506 properties, the Ministry of Land, Infrastructure, Transport, and Tourism created a comprehensive river basin plan. This plan divided a drainage basin into three areas—upper, middle, and lower subbasins—and established sub-basin-specific strategies—dams, retention basins, and channels.

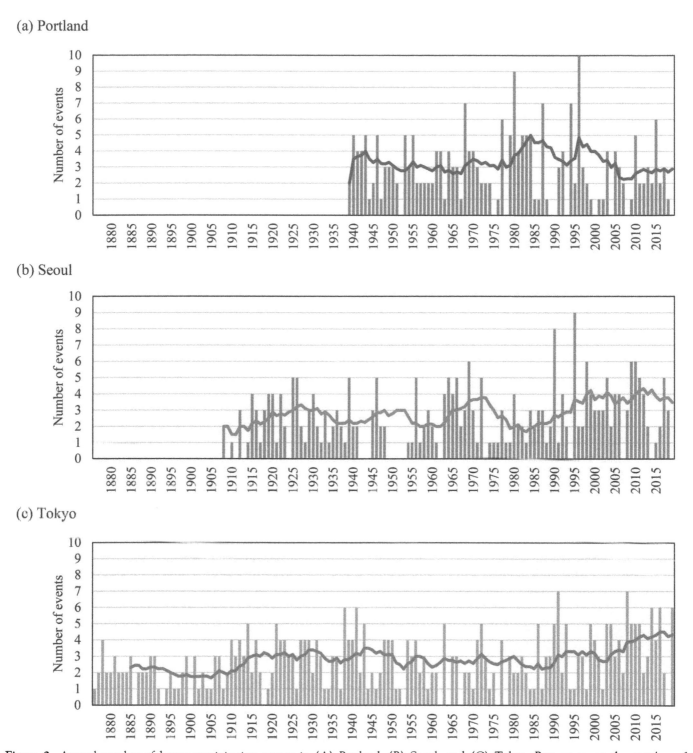

Figure 2. Annual number of heavy precipitation events in (A) Portland, (B) Seoul, and (C) Tokyo. Bars represent the number of events, and the solid lines show the ten-year moving average of the number of events. Daily total precipitation amounts 30 mm, 80 mm, and 65 mm were used for Portland, Seoul, and Tokyo, respectively. Different threshold values were used considering different precipitation characteristics that cause severe urban flooding in each city. Data from National Oceanic Atmospheric Administration (2019), Korea Meteorological Administration (2019), and Japan Meteorological Agency (2019).

The 1958 flood event in the western hilly regions of Tokyo prompted an urgent effort to tackle urban inland floods (Matsuda 2011). The revised River Act in 1964 addressed both flood management and water utilization simultaneously. Realizing the increasing risk of urban flooding with rapid

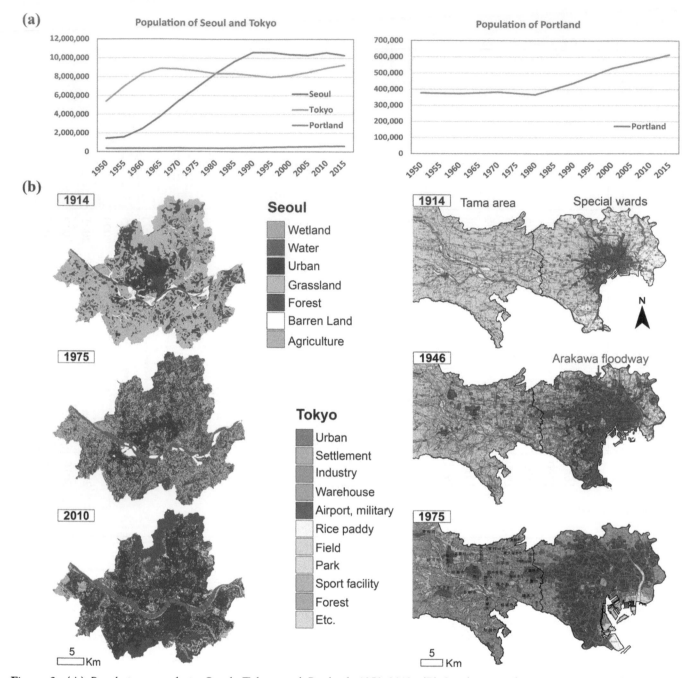

Figure 3. (A) Population growth in Seoul, Tokyo, and Portland, 1950–2019. (B) Land cover change maps in Seoul and Tokyo, 1910s–2010. Data from Korea Ministry of Environment (2019) and Geospatial Information Authority of Japan (1984).

urbanization, the Japanese government initiated "comprehensive flood control measures" in 1977, which embraced environmental conservation while also seeking local communities' opinions on river planning. These measures were intended to regulate lands by increasing infiltration capacity and flood control facilities (Nakamura and Oki 2018). The Tokyo metropolitan government, however, mostly relied on sophisticated gray infrastructures, such as underground detention ponds and underground tunnels, to collect stormwater during massive rainfall events (Bureau of Construction Tokyo Metropolitan Government 2019). These facilities have effectively reduced urban storm runoff in recent years. The underground tunnels, which took thirteen years to complete (1993–2006), are the largest floodwater diversion system in the world and a direct response to the 1991 flood. With pressing concerns about climate change, additional policy shifts occurred at the national level in 2003. Given that four different

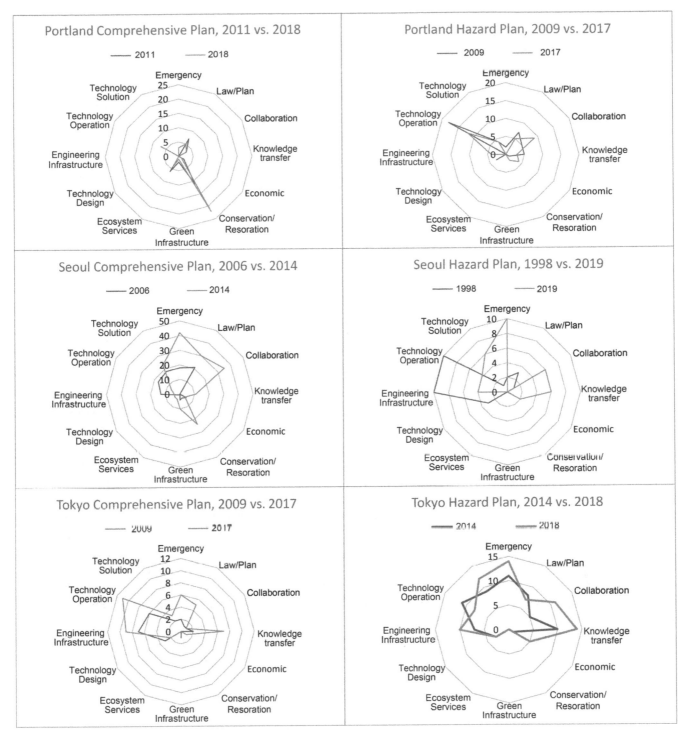

Figure 4. Change in specific social–ecological–technological systems flood risk management strategies mentioned in city comprehensive and hazard mitigation plans in Portland, Seoul, and Tokyo.

laws had been passed related to FRM, a new law, Measures against Inundation Damages in Designated Urban Rivers, was enacted to oversee these individual activities and identify potential synergies when dealing with fluvial and pluvial flooding using both structural and nonstructural measures. This new law has facilitated collaborations among different

institutions to modify FRM with increasing focus on nonstructural measures such as early warning systems, evacuation, public education, and land zoning.

Initially, it appears that all three study cities adapted a reactive or resistant approach in their FRM by emphasizing the technological element of SETS. More recently, however, these cities have embraced more resilient thinking by addressing climate change concerns with a mixture of structural and nonstructural measures that represent other SETS elements (i.e., S, E, and SE).

SETS Dynamics within Governance Documents

Figure 4 shows the relative distribution of SETS elements mentioned in the policy documents for each city over time. In all cities, the hazard mitigation plans predominantly emphasize social elements of SETS, and flood risk reduction strategies mentioned in comprehensive city plans differ by city. The emphasis on social elements in hazard mitigation is not surprising given that hazard mitigation plans tend to focus on people, whereas long-term city plans might seek to achieve some balance among different SETS elements. In Portland, there have not been many changes in the SETS elements in either the comprehensive or hazard mitigation plans. Ecological elements dominated in both versions of the comprehensive plans, whereas emphasis was primarily given to technological and social elements in hazard mitigation plans. In Tokyo, there have been slight changes in the comprehensive plans, with a decreased emphasis on technological elements coinciding with an increased emphasis on social elements. Tokyo's hazard mitigation plans emphasize social and technological elements nearly equally in the earlier document, and social elements became more important in the later document. Seoul's comprehensive plans changed dramatically, with a shift in focus from technological in 2006 to social in 2014. Also notable is an increased emphasis on ecological components in the latter year, pairing with technological elements. As shown in Figure 4, the polygon size became larger in all cities for both documents, suggesting that more strategies were mentioned in the later year documents. Additionally, the shape of the polygons somewhat changed over time, indicating that more diverse SETS strategies were considered as time went on.

The varied and changing emphasis on different SETS elements is likely to be associated with specific pathways each city has taken. Table 4 describes some specific examples of SETS transition elements found in each city's comprehensive plan. As shown in Table 4, some statements include more than one SETS element, which is typically associated with either double-loop learning or triple-loop learning. As shown in Figure 5, Portland has the highest proportion of multiple SETS dimensions (SE, ST, ET, or SET combined strategies), with SE being the most common in the comprehensive plans and ST being the most common in the hazard mitigation plans. Tokyo's multidimensional strategies show heavy emphasis on technological elements; ST strategies slightly strengthened over time in both documents. Seoul's strategies represent a mix of SE and ST. It is noteworthy that the combined SETS strategies are mentioned in recent planning documents in both Portland and Seoul.

Specific Examples of SETS FRM Strategy

Portland's willing seller land acquisition program was designed to permanently relocate residents and businesses out of four flood-prone areas of Johnson Creek, an urban creek that passes through the city (City of Portland 2019b). The city gradually purchased sixty houses over the course of roughly a decade. After completing these purchases, the city converted the site into a natural area. The Foster Floodplain Natural Area features constructed wetlands, floodplain terraces, and other open spaces for holding water during flood events. The restored floodplain is now used for recreational space during low flow periods. In essence, this program was an exemplary case that supported the implementation of the 2001 Johnson Creek Restoration Plan (City of Portland 2019b) and predated a comprehensive city plan that designated Johnson Creek as a special area to reduce chronic floods. A simulation study illustrates the benefits of this floodplain restoration; the restored floodplain effectively minimizes peak flow and retains sediments (Ahilan et al. 2018). In winter 2015, when a storm hit the area, the water level in the creek reached a record peak, but inundation of the adjacent areas of the creek was minimal compared to historical flooding in the area. This program, however, did not consider social displacement of vulnerable people who previously lived in floodplains.

Seoul's Cheonggye Stream restoration project was initiated as part of a downtown revitalization project. The project reopened the 5.8-km stretch of the stream, which was once covered by concrete and inner-city highways and offered new opportunities for economic development and environmental

Table 4. Examples of flood risk reduction strategies social–ecological–technological systems transition elements described in city planning document

	Resistant	Resilient	Transformative
Portland	"Issue Evacuation Orders in accordance with Evacuation Annex" (S) "Protect flood risk reduction system (levee system and drainage infrastructure)" (T)	"Enhance the ability of rivers, streams, wetlands, floodplains, urban forest, habitats, and wildlife to limit and adapt to climate-exacerbated flooding, landslides, wildfire, and urban heat island effects." (E) "Green infrastructure helps minimize risks from flooding and landslides, helps to cool the city—reducing impacts from the urban heat island effect—and creates an overall healthier and more pleasant environment for people." (S, E)	"Promote restoration and protection of floodplains. Prevent development-related degradation of natural systems and associated increases in landslide, wildfire, flooding, and earthquake risks." (S, E) "Improve and maintain the functions of natural and managed drainageways, wetlands, and floodplains to protect health, safety, and property, provide water conveyance and storage, improve water quality, and maintain and enhance fish and wildlife habitat." (S, E, T)
Seoul	"Electric equipment reinforcement in rainwater pumping station" (T) "Lowland basement housing relocation" (S) "Sustainable waterfront planting and vegetation for flood prevention" (E) "Governance expansion for residents' safety and waterfront cleaning movement" (S, E)	"Discharge capacity improvement when over 100 mm/hr (design for 30 years)" (T) "Long-term flood prevention measures: traffic control, strengthening recovery system" (S, T) "Flood Management in conjunction with Urban Planning (land use limitation)" (S) "Enhancement of Natural Flood Prevention System through Ecological River Restoration" (E)	
Tokyo	"Promote constructing realistic and effective warning and evacuation system for each local government" (S) "Improve the awareness of disaster prevention by supporting education and making hazard map with local governments" (S, T)	"Establish flood prevention measures by making fill-up ground for residential and park use in lowlands area" (S, T) "Construct high-standard levee in connection with riverside development projects, and ask the national government for supporting the construction high-standard levee and detention ponds to improve the safety and neighborhood condition" (S, T)	

restoration. Instead of repairing degrading highways and underground channels, the city decided to reopen the stream. The stream now attracts visitors, lowers temperatures in surrounding areas, and helps control urban flooding. The project also separated sewers from channels, improving the water quality of streams. To maintain water during low-flow periods, groundwater was pumped to the beginning of the opened channel, and the streambed was sealed to minimize the seepage of water into the ground. To prevent urban flooding, the channels were deepened with the installation of flood gates. With high land prices and intense development in surrounding areas, the Cheonggye Stream's old floodplains were never completely restored to their natural state. Instead, the flood gates were designed to control high water levels during heavy rainfall events. When the stream water level reaches a certain threshold, they open and transfer water to downstream areas.

Tokyo's super levee was constructed as part of an urban renewal project, creating structures that can adapt to multiple hazards, including earthquakes and floods. Unlike conventional levees that have narrow and tall walls, a super levee is up to thirty times wider than an ordinary levee, making it possible to protect land below sea level from floods caused by seepage failure. The top of the super levee can be used

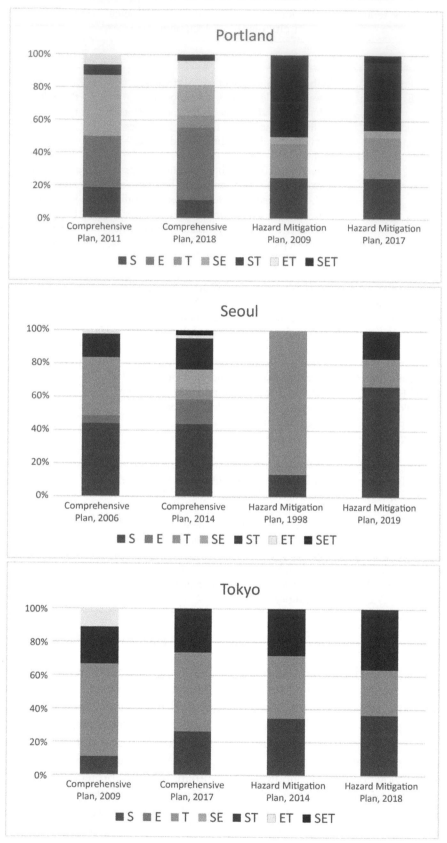

Figure 5. Change in relative proportion of social (S), ecological (E), technological (T), socioecological (SE), sociotechnological (ST), eco-technological (ET), and socio-eco-technological (SET) strategies mentioned in the city comprehensive plan and hazard mitigation plan documents in Portland, Seoul, and Tokyo.

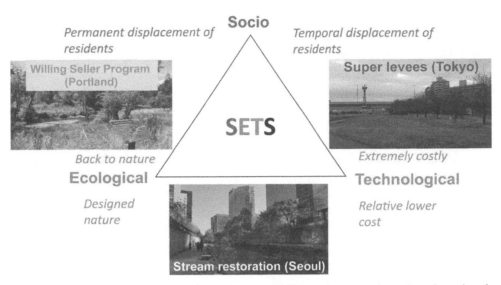

Figure 6. Specific cases of flood risk management strategies representing different elements of social–ecological–technological systems in each city.

as ordinary land, providing a safer area to protect human lives and properties from flood and high tide. High costs and long construction time are the main barriers to the implementation of super levees. Given high property rights in Japan, all residents need to agree on the project and relocate temporarily at their own cost during the construction. As a result of these hurdles, the target area of the super levee was scaled down in 2011. Additional construction of the super levee is currently being reconsidered to prevent the type of seepage that was experienced after Typhoon Hagibis, which caused 140 embankment failures in 2019 ("Pressure for Expanding Construction Investment Due to Successive Disaster" 2019).

Figure 6 summarizes how the aforementioned case studies fit into the SETS framework. Portland's willing seller program emphasizes socioecological elements because people cannot move back to old floodplains anymore. Seoul's Chenggye Stream restoration exemplifies an ecological–technological evolution. It seeks to introduce nature to the built environment, but the water-holding capacity of the restored streams is limited because the river width is constrained by the surrounding buildings. With a heavy initial emphasis on technology, followed by a shift toward social–technological elements, Japan's super levee illustrates that successful construction of super levees requires social consensus as well as enormous budgets for completion.

Discussion

Our SETS-based approach offers insights into each city's FRM strategies in the context of resilience and social learning. In particular, it provides a set of general variables and a conceptual lens by which the diversity of cities' FRM strategies over time (which, we argue here, is a sign of social learning) can be analyzed and interpreted. This conceptual basis is useful for understanding and comparing cities' flood resilience because there is a broad consensus that the level of diversity and learning in a system is closely associated with resilience (Biggs et al. 2012). All study cities appear to primarily rely on the technological components of SETS (e.g., levees and dams) to meet the needs of reducing flood risks in their hazard mitigation plans. Given their land constraints, population density, high land prices, and experience with frequent heavy precipitation events, Seoul and Tokyo have adopted more technological approaches than Portland to reduce flood risk in their comprehensive plans. The emphasis on technological approaches to achieve flood resilience might have been associated with these cities' capital investment strategies that prefer massive construction projects (e.g., super levees). These two cities, however, have increasingly embraced social and ecological strategies to seek resilience (Nishi et al. 2016) or even transformation in response to major flood events. Even though general awareness of flood risk exists, the infeasibility of relocating people out of flood risk zones (due to high land prices and strong property rights) likely caused the city administrators to focus on socially relevant measures such as early warning and evacuation, as clearly shown by the increasing prevalence and diversity of the S elements of SETS within these cities' FRM strategies. Seoul's recent emphasis on ecological elements in FRM appears to be associated with increasing grassroots environmental organizations, which have been involved with river restoration (J. Kim 2013). With their inputs, most stream restoration efforts seek to achieve water quality improvement and flood risk reduction. Similarly, potentially due to its lower density and fewer major flood events, Portland (at least in its comprehensive plans) has emphasized ecologically oriented FRM strategies.

Nonetheless, ecological elements generally appear to be absent or lacking in diversity across all of our case cities. The hazard mitigation plans for all three locations show relatively little scope for ecological elements. In the cases where ecological elements were considered (e.g., Portland and Seoul comprehensive plans), a lack of diversity was observed as nearly all actions focused on conservation, and minimal emphasis was placed on green infrastructure or ecosystem services. Considering the growing popularity and promotion of green infrastructure and ecosystem services (e.g., D. Kim and Song 2019), the apparent shortage of these elements in the cities' FRM strategies is puzzling and worth exploring further. One possible explanation is that green infrastructure and ecosystem services activities might in fact be occurring but in ad hoc or informal ways by different actors that seek different forms of resilience (Grove 2018). That is, these concepts are being implemented to promote resilience but not to the point of being codified into formal planning documents related to flood resilience. Additionally, there might be a disjoint between the theoretical explorations of green infrastructure and ecosystem services and the practical elements of implementing these strategies as part of urban FRM. Regardless of the explanation, the apparent shortage of ecological elements (especially green infrastructure and ecosystem services) highlights future avenues for research, as well as potential room for improvement in FRM strategies.

The findings of our research suggest that future FRM should be tightly coupled with spatial planning for a city to facilitate adaptability and transformability, two dimensions of resilience (Zevenbergen, Gersonius, and Radhakrishan 2020). Given that both risk management and land development plans seek long-term economic, environmental, and social

sustainability, the potential synergies and trade-offs between growth and floodplain management should be examined together when the city plans to develop or conserve lands. The latest planning document in Seoul clearly illustrated the importance of tightly coupled spatial and FRM plans, as locally adapted FRM strategies were suggested for different locations, indicating that spatially well-targeted projects could increase flood resilience (Bertilsson et al. 2019). After the 2019 Typhoon Hagibis in Japan, the Ministry of Finance in Japan noted that it would not be sufficient to rely on structural measures such as dams and levees to cope with increasing flood risk in Japan, and also recommended restrictions to land development in flood-prone areas. Hazardscapes are not created in a vacuum, and institutionalized resilience initiatives could further reinforce uneven power relationships within the city (Grove 2013; Grove, Cox, and Barnett 2020; Webber, Leitner, and Sheppard 2020). Thus, it is imperative to unpack how the local political–economic dynamics produce new development and mitigation strategies in flood-prone areas (Fuchs et al. 2017; Coates and Nygren 2020).

In situating our work within the current debate on resilience, we acknowledge that resilience is essentially a contested concept (Grove 2018). No singular definition of resilience applies to understanding the complex human and flood interactions within cities. Thus, the form of flood resilience that each city has taken is context specific and in line with the inductive critique of resilience initiatives (Chandler 2014; Anderson 2015; Simon and Randalls 2016; Grove 2018; Grove, Barnett, and Cox 2020). Although further study is needed to tease out the contextual specificity of each city, by analyzing the city and hazard planning documents using the SETS learning framework over time, we can infer that some cities continuously promote gray infrastructure projects partly due to the history of development, development density, and the high price of lands. An increasing movement toward ecological and social elements of SETS in recent years speaks to the existence of diversifying perspectives and forms of resilience and their interactions with broader changes in socioeconomics, sociodemographics, and perceptions of climate change in a particular place. Thus, the spatialization of flood resilience using the SETS lens contributes to the conceptual and theoretical innovations in the field.

Conclusions

Understanding the processes of change in FRM and factors that might influence the trajectory of change in FRM is important. In this article, we portrayed a general approach that can be used to analyze the evolution of urban FRM using the SETS learning framework. Based on the case studies of FRM in three cities over the last few decades, we employed the approach to provide a pilot assessment of the characteristics of FRM trajectories that these cities are likely to be on and factors that could be associated with the trajectories. Our findings show that all cities underwent some changes in their focus on different SETS elements in FRM. Historically, Portland has focused on the ecological elements of FRM in its city plan, whereas both Seoul and Tokyo have emphasized technological strategies. These different emphases are likely to be associated with the history of land development and flooding severity in these cities. FRM strategies have been diversifying in recent years, however, particularly in Seoul, with more emphasis on social elements of SETS over conventional technological strategies, indicating that social learning has taken place to shift away from the resistant to resilient or transformative state. Across all cities, FRM systems are lacking in ecological strategies, suggesting room for incorporating nature-based solutions into future planning (Bertilsson et al. 2019). Because diversity is a key condition for building general resilience (Biggs, Schlüter, and Schoon 2015), city managers should consider this aspect when devising future FRM strategies.

Finally, we suggest some directions for future research. First, interview data can be collected from flood experts and city practitioners involved with FRM in the study cities to complement our document-based analysis. Such data can further reveal insights into how and why the regions might have chosen specific strategies and have undergone different pathways through learning. Second, other sources of information, such as newspaper articles or resident surveys, could be used to enrich the context and improve understanding of the potential feedback between public perception of flood risk and flood policy (i.e., link between public preference and codification of SETS strategies in FRM). Third, a modeling approach, such as systems dynamics or agent-based models (e.g., Di Baldassarre et al. 2013), could be employed to better understand the transition of a city from resistant to resilient to transformative phases through social learning. Finally, ample opportunities exist for mutual transdisciplinary learning across the cities through SETS-based comparisons of FRM strategy implementation. Although the SETS learning approach is one of many theorizing resilience and social change, our intent in this study has been to provide a common framework to facilitate this comparative understanding. Balanced SETS strategies could achieve more proactive solutions rather than techno-centric reactive solutions to floods, particularly in the rapidly changing climate and demographic characteristics of many urban regions across the globe.

Acknowledgments

We appreciate Professors Takashi Oguchi, Taikan Oki, and Hyungjun Kim of the University of Tokyo for their constructive comments on the initial phase of the work. Kevin Grove at Florida International University and two anonymous reviewers offered valuable comments that helped strengthen many points of the article. Thanks also go to Yasuyo Makido of Portland State University and Jungbae Kim of Sejong University, who compiled materials used in the article.

Funding

This work was supported by the U.S. National Science Foundation (SES 1444755, CIS-1913665), Korea Environment Industry & Technology Institute through the Advanced Water Management Research Program, funded by the Korea Ministry of Environment (Grant 83079), a grant from the Abe Fellowship Program administered by the Social Science Research Council in cooperation with and with funds provided by the Japan Foundation Center for Global Partnership, and the Center for the Environment at Purdue University through its C4E seed grant program. Any opinions, findings, and conclusions or recommendations expressed in this material are those of the authors and do not necessarily reflect the views of the sponsoring agencies.

ORCID

Heejun Chang http://orcid.org/0000-0002-5605-6500
David J. Yu http://orcid.org/0000-0001-9929-1933
Samuel A. Markolf http://orcid.org/0000-0003-4744-0006
Chang-yu Hong http://orcid.org/0000-0001-5368-8855

Sunyong Eom ⓘ http://orcid.org/0000-0002-8164-7097

Wonsuh Song ⓘ http://orcid.org/0000-0001-6588-495X

Deghyo Bae ⓘ http://orcid.org/0000-0002-0429-1154

References

Aerts, J., and W. Botzen. 2014. Adaptation: Cities' response to climate risks. *Nature Climate Change* 4 (9):759–60. doi: 10.1038/nclimate2343.

Ahilan, S., M. Guan, A. Sleigh, N. Wright, and H. Chang. 2018. The influence of floodplain restoration on flow and sediment dynamics in an urban river. *Journal of Flood Risk Management* 11 (Suppl. 2):S986–S1001. doi: 10.1111/jfr3.12251.

Allen, T. R., T. Crawford, B. E. Montz, J. Whitehead, S. Lovelace, A. D. Hanks, A. R. Christensen, and G. D. Kearney. 2019. Linking water infrastructure, public health, and sea level rise: Integrated assessment for flood resilience in coastal cities. *Public Works Management and Policy* 24 (1):110–39. 10.1177/1087724X18798380.

Anderson, B. 2015. What kind of thing is resilience? *Politics* 35 (1):60–66. doi: 10.1111/1467-9256.12079.

Argyris, C., and D. Schön. 1978. *Organizational learning*. Reading, MA: Addison-Wesley.

Bae, S., and H. Chang. 2019. Urbanization and floods in the Seoul metropolitan area of South Korea: What old maps tell us. *International Journal of Disaster Risk Reduction* 37:101186. doi: 10.1016/j.ijdrr.2019.101186.

Bertilsson, L., K. Wiklund, I. de Moura Tebaldi, O. M. Rezende, A. P. Verol, and M. G. Miguez. 2019. Urban flood resilience—A multi criteria index to integrate flood resilience into urban planning. *Journal of Hydrology* 573:970–82. doi: 10.1016/j.hydrol.2018.06.05.

Biggs, R., M. Schlüter, D. Biggs, E. L. Bohensky, S. BurnSilver, G. Cundill, V. Dakos, T. M. Daw, L. S. Evans, K. Kotschy, A. M. Leitch, C. Meek, A. Quinlan, C. Raudsepp-Hearne, M. D. Robards, M. L. Schoon, L. Schultz, and P. C. Miguez. 2012. Toward principles for enhancing the resilience of ecosystem services. *Annual Review of Environment and Resources* 37 (1):421–48. http://www.annualreviews.org/doi/10.1146/annurev-environ-051211-123836.

Biggs, R., M. Schlüter, and M. L. Schoon. 2015. *Principles for building resilience: Sustaining ecosystem services in social–ecological systems*. Cambridge, UK: Cambridge University Press.

Bureau of Construction, Tokyo Metropolitan Government. 2019. Retention basin of Tokyo. Accessed December 5, 2019. http://www.kensetsu.metro.tokyo.jp/kasenbu0035.html.

Chan, F. K. S., C. J. Chuah, A. D. Ziegler, M. Dąbrowski, and O. Varis. 2018. Towards resilient flood risk management for Asian coastal cities: Lessons learned from Hong Kong and Singapore. *Journal of Cleaner Production* 187:576–89. doi: 10.1016/j.jclepro.2018.03.217.

Chandler, D. 2014. *Resilience: The art of governing complexity*. London and New York: Routledge.

Chang, H., and J. Franczyk. 2008. Climate change, land use change, and floods: Toward an integrated assessment. *Geography Compass* 2 (5):1549–79. doi: 10.1111/j.1749-8198.2008.00136.x.

Chang, H., M. Lafrenz, I.-W. Jung, M. Figliozzi, D. Platman, and C. Pederson. 2010. Potential impacts of climate change on flood-induced travel disruption: A case study of Portland in Oregon, USA. *Annals of the Association of American Geographers* 100 (4):938–52. doi: 10.1080/00045608.2010.497110.

Cho, S. 2011. Urban transformation of Seoul and Tokyo by legal redevelopment project. *ITU AZ* 8 (1):169–83.

City of Portland. 1991. Johnson Creek Basin protection plan. Accessed December 11, 2019. https://www.portlandoregon.gov/bes/article/214364.

City of Portland. 2011. *Comprehensive plan goals and policies*. Portland, OR: City of Portland. Accessed December 11, 2019. https://www.portlandonline.com/bps/Comp_Plan_Nov2011.pdf.

City of Portland. 2018. 2035 Comprehensive plan, July 2018 updated. Portland, OR: City of Portland. Accessed December 11, 2019. https://www.portland.gov/bps/comp-plan/2035-comprehensive-plan-and-supporting-documents

City of Portland. 2019a. Flood insurance resources. Accessed December 11, 2019. https://www.portlandoregon.gov/phb/74674.

City of Portland. 2019b. Willing seller program. Accessed December 11, 2019. https://www.portlandoregon.gov/bes/article/106234.

Coates, R., and A. Nygren. 2020. Urban floods, clientelism, and the political ecology of the state in Latin America. *Annals of the American Association of Geographers* 110 (5): 1301–17.

Cole, D. H., and M. D. McGinnis. 2017. *Elinor Ostrom and the Bloomington School of Political Economy: A framework for policy analysis*. Vol. 3. Lexington, MA: Lexington Books.

den Boer, J., C. Dieperink, and F. Mukhtarov. 2019. Social learning in multilevel flood risk governance: Lessons from the Dutch Room for the River program. *Water* 11 (10). doi: 10.3390/w11102032.

Di Baldassarre, G. D., A. Viglione, G. Carr, L. Kuil, J. L. Salinas, and G. Blöschl. 2013. Socio-hydrology: Conceptualising human–flood interactions. *Hydrology and Earth System Sciences* 17 (8):3295–303. doi: 10.5194/hess-17-3295-2013.

Fahy, B., E. Brenneman, H. Chang, and V. Shandas. 2019. Spatial analysis of urban floods and extreme heat potential in Portland, OR. *International Journal of Disaster Risk Reduction* 39:101117. doi: 10.1016/j.ijdrr.2019.101117.

Folke, C. 2006. Resilience: The emergence of a perspective for social–ecological systems analyses. *Global Environmental Change* 16 (3):253–67. doi: 10.1016/j.gloenvcha.2006.04.002.

Fuchs, S., V. Röthlisberger, T. Thaler, A. Zischg, and M. Keiler. 2017. Natural hazard management from a coevolutionary perspective: Exposure and policy response in the European Alps. *Annals of the American Association of Geographers* 107 (2):382–92. doi: 10.1080/24694452.2016.1235494.

Geospatial Information Authority of Japan. 1984. Regional planning atlas of Japan. Accessed December 10, 2019. https://www.gsi.go.jp/atlas/kokudo-etsuran.html

Goodchild, B., R. Sharpe, and C. Hanson. 2018. Between resistance and resilience: A study of flood risk management in the Don catchment area (UK). *Journal of Environmental Policy & Planning* 20 (4):434–49.

Grove, K. J. 2013. From emergency management to managing emergence: A genealogy of disaster management in Jamaica. *Annals of the Association of American Geographers* 103 (3):570–88. doi: 10.1080/00045608.2012.740357.

Grove, K. J. 2018. *Resilience*. London and New York: Routledge.

Grove, K. J., A. Barnett, and S. Cox. 2020. Designing justice? Race and the limits of recognition in Greater Miami resilience planning. *Geoforum* 117:134–43. doi: 10.1016/j.geoforum.2020.09.014.

Grove, K. J., S. Cox, and A. Barnett. 2020. Racializing resilience: Assemblage, critique, and contested futures in Greater Miami resilience planning. *Annals of the American Association of Geographers* 110 (5):1613–18. doi: 10.1080/24694452.2020.1715778.

Helaire, L. T., S. A. Talke, D. A. Jay, and H. Chang. 2020. Present and future flood hazard in the Lower River Columbia River estuary. *Geophysical Research Letters-Oceans*. doi: 10.1029/2019JC015928.

Intergovernmental Panel on Climate Change. 2012. *Managing the risks of extreme events and disasters to advance climate change adaptation*, ed. C. B. Field, V. Barros, T. F. Stocker, D. Qin, D. J. Dokken, K. L. Ebi, M. D. Mastrandrea, K. J. Mach, G.-K. Plattner, S. K. Allen, et al. Cambridge, UK: Cambridge University Press.

Japan Meteorological Agency. 2019. Accessed December 7, 2019. http://www.data.jma.go/jp/gmd/risk/obsdl/index.php

Johannessen, Å., Å. Gerger Swartling, C. Wamsler, K. Andersson, J. T. Arran, D. I. Hernández Vivas, and T. A. Stenström. 2019. Transforming urban water governance through social (triple-loop) learning. *Environmental Policy and Governance* 29 (2):144–54. doi: 10.1002/eet.1843.

Kim, D., and S. Song. 2019. The multifunctional benefits of green infrastructure in community development: An analytical review based on 447 cases. *Sustainability* 11 (14):3917. doi: 10.3390/su11143917.

Kim, J. 2013. Desirable use and conservation of riverine environment in Seoul. Working paper 2013-PR-59, The Seoul Institute, Seoul, South Korea.

Korea Meteorological Administration. 2019. Daily precipitation data. Accessed November 18, 2019. https://data.kma.go.kr/cmmn/main.do.

Korea Ministry of Environment. 2019. Environmental Spatial Information Service. Accessed November 17, 2019. https://egis.me.go.kr/api/land.do.

Liao, K.-H. 2012. A theory on urban resilience to floods: A basis for alternative planning practices. *Ecology and Society* 17 (4):15. doi: 10.5751/ES-05231-170448.

Luo, P., B. He, K. Takara, Y. E. Xiong, D. Nover, W. Duan, and K. Fukushi. 2015. Historical assessment of Chinese and Japanese flood management policies and implications for managing future floods. *Environmental Science & Policy* 48:265–77. doi: 10.1016/j.envsci.2014.12.015.

Markolf, S., M. Chester, D. Eisenberg, D. Iwaniec, C. Davidson, R. Zimmerman, T. Miller, B. Ruddell, and H. Chang. 2018. Interdependent infrastructure as linked social, ecological, and technological systems (SETSs) to address lock-in and enhance resilience. *Earth's Future* 6 (12):1638–59. doi: 10.1029/2018EF000926.

Matsuda, I. 2011. *Land development and disaster mitigation: Disaster and land history of Edo-Tokyo. Suggestions from physical geography*. Tokyo: Imagine.

McPhearson, T., D. Haase, S. Kabisch, and A. Gren. 2016. Advancing understanding of the complex nature of urban systems. *Ecological Indicators* 70:566–73. doi: 10.1016/j.ecolind.2016.03.054.

Merz, B., S. Vorogushyn, U. Lall, A. Viglione, and G. Blöschl. 2015. Charting unknown waters—On the role of surprise in flood risk assessment and management. *Water Resources Research* 51 (8):6399–416. doi: 10.1002/2015WR017464.

Morrison, A., B. F. Noble, and C. J. Westbrook. 2018. Flood risk management in the Canadian Prairie provinces: Defaulting toward flood resistance and recovery versus resilience. *Canadian Water Resources Journal/ Revue Canadienne des Ressources Hydriques* 43 (1):33–46. doi: 10.1080/07011784.2018.1428501.

Morrison, A., C. J. Westbrook, and B. F. Noble. 2018. A review of the flood risk management governance and resilience literature. *Journal of Flood Risk Management* 11 (3):291–304. doi: 10.1111/jfr3.12315.

Multnomah County. 2009. Multnomah County Hazard Mitigation Plan, Portland, OR.

Multnomah County. 2017. Multi-Jurisdictional Natural Hazards Mitigation Plan, Portland, OR.

Nakamura, S., and T. Oki. 2018. Paradigm shifts on flood risk management in Japan: Detecting triggers of design flood revisions in the modern era. *Water Resources Research* 54 (8):5504–15. doi: 10.1029/2017WR022509.

National Academies of Sciences, Engineering, and Medicine. 2019. *Framing the challenge of urban flooding in the United States*. Washington, DC: The National Academies Press. 10.17226/25381.

National Oceanic Atmospheric Administration. 2019. Climate data online. Accessed December 8, 2019. https://www.ncdc.noaa.gov/cdo-web/.

Nishi, M., M. Pelling, M. Yamamuro, W. Solecki, and S. Kraines. 2016. Risk management regime and its scope for transition in Tokyo. *Journal of Extreme Events* 3 (3):1650011. doi: 10.1142/S2345737616500111.

Organization for Economic Co-operation and Development. 2016. *Financial management of flood risk*. Paris: OECD Publishing.

Ostrom, E. 2005. *Understanding institutional diversity*. Princeton, NJ: Princeton University Press.

Ostrom, E. 2009. A general framework for analyzing sustainability of social-ecological systems. *Science* 325 (5939):419–22. doi: 10.1126/science.1172133.

Pahl-Wostl, C. 2009. A conceptual framework for analysing adaptive capacity and multi-level learning processes in resource governance regimes. *Global Environmental Change* 19 (3):354–65. doi: 10.1016/j.gloenvcha.2009.06.001.

Pelling, M. 2011. *Adaptation to climate change: From resilience to transformation*. London and New York: Routledge.

Pelling, M., C. High, J. Dearing, and D. Smith. 2008. Shadow spaces for social learning: A relational understanding of adaptive capacity to climate change within organisations. *Environment and Planning A* 40 (4):867–84.

Pelling, M., K. O'Brien, and D. Matyas. 2015. Adaptation and transformation. *Climatic Change* 133 (1):113–27. doi: 10.1007/s10584-014-1303-0.

Pressure for expanding construction investment due to successive disaster. 2019. *The Nikkei*, November 23.

Reed, M. S., A. C. Evely, G. Cundill, I. Fazey, J. Glass, A. Laing, J. Newig, B. Parrish, C. Prell, C. Raymond, et al. 2010. What is social learning? *Ecology and Society* 15 (4):r1. doi: 10.5751/ES-03564-1504r01.

Simon, S., and S. Randalls. 2016. Geography, ontological politics and the resilient future. *Dialogues in Human Geography* 6 (1):3–18. doi: 10.1177/2043820615624047.

Solecki, M., M. Pelling, and M. Garschagen. 2017. Transitions between risk management regimes in cities. *Ecology and Society* 22 (2):38. doi: 10.5751/ES-09102-220238.

Syed, M., and S. Nelson. 2015. Guidelines for establishing reliability when coding narrative data. *Emerging Adulthood* 3 (6):375–87. doi: 10.1177/2167696815587648.

Tokyo Metropolitan Government. 2018. Land use in Tokyo (Special-ward area 2016 edition). Accessed December 8, 2019. http://www.toshiseibi.metro.tokyo.jp/seisaku/tochi_c/tochi_5.html.

Tokyo Metropolitan Government. 2019. Land use in Tokyo (Tama and Island areas 2017 edition). Accessed December 8, 2019. http://www.toshiseibi.metro.tokyo.jp/seisaku/tochi_c/tochi_6.html.

Webber, S., H. Leitner, and E. Sheppard. 2020. Wheeling out urban resilience: Philanthrocapitalism, marketization, and local practice. *Annals of the American Association of Geographers*. Advance online publication. doi:10.1080/24694452.2020.1774349.

Yu, D. J., M. L. Schoon, J. K. Hawes, S. Lee, J. Park, P. S. C. Rao, L. K. Siebeneck, and S. V. Ukkusuri. 2020. Toward general principles for resilience engineering. *Risk Analysis: An Official Publication of the Society for Risk Analysis* 40 (8):1509–37. doi: 10.1111/risa.13494.

Yu, D. J., H. C. Shin, I. Pérez, J. M. Anderies, and M. A. Janssen. 2016. Learning for resilience-based management: Generating hypotheses from a behavioral study. *Global Environmental Change* 37:69–78. doi: 10.1016/j.gloenvcha.2016.01.009.

Yu, D. J., N. Sangwan, K. Sung, X. Chen, and V. Merwade. 2017. Incorporating institutions and collective action into a sociohydrological model of flood resilience. *Water Resources Research* 53 (2):1336–53. doi: 10.1002/2016WR019746.

Zevenbergen, C., B. Gersonius, and M. Radhakrishan. 2020. Flood resilience. *Philosophical Transactions: Series A. Mathematical, Physical, and Engineering Sciences* 378 (2168). doi: 10.1098/rsta.2019.0212.

Document title	Date	Document type	Location	Link
Multnomah County Multi-Hazard Mitigation Plan	2009	Hazard-specific	Multnomah County, OR	https://scholarsbank.uoregon.edu/xmlui/handle/1794/3186
Multnomah County Multi-Jurisdictional Natural Hazards Mitigation Plan	2017	Hazard-specific	Multnomah County, OR	https://multco.us/file/65292/download
Comprehensive Plan Goals and Policies	2011[a]	General plan	Portland, OR	https://www.portlandonline.com/bps/Comp_Plan_Nov2011.pdf
2035 Comprehensive Plan	2018	General plan	Portland, OR	https://beta.portland.gov/bps/comp-plan/2035-comprehensive-plan-and-supporting-documents#toc-2035-comprehensive-plan-as-amended-through-march-2020
1998 Seoul Floods and Mitigation Policy	1998	Hazard-specific	Seoul, South Korea	City of Seoul library
City of Seoul, Flood Mitigation Policy	2019	Hazard-specific	Seoul, South Korea	City of Seoul library
City of Seoul, Comprehensive Plan	1997	General plan	Seoul, South Korea	http://urban.seoul.go.kr/4DUPIS/sub3/sub3_1.jsp
City of Seoul, Comprehensive Plan	2006	General plan	Seoul, South Korea	http://urban.seoul.go.kr/4DUPIS/sub3old/concept_3_1_6_2020BU.jsp
City of Seoul, Comprehensive Plan	2014	General plan	Seoul, South Korea	http://urban.seoul.go.kr/4DUPIS/sub3/sub3_1.jsp
Tokyo Disaster Prevention Plan	2014	Hazard-specific	Tokyo, Japan	https://www.bousai.metro.tokyo.lg.jp/_res/projects/default_project/_page_/001/005/784/1.pdf
Safe City Tokyo Disaster Prevention Plan	2018	Hazard-specific	Tokyo, Japan	https://www.bousai.metro.tokyo.lg.jp/_res/projects/default_project/_page_/001/005/783/NEW_PLAN.pdf
Tokyo Urban Planning Vision	2009	General plan	Tokyo, Japan	https://www.toshiseibi.metro.tokyo.lg.jp/kanko/mnk/
The Grand Design for Urban Development	2017	General plan	Tokyo, Japan	http://www.toshiseibi.metro.tokyo.jp/keikaku_chousa_singikai/pdf/grand_design_42.pdf

[a]The 2011 Comprehensive Plan includes all actions and strategies that have been included in city comprehensive plans since 1980.

Part 5

The Environment and Environmental Degradation

Reframing Pre-European Amazonia through an Anthropocene Lens

Antoinette M. G. A WinklerPrins ⓘ and Carolina Levis ⓘ

This article examines three intertwined forms of human transformation of Amazonia's landscapes: (1) anthrosols, (2) cultural or domesticated forests, and (3) anthropogenic earthworks. By acknowledging the extent to which landscapes are humanized, an Anthropocene lens provides an opportunity to examine Amazonia as an Anthropogenic space (anthrome), providing a more realistic approach to understanding the region's past and for guiding its conservation. *Key Words: Amazonia, anthrosols, domesticated forests, earthworks, landesque capital.*

The significance of the Anthropocene resides in its role as a new lens through which age-old narratives and philosophical questions are being revisited and rewritten.

—Ellis (2018, 4)

The popular imagery of Amazonia continues to conjure up two extreme views: rampant deforestation and environmental destruction on the one hand and intact or pristine wilderness on the other. Neither is a correct representation of the region, yet persistence of these imageries hinders a more realistic approach to understanding and conserving the region. The Anthropocene lens, which acknowledges that the Earth has long been significantly transformed by the actions of humans, permits a redress of this popular imagery (Ellis and Ramankutty 2008; Ellis and Ramankutty 2008; Ellis 2015, 2018). Earlier reframing of the Neotropics, including Amazonia, as anthropogenic spaces came through the historical ecology research program, which is concerned with the interactions through time between human societies, environments, plants, and animals and the consequences of these interactions for understanding the formation of current landscapes (Balée 2006; Balée et al. 2020). For Amazonia specifically, historical ecologists take the perspective that indigenous people "did not adapt to nature but rather they created what they wanted through human creativity, technology and engineering, and cultural institutions," which resulted in a widespread distribution of domesticated landscapes

across the region (Erickson 2003, 456). This perspective contrasts with still persistent ideas of environmental determinism (Meggers [1971] 1996) and also with traditional ecological and land use change research that treats Amazonia ahistorically and visualizes it as a demographic void. These approaches typically do not acknowledge or discuss past human action or treat it as inconsequential for conservation planning (e.g., Barlow et al. 2012).

This article considers Amazonia through an Anthropocene lens by examining three intertwined human transformations of Amazonian landscapes: (1) anthrosols (2) cultural or domesticated forests, and (3) anthropogenic earthworks. The first two are legacies of long-term and cumulative activities of Native Amazonians (Denevan 2007; Clement et al. 2015a; Levis et al. 2018), and the third is a form of landesque capital, because earthworks were intentionally produced and their creation involves permanent changes to the landscape in accordance with economic, social, and ritual purposes (Håkansson and Widgren 2014; Arroyo-Kalin 2016). The interactions of these three types of domesticated landscapes result in feedback mechanisms that are not yet fully understood but are the topic of ongoing research regarding their persistence through time (Levis et al. 2020). Anthropogenic earthworks were identified before anthrosols, but research on anthrosols in turn contributed to the further identification of earthworks. Both influenced forest composition in the region long before the arrival of Europeans. Some of these

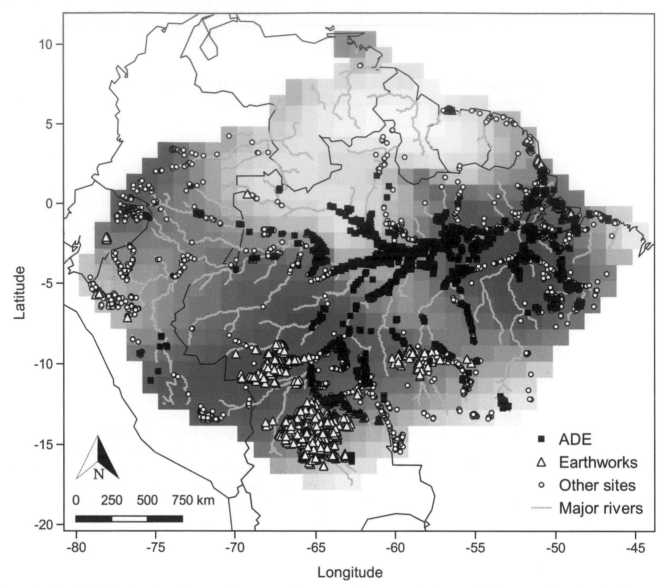

Figure 1. Spatial distribution of pre-European human transformations of Amazonian landscapes. Black squares show the spatial extent of known locations of anthrosols (Amazonian Dark Earth [ADE]), white triangles show the spatial extent of known locations of earthworks, and white circles indicate other archaeological sites. *Note:* There are likely many more archaeological sites; this is not meant to be a comprehensive map. Archaeological data were obtained from the Amazon Archaeological Sites Network (2020), Instituto do Patrimônio Histórico e Artístico Nacional (http://portal.iphan.gov.br/pagina/detalhes/1701/), Lombardo et al. (2020), and the second author's own research. The white-red background shows the interpolation of the observed values of the total number of domesticated species (richness) in each Amazon Tree Diversity Network (http://atdn.myspecies.info/) forest plot modeled as a function of latitude and longitude on a 1° grid cell scale by use of loess spatial interpolation (modified after Levis et al. 2017). The major river network was obtained from the HydroSHEDS data set (http://hydrosheds.cr.usgs.gov). Map was created in QGIS 2.18.25 by Carolina Levis. ADE = Amazonian Dark Earth.

domesticated forests, such as forest islands, have been dated to as early as 10,850 calibrated years before present (cal. yr BP) in the Bolivian Amazonia (Lombardo et al. 2020). Pre-European transformations persist in hyperdiverse Amazonian forests and their resources are maintained and re-created by indigenous and traditional peoples (Roberts et al. 2017; Levis et al. 2018). These landscape transformations are widespread across the region (Figure 1).

Anthrosols

Anthrosols are the most visible legacy of widespread pre-European human settlements in Amazonia. They are a continuum of fertile soils found in patches (1 to 300 ha) throughout the Amazon basin (Kern et al. 2003; WinklerPrins 2014). The most consistently used and inclusive term for these soils is Amazonian Dark Earths (ADEs) (Woods and

McCann 1999). *Terra Preta* (shorthand for the Portuguese *Terra Preta do Índio*, black earth of the Indians) is also used (Lehmann et al. 2003; Glaser and Woods 2004; Teixeira et al. 2009; Woods et al. 2009). The anthrosol continuum ranges from "true" black *Terra Preta* (TP), with embedded ceramics to *Terra Marrom* (TM),[1] which has shades of brown and covers a much larger area than TP (Denevan 2004). TM usually does not contain ceramics, yet carries charcoal and chemical signatures that indicate human origin (Fraser et al. 2011).

The dark color of Amazonian anthrosols contrasts with the lighter colored, usually yellow and red, dominant tropical soils of the region. The dark color is due to high levels of soil organic matter (SOM) and charcoal, particularly pyrogenic carbon (Glaser and Birk 2012). The dark color and associated SOM persist over time, challenging traditional understanding of tropical soils wherein organic matter is thought to be rapidly degraded and leached out of the system. The persistence of SOM and soil nutrients (e.g., phosphorus and calcium) is because much of the carbon is pyrogenic, which is highly recalcitrant and resists weathering in the high temperatures and precipitation regimes typical of Amazonia (Glaser, Lehmann, and Zech 2002). This pyrogenic carbon is formed during slow, cool burns with smoldering, a process identified as variations on "slash and char" in contrast to "slash and burn" (Steiner, Teixeira, and Zech 2004; WinklerPrins 2009; Arroyo-Kalin 2012). Soils with pyrogenic carbon absorb and retain nutrients and moisture better, and yield more plant-available nutrients over the long term, a combination that contributes to their fertility.

The texture and mineralogy of these anthrosols are generally similar to those of surrounding not-anthrosols, which confirms that they were formed in situ via additions from above but in the same parent material (Glaser and Birk 2012). Current thinking about anthrosol formation is that TP was formed as a result of refuse accumulation from long-term habitation; hence the accumulation of ceramics (Glaser and Birk 2012; Schmidt et al. 2014). Although debated, TM was likely created as a result of semi-intensive active soil management, including the addition of organic inputs and charcoal (Woods and McCann 1999; Denevan 2004, 2006). These forms of management, light in-field burning and smoldering, have been noted in the ethnographic record and can be observed today among indigenous and traditional

Amazonian villages (Hecht 2003; Heckenberger et al. 2003; Heckenberger et al. 2007; Heckenberger et al. 2008; WinklerPrins 2009; WinklerPrins and Falcão 2010; Schmidt et al. 2014). Thus, Amazonian anthrosols formed as a result of long-term human occupancy and active land management.

It is becoming increasingly evident that these anthrosols are deeply intertwined with vegetation patterns in Amazonia and illustrate the degree to which forests are domesticated. Changes in soil properties due to past human activities lead to differentiation in vegetation patterns of contemporary home gardens, swiddens, secondary, and old-growth forests (Junqueira, Shepard, and Clement 2010; Junqueira et al. 2011; Lins et al. 2015; Quintero-Vallejo et al. 2015; Junqueira et al. 2016; Junqueira et al. 2017; Maezumi et al. 2018; Levis et al. 2020). For example, Junqueira, Shepard, and Clement (2010; Junqueira et al. 2017) demonstrated that secondary forest growth on anthrosols conserves agrobiodiversity and that old-growth forests on anthrosols concentrate plant species domesticated to some degree. Lins et al. (2015) found greater plant diversity of native species in home gardens where there is evidence of multiple occupancies in pre-European times. Maezumi et al. (2018) demonstrated evidence of 4,500 years of polycultural agroforestry in the lower Tapajós basin and documented enrichment of fruit-bearing forest species. Levis et al. (2020) demonstrated that forest enrichment with such species is associated with pre-European soil fertilization and this legacy might extend far beyond localized former occupation sites as evidenced by TM.

Most ADEs identified thus far are located along or near bluffs or close to a source of perennial water reflecting past Amerindian settlement patterns that were predominantly on bluffs (Denevan 1996; see Figure 1). Recent mapping efforts and modeling, however, have revealed that these soils are increasingly common along minor perennial and temporary rivers that are very abundant across the region (see maps in Levis et al. [2014] and Palace et al. [2017]). Their predominance along bluffs might also reflect modern-day accessibility. The extent to which ADEs are found in interfluves, away from major perennial water sources (e.g., McMichael et al. 2012; Bush et al. 2015; Piperno, McMichael, and Bush 2015), is still debated, but the increasing evidence is that they are found in the interfluves and throughout the region (Franco-Moraes et al. 2019; AmazonArch

2020; Levis et al. 2020). Although the spatial extent of TP has been intensively studied and is estimated to cover from 0.1 percent up to 3 percent of the Amazon basin (Madari et al. 2004; McMichael et al. 2014), the spatial extent of TM has not yet been evaluated.

Domesticated Forests

A widespread and common legacy of human transformation of Amazonian landscapes is the forests themselves. Although long thought to be the result of ecological and evolutionary processes with limited influence by humans (Meggers [1971] 1996; Barlow et al. 2012; McMichael et al. 2012; Bush et al. 2015; Piperno, McMichael, and Bush 2015), substantial research since 2000, building on earlier scholarship (e.g., Sauer 1963; Denevan and Padoch 1987; Balée 1989; Denevan 1992; Neves 1998) and undertaken from a variety of disciplinary perspectives (Balée 2006), demonstrates that Native Amazonians were active managers of those forests, intentionally or not, and to varying degrees (Heckenberger et al. 2007; Clement et al. 2015a; Piperno, McMichael, and Bush 2015; Roberts et al. 2017; Levis et al. 2018; McKey 2019). Peters (2000) argued that "managed forest systems are subtle, but they can produce lasting changes" (213) and that "what is overlooked in t[he] historical treatment of tropical silviculture is the fact that the indigenous population … [has] been using, manipulating, and managing tropical forests for several thousand years" (203). Researchers demonstrate that various forms and combinations of incidental and active forest management and arboriculture (forest, garden, and swidden mosaics, as well as longer term forest management combined with semi-intensive agroforestry) were practiced for millennia in Amazonia before European arrival and sustained substantial and increasing populations (e.g., Denevan 1992, 2007, 2014; Heckenberger et al. 2003; Erickson 2006; Heckenberger et al. 2007; Heckenberger et al. 2008; Neves 2013; Roosevelt 2013; Clement et al. 2015a; Levis et al. 2017; Levis et al. 2018; Clement et al. 2020; Iriarte, Elliott, et al. 2020). Today, "domesticated forests are recognizable by the presence of forest patches dominated by one or a few useful species favored by long-term human activities" (Levis et al. 2018, 1).

Forests on and around anthrosols were initially described as "cultural" or "anthropogenic" forests (Balée 1989, 2013; Denevan 1992; Peters 2000; Shepard et al. 2020), in which "species … [were] manipulated, often without a reduction in natural diversity" (Denevan 1992, 374). Recent research by interdisciplinary teams of archaeologists, ecologists, geographers and others revealed that Amazonian forests are not just cultural forests but domesticated ones (e.g., Erickson 2006; Clement et al. 2015a; Hecht 2016; Levis et al. 2017; de Souza et al. 2018; Levis et al. 2018; Franco-Moraes et al. 2019; McKey 2019; Clement et al. 2020). There is significant evidence of human management on species distribution and abundance: "many present Amazonian forests, while seemingly natural, are domesticated to varying degrees in terms of altered plant distributions and densities" (Clement et al. 2015a, 2). Plant species with utility and domesticated to some extent by Native Amazonians occur in high densities in and around archaeological sites across Amazonia (Levis et al. 2017).

Domesticated forests demonstrate the degree to which Amazonians worked with ecological processes to make their landscapes more productive than the natural endowment provided them (Levis et al. 2018). Through subtle intentional and unintentional actions, including managing, cultivating, fishing, and hunting, Amazonian forests and other ecosystems were to some degree transformed by the activities of indigenous and traditional peoples. Shepard et al. (2020) recently urged a turn away from the term *agriculture* and the concept of *farming*, because these carry cultural history and baggage with them and this bias hinders the ability to see the entirety of what is really a food production system that has been practiced in the region for millennia. A food production system is more appropriate because it encompasses the broad continuum of the varied activities that Native Amazonians engaged with to produce food. The continued attempt to fit what Amazonian people did in the past into the agriculture and farming mold does not do justice to their landscape management (Neves 2013).

Although the full scale and degree of Amazonian domestication remains a topic of debate (e.g., Clement et al. 2015b; McMichael et al. 2015; Junqueira et al. 2017; McMichael et al. 2017; Piperno, McMichael, and Bush 2017; Watling et al. 2017b) and requires more attention, what is clear from research to date is that Amazonia, long thought to be an intact, pristine rainforest, the epitome of wilderness and untrammeled nature, is actually an anthrome. Active landscape management and transformation by pre-European peoples resulted in changes to varying degrees in forest structure and composition that are still discernable today across the region, especially where archaeological sites are found (Levis et al. 2017; Levis et al. 2018; AmazonArch 2020).

Earthworks

The most intentionally produced forms of human transformation of Amazonian landscapes are anthropogenic earthworks. These earthworks can be regarded as landesque capital, because the permanent changes to the landscape generated by humans are the result of intentional actions produced to endure economic, social, and ritual organizations (Håkansson and Widgren 2014; Arroyo-Kalin 2016). Although numerous anthropogenic earthworks in Amazonia were identified more than fifty years ago in the Llanos de Mojos of Bolivia (Denevan 1966), their variety and ubiquity are becoming increasingly apparent through greater visibility due to land clearing and the advances of remote sensing techniques such as LiDAR (e.g., de Souza et al. 2018; Stenborg, Schaan, and Figueiredo 2018; Iriarte, Robinson, et al. 2020; Lombardo et al. 2020). Although earthworks have been identified throughout Amazonia, most sites are in the periphery of the region, and it is likely that there are many more (Figure 1). These include raised fields, mounds, ditches, fish weirs, causeways, canals, moats, embankments, forest islands, and geoglyphs/ring ditches (e.g., Roosevelt 1991, 2013; Heckenberger, Petersen, and Neves 1999; Erickson 2000, 2006; Heckenberger et al. 2003; Heckenberger et al. 2008; Pärssinen, Schaan, and Ranzi 2009; McKey et al. 2010; Schmidt et al. 2014; Schaan 2016; Watling et al. 2017a; de Souza et al. 2018; Iriarte, Robinson, et al. 2020; Lombardo et al. 2020). These earthworks had many functions, although these are not yet well understood, but most concentrated or provided access to food resources, eased transportation and communication between communities, and were likely used for ceremonial functions. Across southern Amazonia, earthworks are organized in to complex networks of villages, suggesting that this part of the region sustained a high population density in the late Holocene (de Souza

et al. 2018). In southwestern Amazonia, thousands of anthropic forest islands were constructed within a seasonally flooded savannah starting in the early Holocene (about 10,850 cal. yr BP) and continuing up to 2,300 cal. yr BP, indicating significant human transformation of landscapes much earlier than previously thought (Iriarte, Elliott, et al. 2020; Lombardo et al. 2020). Since the early Holocene and throughout this epoch, forest builders cultivated domesticated plant species, such as squash, manioc, and many palms. Today, anthropogenic forest islands concentrate edible plants that feed not only local communities (Balée and Erickson 2006) but also critically endangered bird species (Lombardo et al. 2020).

The construction of earthworks, along with anthrosols and domesticated forests, is a result of societal development and ecosystem engineering techniques that increased habitat heterogeneity and the productivity of Amazonian landscapes. According to Ellis's framing of anthroecological change, a general causal theory that "explain[s] why human societies gained the capacity to globally alter the patterns, processes, and dynamics of ecology" (Ellis 2015, 287), these domesticated landscapes contribute to an understanding of the origin of anthromes. Recent work by Ellis, Beusen, and Klein Goldewijk (2020) acknowledges human transformations in pre-European Amazonia, although their study continues to treat the region homogenously. Recognizing the heterogeneity of the region's domesticated landscapes is key to understanding how extensive and intense these transformations were and will better inform conservation efforts.

Conservation of Amazonia through an Anthropocene Lens

Widespread anthrosols, domesticated forests, and earthworks reveal significant human transformations, an anthrome, in contrast to a perceived "intact" or "pristine" Amazonian forest. Understanding Amazonia through this Anthropocene lens challenges the definition of what is cultural and what is natural as "the separation between the human and the non-human … has grown increasingly fuzzy, to the point that it is rendered almost meaningless" (Kawa 2016, 19).

Developers continue to see the region as tabula rasa, a vast storehouse of riches for exploitation: soybean production, cattle ranching, lumbering, mining, and other natural resource extraction. Technocrats and large land owners tend to see indigenous and traditional people and their activities as getting in the way of progress and with little value besides the labor they provide. Conservationists consider the region essential for global ecosystem services (e.g., climate regulation), a vast storehouse of yet undiscovered biodiversity, and one of the last wildernesses on Earth. They urge maximum conservation of the forest, ideally without people, to conserve intact ecosystems (Watson et al. 2018). Both perspectives perpetuate the belief that "wilderness areas are … the only places that contain mixes of species at near-natural levels of abundance" (Watson et al. 2018, 28). This goes along with the persistent and pernicious myth of environmental constraints on forest people that together with a still perpetuated textbook trope of demographic emptiness of Amazonia needs to be moved beyond (Neves 1998). The Anthropocene lens has the potential to deconstruct this old model and open the way to a new framework for how conservation is approached in the region.

When significant findings by Heckenberger, Petersen, and Neves (1999) about Kuikuru landscape transformations were first published, Meggers (2001) wrote a rebuttal in which she stated that "uncritical acceptance of the conclusions of [revisionist assessments] not only conflicts with ecological and archaeological evidence, *but provides support for the unconstrained deforestation of the region*" (304, italics added). Meggers was upset that her theory regarding limited cultural potential in the region was being challenged (Woods 2013), but she also raised concerns that breaking the belief of Amazonia as a wilderness with humans as minimalist interlopers would open up the region to unprecedented development. Essentially, she and many others do not want to acknowledge that humans were significant landscape agents in Amazonia because that runs counter to the conservationist approach to conserving the forest (Denevan 2011). There are few who want to see the wholesale destruction of the forest, but by ignoring the *longue durée* of sustainable human use of the region, as evidenced by the existing intertwined landscape transformations, scientists are missing an engagement with the instructive ways in which Amazonians have managed and transformed ecosystems (Clement et al. 2020).

Ziegler (2019) observed that conservation in the Anthropocene is "wrought with tension" (274), because it forces an acceptance that there is not a nature–culture divide and that there are no intact,

unhumanized places on Earth. What needs to become accepted—and this is what makes the Anthropocene lens a constructive framing—is that domesticated landscapes are just as worthy of conservation as apparently "pristine" intact ones, because biodiversity exists in a domesticated forest just as it does in a less domesticated forest (Balée 2013). In fact, beta diversity of some life forms, such as the spatial turnover of plant species, increases in Amazonian landscapes with different types and degrees of pre-European human transformations (Lins et al. 2015; Odonne et al. 2019). Similarly, the provisioning of ecosystem services increases in domesticated forests because these forests concentrate agrobiodiversity and food resources highly valued by modern societies and wildlife (Junqueira, Shepard, and Clement 2010; Levis et al. 2020). Such concentrations form an essential component of participatory conservation and community-based management approaches, because they can promote socioeconomic benefit to local communities from the sustainable management of nontimber forest products such as Brazil nuts (Guariguata et al. 2017). It is increasingly understood that the best conservation policy in a region such as Amazonia is a participatory conservation approach, working with local people, because people-less "set-aside" spaces efface the rich history of the region (Katz 2005; Cámara-Leret, Fortuna, and Bascompte 2019; Intergovernmental Science-Policy Platform on Biodiversity and Ecosystem Services (IPBES) 2019; Clement et al. 2020). Research also demonstrates that indigenous territories protect better against rampant development than other forms of protection such as national parks (Nepstad et al. 2006; R. Walker et al. 2009; Garnett et al. 2018; Balée et al. 2020; W. S. Walker et al. 2020).

The landscape legacies and landesque capital evident today in Amazonia demonstrate that indigenous peoples, since before European conquest, practiced landscape transformations that enhanced the physical environment for humans while maintaining ecosystem functioning and its ecological integrity, in ways that current industrial-scale land use does not (e.g., McKey et al. 2010; Watling et al. 2017a). The local knowledge to transform soils, forest assemblages, and the land itself to improve its utility for human use, while sustaining, even improving, ecosystems services, represents a promising alternative to ensure the conservation of Amazonian ecosystems and to promote the rights and livelihoods of indigenous and traditional peoples.

Acknowledgments

Earlier versions of this article were presented at various talks delivered by the first author, and we thank those who provided suggestions for refinement of the argument. We also thank Bill Denevan, the reviewers, and the editor for their careful reading of the article and thoughtful suggestions for its improvement. We thank AmazonArch for providing the archeological data presented in Figure 1.

Funding

The writing of this article was supported by the U.S. National Science Foundation while the first author worked at the Foundation. Any opinions, findings, conclusions, or recommendations expressed in this material are those of the authors and do not necessarily reflect the views of the National Science Foundation. The second author acknowledges support from and thanks CNPq and CAPES for a post-doctoral fellowship (CNPq 159440/2018-1; CAPES 88887.474572/2020-00) that supported her research.

Note

1. *Terra Marrom* was formerly known as *Terra Mulata* and is now called Amazonian Brown Earth (Iriarte, Elliott, et al. 2020).

ORCID

Antoinette M. G. A WinklerPrins (iD) https://orci-d.org/0000-0003-1076-9019
Carolina Levis (iD) https://orcid.org/0000-0002-8425-9479

References

AmazonArch. 2020. Home page. Accessed September 26, 2020. https://sites.google.com/view/amazonarch/home.

Arroyo-Kalin, M. 2012. Slash-burn-and-churn: Landscape history and crop cultivation in pre-Columbian Amazonia. *Quaternary International* 249:4–18.

Arroyo-Kalin, M. 2016. Landscaping, landscape legacies, and landesque capital in pre-Columbian Amazonia. In *The Oxford handbook of historical ecology and applied archaeology*, ed. C. Isendahl and D. Stump. Oxford, UK: Oxford University Press.

Balée, W. 1989. The culture of Amazonian forests. *Advances in Economic Botany* 7:1–21.

Balée, W. 2006. The research program of historical ecology. *Annual Review of Anthropology* 35 (1):75–98. doi: 10.1146/annurev.anthro.35.081705.123231.

Balée, W. 2013. *Cultural forests of the Amazon: A historical ecology of people and their landscapes.* Tuscaloosa: University of Alabama Press.

Balée, W., V. H. de Oliveira, R. dos Santos, M. Amaral, B. Rocha, N. Guerrero, S. Schwartzman, M. Torres, and J. Pezzuti. 2020. Ancient transformation, current conservation: Traditional forest management on the Iriri River, Brazilian Amazonia. *Human Ecology* 48 (1):1–15. doi: 10.1007/s10745-020-00139-3.

Balée, W., and C. L. Erickson, eds. 2006. *Time and complexity in the neotropical lowlands: Studies in historical ecology.* New York: Columbia University Press.

Barlow, J., T. A. Gardner, A. C. Lees, L. Parry, and C. A. Peres. 2012. How pristine are tropical forests? An ecological perspective on the pre-Columbian human footprint in Amazonia and implications for contemporary conservation. *Biological Conservation* 151 (1):45–49. doi: 10.1016/j.biocon.2011.10.013.

Bush, M. B., C. H. McMichael, D. R. Piperno, M. R. Silman, J. Barlow, C. A. Peres, M. Power, and M. W. Palace. 2015. Anthropogenic influence on Amazonian forests in pre-history: An ecological perspective. *Journal of Biogeography* 42 (12):2277–88.

Cámara-Leret, R., M. A. Fortuna, and J. Bascompte. 2019. Indigenous knowledge networks in the face of global change. *Proceedings of the National Academy of Sciences of the United States of America* 116 (20):9913–18. doi: 10.1073/pnas.1821843116.

Clement, C. R., W. M. Denevan, M. J. Heckenberger, A. B. Junqueira, E. G. Neves, W. G. Teixeira, and W. I. Woods. 2015a. The domestication of Amazonia before European conquest. *Proceedings of the Royal Society B-Biological Sciences* 282: 20150813.

Clement, C. R., W. M. Denevan, M. J. Heckenberger, A. B. Junqueira, E. G. Neves, W. G. Teixeira, and W. I. Woods. 2015b. Response to comment by McMichael, Piperno and Bush. *Proceedings of the Royal Society B: Biological Sciences* 282 (1821):20152459. doi: 10.1098/rspb.2015.2459.

Clement, C. R., C. Levis, J. Franco-Moraes, and A. B. Junqueira. 2020. Domesticated nature: The culturally constructed niche of humanity. In *Participatory biodiversity conservation*, ed. C. Baldauf, 35–51. Cham, Switzerland: Springer Nature.

Denevan, W. M. 1966. *The aboriginal cultural geography of the Llanos de Mojos of Bolivia, Ibero Americana 48.* Berkeley: University of California Press.

Denevan, W. M. 1992. The pristine myth: The landscape of the Americas in 1492. *Annals of the Association of American Geographers* 82 (3):369–85. doi: 10.1111/j.1467-8306.1992.tb01965.x.

Denevan, W. M. 1996. A bluff model of riverine settlement in prehistoric Amazonia. *Annals of the Association of American Geographers* 86 (4):654–81. doi: 10.1111/j.1467-8306.1996.tb01771.x.

Denevan, W. M. 2004. Semi-intensive pre-European cultivation and the origins of anthropogenic dark earths in Amazonia. In *Amazonian dark earths: Explorations in space and time*, ed. B. Glaser and W. I. Woods, 135–43. Berlin: Springer.

Denevan, W. M. 2006. Pre-European forest cultivation in Amazonia. In *Time and complexity in the neotropical lowlands: Studies in historical ecology*, ed. W. Balée and C. L. Erickson, 153–64. New York: Columbia University Press.

Denevan, W. M. 2007. Pre-European human impacts on tropical lowland environments. In *The physical geography of South America*, ed. T. T. Veblen, K. R. Young, and A. R. Orme, 265–78. Oxford, UK: Oxford University Press.

Denevan, W. M. 2011. The "pristine myth" revisited. *Geographical Review* 101 (4):576–91. doi: 10.1111/j.1931-0846.2011.00118.x.

Denevan, W. M. 2014. Estimating Amazonian Indian numbers in 1492. *Journal of Latin American Geography* 13 (2):207–17. doi: 10.1353/lag.2014.0036.

Denevan, W. M., and C. Padoch. 1987. Swidden-fallow agroforestry in the Peruvian Amazon. *Advances in Economic Botany* 5:1–102.

de Souza, J. G., D. P. Schaan, M. Robinson, A. D. Barbosa, L. E. O. C. Aragão, B. H. Marimon, B. S. Marimon, I. B. da Silva, S. S. Khan, F. R. Nakahara, et al. 2018. Pre-Columbian earth-builders settled along the entire southern rim of the Amazon. *Nature Communications* 9: 1125. doi: 10.1038/s41467-018-03510-7.

Ellis, E. C. 2015. Ecology in an anthropogenic biosphere. *Ecological Monographs* 85 (3):287–331.

Ellis, E. C. 2018. *Anthropocene: A very short introduction.* Oxford, UK: Oxford University Press.

Ellis, E. C., A. H. W. Beusen, and K. Klein Goldewijk. 2020. Anthropogenic biomes: 10,000 BCE to 2015 CE. *Land* 9: 129. doi: 10.3390/land9050129.

Ellis, E. C., and N. Ramankutty. 2008. Putting people in the map: Anthropogenic biomes of the world. *Frontiers in Ecology and Environment* 6 (8):439–47.

Erickson, C. L. 2000. An artificial landscape-scale fishery in the Bolivian Amazon. *Nature* 408 (6809):190–93. doi: 10.1038/35041555.

Erickson, C. L. 2006. The domesticated landscape of the Bolivian Amazon. In *Time and complexity in the neotropical lowlands: Studies in historical ecology*, ed. W. Balée and C. L. Erickson, 236–78. New York: Columbia University Press.

Erickson, C. L. 2003. Historical ecology and future explorations. In *Amazonian dark earths: Wim Sombroek's vision*, ed. J. Lehmann, 455–500. Dordrecht, The Netherlands: Springer.

Franco-Moraes, J., A. F. M. B. Baniwa, F. R. C. Costa, H. P. Lima, C. R. Clement, and G. H. Shepard, Jr. 2019. Historical landscape domestication in ancestral forests with nutrient-poor soils in northwestern Amazonia. *Forest Ecology and Management* 446:317–30.

Fraser, J., W. Teixeira, N. P. S. Falcão, W. I. Woods, J. Lehmann, and A. B. Junqueira. 2011. Anthropogenic soils in the Central Amazon: From categories to a continuum. *Area* 43 (3):264–73.

Garnett, S. T., N. D. Burgess, J. E. Fa, Á. Fernández-Llamazares, Z. Molnár, C. J. Robinson, J. E. M. Watson, K. K. Zander, B. Austin, E. S. Brondizio, et al. 2018. A spatial overview of the global importance of Indigenous lands for conservation. *Nature Sustainability* 1 (7):369–74.

Glaser, B., and J. J. Birk. 2012. State of the scientific knowledge on the properties and genesis of anthropogenic dark earths in Central Amazonia (*terra preta do índio*). *Geochimica et Cosmochimica Acta* 82:39–51.

Glaser, B., J. Lehmann, and W. Zech. 2002. Ameliorating physical and chemical properties of highly weathered soils in the tropics with charcoal—A review. *Biology and Fertility of Soils* 35 (4):219–30.

Glaser, B., and W. I. Woods, eds. 2004. *Amazonian dark earths: Explorations in space and time.* Berlin: Springer.

Guariguata, M. R., P. Cronkleton, A. E. Duchelle, and P. A. Zuidema. 2017. Revisiting the "cornerstone of Amazonian conservation": A socioecological assessment of Brazil nut exploitation. *Biodiversity and Conservation* 26 (9):2007–27.

Håkansson, N. T., and M. Widgren, eds. 2014. *Landesque capital: The historical ecology of enduring landscape modifications.* Walnut Creek, CA: Left Coast Press.

Hecht, S. B. 2003. Indigenous soil management and the creation of Amazonian dark earths: Implications of Kayapó practices. In *Amazonian dark earths: Origin, properties, management*, ed. J. Lehmann, 355–72. Dordrecht, The Netherlands: Kluwer.

Hecht, S. B. 2016. Domestication, domesticated landscapes, and tropical natures. In *The Routledge companion to the environmental humanities*, ed. U. Kitteise, J. Christensen, and M. Niemann, 21–34. London and New York: Routledge.

Heckenberger, M. J., A. Kuikuro, U. T. Kuikuro, J. C. Russell, M. Schmidt, C. Fausto, and B. Franchetto. 2003. Amazonia 1492: Pristine forest or cultural parkland? *Science* 301 (5640):1710–14. doi: 10.1126/science.1086112.

Heckenberger, M. J., J. B. Petersen, and E. G. Neves. 1999. Village size and permanence in Amazonia: Two archaeological examples from Amazonia. *Latin American Antiquity* 10 (4):353–76.

Heckenberger, M. J., J. C. Russell, C. Fausto, J. R. Toney, M. J. Schmidt, E. Pereira, B. Franchetto, and A. Kuikuro. 2008. Pre-Columbian urbanism, anthropogenic landscapes, and the future of the Amazon. *Science* 321 (5893):1214–17. doi: 10.1126/science.1159769.

Heckenberger, M. J., J. C. Russell, J. R. Toney, and M. J. Schmidt. 2007. The legacy of cultural landscapes in the Brazilian Amazon: Implications for biodiversity. *Philosophical Transactions of the Royal Society of London, Series B: Biological Sciences* 362 (1478):197–208. doi: 10.1098/rstb.2006.1979.

IPBES. 2019. *Global assessment report on biodiversity and ecosystem services of the Intergovernmental Science-Policy Platform on Biodiversity and Ecosystem Services.*

Iriarte, J., S. Elliott, S. Y. Maezumi, D. Alves, R. Gonda, M. Robinson, J. Gregorio de Souza, J. Watling, and J. Handley. 2020. The origins of Amazonian landscapes: Plant cultivation, domestication and the spread of food production in tropical South America. *Quaternary Science Reviews* 248:106582. doi: 10.1016/j.quascirev.2020.106582.

Iriarte, J., M. Robinson, J. de Souza, A. Damasceno, F. da Silva, F. Nakahara, A. Ranzi, and L. Aragao. 2020. Geometry by design: Contributions of LIDAR to the understanding of settlement patterns of the mound villages of S.W. Amazonia. *Journal of Computer Applications in Archaeology* 3 (1):151–69. doi: 10.5334/jcaa.45.

Junqueira, A. B., C. Levis, F. Bongers, M. Peña-Claros, C. R. Clement, F. Costa, and H. ter Steege. 2017. Response to comment on "Persistent effects of pre-Columbian plant domestication on Amazonian forest composition." *Science* 358 (6361):eaan8837. doi: 10.1126/science.aan8837.

Junqueira, A. B., G. H. Shepard, Jr., and C. R. Clement. 2010. Secondary forests on anthropogenic soils in Brazilian Amazonia conserve agrobiodiversity. *Biodiversity and Conservation* 19 (7):1933–61.

Junqueira, A. B., N. B. Souza, T. J. Stomph, C. J. Almekinders, C. R. Clement, and P. C. Struik. 2016. Soil fertility gradients shape the agrobiodiversity of Amazonian homegardens. *Agriculture, Ecosystems & Environment* 221:270–81.

Junqueira, A. B., T. J. Stomph, C. R. Clement, and P. C. Struik. 2011. Variation in soil fertility influences cycle dynamics and crop diversity in shifting cultivation systems. *Agriculture, Ecosystems & Environment* 215:122–32.

Katz, C. 2005. Whose nature, whose culture? Private production of space and the "preservation" of nature. In *Remaking reality: Nature at the millennium*, ed. B. Braun and N. Castree, 46–63. London and New York: Taylor & Francis.

Kawa, N. C. 2016. *Amazonia in the Anthropocene: People, soils, plants, forests*. Austin: University of Texas Press.

Kern, D. C., G. D'aquino, T. E. Rodrigues, F. J. Lima Frazao, W. Sombroek, T. P. Myers, and E. G. Neves. Distribution of Amazonian dark earths in the Brazilian Amazon. In *Amazonian dark earths: Origin, properties, management*, ed. J. Lehmann, 51–75. Dordrecht, The Netherlands: Kluwer.

Lehmann, J., D. C. Kern, B. Glaser, and W. I. Woods, eds. 2003. *Amazonian dark earths: Origin, properties, management*. Dordrecht, The Netherlands: Kluwer.

Levis, C., M. S. Silva, M. A. Silva, C. P. Moraes, E. K. Tamanaha, B. M. Flores, E. G. Neves, and C. R. Clement 2014. What do we know about the distribution of Amazonian Dark Earth along tributary rivers in Central Amazonia? In *Antes de Orellana: Actas del 3er Encuentro Internacional de Arqueología Amazónica*, ed. R. Stéphen, pp. 305–12. Lima, Peru: Instituto Francés de Estudios Andinos.

Levis, C., F. R. C. Costa, F. Bongers, M. Peña-Claros, C. R. Clement, A. B. Junqueira, E. G. Neves, E. K. Tamanaha, F. O. G. Figueiredo, R. P. Salomão, et al. 2017. Persistent effects of pre-Columbian plant domestication on Amazonian forest composition. *Science* 355 (6328):925–31. doi: 10.1126/science.aal0157.

Levis, C., B. M. Flores, P. A. Moreira, B. G. Luize, R. P. Alves, J. Franco-Moraes, J. Lins, E. Konings, M. Peña-Claros, F. Bongers, et al. 2018. How people

domesticated Amazonian forests. *Frontiers in Ecology and Evolution* 5:1–21. doi: 10.3389/fevo.2017.00171.

Levis, C., M. Peña-Claros, C. R. Clement, F. R. C. Costa, R. P. Alves, M. J. Ferreira, C. G. Figueiredo, and F. Bongers. 2020. Pre-Columbian soil fertilization and current management maintain food resource availability in old-growth Amazonian forests. *Plant and Soil* 450 (1–2):29–48. doi: 10.1007/s11104-020-04461-z.

Lins, J., H. P. Lima, F. B. Baccaro, V. F. Kinupp, G. H. Shepard, Jr., and C. R. Clement. 2015. Pre-Columbian floristic legacies in modern homegardens of Central Amazonia. *PLoS ONE* 10 (6):e0127067. doi: 10.1371/journal.pone.0127067.

Lombardo, U., J. Iriarte, L. Hilbert, J. Ruiz-Pérez, J. M. Capriles, and H. Veit. 2020. Early Holocene crop cultivation and landscape modification in Amazonia. *Nature* 581 (7807):190–93. doi: 10.1038/s41586-020-2162-7.

Madari, B. E., W. G. Sombroek, and W. I. Woods 2004. Research on anthropogenic dark earth soils: Could it be a solution to sustainable development in the Amazon? In *Amazonian Dark Earths: Explorations in space and time*, ed. B. Glaser and W. I. Woods, pp. 169–81. Berlin, DE: Springer.

Maezumi, S. Y., D. Alves, M. Robinson, J. G. de Souza, C. Levis, R. L. Barnett, E. Almeida de Oliveira, D. Urrego, D. Schaan, and J. Iriarte. 2018. The legacy of 4,500 years of polyculture agroforestry in the eastern Amazon. *Nature Plants* 4 (8):540–47. doi: 10.1038/s41477-018-0205-y.

McKey, D. 2019. Pre-Columbian human occupations of Amazonia and its influence on current landscapes and biodiversity. *Anais da Academia Brasileira de Ciências* 91 (Suppl. 3):e20190087. doi: 10.1590/0001-3765201920190087.

McKey, D., S. Rostain, J. Iriarte, B. Glaser, J. J. Birk, I. Holst, and D. Renard. 2010. Pre-Columbian agricultural landscapes, ecosystem engineers, and self-organized patchiness in Amazonia. *Proceedings of the National Academy of Sciences of the United States of America* 107 (17):7823–28. doi: 10.1073/pnas.0908925107.

McMichael, C. H., K. J. Feeley, C. W. Dick, D. R. Piperno, and M. B. Bush. 2017. Comment on "Persistent effects of pre-Columbian plant domestication on Amazonian forest composition." *Science* 358 (6361):eaan8347.

McMichael, C. H., D. R. Piperno, and M. B. Bush. 2015. Comment on Clement et al. 2015. "The domestication of Amazonia before European conquest." *Proceedings of the Royal Society B: Biological Sciences* 282 (1821):20151837. doi: 10.1098/rspb.2015.1837.

McMichael, C. H., D. R. Piperno, M. B. Bush, M. R. Silman, A. R. Zimmerman, M. F. Raczka, and L. C. Lobato. 2012. Sparse pre-Columbian human habitation in western Amazonia. *Science* 336 (6087):1429–31. doi: 10.1126/science.1219982.

McMichael, C. H., M. W. Palace, M. B. Bush, B. Braswell, S. Hagen, E. G. Neves, M. R. Silman, E. K. Tamanaha, and C. Czarnecki. 2014. Predicting pre-Columbian anthropogenic soils in Amazonia.

Proceeding of the Royal Society B: Biological Sciences. 281:20132475. doi: 10.1098/rspb.2013.2475.

Meggers, B. J. [1971] 1996. *Amazonia: Man and culture in a counterfeit paradise.* Washington, DC: Smithsonian.

Meggers, B. J. 2001. The continuing quest of El Dorado: Round two. *Latin American Antiquity* 12 (3):304–25. doi: 10.2307/971635.

Nepstad, D., S. Schwartzman, B. Bamberger, M. Santilli, D. Ray, P. Schlesinger, P. Lefebvre, A. Alencar, E. Prinz, G. Fiske, et al. 2006. Inhibition of Amazon deforestation and fire by parks and indigenous lands. *Conservation Biology: The Journal of the Society for Conservation Biology* 20 (1):65–73. doi: 10.1111/j.1523-1739.2006.00351.x.

Neves, E. G. 1998. Twenty years of Amazonian archaeology in Brazil (1977–1997). *Antiquity* 72 (277):625–32.

Neves, E. G. 2013. Was agriculture a key productive activity in pre-colonial Amazonia? The stable productive basis for social equality in the Central Amazon. In *Human–environmental interactions: Current and future directions*, ed. E. S. Brondízio and E. F. Moran, 371–88. Dordrecht, The Netherlands: Springer.

Odonne, G., M. van den Bel, M. Burst, O. Brunaux, M. Bruno, E. Dambrine, D. Davy, M. Desprez, J. Engel, B. Ferry, et al. 2019. Long-term influence of early human occupations on current forests of the Guiana Shield. *Ecology* 100 (10):e02806. doi: 10.1002/ecy.2806.

Palace, M. W., C. N. H. McMichael, B. H. Braswell, S. C. Hagen, M. B. Bush, E. Neves, E. Tamanaha, C. Herrick, and S. Frolking. 2017. Ancient Amazonian populations left lasting impacts on forest structure. *Ecosphere* 8 (12):e02035.

Pärssinen, M., D. P. Schaan, and A. Ranzi. 2009. Pre-Columbian geometric earthworks in the upper Psurus: A complex society in western Amazonia. *Antiquity* 83 (322):1084–95.

Peters, C. M. 2000. Pre-Columbian silviculture and indigenous management of neotropical forests. In *Imperfect balance: Landscape transformations in the pre-Columbian Americas*, ed. D. L. Lentz, 203–22. New York: Columbia University Press.

Piperno, D. R., C. H. McMichael, and M. B. Bush. 2015. Amazonia and the Anthropocene: What was the spatial extent and intensity of human landscape modification in the Amazon Basin at the end of prehistory? *The Holocene* 25 (10):1588–97.

Piperno, D. R., C. H. McMichael, and M. B. Bush. 2017. Further evidence for localized, short-term anthropogenic forest alterations across pre-Columbian Amazonia. *Proceedings of the National Academy of Sciences* 114 (21):E4118–E4119. doi: 10.1073/pnas.1705585114.

Quintero-Vallejo, E., Y. Klomberg, F. Bongers, L. Poorter, M. Toledo, and M. Peña-Claros. 2015. Amazonian dark earth shapes the understory plant community in a Bolivian forest. *Biotropica* 47 (2):152–61.

Roberts, P., C. Hunt, M. Arroyo-Kalin, D. Evans, and N. Boivin. 2017. The deep human prehistory of global tropical forests and its relevance for modern conservation. *Nature Plants* 3:17093. doi: 10.1038/nplants.2017.93.

Roosevelt, A. C. 1991. *Moundbuilders of the Amazon: Geophysical archaeology on Marajó Island, Brazil.* San Diego, CA: Academic Press.

Roosevelt, A. C. 2013. The Amazon and the Anthropocene: 13,000 years of human influence in a tropical rainforest. *Anthropocene* 4:69–87. doi: 10.1016/j.ancene.2014.05.001.

Sauer, C. O. [1958] 1963. Man in the ecology of tropical America. In *Land and life: A selection from the writings of Carl Ortwin Sauer*, ed. J. Leighly, 182–93. Berkeley: University of California Press.

Schaan, D. P. 2016. *Sacred geographies of ancient Amazonia: Historical ecology and social complexity.* London and New York: Routledge.

Schmidt, M. J., A. Rapp Py-Daniel, C. de Paula Moraes, R. B. M. Valle, C. F. Caromano, W. G. Texeira, C. A. Barbosa, J. A. Fonseca, M. P. Magalhães, D. Silva do Carmo Santos, et al. 2014. Dark earths and the human built landscape in Amazonia: A widespread pattern of anthrosol formation. *Journal of Archaeological Research* 42:152–65.

Shepard, G. H., Jr., E. G. Neves, C. R. Clement, H. Lima, C. Moraes, and G. Mendes dos Santos. 2020. Ancient and traditional agriculture in South America: Tropical lowlands. In *Oxford research encyclopedia: Environmental sciences.* New York: Oxford University Press.

Steiner, C., W. G. Teixeira, and W. Zech. 2004. Slash and char: An alternative to slash and burn practiced in the Amazon Basin. In *Amazonian dark earths: Explorations in space and time*, ed. B. Glaser and W. I. Woods, 183–93. Berlin: Springer.

Stenborg, P., D. P. Schaan, and C. G. Figueiredo. 2018. Contours of the past: LiDAR data expands the limits of late pre-Columbian human settlement in the Santarém Region, Lower Amazon. *Journal of Field Archaeology* 43 (1):44–57.

Teixeira, W. G., D. C. Kern, B. E. Madari, H. N. Lima, and W. I. Woods, eds. 2009. *As Terras Pretas de Índio da Amazônia: Sua caracterização e uso deste conhecimento na criação de novas areas.* Manaus, Brazil: Embrapa Amazônia Ocidental.

Walker, R., N. J. Moore, E. Arima, S. Perz, C. Simmons, M. Caldas, D. Vergara, and C. Bohrer. 2009. Protecting the Amazon with protected areas. *Proceedings of the National Academy of Sciences of the United States of America* 106 (26):10582–86. doi: 10.1073/pnas.0806059106.

Walker, W. S., S. R. Gorelik, A. Baccini, J. L. Aragon-Osejo, C. Josse, C. Meyer, M. N. Macedo, C. Augusto, S. Rios, T. Katan, et al. 2020. The role of forest conversion, degradation, and disturbance in the carbon dynamics of Amazon indigenous territories and protected areas. *Proceedings of the National Academy of Sciences* 117 (6):3015–25. doi: 10.1073/pnas.1913321117.

Watling, J., J. Iriarte, F. E. Mayle, D. Schaan, L. C. R. Pessenda, N. J. Loader, F. A. Street-Perrott, R. E. Dickau, A. Damasceno, A. Ranzi, et al. 2017a. Impact of pre-Columbian "geoglyph" builders on

Amazonian forests. *Proceedings of the National Academy of Sciences* 114 (8):1868–73. doi: 10.1073/pnas.1614359114.

Watling, J., J. Iriarte, F. E. Mayle, D. Schaan, L. C. R. Pessenda, N. J. Loader, F. A. Street-Perrott, R. E. Dickau, A. Damasceno, A. Ranzi, et al. 2017b. Reply to Piperno et al., "It is too soon to argue for localized, short-term human impacts in interfluvial Amazonia." *Proceedings of the National Academy of Sciences* 114 (21):E4120–E4121. doi: 10.1073/pnas.1705697114.

Watson, J. E. M., O. Venter, J. Lee, K. R. Jones, J. G. Robinson, H. P. Possingham, and J. R. Allan. 2018. Protect the last of the wild. *Nature* 563 (7729):27–30. doi: 10.1038/d41586-018-07183-6.

WinklerPrins, A. M. G. A. 2009. Sweep and char and the creation of Amazonian dark earths in homegardens. In *Amazonian dark earths: Wim Sombroek's vision*, ed. W. I. Woods, 205–11. Dordrecht, The Netherlands: Springer.

WinklerPrins, A. M. G. A. 2014. *Terra preta*—The mysterious soils of the Amazon. In *The soil underfoot: Infinite possibilities for a finite resource*, ed. G. J. Churchman and E. R. Landa, 235–45. Boca Raton, FL: CRC.

WinklerPrins, A. M. G. A., and N. P. S. Falcão. 2010. Soil fertility management and its contribution to the formation of Amazonian dark earth in urban homegardens, Santarém, Pará, Brazil. In *Transactions of the Nineteenth World Congress of Soil Science, Brisbane, Australia.*

Woods, W. I. 2013. Betty Meggers: Her later years. *Andean Past* 11:15–20.

Woods, W. I., and J. M. McCann. 1999. The anthropogenic origin and persistence of Amazonian dark earths. *Yearbook of the Conference of Latin American Geographers* 25:7–14.

Woods, W. I., W. G. Teixeira, J. Lehmann, C. Steiner, A. M. G. A. WinklerPrins, and L. Rebellato, eds. 2009. *Amazonian dark earths: Wim Sombroek's vision.* Dordrecht, The Netherlands: Springer.

Ziegler, S. Z. 2019. The Anthropocene in geography. *Geographical Review* 109 (2):271–80. doi: 10.1111/gere.12343.

Forests in the Anthropocene

Jaclyn Guz and Dominik Kulakowski

Disturbances have shaped most terrestrial ecosystems for millennia and are natural and essential components of ecological systems. However, direct and indirect human activities during the Anthropocene have amplified disturbances globally. This amplification, coupled with increasingly unfavorable post-disturbance climatic conditions or ecosystem management that intensifies the initial disturbance, is compromising the resilience of some ecosystems, with cascading effects on Earth system function and ecosystem services. Such dynamics are especially prevalent in forests, which are one of the most important ecosystems on Earth and provide countless ecosystem services for people and nonhuman species. Although climate change and its effects are ubiquitous, they do vary spatially in their intensity, and many ecological systems are more affected by changing land use than by changing climate. Understanding the geographic variation in relationships and feedbacks among climate, vegetation, disturbances, regeneration, and human activity is necessary for developing management strategies that will promote forest resilience (i.e., facilitate ecosystems tolerating and recovering from novel or intensified perturbations without shifting to alternative states controlled by different processes). Successful management strategies will vary geographically depending on the degree of departure from the ecological dynamics that preceded the Anthropocene and the spatial variability in drivers of change. As global environmental change accelerates, conservation areas, and the species and ecosystem services that rely on them, are particularly vulnerable. Where disturbances increase, expanding the size of protected areas and minimizing secondary anthropogenic disturbances are likely to be the only ways to maintain the minimum dynamic area that will cultivate adequate resilience. *Key Words: Anthropocene, disturbances, ecology, ecosystem management, forest resilience.*

The Anthropocene is a period of unprecedented human impact on ecosystems worldwide (Ellis et al. 2013). Major human impacts are ubiquitous across the Earth system—with documented influence on the atmosphere, hydrosphere, and biosphere. In terrestrial ecosystems, the Anthropocene is characterized by new and novel, human- and climate-induced disturbances that are exceeding the resilience mechanisms that have historically governed disturbed ecosystems. This perfect storm is driving rapid change that is pushing, and will likely continue to push, ecosystems into fundamentally different structure and function compared to that of past decades and centuries. This is especially so in biophysical settings in which post-disturbance climate conditions are increasingly unfavorable and where post-disturbance management intensifies the initial disturbance. Importantly, these processes vary geographically, and understanding the variation in relationships and feedbacks among climate, vegetation, disturbances, regeneration, and human activity is necessary for developing management strategies that will maintain or promote future forest resilience.

Although forest disturbances are among the most spectacular manifestations of ecological change, it has long been understood that ecological systems exist in one of only two states: gradual change and abrupt change. The latter is driven by disturbances such as fires, insect outbreaks, and windstorms, which have shaped most terrestrial ecosystems for millennia and are normal and essential components of ecological systems. The most common and generally accepted definition of disturbance is "any relatively discrete event in time that disrupts ecosystem, community or population structure and changes resources, substrate availability, or the physical environment" (Pickett and White 1985, 7). The focus on disturbances as relatively discrete events implies that gradual climate change (e.g., changes in temperature and precipitation) is not considered a disturbance in and of itself. Rather, climate change indirectly makes forests more susceptible to disturbances and has the potential to inhibit post-disturbance recovery. Disturbances have exerted an evolutionary force and native species are typically well adapted to the characteristic disturbance regimes of the ecosystems in which they occur

(Kulakowski et al. 2019). Although there is ongoing debate as to the exact role of disturbances in driving species-level genetic adaptation (Poulos et al. 2018), there is little doubt that at the ecosystem scale, disturbances are integral to normal ecological structure and function (Kulakowski et al. 2019).

Disturbance processes maintain the structure and function of ecosystems by creating and responding to landscape heterogeneity, adding a dimension to the underlying spatial heterogeneity imposed by differences in species composition and physical setting (Mori, Isbell, and Seidl 2018). Further complexity is introduced by intermediate or mixed severity disturbances and multiple disturbance types, all of which can vary across space and time (Peterson 2019). Complex physical and compositional structure, in turn, is generally positively correlated with biodiversity (Camargo, Sano, and Vieira 2018), carbon storage, and long-term resource use efficiency (Bartowitz et al. 2019).

Although the occurrence of disturbances, even large and severe ones, does not necessarily signal an environmental problem, direct and indirect human activities have amplified disturbances globally during the Anthropocene (Cattau et al. 2020). Deviation from historic forest disturbance regimes during the Anthropocene has been documented based on dendroecology (Kim, Millington, and Lafon 2020), paleoecology (Gillson, Whitlock, and Humphrey 2019), historical aerial images (Lydersen and Collins 2018), and satellite data (Cattau et al. 2020). The combined influence of human activity and human-induced climate change has the potential to compromise forest resilience (Figure 1).

Natural (e.g., wind) and anthropogenic (e.g., logging) disturbances differ in their ecological effects and in the degree to which forest species are adapted to them. During the Anthropocene, most so-called natural disturbances are influenced by anthropogenic climate change, which can strengthen the disturbance agent as well as reduce the resistance of trees to that agent. The resulting amplified disturbance regime, sometimes coupled with unfavorable post-disturbance climatic conditions or post-disturbance ecosystem management that intensifies the effect of the initial disturbance, is inhibiting post-disturbance regeneration in some systems, with cascading effects on Earth system function and ecosystem services. The compounded effects of climate change and intense human activity are overwhelming ecological resilience and tipping some forests to alternative states. Importantly, although climate change and its effects are ubiquitous during the Anthropocene, they do vary spatially in their intensity. At the global scale, forest disturbance patterns are variable, but high disturbance activity is most correlated with climate (Sommerfeld et al. 2018), although the relationship with climate is weaker in areas of high direct anthropogenic pressures. For example, many ecological systems are more affected by changing land use than by changing climate. Understanding the geographic variability in the drivers of ecological change is necessary for understanding ecological dynamics and formulating appropriate management strategies (Kulakowski, Bebi, and Rixen 2011).

As disturbances increase in the size, frequency, and severity under climate change, forests and other ecosystems are increasingly being affected by multiple disturbances over relatively short periods, with major effects on ecosystem patterns and processes. For example, disturbances exert a strong control on subsequent disturbance probabilities (e.g., Buma 2015), often by limiting biomass or influencing structure or composition to increase or decrease the likelihood, extent, or severity of a second disturbance event. Resistance to disturbances conferred by preceding disturbances, however, could be reduced as climate change increases the intensity of disturbances, overwhelming the influence of predisturbance conditions in modulating disturbance behavior and effects. Whereas in some cases, climate change can dampen the interactions between linked disturbances, in other cases those interactions and the resulting ecological effects can be intensified as in windstorms creating habitat for some tree-killing insects such as the European spruce bark beetle (*Ips typographus*), whose populations can grow rapidly under warmer conditions (e.g., Potterf and Bone 2017). Finally, climate can serve as an underlying driver of multiple disturbances in the same landscape (e.g., Williams et al. 2019). For example, over the past decades, occurrence of large wildfires across the Western United States has been driven primarily by warm and dry climate conditions. At the same time, extensive outbreaks of mountain pine beetle (*Dendroctonus ponderosae*) have killed trees over extensive areas across the same region and have also primarily been driven by climatic conditions that promote the growth of beetle populations and reduce host resistance. Despite the cooccurrence of these two disturbances in the same landscape, this

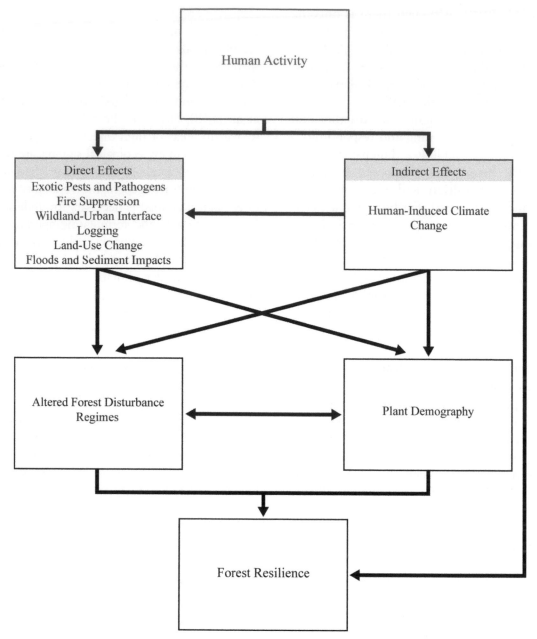

Figure 1. Conceptual summary of the relationship among human-induced disturbances, human-induced climate change, and forest resilience.

manifestation is due to a common climatic driver, rather than interactions between these disturbances (Mietkiewicz and Kulakowski 2016). Identifying the true drivers of ecological change, the geographic variation in their causes and effects, as well as the conditions under which ecosystems are resilient to direct and indirect human pressures is critical to understanding and managing forests and other ecosystems during the Anthropocene. Without detailed, site-specific research, management of forest landscapes

could further compound climatically driven stresses and disturbances or miss necessary restoration efforts.

Ecological Resilience during the Anthropocene

The concept of resilience has become central in discussions of global environmental change. Most generally, resilience is defined by the capacity of a system to be disturbed without shifting to a different state controlled by different processes (Holling 1973). However, "state" can refer to tree size, forest structure, forest type, vegetation type, or even broader categories. Given this breadth of possible criteria, it follows that our definition of resilience affects our assessment of ecological change. The broader the range of conditions that we consider normal, natural, or acceptable, the more resilient we will understand a system to be (Kulakowski et al. 2017). Nevertheless, questions of whether disturbances threaten or promote resilience, whether recent disturbances are unprecedented or simply infrequent, and what characterizes natural ecosystems require a fundamental and often long-term understanding of ecological dynamics (Johnstone et al. 2016).

Earlier in the Anthropocene, the historic range of variability (HRV, which describes the range of critical patterns and processes prior to major human modification) was advocated as a compelling model of ecosystem management, because it described the conditions under which ecosystems evolved and sustained themselves. The concept is still relevant, but now the past is less a guide to managing for the future and more of a guide for understanding departures from characteristic conditions (Keane et al. 2018). Indeed, we can only know whether ecosystem behavior and conditions are unprecedented if we know the precedent. Views of ecosystem dynamics over short periods tend to minimize the true range of characteristic conditions and predispose the observer to conclude that an observed event or state is novel and unprecedented. The longer the baseline, the more reliable a barometer it is for determining truly uncharacteristic ecosystem behavior (Gillson, Whitlock, and Humphrey 2019). Ecosystems are generally well adapted to disturbance frequencies, sizes, and severities that are characteristic of the respective systems. If these parameters are altered then post-disturbance regeneration might fail (Johnstone et al. 2016).

Resilience mechanisms are fundamentally properties and functions of a species or ecosystem that allow it to regenerate after disturbance (Buma and Wessman 2011). Consequently, a mechanistic understanding of interactions among climate, disturbance, and vegetation that shape forest demography is critical (Gillson, Whitlock, and Humphrey 2019), particularly regarding those dynamics that affect the processes, time, and limitations of ecosystem regeneration after disturbance (Reyer et al. 2015).

Field-based studies and ecological models have highlighted how some ecological mechanisms are not resilient to changes in disturbance regimes under an altered climate. For example, contrasting conifer species illustrates how resilience mechanisms influence the success or failure of forest regeneration. After compounded insect outbreak and fire, Douglas fir (*Pseudotsuga menziesii*) and Engelmann spruce (*Picea engelmannii*) rely on mature trees from unburned patches to produce seeds that can disperse to burned areas (Harvey et al. 2013). As disturbance size increases and high-severity patches become larger, disturbances can exceed the capacity of the resilience mechanisms that had maintained affected systems; that is, dispersal distance increases and successful post-disturbance seedling establishment and survival become more difficult. In contrast, lodgepole pines (*Pinus contorta*) produce serotinous cones that create a large seed bank after disturbances but with important spatiotemporal variability (Turner et al. 2016). Where they are abundant, serotinous cones reduce the need for long-distance seed dispersal. However, this resilience mechanism may be compromised as fire frequency falls below the time needed for lodgepole pine to reach reproductive age. Resilience mechanisms can be further inhibited by an altered post-disturbance climate (Kemp et al. 2019).

Climate change will continue to magnify and intensify natural disturbances, as well as effects of human activities such as land abandonment, agricultural practices, deforestation, introduction of exotic species, and increased populations in the wildland–urban interface (WUI; Johnstone et al. 2016). As ecosystems diverge from historical trends, managing ecosystem resilience based on HRV is not always feasible, particularly where disturbances are inevitable and preventative management is costly, dangerous, or impossible. In some cases, forests might be continuing to function as healthy systems and disturbances might not compromise resilience or might promote adaptive resilience (Gill et al. 2017; Gillson, Whitlock, and Humphrey 2019). In other cases, the combination of altered disturbance regimes and unfavorable post-disturbance climate could overwhelm the resilience of forest systems. In such cases, managers might know the HRV of a forest but make decisions that will allow a forest to persist, although in an altered state (Falk and Millar 2016).

Management of Protected versus Intensely Used Forests

Forests globally have been compromised by the combination of altered disturbance regimes and climate change, but important geographic variability exists in these pressures and in the consequences for resilience. Thus, managing for ecosystem resilience requires different strategies based on forest proximity to intensive human land use (Morales-Hidalgo, Oswalt, and Somanathan 2015) and the degree to which forests have been altered. Many forests have been shown to be resilient to climate change and altered natural disturbance regimes thus far but tip to alternative states when exposed to the additional compounded effects of human-induced disturbances. In general, although protected areas (PAs) and intact forests (that have been governed by natural or close-to-natural dynamics) tend to be more resilient, the pressures of the Anthropocene put disproportionate pressure on these same ecosystems. On the other hand, disturbances in human-dominated landscapes tend to be larger, more severe, and less complex than disturbances in PAs (Sommerfeld et al. 2018) and resilience mechanisms in human-dominated landscapes respond more slowly than in PAs (Leemput et al. 2018).

Protected Ecosystems

Reserves and other protected ecosystems provide refuge from direct human impacts, provide critical information about ecosystem dynamics, and provide rich research opportunities for disentangling the complex web of influences on ecological patterns and processes. Large, intact, protected forests are expected to be relatively more resilient during the Anthropocene (Schmitz et al. 2015). However, management strategies in PAs must adapt to accommodate for changing disturbance regimes and potentially compromised resilience mechanisms.

Although formally PAs have increased 50 percent from 1990 to 2015 (Morales-Hidalgo, Oswalt, and Somanathan 2015), many nature reserves are relatively small and fragmented from other PAs. Because every ecosystem is dynamic, each requires a respective minimum dynamic area (i.e., an area large enough to contain within it multiple patches in various stages of disturbance or recovery such that internal recolonization contributes to the maintenance of the overall ecosystem; sensu Pickett and Thompson 1978). At a minimum, no nature reserve should be smaller than the largest disturbance that is likely to affect that ecosystem. As sizes of disturbances increase during the Anthropocene, requisite minimum dynamic area also increases. It follows that the size of PAs must increase to facilitate maintenance of basic ecological function. However, management responses to increasing disturbances frequently involve post-disturbance logging, even in areas in which logging is otherwise excluded. Consequently, as climatically driven disturbances increase, so does post-disturbance logging. Often motivated by economic factors or an attempt to promote post-disturbance regeneration, the net effect of post-disturbance logging is frequently that of a compounded disturbance that increases the size or severity of the initial disturbance and further impinges on the minimum dynamic area, which is already under pressure (Thorn et al. 2017; Lindenmayer et al. 2020).

Managing PAs in the Anthropocene requires increasing the size, amount, and connectivity of protected ecosystems, creating buffers between protected ecosystems and intense human activity, and preventing secondary human-induced disturbances. Where possible, expanding PAs should cover a range of environmental gradients, which give ecosystems the opportunity to shift in response to climate change, although in some locations, physiographic limitations or variable species requirements would make this unrealistic. PAs ideally contain diversity in stand age (Lindenmayer 2019) and the ability to regenerate without external colonizers (Pickett and Thompson 1978). We suggest that at a minimum, the size of PAs should be increased proportional to the increasing size of potential future disturbances. Others, though, have called for much larger expansion, including the "Half Earth" proposal to set aside and protect at least half of the Earth's surface (Wilson 2016; Guo et al. 2020).

In addition to increasing the size of PAs, increasing connectivity among PAs is critical to maintaining ecosystem resilience (Lapola et al. 2020). Corridors increase the rate of immigration to PAs, and animals traveling among PAs can promote overall ecosystem resilience. Connectivity can enhance the resilience mechanisms of seed-dispersing trees and increase potential recolonization of disturbed patches (Gardner et al. 2019).

It is important to remember that even strictly protected areas do not function in isolation of human influence (Cumming 2016). Currently one third of PAs globally are affected by nearby human activity (Jones et al. 2018). This percentage is expected to increase as exposure of PAs to human activity continues to grow (Geldmann et al. 2019). Ecosystems immediately outside PAs have a large impact on the resilience of protected ecosystems. Thus, designating buffers between PAs and intense human activity potentially ameliorates degradation of protected ecosystems (Lindenmayer 2019).

Many forests globally have been compromised by the combination of altered disturbance regimes and climate change, but many others are so far resilient to these two stressors but vulnerable when exposed to the additional compounded effects of direct human-induced pressures. For example, some forests in the Amazon basin have been shown to be resilient to increased drought and natural disturbances but not the additional compound impact of repeated fires or prolonged land clearing for pastures (e.g., Bullock et al. 2020). Additionally, post-disturbance logging often creates compounded disturbance effects that reduce ecosystem heterogeneity and biodiversity, remove ecological legacies (Thorn et al. 2017), and slow ecosystem development (Beudert et al. 2015). Thus, eliminating secondary human-induced disturbances should be prioritized in all PAs.

Management of Human-Modified Ecosystems

The majority of the Earth's forests are affected by direct human activity (Potapov et al. 2017). Forests in human-modified landscapes are subject to a cocktail of climatic effects, altered vegetation land cover, novel fuel configurations, and human-induced disturbances (Godoy et al. 2019). Ecological dynamics in these areas are altered by roads, exotic species, habitat fragmentation, impervious surfaces, urban heat islands, and other pressures to which most ecosystems are not well adapted (e.g., Wittenberg et al. 2020). These effects are intensified in densely populated urban landscapes (Modugno et al. 2016). Even by the single virtue of urban areas continuing to expand during the Anthropocene, the rate at which human activity degrades forest is expected to increase (Venter et al. 2016) under business-as-usual scenarios.

Low-Elevation Forests. Low-lying forests are widely influenced by agricultural practices (Fernandez et al. 2018), with expansion especially pronounced in rainforests. Between 1980 and 2000, about 83 percent of new agricultural lands were close to rainforests (Gibbs et al. 2010). In contrast, land abandonment has been dramatic in other low-elevation forests where agriculture has ceased being economically viable. For example, European forests have fewer sheep and cattle and less agriculture than in prior centuries (Navarro and Pereira 2015), and changes in land use have contributed to expanded forest area. A similar trend in forest expansion has occurred in Central Asia after the collapse of the Soviet Union. Large amounts of land in northern Kazakhstan were abandoned and returned to woody grasslands and forests (Zhou et al. 2019). In general, as forests expand, so do natural disturbances in the same areas (Kulakowski et al. 2017). Passive forest expansion, especially if coupled with close-to-nature management or rewilding, provides an opportunity to offset forest losses elsewhere.

The Wildland–Urban Interface. The WUI is an area where anthropogenic land use and forests frequently intersect (Modugno et al. 2016). Fires and other disturbances in the WUI threaten human life and property (Manzello et al. 2018) and are also more likely to occur because fires can be ignited by lightning but also by human-caused ignition or natural disturbances damaging infrastructure, such as windstorms blowing down electrical lines and igniting fires (Radeloff et al. 2018). Management strategies in the WUI center around risk reduction, creating strategies to mitigate the impact of disturbances, and maintaining resilient landscapes (Godoy et al. 2019), but ecosystem restoration is only viable where it is consistent with other land uses. Maintaining resilient forests in the WUI comes with unique challenges that can be optimized depending on density of houses and infrastructure.

The Urban Forest. By definition, urban forests are intensely modified systems. Resilience in urban forests is not constrained to ecological context (Pickett et al. 2014) and can be conceived of as the ability to absorb a disturbance, recover from management mistakes, and integrate social engagement and institutional support (Steenberg, Duinker, and Nitoslawski 2019). The resilience mechanisms in urban forests rely on human actors and, as such, maintaining and promoting resilience in urban landscapes requires strong partnerships between local governments, ecologists, and urban designers (Geron et al. 2019; Hersh et al. 2020). Management strategies to increase the future resilience of urban forests have focused on increasing the sense of ownership in communities with urban trees, planting species that are adapted to increasing temperatures, and diversifying trees as a strategy to mitigate the intensity of exotic insect outbreaks (Elmes et al. 2018).

Promoting Resilience in the Anthropocene

Forests provide multiple ecosystem services, including those related to biodiversity, carbon, water, timber, protection from natural hazards, recreation, and spiritual values. Fortunately, many of these services are mutually compatible, and management often can promote multiple ecosystem services in the same forest. Seizing these opportunities will be crucial during the Anthropocene, as human pressures expand and intensify and forest management decisions continue to become increasingly complex. Debates around protecting human life and infrastructure, managing ecosystem resilience, and appropriate management tools will be ongoing. Ultimately, successful forest management depends on ecological and social spatiotemporal context, the former referring to particular requirements of species and ecosystems and the latter to social desires for the ecosystem in question (e.g., wilderness, timber production, etc.). Additionally, understanding the geographic variability in the mechanisms and feedbacks that have maintained resilience will continue to be essential for managing forests in the Anthropocene.

Natural forest disturbances are key players in maintaining ecological resilience and long-term forest health (Cumming 2016). Ecological disturbances can create spatial heterogeneity that can promote biodiversity, primary production, wildlife habitat, hydrogeologic protection, and ecosystem resilience (Johnstone et al. 2016). Furthermore, variability of tree, stand, and landscape patch conditions can modulate subsequent disturbance size and severity and increase the likelihood of survival of individual trees or groups of trees that can be important in post-disturbance regeneration (Buma and Wessman 2011). Spatial heterogeneity also can facilitate ecological adaptation to future environmental change and help sustain important ecosystem services. Consequently, a common goal of recent management is to increase

structural diversity and other attributes of heterogeneity, in part as a safeguard for future conditions (Lindenmayer et al. 2020).

Although disturbances can promote heterogeneity, if disturbances are too large, severe, or frequent, legacies of the predisturbance system might be lost and the ability of affected ecosystems to regenerate could be compromised (e.g., Kuuluvainen et al. 2017). This loss can shift ecosystems to alternative stable states, sometimes over extensive areas (Johnstone et al. 2016). Consequently, it is essential to understand the range of variability of disturbances that promote resilience, as well as identify the threshold beyond which disturbances might compromise it. Similarly, it is important to understand how resilience, and its underlying mechanisms, could be changing under climate change. Tipping points are most likely to be crossed as a result of extreme climate events that increase the size, frequency, and intensity of disturbances (Allen et al. 2010), compounding events (Buma and Wessman 2011), post-disturbance climatic conditions that hinder post-disturbance regeneration (Harvey, Donato, and Turner 2016), and post-disturbance logging that overrides disturbance-created heterogeneity (Thorn et al. 2017; Lindenmayer et al. 2020) and reduces post-disturbance regeneration. The relative importance of all of these drivers varies geographically, and understanding the spatiotemporal nature of that variability is a critical contemporary research priority.

Although the ecological benefits of disturbances have been extensively documented, disturbances do pose difficult management challenges—challenges that are long-standing but that are intensified in the Anthropocene. For example, potential risks to human populations associated with natural disturbances have increased as urban and suburban development expands into forested areas and also as forests expand into some urban areas. Consequently, despite a collective improved understanding of the ecological roles of disturbances, the combination of sometimes actual risks to human endeavors coupled with entrained views results in pressure to attempt to suppress disturbances and tightly control forest dynamics (Fares et al. 2015). This command-and-control approach to forest management is increasingly untenable as the Anthropocene unfolds and increases the amplitude of ecological patterns and processes. In this light, conversations become more urgent about how to work with, rather than against, natural processes in managing forests (Moritz et al. 2014). Given the inevitability of disturbances in forest ecosystems and the ecological benefits of many disturbances, a key opportunity is to identify areas in which it is feasible to allow disturbances to operate as natural ecological processes and, in doing so, to optimize their ecological benefit while protecting human safety and well-being and maintaining other desired ecosystem services. Acceptance and appreciation of natural disturbances by the general public, especially in tightly coupled human–natural systems, would facilitate land managers and policymakers maintaining forests in a more natural state (Miller et al. 2018).

Ecosystems altered by extreme land use change will continue to be evermore difficult to manage. Human activity and infrastructure development will continue to increase the potential for large disturbances and reduce the proficiency of resilience mechanisms. In intensely used ecosystems, restoration efforts should focus on reducing high-severity disturbances and maintaining heterogenous landscapes resilient to climate change. Risk management should consider the proximity of human infrastructure to forests areas and ecosystem management should aim to maintain natural ecological dynamics to the degree possible while protecting human life and infrastructure.

Although many ecosystem services are mutually compatible, strictly conserved areas are exceptionally valuable as heritage sites, integral components of healthy Earth system function, and places where non-human species have a say. Protected ecosystems will continue to define baselines of variability, function as islands of biodiversity, and provide other unique ecosystem services. To maintain the value and viability of protected forests, we suggest that the size of PAs should be increased proportional to the increasing size of potential future disturbances. Additionally, buffers should dampen the amount and intensity of human activity, and connectivity among PAs should be prioritized. Within reserves and other protected ecosystems, secondary anthropogenic disturbances (e.g., post-disturbance logging) should be eliminated to avoid compounding disturbance effects (Geldmann et al. 2019).

Acknowledgments

We thank two anonymous referees who reviewed a previous draft of this article. All authors contributed critically to the drafts and gave final approval for publication.

Funding

This work was supported by the Society of Woman Geographers, Evelyn L. Pruitt National Fellowship for Dissertation Research.

References

Allen, C. D., A. K. Macalady, H. Chenchouni, D. Bachelet, N. McDowell, M. Vennetier, T. Kitzberger, A. Rigling, D. D. Breshears, E. H. Hogg, et al. 2010. A global overview of drought and heat-induced tree mortality reveals emerging climate change risks for forests. *Forest Ecology and Management* 259 (4):660–84. doi: 10.1016/j.foreco.2009.09.001.

Bartowitz, K. J., P. E. Higuera, B. N. Shuman, K. K. McLauchlan, and T. W. Hudiburg. 2019. Post-fire carbon dynamics in subalpine forests of the Rocky Mountains. *Fire* 2 (4):58. doi: 10.3390/fire2040058.

Beudert, B., C. Bässler, S. Thorn, R. Noss, B. Schröder, H. Dieffenbach-Fries, N. Foullois, and J. Müller. 2015. Bark beetles increase biodiversity while maintaining drinking water quality. *Conservation Letters* 8 (4):272–81. doi: 10.1111/conl.12153.

Bullock, E. L., C. E. Woodcock, C. Souza, Jr., and P. Olofsson. 2020. Satellite-based estimates reveal widespread forest degradation in the Amazon. *Global Change Biology* 26 (5):2956–69. doi: 10.1111/gcb.15029.

Buma, B. 2015. Disturbance interactions: Characterization, prediction, and the potential for cascading effects. *Ecosphere* 6 (4):1–15. doi: 10.1890/ES15-00058.1.

Buma, B., and C. A. Wessman. 2011. Disturbance interactions can impact resilience mechanisms of forests. *Ecosphere* 2 (5):art64. doi: 10.1890/ES11-00038.1.

Camargo, N. F. d., N. Y. Sano, and E. M. Vieira. 2018. Forest vertical complexity affects alpha and beta diversity of small mammals. *Journal of Mammalogy* 99 (6):1444–54. doi: 10.1093/jmammal/gyy136.

Cattau, M. E., C. Wessman, A. Mahood, and J. K. Balch. 2020. Anthropogenic and lightning-started fires are becoming larger and more frequent over a longer season length in the USA. *Global Ecology and Biogeography* 29 (4):668–81. doi: 10.1111/geb.13058.

Cumming, G. S. 2016. The relevance and resilience of protected areas in the Anthropocene. *Anthropocene* 13:46–56. doi: 10.1016/j.ancene.2016.03.003.

Ellis, E. C., D. Q. Fuller, J. O. Kaplan, and W. G. Lutters. 2013. Dating the anthropocene: towards an empirical global history of human transformation of the terrestrial biosphere. *Elementa: Science of the Anthropocene* 1 (December): 000018. https://doi.org/10.12952/journal.elementa.000018.

Elmes, A., J. Rogan, L. A. Roman, C. A. Williams, S. J. Ratick, D. J. Nowak, and D. G. Martin. 2018. Predictors of mortality for juvenile trees in a residential urban-to-rural cohort in Worcester, MA. *Urban*

Forestry & Urban Greening 30 (March):138–51. doi: 10.1016/j.ufug.2018.01.024.

Falk, D. A., and C. I. Millar. 2016. The influence of climate variability and change on the science and practice of restoration ecology. In Foundations of restoration ecology, 484–513. Washington, DC: Island Press.

Fares, S., G. Scarascia Mugnozza, P. Corona, and M. Palahí. 2015. Sustainability: Five steps for managing Europe's forests. Nature News 519 (7544):407–9. doi: 10.1038/519407a.

Fernandez, M., J. Williams, G. Figueroa, G. Graddy Lovelace, M. Machado, L. Vasquez, N. Perez, L. Casimiro, G. Romero, and F. Funes Aguilar. 2018. New opportunities, new challenges: Harnessing Cuba's advances in agroecology and sustainable agriculture in the context of changing relations with the United States. Elementa: Science of the Anthropocene 6 (1):76. doi: 10.1525/elementa.337.

Gardner, C. J., J. E. Bicknell, W. Baldwin-Cantello, M. J. Struebig, and Z. G. Davies. 2019. Quantifying the impacts of defaunation on natural forest regeneration in a global meta-analysis. Nature Communications 10 (1):1–7. doi: 10.1038/s41467-019-12539-1.

Geldmann, J., A. Manica, N. D. Burgess, L. Coad, and A. Balmford. 2019. A global-level assessment of the effectiveness of protected areas at resisting anthropogenic pressures. Proceedings of the National Academy of Sciences of the United States of America 116 (46):23209–15. doi: 10.1073/pnas.1908221116.

Geron, N. A., J. Rogan, D. Martin, and M. Healy. 2019. The impact of tree planting program governance structure on tree survivorship and vigor: A case study using the Massachusetts Greening the Gateway Cities Program. Proceedings of the Fábos Conference on Landscape and Greenway Planning 6 (1):61.

Gibbs, H. K., A. S. Ruesch, F. Achard, M. K. Clayton, P. Holmgren, N. Ramankutty, and J. A. Foley. 2010. Tropical forests were the primary sources of new agricultural land in the 1980s and 1990s. Proceedings of the National Academy of Sciences of the United States of America 107 (38):16732–37. doi: 10.1073/pnas.0910275107.

Gill, N. S., F. Sangermano, B. Buma, and D. Kulakowski. 2017. Populus tremuloides seedling establishment: An underexplored vector for forest type conversion after multiple disturbances. Forest Ecology and Management 404:156–64. doi: 10.1016/j.foreco.2017.08.008.

Gillson, L., C. Whitlock, and G. Humphrey. 2019. Resilience and fire management in the Anthropocene. Ecology and Society 24 (3). doi: 10.5751/ES-11022-240314.

Godoy, M. M., S. Martinuzzi, H. A. Kramer, G. E. Defossé, J. Argañaraz, and V. C. Radeloff. 2019. Rapid WUI growth in a natural amenity-rich region in central-western Patagonia, Argentina. International Journal of Wildland Fire 28 (7):473–84. doi: 10.1071/WF18097.

Guo, W.-Y., J. M. Serra-Diaz, F. Schrodt, W. L. Eiserhardt, B. S. Maitner, C. Merow, C. Violle, M. Anand, M. Belluau, H. H. Bruun, et al. 2020. Half of the world's tree biodiversity is unprotected and is increasingly threatened by human activities. Ecology. 10.1101/2020.04.21.052464.

Harvey, B. J., D. C. Donato, W. H. Romme, and M. G. Turner. 2013. Influence of recent bark beetle outbreak on fire severity and postfire tree regeneration in montane Douglas-fir forests. Ecology 94 (11):2475–86. doi: 10.1890/13-0188.1.

Harvey, B. J., D. C. Donato, and M. G. Turner. 2016. High and dry: Post-fire tree seedling establishment in subalpine forests decreases with post-fire drought and large stand-replacing burn patches. Global Ecology and Biogeography 25 (6):655–69. doi: 10.1111/geb.12443.

Hersh, J., D. G. Martin, N. B. Geron, and J. Rogan. 2020. A relational theory of risk: A case study of the Asian longhorned beetle infestation in Worcester, MA. Journal of Risk Research 23 (6):781–815. doi: 10.1080/13669877.2019.1628091.

Holling, C. S. 1973. Resilience and stability of ecological systems. Annual Review of Ecology and Systematics 4 (1):1–23. doi: 10.1146/annurev.es.04.110173.000245.

Johnstone, J. F., C. D. Allen, J. F. Franklin, L. E. Frelich, B. J. Harvey, P. E. Higuera, M. C. Mack, R. K. Meentemeyer, M. R. Metz, G. L. W. Perry, et al. 2016. Changing disturbance regimes, ecological memory, and forest resilience. Frontiers in Ecology and the Environment 14 (7):369–78. doi: 10.1002/fee.1311.

Jones, K. R., O. Venter, R. A. Fuller, J. R. Allan, S. L. Maxwell, P. J. Negret, and J. E. Watson. 2018. One-third of global protected land is under intense human pressure. Science 360 (6390):788–91. doi: 10.1126/science.aap9565.

Keane, R. E., R. A. Loehman, L. M. Holsinger, D. A. Falk, P. Higuera, S. M. Hood, and P. F. Hessburg. 2018. Use of landscape simulation modeling to quantify resilience for ecological applications. Ecosphere 9 (9):e02414. doi: 10.1002/ecs2.2414.

Kemp, K. B., P. E. Higuera, P. Morgan, and J. T. Abatzoglou. 2019. Climate will increasingly determine post-fire tree regeneration success in low-elevation forests. Ecosphere 10 (1):e02568. doi: 10.1002/ecs2.2568.

Kim, D., A. C. Millington, and C. W. Lafon. 2020. Disturbance after disturbance: Combined effects of two successive hurricanes on forest community structure. Annals of the American Association of Geographers 110 (3):571–85. doi: 10.1080/24694452.2019.1654844.

Kulakowski, D., P. Bebi, and C. Rixen. 2011. The interacting effects of land use change, climate change and suppression of natural disturbances on landscape forest structure in the Swiss Alps. Oikos 120 (2):216–25. doi: 10.1111/j.1600-0706.2010.18726.x.

Kulakowski, D., B. Buma, J. Guz, and K. Hayes. 2019. The ecology of forest disturbances. In Encyclopedia of the World's Biomes, ed. M. I. Goldstein and D. A. DellaSala, 35–46. Oxford: Elsevier. https://doi.org/10.1016/B978-0-12-409548-9.11878-0.

Kulakowski, D., R. Seidl, J. Holeksa, T. Kuuluvainen, T. A. Nagel, M. Panayotov, M. Svoboda, S. Thorn, G. Vacchiano, C. Whitlock, et al. 2017. A walk on the wild side: Disturbance dynamics and the conservation and management of European mountain forest ecosystems. Forest Ecology and Management 388 (March):120–31. doi: 10.1016/j.foreco.2016.07.037.

Kuuluvainen, T., A. Hofgaard, T. Aakala, and B. G. Jonsson. 2017. Reprint of: North Fennoscandian mountain forests: History, composition, disturbance dynamics and the unpredictable future. *Forest Ecology and Management* 388 (March):90–99. doi: 10.1016/j.foreco.2017.02.035.

Lapola, D. M., J. M. C. da Silva, D. R. Braga, L. Carpigiani, F. Ogawa, R. R. Torres, L. C. Barbosa, J. P. Ometto, and C. A. Joly. 2020. A climate-change vulnerability and adaptation assessment for Brazil's protected areas. *Conservation Biology: The Journal of the Society for Conservation Biology* 34 (2):427–37. doi: 10.1111/cobi.13405.

Leemput, I. A. V. D., V. Dakos, M. Scheffer, and E. H. van Nes. 2018. Slow recovery from local disturbances as an indicator for loss of ecosystem resilience. *Ecosystems* 21 (1):141–52. doi: 10.1007/s10021-017-0154-8.

Lindenmayer, D. B. 2019. Integrating forest biodiversity conservation and restoration ecology principles to recover natural forest ecosystems. *New Forests* 50 (2):169–81. doi: 10.1007/s11056-018-9633-9.

Lindenmayer, D. B., R. M. Kooyman, C. Taylor, M. Ward, and J. E. Watson. 2020. Recent Australian wildfires made worse by logging and associated forest management. *Nature Ecology & Evolution* 4 (7):898–900. doi: 10.1038/s41559-020-1195-5.

Lydersen, J. M., and B. M. Collins. 2018. Change in vegetation patterns over a large forested landscape based on historical and contemporary aerial photography. *Ecosystems* 21 (7):1348–63. doi: 10.1007/s10021-018-0225-5.

Manzello, S. L., K. Almand, E. Guillaume, S. Vallerent, S. Hameury, and T. Hakkarainen. 2018. FORUM position paper: The growing global wildland urban interface (WUI) fire dilemma: Priority needs for research. *Fire Safety Journal* 100. doi: 10.1016/j.firesaf.2018.07.003.

Mietkiewicz, N., and D. Kulakowski. 2016. Relative importance of climate and mountain pine beetle outbreaks on the occurrence of large wildfires in the Western USA. *Ecological Applications* 26 (8):2525–37. doi: 10.1002/eap.1400.

Miller, S., R. Addington, G. Aplet, M. Battaglia, T. Cheng, J. Feinstein, and J. Underhill. 2018. Back to the future: Building resilience in Colorado Front Range forests using research findings and a new guide for restoration of ponderosa and dry-mixed conifer landscapes. *Science You Can Use Bulletin* 28:15. Fort Collins, CO: Rocky Mountain Research Station.

Modugno, S., H. Balzter, B. Cole, and P. Borrelli. 2016. Mapping regional patterns of large forest fires in wildland–urban interface areas in Europe. *Journal of Environmental Management* 172 (May):112–26. doi: 10.1016/j.jenvman.2016.02.013.

Morales-Hidalgo, D., S. N. Oswalt, and E. Somanathan. 2015. Status and trends in global primary forest, protected areas, and areas designated for conservation of biodiversity from the Global Forest Resources Assessment 2015. *Forest Ecology and Management* 352 (September):68–77. doi: 10.1016/j.foreco.2015.06.011.

Mori, A. S., F. Isbell, and R. Seidl. 2018. β-Diversity, community assembly, and ecosystem functioning. *Trends in Ecology & Evolution* 33 (7):549–64. doi: 10.1016/j.tree.2018.04.012.

Moritz, M. A., E. Batllori, R. A. Bradstock, A. M. Gill, J. Handmer, P. F. Hessburg, J. Leonard, S. McCaffrey, D. C. Odion, T. Schoennagel, et al. 2014. Learning to coexist with wildfire. *Nature* 515 (7525):58–66. doi: 10.1038/nature13946.

Navarro, L. M., and H. M. Pereira. 2015. Rewilding abandoned landscapes in Europe. In *Rewilding European landscapes*, 3–23. Cham, Switzerland: Springer.

Peterson, C. J. 2019. Damage diversity as a metric of structural complexity after forest wind disturbance. *Forests* 10 (2):85. doi: 10.3390/f10020085.

Pickett, S. T. A., B. McGrath, M. L. Cadenasso, and A. J. Felson. 2014. Ecological resilience and resilient cities. *Building Research & Information* 42 (2):143–57. doi: 10.1080/09613218.2014.850600.

Pickett, S. T. A., and J. N. Thompson. 1978. Patch dynamics and the design of nature reserves. *Biological Conservation* 13 (1):27–37. doi: 10.1016/0006-3207(78)90016-2.

Pickett, S. T. A., and P. S. White. 1985. Patch dynamics: A synthesis. In *The ecology of natural disturbance and patch dynamics*, ed. S. T. A. Pickett and P. S. White, 371–84. San Diego, CA: Academic Press.

Potapov, P., M. C. Hansen, L. Laestadius, S. Turubanova, A. Yaroshenko, C. Thies, W. Smith, et al. 2017. The last frontiers of wilderness: Tracking loss of intact forest landscapes from 2000 to 2013. *Science Advances* 3 (1):e1600821. doi: 10.1126/sciadv.1600821.

Potterf, M., and C. Bone. 2017. Simulating bark beetle population dynamics in response to windthrow events. *Ecological Complexity* 32:21–30. doi: 10.1016/j.ecocom.2017.08.003.

Poulos, H. M., A. M. Barton, J. A. Slingsby, and D. M. J. S. Bowman. 2018. Do mixed fire regimes shape plant flammability and post-fire recovery strategies? *Fire* 1 (3):39. doi: 10.3390/fire1030039.

Radeloff, V. C., D. P. Helmers, H. A. Kramer, M. H. Mockrin, P. M. Alexandre, A. Bar-Massada, V. Butsic, T. J. Hawbaker, S. Martinuzzi, A. D. Syphard, et al. 2018. Rapid growth of the U.S. wildland–urban interface raises wildfire risk. *Proceedings of the National Academy of Sciences of the United States of America* 115 (13):3314–19. doi: 10.1073/pnas.1718850115.

Reyer, C. P. O., A. Rammig, N. Brouwers, and F. Langerwisch. 2015. Forest resilience, tipping points and global change processes. *Journal of Ecology* 103 (1):1–4. doi: 10.1111/1365-2745.12342.

Schmitz, O. J., J. J. Lawler, P. Beier, C. Groves, G. Knight, D. A. Boyce, J. Bulluck, K. M. Johnston, M. L. Klein, K. Muller, et al. 2015. Conserving biodiversity: Practical guidance about climate change adaptation approaches in support of land use planning. *Natural Areas Journal* 35 (1):190–203. doi: 10.3375/043.035.0120.

Sommerfeld, A., C. Senf, B. Buma, A. W. D'Amato, T. Després, I. Díaz-Hormazábal, S. Fraver, L. E. Frelich, A. G. Gutierrez, S. J. Hart, et al. 2018. Patterns and drivers of recent disturbances across the temperate forest biome. *Nature Communications* 9:4355. doi: 10.1038/s41467-018-06788-9.

Steenberg, J. W., P. N. Duinker, and S. A. Nitoslawski. 2019. Ecosystem-based management revisited: Updating the concepts for urban forests. *Landscape and Urban Planning* 186:24–35. doi: 10.1016/j.landurbplan.2019.02.006.

Thorn, S., C. Bässler, M. Svoboda, and J. Müller. 2017. Effects of natural disturbances and salvage logging on biodiversity—Lessons from the Bohemian forest. *Forest Ecology and Management* 388 (March):113–19. doi: 10.1016/j.foreco.2016.06.006.

Turner, M. G., T. G. Whitby, D. B. Tinker, and W. H. Romme. 2016. Twenty-four years after the Yellowstone Fires: Are postfire lodgepole pine stands converging in structure and function? *Ecology* 97 (5):1260–73. doi: 10.1890/15-1585.1.

Venter, O., E. W. Sanderson, A. Magrach, J. R. Allan, J. Beher, K. R. Jones, H. P. Possingham, W. F. Laurance, P. Wood, B. M. Fekete, et al. 2016. Sixteen years of change in the global terrestrial human footprint and implications for biodiversity conservation. *Nature Communications* 7 (1):12558–11. doi: 10.1038/ncomms12558.

Williams, A. P., J. T. Abatzoglou, A. Gershunov, J. Guzman-Morales, D. A. Bishop, J. K. Balch, and D. P. Lettenmaier. 2019. Observed impacts of anthropogenic climate change on wildfire in California. *Earth's Future* 7 (8):892–910. doi: 10.1029/2019EF001210.

Wilson, E. O. 2016. *Half-Earth: Our planet's fight for life.* New York: Norton.

Wittenberg, L., H. van der Wal, S. Keesstra, and N. Tessler. 2020. Post-fire management treatment effects on soil properties and burned area restoration in a wildland–urban interface, Haifa fire case study. *Science of the Total Environment* 716:135190. doi: 10.1016/j.scitotenv.2019.135190.

Zhou, Y., L. Zhang, J. Xiao, C. A. Williams, I. Vitkovskaya, and A. Bao. 2019. Spatiotemporal transition of institutional and socioeconomic impacts on vegetation productivity in central Asia over last three decades. *Science of the Total Environment* 658 (March):922–35. doi: 10.1016/j.scitotenv.2018.12.155.

Abandoning Holocene Dreams: Proactive Biodiversity Conservation in a Changing World

Kenneth R. Young and Sisimac Duchicela

Although species have always shifted their ranges, the rapid pace of current biophysical changes and the further complications imparted by human land use provide unprecedented challenges for biodiversity conservation. As a result, goals, methods, and strategies are being reconceptualized. For example, the terms *conservation* and *wilderness protection*, and their associated practices, seem to be static and simplistic compared to the challenges of managing novel species assemblages in unique climatic and disturbance regimes. A more proactive approach is developing that builds in the needed adaptations as biophysical change progressed; that would need to consider possible nonlinear ecosystem changes, with threshold effects and ecological surprises; that might force a reconsideration of the goals of ecological restoration; and that might require management of lands today for goals that could be quite different in 50 to 100 years. As examples, this could require such potentially controversial activities as planting trees outside their ranges so that they could serve as part of wildlife habitat in several decades, using prescribed burns, or bringing some species into captivity or botanical gardens until reintroduction becomes feasible. With the Anthropocene providing a potential new label for the current epoch and ongoing research providing new insights into disturbance regimes and successful conservation practices, it is appropriate to rethink implications for sustainability and human–nature relationships in general and for biodiversity in particular. *Key Words: Anthropocene, biodiversity, climate change, conservation.*

Conservation of biological diversity has traditionally taken as goals reference points derived from idealized or past time periods; that is, "nature without people." More concretely, conservation objectives could involve what are judged to be (1) "pre-European" conditions; (2) "road-less areas," which then can be mapped and used to assess habitat quality near and far from direct human influences; or (3) "historic fire regimes," which theoretically can be (re)created through judicious fire management. These biophysical settings are routinely used as restoration goals, and are used to judge how to undertake remediation following environmental degradation (Hobbs and Cramer 2008). They can also serve as targets for preservation, in cases where previous human impact is judged minimal ("wilderness"). Sometimes those targets are revealed to be based more on illusion and tradition than on empirical measures of success (e.g., Lave 2012). The full repertoire of biodiversity conservation strategies includes hotspot identification, systematic planning of protected area systems, landscape and corridor management, targeted endangered species programs, and the recognition and quantification of ecosystem services (Brooks et al. 2006; Zimmerer 2006; Brown et al. 2013).

Yet, the totality of this conservation planning has been developed and implemented in the late Holocene. The advent of interest in the Anthropocene, whether as a potential new designation for our geological epoch or as a fulcrum for discussion and debate about the place of *Homo sapiens* in the world today (Davies 2016), suggests the need for new thinking, especially as climate change continues to occur and requires the planning of adaptive approaches (e.g., Hansen et al. 2016). If recent experiences provide insufficient information for planning for future challenges, then one option is to examine the past, looking for useful analogues (Williams, Jackson, and Kutzbach 2007). For example, paleoconservation biogeography can be used to get rates that species can potentially shift their distributions in the future (based on location and timing data extracted from paleorecords) or to evaluate how the movement of species range shifts compares to those of the climate isobars and associated ecological change (Svenning and Sandel 2013).

Here we argue that the Anthropocene can also be viewed as an opportunity to reassess biodiversity conservation, making it more mission directed and more proactive, commensurate with the scope and magnitude of likely future challenges. This is an active area of research by conservation scientists and conservationists (e.g., Hobbs, Higgs, and Harris 2009; Barnosky et al. 2017). In this article, we highlight three research foci, namely, disturbance regimes, species gains and losses, and ecological restoration, along with their associated practices, which are being rethought and are also of great interest to broader societal needs. Many of these approaches require holistic or interdisciplinary approaches that meld the social with the ecological, providing opportunities for engagement with geographers and geography.

Disturbance Regimes of the Future

Disturbance regimes are used as conceptual frameworks for quantifying the timing and magnitude or intensity of disturbance events, recovery rates post-disturbance, and, by extension, spatiotemporal change over landscapes or across environmental gradients. Hence, they embed both time and space considerations. They in turn can be related by researchers and land managers to further synergistic disturbance dynamics, to the use of habitat by wildlife, and to management prescriptions that tie extraction for human needs to the recovery rates of the ecological systems involved. The dynamics of vegetation and of landscapes is a core research focus in biogeography (Glenn-Lewin, Peet, and Veblen 1992; Millington, Blumler, and Schickhoff 2011), hence linking these concepts and practices to other parts of geography.

Regimes change in their modes, so the vagaries in the pacing or severity of disturbance events lead to altered outcomes and change underlying processes. Those shifts might be especially responsive to directional changes imposed by new climatic regimes and anthropogenic fire sources, for example, with longer fire seasons or extended dry seasons; such shifts will likely produce novel regimes (Balch et al. 2017). There are also historical legacies of fire suppression that have left their mark on current disturbance regimes, with many implications for the dynamics of the wildland–urban interface, which increasingly characterizes large parts of North America and Europe (Davis et al. 2019; Hessburg et al. 2019). Regime shifts can also be quantified in other contexts, especially using physical geography approaches. For example, Ramos-Scharrón and Arima (2019)

recently assessed the intensity of Hurricane Maria in Puerto Rico, finding that the storm was stronger than any other measured yet might represent a that new norm, given increasing sea surface temperatures and hurricane intensities. Similarly, an increase in 100-year floods occurring one after another suggests that new hydrological and storm regimes are in place (Milly et al. 2008), requiring different responses for natural disaster prediction, recovery, and mitigation.

In ecological systems, unprecedented disturbances or its opposite, suppression of natural disturbances, could result in change in biotic configurations, including as examples increases in the dominance of certain species or extinctions of others, with implications for ecosystem functions. It might be necessary to have aggressive habitat management for species of special concern to prevent losses. As one example, severe-fire-adapted shrubs might not persist under altered fire regimes (Le Breton et al. 2020). Management of the colonization and spread of invasive species adapted to altered regimes might be needed to deter tendencies for biotic homogenization (Baiser et al. 2012), which lessens native diversity and could affect ecosystem functioning. Disturbances also affect management of natural and utilized landscapes for carbon services and for calibrating values for scenic beauty versus the need for human and infrastructure safety.

Climatic and environmental surprises are to be expected, with feedbacks creating new dynamism and hence novel outcomes. Some of these can be assessed using scenario testing, in the manner of what has been done with Intergovernmental Panel on Climate Change methodologies or as adapted by the new IPBES (Intergovernmental Science-Policy Platform on Biodiversity and Ecosystem Services) approach (Díaz et al. 2015). More practically, at landscape levels, there need be predictive conservation frameworks based on predicted future biophysical parameters that can be applied to given sites or management modes and that can be evaluated in regard to the ways to connect ecological sustainability to its other dimensions. In practice, such modeling might be too simplistic, so experimentation and adaptive management that reassesses its goals could be part of new conservation norms.

Conservation Prescriptions for Shifting Species

Prescribed burns are a standard tool used for manipulating or influencing fire regimes. They are used for fire control by preemptively reducing fuel loads and by altering the spatial connectivity that would permit larger fires. Ecologically, they are valuable for habitat management for species that need burned or open habitats and for limiting nuisance species. More generally, they are prescriptions meant to right a wrong or to guide systems toward a different outcome. As such, they appear to offer a conceptual route to further interrogate Anthropocene considerations in biodiversity conservation. Other such conservation prescriptions could include species reintroductions, fertilization or other soil remediations, and the thinning of trees or targeted harvesting of wildlife (Whisenant 1999).

Nevertheless, here we also offer a more provocative concern: Should we add "prescribed extinctions" to the conservation toolkit? Smallpox, rinderpest, and soon polio and Guinea worm will have been guided deliberately into extinction given their role as disease and parasitic vectors that severely harm people (Klepac et al. 2013). Genetic engineering is now being used to develop mosquitos that could result in their reduction in the wild, or maybe even their extinction. What kind of societal debates and impact assessments should be taking place about these planned extinctions, and who should decide whether these are wise actions to pursue? Is the Anthropocene meant to not include species harmful to our social, economic, or health interests? These are some of the motivating questions for researchers but for which answers will likely only come following debates across all societal realms.

Alternatively, consider "de-extinction" efforts, which have already received development of possible protocols (Valdez et al. 2019) for deciding whether bringing species back from extinction makes ecological sense, once the technical challenges are met, as they surely will be (Monbiot 2014; Shapiro 2015). Current debates include asking, as an example, whether reconstituted and rewilded mammoths would add missing kinds of ecosystem management in places where extinct megafauna were once significant ecological actors. Presumably, the difficulty of such efforts would tend to make them be designated as showcases for charismatic species. Nevertheless, a more general approach of rewilding could consider all sorts of reintroductions and reestablishments, including for the uncharismatic microbes, invertebrates, or fungi that actually do most ecosystem functions but whose distributions might have been reduced by human actions in the past.

As another set of examples, contemplate the scope of some of the challenges: The páramo/tropical alpine ecoregion found in high mountains of northwest South America has some 3,500 vascular plant species (Madriñán, Cortés, and Richardson 2013), meaning that its projected shift (Tovar et al. 2013) to woody vegetation could result in extinctions of half or more of the species that depend on cool, open habitats. Simultaneously, the more than 15,000 species of Amazonian plants would be placed into "novel climate regimes" but situated hundreds of kilometers from the Andean uplands (Feeley and Silman 2009), meaning that they must adapt in place, persist and tolerate, or go extinct. It does not appear feasible that current institutions could bring those thousands of plant species into conservation programs: Planning the use of seed banks or botanical gardens for this task appears to be so daunting that there are few organized efforts in place. The institutional challenges are at least as difficult in thinking about animals, given that the use of zoos, aquaria, and captive breeding programs would have to be dramatically upscaled and made more proactive, predictive, interventionist, and better funded than they are today.

Directed, facilitated, and passive means of shifting species ranges are encapsulated in the existing "assisted migration" initiatives (Hoegh-Guldberg et al. 2008; Scheffers and Pecl 2019). There are nascent attempts to institutionalize these concerns, including efforts to surmount the bureaucratic, legal, and practical challenges. In turn, these efforts require improvements in the art of species distribution modeling, combining future range predictions using climatic inferences with the capabilities of dispersal by the respective taxa (Miller and Holloway 2015). The future status quo of biodiversity conservation, as a human-led operation, will be either assisted migration or a "benign neglect" or "business as usual" approach that documents extinction but chooses to not intervene. Note that species range shifts also carry along the respective species traits, meaning that the future will often include ecosystems with novel functions.

Anticipatory Ecological Restoration

In the 1990s some geographers, including Kellman (1996), began to worry that conservation was becoming a nostalgic practice, where maintaining ecosystems in their intact or original states was

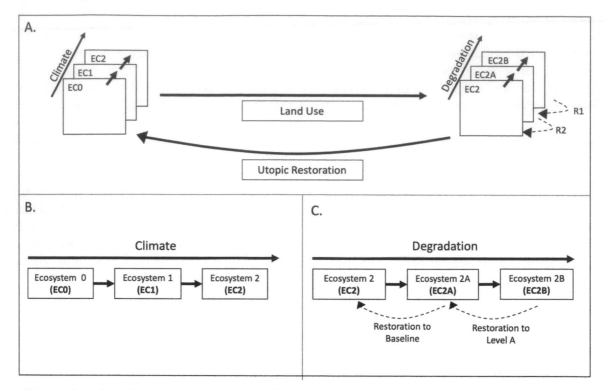

Figure 1. Conceptual model of current and proposed ecological restoration practices, showing changes expected through time. (A) Overall representation of aims for restoration projects using experimental or projected data on ecosystems, plus recognition of land use effects. Climate is expected to modify ecosystems, for which utopic ecological restoration projects could use predicted ecosystem conditions as their targets (EC2, etc.). (B) Detailed view of how climate change will affect ecosystems, which might change at different rates than climate drivers due to lag effects and differing rates of species colonization. (C) Detailed view of how restoring ecosystem functioning might be considered "business as usual" restoration where restoration is reverting to past ecosystem conditions, without taking into account the future.

seemingly unrealistic. He argued that "the fragmented world of biological communities in the future will be so different from that of the past, that we must reformulate preservation strategies to forms that go beyond thinking only of preserving microcosms of the original community types" (Kellman 1996, 115). He also discussed the possibility that ecosystems might be composed of species assemblies never seen before. Around that same time, restoration ecology became institutionalized (Jordan and Lubick 2011) and defined as the study of interrelationships among organisms and their environment while assisting in their recovery, using ecological theory and practice to guide a degraded or disturbed ecosystem to a less perturbed state. The Society for Ecological Restoration International Science and Policy Working Group (2004, 3) defined this as "the process of assisting the recovery of an ecosystem that has been damaged, degraded or destroyed." Although elements of restoration had been practiced since the 1930s, as famously depicted by Aldo Leopold (1949) in *A Sand County Almanac* (Leopold 1949), it was

not until the 1970s that ecological restoration began to be formally practiced (Jordan and Lubick 2011; Palmer, Zedler, and Falk 2016).

Ecological restoration is said to be "rooted in ecological history" (Jackson and Hobbs 2009, 567). Specifically, information on the original ecosystem and its land use history is necessary for identifying the goals for restoration and to return an ecosystem to its historical trajectory (Society for Ecological Restoration International Science and Policy Working Group 2004). Although an ecosystem might be set on a trajectory to return to a historical, undegraded condition, climatic pressures (and associated disturbance regimes) might not allow that trajectory to occur (Figure 1). In those cases, climatic predictions via models or experimentation should be incorporated into ecological restoration plans in the development of goals and success metrics (Bastin et al. 2019). Thus, the current rates of climate change call for a reevaluation not only of historic conservation practices but also of these newer restoration criteria. Furthermore, the term *ecological restoration* might no longer encompass the methods, goals, and strategies needed and should perhaps be referred to as *anticipatory ecological restoration* or *ecological rehabilitation*.

Through the use of anticipatory or predictive methods, appropriate metrics for ecological restoration practices that account for the future conditions of the ecosystem being restored will jump-start the trajectory that that ecosystem will follow as climatic conditions begin to change. Climate is an important driver for the interplay of biotic factors and up to a certain point might be deterministic for ecosystem processes and organismal interactions (Pörtner and Farrell 2008). Therefore, as climate continues to change, the baseline for the undegraded condition of the target ecosystem also changes. In some cases, instead of estimating future conditions, a current reference ecosystem could be used if that reference also changed along with climate. In mountains, sampling at lower elevations would reflect possible future change with warming; however, for other cases, warming experiments or reciprocal transplants might be better research designs.

We propose that ecological restoration follow an anticipatory path driven by predictive modeling, experimentation, or both (e.g., Le Breton et al. 2020). Figure 1 conceptually shows three elements of current and proposed ecological restoration projects, including the climatic impacts, land use effects, and a proposed "utopic" restoration initiative acting on ecosystems. As climate changes, ecosystems shift as well, but this might not occur at the same rate as overall climate change (Figure 1B). The effects of past land use (Figure 1C) could result in a degraded ecosystem, which is addressed in typical or "business as usual" restoration, where ecosystem functioning is restored but ecosystems are not equipped for future changes. Degradation in this case implies the simplification of an ecosystem when biotic homogenization has occurred, which, in turn, might affect ecosystem functioning; the more intense the land uses, the less functionally diverse it is (represented as EC2 to EC2A to EC2B in Figure 1). Mitigating or removing the degrading influence (labeled R1 in Figure 1) recovers ecosystem functions but not necessarily the original species composition, unless the past influence of land use is accounted for and missing species are reintroduced (R2). Figure 1A shows that desired restoration must account for both ongoing and future climate change (Figure 1B), perhaps through

modeling, monitoring, or experimentation, also considering land use effects and recovery from those influences. Decisions need to include desired species composition, their interrelationships, ecosystem functions, needs for given land use practices, and ongoing climate change. There will often be many additional societal concerns that also must be accommodated in terms of costs and benefits or inequalities.

Some current ecological restoration projects are considering changing baselines for restoration projects: Duchicela et al. (2019) recently assessed success in restoration practices installed in four sites in highland Peru by comparing specific ecological indicators of the restored plots to a shifting control. The goal of our restoration trials was to increase vegetation coverage of palatable species for alpaca grazing and the indicators included diversity, biomass, and soil variables. No historic reference was used as a target; instead, success was determined by comparing the restoration treatments among themselves and from unrestored control plots that were monitored for changes throughout the years. The Andes have a long, intense land use history where undisturbed sites are either not accessible or cannot be found (Young 2009; Sylvester, Sylvester, and Kessler 2014). Additionally, climate change is already having strong effects on water availability, temperature, and seasonality in this region (e.g., Postigo, Young, and Crews 2008), making it imperative to develop applied conservation practices that incorporate this environmental dynamism.

These complications, in addition to ecosystem dynamics and further genetic considerations, all imply that "business as usual" restoration based on goals determined by historic land use and past conditions would not produce resilient ecosystems under future change. Setting targets that adapt to changes in climatic predictions might enable the resilience needed to assist the ecosystem into future conditions (EC2 in Figure 1A). What we call utopic restoration in Figure 1 might need to become standard practice. Some examples of setting ecosystems into future ecological trajectories include determining future species ranges and using assisted migration to move species outside of their current range toward their future range. Roberts and Hamann (2016) modeled the future distributional ranges of twenty-four common western North American trees and determined the velocity of migration needed for these trees to reach the nearest climate refugia. Findings from their study showed that for some species, the rate of climate change was higher than their migrating velocity. Thus, ecological restoration interventions would need to plant trees outside of their current distributional ranges, ahead of the range shifts of targeted wildlife species that require a forested environment. Another intervention could consist of using artificial selection for extreme traits; for example, selecting for plants that will be more drought tolerant or fire resistant in areas where drastic humidity loss is predicted (e.g., Exposito-Alonso et al. 2019).

Ecological restoration, whether referred to as anticipatory ecological restoration or ecological rehabilitation, can be used for mitigating and increasing resilience from the effects of climate change (Harris et al. 2006; McDonald et al. 2016; Palmer, Zedler, and Falk 2016). Because temperature and other climatic conditions constantly shape vegetation zones, considering climate change is unavoidable for restoration ecology in practice. Therefore, discussions around how to "future-proof" ecosystems should be at the forefront of conservation and management policies. Studies in different systems (e.g., Urrutia and Vuille 2009; Elmendorf et al. 2012; Morueta-Holme et al. 2015) are attempting to define or quantify the effects of climate change on vegetation communities and, in turn, how these changes in plant composition and distribution will affect the adaptability and land-use systems of human populations living nearby (e.g., Young and Lipton 2006). The interdisciplinary and transdisciplinary strategies required also would provide new foci for human–environment studies. Lessons learned from the study of fire regimes and other disturbance-focused approaches might be especially useful.

Conclusions

The title of this article uses the gerund form of the verb *abandon* to highlight the efforts in the past few decades to develop proactive and anticipatory approaches to biodiversity conservation (Seastedt, Hobbs, and Suding 2008; Zedler, Doherty, and Miller 2012), which in turn requires societal participation. The Holocene is anomalous compared to the previous 100,000 years (Archer 2016); lessons derived in the past two centuries of Western science might not be sufficient for the novel or directional changes of the future. Biodiversity conservation

might require proactive approaches that go beyond the recipes developed to date. As examples, ecological restoration projects have long used reference ecosystems as targets for remediation of degraded sites, but this is a practice that loses its clarity in a world affected by directional changes. Alternatively, consider the endangered species affected by range shifts that move them out of the very landscapes managed for their benefit, requiring strategies for land acquisition for the biological corridors needed in the future.

The Anthropocene might be useful as a label for this effort, by freeing us from prevailing conservation prescriptions and making the biodiversity target a shifting one, which will depend on the time horizon used for the planning and the kinds of goals that will suffice. We suggest that idealism and utopic goal setting would be helpful in organizing conservation efforts, given other competing societal needs and priorities. We further suggest that ethics protocols might be useful devices for both researchers and decision makers, otherwise adrift in the complexities of possible futures. As an example, the bold suggestion of Wilson (2016) to dedicate one half of the planet to people's use and one half to all other species was recently formalized with spatial analyses that indeed imply that this might be a feasible vision to use as a goal (Dinerstein et al. 2019; Schleicher et al. 2019). Doubtless, biodiversity conservation is but one concern for Future Earth, but there could be many means to combine concerns for human needs and inequalities with other innovative approaches (Shoreman-Ouimet and Kopnina 2015).

Ecological restoration goals are perplexing given directional climate change affecting ecosystem dynamism. The term itself begins to lose meaning in this context; in this article we add adjectives such as *anticipatory* and *utopic* to suggest possible ways to improve communication. Coincident socioeconomic transitions can alter, augment, or decrease ecological processes of change in land use. Thus, we propose that the rehabilitation of degraded ecosystems should continue but might need to be categorized as a "business as usual" approach, not sufficient in terms of the institutional challenges of also addressing likely future change trajectories (Figure 1). Typically, goal setting for ecological restoration projects entails obtaining a reference site that is either considered an unaltered state of the targeted degraded ecosystem or that is meant to be a historical portrayal of an ecosystem that once occurred in that same location. Instead, we observe that climate change creates shifting baselines for ecological restoration, and scenario-building experiments might need to be used to predict future baselines. Ecological restoration needs to be adaptable to changes in climatic and ecological conditions in specific landscape contexts. Doing so requires that predictive models, experimentation, or both be done as part of goal setting.

Many of the concerns raised here must be addressed by a broader cross section of society, but researchers can continue to provide policy-relevant findings. Especially challenging is the need for simultaneously evaluating land use and shifting environmental baselines, with attention paid to the social, educational, legal, and economic implications of those approaches. These challenges also represent research opportunities, especially with geography's strengths in building linkages from the biophysical to the various human dimensions. Others have noted the complexities of evaluating natural resource management and preparing for natural disasters given directional biophysical changes (Siders, Hino, and Mach 2019). Similarly, the challenges for nature conservation represent yet another existential test for the planet's biodiversity.

Funding

Kenneth R. Young acknowledges funding from the National Science Foundation (DEB1617429). Sisimac Duchicela acknowledges financial support from the National Geographic Society (Grant No. EC-56298C-19).

References

Archer, D. 2016. *The long thaw: How humans are changing the next 100,000 years of Earth's climate.* 2nd ed. Princeton, NJ: Princeton University Press.

Baiser, B., J. D. Olden, S. Record, J. L. Lockwood, and M. L. McKinney. 2012. Pattern and process of biotic homogenization in the New Pangaea. *Proceedings. Biological Sciences* 279 (1748):4772–77. doi: 10.1098/rspb.2012.1651.

Balch, J. K., B. A. Bradley, J. T. Abatzoglou, R. C. Nagy, E. J. Fusco, and A. L. Mahood. 2017. Human-started wildfires expand the fire niche across the United States. *Proceedings of the National Academy of Sciences of the United States of America* 114 (11):2946–51. doi: 10.1073/pnas.1617394114.

Barnosky, A. D., E. A. Hadly, P. Gonzalez, J. Head, P. D. Polly, A. M. Lawing, J. T. Eronen, D. D. Ackerly, K.

Alex, E. Biber, et al. 2017. Merging paleobiology with conservation biology to guide the future of terrestrial ecosystems. *Science* 355 (6325) eaah4787. doi: 10.1126/science.aah4787.

Bastin, J.-F., Y. Finegold, C. Garcia, D. Mollicone, M. Rezende, D. Routh, C. M. Zohner, and T. W. Crowther. 2019. The global tree restoration potential. *Science* 365 (6448):76–79. doi: 10.1126/science.aax0848.

Brooks, T. M., R. A. Mittermeier, G. A. B. da Fonseca, J. Gerlach, M. Hoffmann, J. F. Lamoreux, C. G. Mittermeier, J. D. Pilgrim, and A. S. L. Rodrigues. 2006. Global biodiversity conservation priorities. *Science* 313 (5783):58–61. doi: 10.1126/science.1127609.

Brown, D. G., D. T. Robinson, N. H. F. French, and B. C. Reed, eds. 2013. *Land use and the carbon cycle: Advances in integrated science, management, and policy.* Cambridge, UK: Cambridge University Press.

Davies, J. 2016. *The birth of the Anthropocene.* Oakland: University of California Press.

Davis, K. T., S. Z. Dobrowski, P. E. Higuera, Z. A. Holden, T. T. Veblen, M. T. Rother, S. A. Parks, A. Sala, and M. P. Maneta. 2019. Wildfires and climate change push low-elevation forests across a critical climate threshold for tree regeneration. *Proceedings of the National Academy of Sciences of the United States of America* 116 (13):6193–98. doi: 10.1073/pnas.1815107116.

Díaz, S., S. Demissew, J. Carabias, C. Joly, M. Lonsdale, N. Ash, A. Larigauderie, J. R. Adhikari, S. Arico, A. Báldi, et al. 2015. The IPBES conceptual framework—Connecting nature and people. *Current Opinion in Environmental Sustainability* 14:1–16. doi: 10.1016/j.cosust.2014.11.002.

Dinerstein, E., C. Vynne, E. Sala, A. R. Joshi, S. Fernando, T. E. Lovejoy, J. Mayorga, D. Olson, G. P. Asner, J. E. M. Baillie, et al. 2019. A global deal for nature: Guiding principles, milestones, and targets. *Science Advances* 5 (4), eaaw2869. doi: 10.1126/sciadv.aaw2869.

Duchicela, S. A., F. Cuesta, E. Pinto, W. D. Gosling, and K. R. Young. 2019. Indicators for assessing tropical alpine rehabilitation practices. *Ecosphere* 10 (2): e02595. doi: 10.1002/ecs2.2595.

Elmendorf, S. C., G. H. R. Henry, R. D. Hollister, R. G. Björk, A. D. Bjorkman, T. V. Callaghan, L. S. Collier, E. J. Cooper, J. H. C. Cornelissen, T. A. Day, et al. 2012. Global assessment of experimental climate warming on tundra vegetation: Heterogeneity over space and time. *Ecology Letters* 15 (2):164–75. doi: 10.1111/j.1461-0248.2011.01716.x.

Exposito-Alonso, M., H. A. Burbano, O. Bossdorf, R. Nielsen, D. Weigel, and the 500 Genomes Field Experiment Team. 2019. Natural selection on the *Arabidopsis thaliana* genome in present and future climates. *Nature* 573 (7772):126–29. doi: 10.1038/s41586-019-1520-9.

Feeley, K. J., and M. R. Silman. 2009. Extinction risks of Amazonian plant species. *Proceedings of the National Academy of Sciences of the United States of America* 106 (30):12382–87. doi: 10.1073/pnas.0900698106.

Glenn-Lewin, D. C., R. K. Peet, and T. T. Veblen, eds. 1992. *Plant succession: Theory and prediction.* London: Chapman and Hall.

Hansen, A. J., W. B. Monahan, D. M. Theobald, and S. T. Olliff, eds. 2016. *Climate change in wildlands: Pioneering approaches to science and management.* Washington, DC: Island.

Harris, J. A., R. J. Hobbs, E. Higgs, and J. Aronson. 2006. Ecological restoration and global climate change. *Restoration Ecology* 14 (2):170–76. doi: 10.1111/j.1526-100X.2006.00136.x.

Hessburg, P. F., C. L. Miller, S. A. Parks, N. A. Povak, A. H. Taylor, P. E. Higuera, Prichard SJ, North MP, Collins BM, Hurteau MD, et al. 2019. Climate, environment, and disturbance history govern resilience of western North American forests. *Frontiers in Ecology and Evolution* 7:239. doi: 10.3389/fevo.2019.00239.

Hobbs, R. J., and V. A. Cramer. 2008. Restoration ecology: Interventionist approaches for restoring and maintaining ecosystem function in the face of rapid environmental change. *Annual Review of Environment and Resources* 33 (1):39–61. doi: 10.1146/annurev.environ.33.020107.113631.

Hobbs, R. J., E. Higgs, and J. A. Harris. 2009. Novel ecosystems: Implications for conservation and restoration. *Trends in Ecology & Evolution* 24 (11):599–605. doi: 10.1016/j.tree.2009.05.012.

Hoegh-Guldberg, O., L. Hughes, S. McIntyre, D. B. Lindenmayer, C. Parmesan, H. P. Possingham, and C. D. Thomas. 2008. Ecology: Assisted colonization and rapid climate change. *Science* 321 (5887):345–46. doi: 10.1126/science.1157897.

Jackson, S. T., and R. J. Hobbs. 2009. Ecological restoration in the light of ecological history. *Science* 325 (5940):567–69. doi: 10.1126/science.1172977.

Jordan, W. R., and G. M. Lubick. 2011. *Making nature whole: A history of ecological restoration.* Washington, DC: Island.

Kellman, M. 1996. Redefining roles: Plant community reorganization and species preservation in fragmented systems. *Global Ecology and Biogeography Letters* 5 (3):111–16. doi: 10.2307/2997393.

Klepac, P., C. J. E. Metcalf, A. R. McLean, and K. Hampson. 2013. Towards the endgame and beyond: Complexities and challenges for the elimination of infectious diseases. *Philosophical Transactions of the Royal Society B Biological Sciences* 368(1623): 20120137. doi: 10.1098/rstb.2012.0137.

Lave, R. 2012. *Fields and streams: Stream restoration, neoliberalism, and the future of environmental science.* Athens: University of Georgia Press.

Le Breton, T. D., S. Natale, K. French, B. Gooden, and M. J. K. Ooi. 2020. Fire-adapted traits of threatened shrub species in riparian refugia: Implications for fire regime management. *Plant Ecology* 221 (1):69–81. doi: 10.1007/s11258-019-00993-2.

Leopold, A. 1949. *A Sand County almanac and sketeches here and there.* London: Oxford University Press.

Madriñán, S., A. J. Cortés, and J. E. Richardson. 2013. Páramo is the world's fastest evolving and coolest biodiversity hotspot. *Frontiers in Genetics* Vol. 4 (2): 1–7. doi: 10.3389/fgene.2013.00192.

McDonald, T., G. Gann, J. Jonson, and K. Dixon. 2016. *International standards for the practice of ecological restoration—Including principles and key concepts.* Washington, DC: Society for Ecological Restoration.

Miller, J. A., and P. Holloway. 2015. Incorporating movement in species distribution models. *Progress in Physical Geography: Earth and Environment* 39 (6):837–49. doi: 10.1177/0309133315580890.

Millington, A. C., M. A. Blumler, and U. Schickhoff, eds. 2011. *The Sage handbook of biogeography.* London: Sage.

Milly, P. C. D., J. Betancourt, M. Falkenmark, R. M. Hirsch, Z. W. Kundzewicz, D. P. Lettenmaier, and R. J. Stouffer. 2008. Climate change. Stationarity is dead: Whither water management? *Science* 319 (5863):573–74. doi: 10.1126/science.1151915.

Monbiot, G. 2014. *Feral: Rewilding the land, the sea and human life.* Chicago: University of Chicago Press.

Morueta-Holme, N., K. Engemann, P. Sandoval-Acuña, J. D. Jonas, R. M. Segnitz, and J.-C. Svenning. 2015. Strong upslope shifts in Chimborazo's vegetation over two centuries since Humboldt. *Proceedings of the National Academy of Sciences of the United States of America* 112 (41):12741–45. doi: 10.1073/pnas.1509938112.

Palmer, M. A., J. B. Zedler, and D. A. Falk. 2016. *Foundations of restoration ecology.* Washington, DC: Island.

Pörtner, H. O., and A. P. Farrell. 2008. Ecology, physiology and climate change. *Science* 322 (5902):690–92. doi: 10.1126/science.1163156.

Postigo, J. C., K. R. Young, and K. A. Crews. 2008. Change and continuity in a pastoralist community in the high Peruvian Andes. *Human Ecology* 36 (4):535–51. doi: 10.1007/s10745-008-9186-1.

Ramos-Scharrón, C. E., and E. Arima. 2019. Hurricane María's precipitation signature in Puerto Rico: A conceivable presage of rains to come. *Scientific Reports* 9 (1):15612. doi: 10.1038/s41598-019-52198-2.

Roberts, D. R., and A. Hamann. 2016. Climate refugia and migration requirements in complex landscapes. *Ecography* 39 (12):1238–46. doi: 10.1111/ecog.01998.

Scheffers, B. R., and G. Pecl. 2019. Persecuting, protecting or ignoring biodiversity under climate change. *Nature Climate Change* 9 (8):581–86. doi: 10.1038/s41558-019-0526-5.

Schleicher, J., J. G. Zaehringer, C. Fastré, B. Vira, P. Visconti, and C. Sandbrook. 2019. Protecting half of the planet could directly affect over one billion people. *Nature Sustainability* 2 (12):1094–96. doi: 10.1038/s41893-019-0423-y.

Seastedt, T. R., R. J. Hobbs, and K. N. Suding. 2008. Management of novel ecosystems: Are novel approaches required? *Frontiers in Ecology and the Environment* 6 (10):547–53. doi: 10.1890/070046.

Shapiro, E. 2015. *How to clone a mammoth: The science of de-extinction.* Princeton, NJ: Princeton University Press.

Shoreman-Ouimet, E., and H. Kopnina. 2015. Reconciling ecological and social justice to promote biodiversity conservation. *Biological Conservation* 184:320–26. doi: 10.1016/j.biocon.2015.01.030.

Siders, A. R., M. Hino, and K. J. Mach. 2019. The case for strategic and managed climate retreat. *Science* 365 (6455):761–63. doi: 10.1126/science.aax8346.

Society for Ecological Restoration International Science and Policy Working Group. 2004. *The SER international primer on ecological restoration*. Tuscon, AZ: Society for Ecological Restoration.

Svenning, J.-C., and B. Sandel. 2013. Disequilibrium vegetation dynamics under future climate change. *American Journal of Botany* 100 (7):1266–86. doi: 10.3732/ajb.1200469.

Sylvester, S. P., M. D. P. V. Sylvester, and M. Kessler. 2014. Inaccessible ledges as refuges for the natural vegetation of the high Andes. *Journal of Vegetation Science* 25 (5):1225–34. doi: 10.1111/jvs.12176.

Tovar, C., C. A. Arnillas, F. Cuesta, and W. Buytaert. 2013. Diverging responses of tropical Andean biomes under future climate conditions. *PLOS ONE* 8 (5): e63634. doi: 10.1371/journal.pone.0063634.

Urrutia, R., and M. Vuille. 2009. Climate change projections for the tropical Andes using a regional climate model: Temperature and precipitation simulations for the end of the 21st century. *Journal of Geophysical Research* 114 (D2): D02108. doi: 10.1029/2008JD011021.

Valdez, R. X., J. Kuzma, C. L. Cummings, and M. N. Peterson. 2019. Anticipating risks, governance needs, and public perceptions of de-extinction. *Journal of Responsible Innovation* 6 (2):211–31. doi: 10.1080/23299460.2019.1591145.

Whisenant, S. G. 1999. *Repairing damaged wildlands: A process-oriented, landscape-scale approach*. Cambridge, UK: Cambridge University Press.

Williams, J. W., S. T. Jackson, and J. E. Kutzbach. 2007. Projected distributions of novel and disappearing climates by 2100 AD. *Proceedings of the National Academy of Sciences of the United States of America* 104 (14):5738–42. doi: 10.1073/pnas.0606292104.

Wilson, E. O. 2016. *Half-Earth: Our planet's fight for life*. New York: Norton.

Young, K. R. 2009. Andean land use and biodiversity: Humanized landscapes in a time of change. *Annals of the Missouri Botanical Garden* 96 (3):492–507. doi: 10.3417/2008035.

Young, K. R., and J. K. Lipton. 2006. Adaptive governance and climate change in the tropical highlands of western South America. *Climatic Change* 78 (1):63–102. doi: 10.1007/s10584-006-9091-9.

Zedler, J. B., J. M. Doherty, and N. A. Miller. 2012. Shifting restoration policy to address landscape change, novel ecosystems, and monitoring. *Ecology and Society* 17(4): 1–36 doi: 10.5751/ES-05197-170436.

Zimmerer, K. S., ed. 2006. *Globalization and new geographies of conservation*. Chicago: University of Chicago Press.

Re-envisioning the Toxic Sublime: National Park Wilderness Landscapes at the Anthropocene

Nicolas T. Bergmann ⓘ and Robert M. Briwa ⓘ

The Anthropocene concept stimulates much debate among geographers. This wider conversation often neglects the role that visual imagery plays in shaping geographical imaginations of the Anthropocene. This article examines artist Hannah Rothstein's revisionist collection *National Parks 2050* to better understand the intersections of visual imagery and the Anthropocene. Rothstein's collection draws on Works Progress Administration–style artwork to visualize a bleak future. In particular, her artwork mobilizes the aesthetic of the toxic sublime. To assess *National Parks 2050*'s use of the toxic sublime, we conducted a visual analysis that found Rothstein's use of the concept innovative in two important ways. First, Rothstein's toxic sublime is derived from deeper traditions of the romantic sublime, which diverges from existing understandings of the toxic sublime as the counterpart to a technological sublime. Second, it brings two underexamined themes of the toxic sublime to the fore. The first theme is death and disappearance; the second is scale and the toxic. We argue that Rothstein's toxic sublime re-instills Burkean sublime's heightened awareness—here understood as horror and despair—into the romantic natural sublime of U.S. national park wilderness landscapes. *Key Words: Anthropocene, geographical imagination, national parks, toxic sublime, wilderness.*

The twenty-first-century emergence of the Anthropocene concept has evoked much conceptual debate among geographers (Castree 2014; Lorimer 2017; Ziegler and Kaplan 2019). Although it provides impetus for furthering empirical understandings of the climate crisis (Steffen et al. 2011), many critical scholars remain uncomfortable with the concept's power-laden narrative and human-centered terminology (Moore 2016). This wider conversation increasingly includes analysis of visual culture (Mirzoeff 2014; Demos 2017; Jacobson 2019), but visual representations of the Anthropocene— especially national park landscapes—remain understudied. Representing a field with an active tradition of visual analysis (Cosgrove 1994; Schwartz and Ryan 2003; Rose 2016), geographers are well suited to critically analyze artistic expressions of national parks within the context of the Anthropocene.

Diverse geographic park scholarship exists at many scales (Byrne and Wolch 2009). Geographers view U.S. national parks as essential environments for field research and as objects worthy of study in their own right (Dilsaver 2009). Recent works examine how national parks fit within the larger public lands system (Wilson 2020), assess relationships between national parks and wilderness preservation

(Dilsaver 2014), identify causes of racial inequality in national park usage (Weber and Sultana 2013; Davis 2019), and discuss risk and hazards within park spaces (Youngs 2018). Furthermore, national parks remain powerful shapers of national identity (Runte 2010). Because of their environmental, aesthetic, and touristic values, a great deal of visual imagery exists depicting various national park landscapes. From examining railroad promotional imagery (Wyckoff and Dilsaver 1997) to historical photographs of fire towers (Butler 2014) and souvenir postcards (Youngs 2012), geographers investigate ways in which imagery shapes perceptions of national park landscapes (Dunaway 2005).

Artist Hannah Rothstein's provocative collection of national park poster art entitled *National Parks 2050* (Figure 1) provides opportunities to critically engage with Anthropocene visual culture and to further geographic understanding of wilderness perceptions in an era of unprecedented socioecological change. Critical geographers have systematically deconstructed and denaturalized the wilderness concept (Cronon 1995). Despite revised understandings of wilderness as a deeply social and historical construction, it remains a powerful trope within the visual rhetoric of national parks (Briwa and

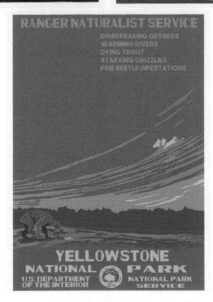

Figure 1. Hannah Rothstein's *National Parks 2050*. Copyright Hannah Rothstein, 2017. All rights reserved. Used with permission.

Bergmann 2020). Rothstein's *National Parks 2050* challenges this paradigm. Far from invoking cherished notions of sacred wilderness, Rothstein's poster art represents national park landscapes as toxic wastelands filled with starving grizzlies, burned-out redwoods, and melted permafrost.

This article places Rothstein's collection within larger scholarly discussions involving environmental reform, national park wilderness, and the aesthetic tradition of the sublime. Beginning with understandings of the toxic sublime related to landscape photography and industrial ruins (Peeples 2011; Gatlin 2015; Ray 2016; Kane 2018; Balayannis 2019), we extend its application to national park landscapes. Employing visual analysis, we argue that Rothstein's collection—through its close association between the toxic and natural sublime—challenges previous interpretations. Specifically, we find that in the context of *National Parks 2050*, the toxic sublime is not defined by tensions between toxicity and awe-inspiring magnificence (Peeples 2011) but instead by its ability to re-instill a Burkean sense of horror, fear, and despair into the romanticized natural sublime of the late nineteenth, twentieth, and early twenty-first centuries. Through this analysis, we complicate existing understandings of the toxic sublime and demonstrate how Rothstein's artwork destabilizes popular understandings of national park wilderness landscapes.

(Re)Viewing the Anthropocene's Images

Scholarly analysis of Anthropocene visual imagery occurs in several ways. Perhaps the most explicit type is an image categorization known as "Anthropocene visualization." Demos (2017) characterizes Anthropocene visualization through its general reliance on cutting-edge technologies to visualize previously invisible data. Instead of employing more traditional media such as photography, he argues that Anthropocene visualization relies on high-resolution satellite imagery. Thus, technological advancement and innovation allows for visualization of the material world's previously imperceptible aspects. Demos claims, however, that producers of these visualizations provide "hyper-legible" images that already interpret complex data for their viewers. He finds this especially problematic because these visualizations are "carefully edited in order to show generally positive examples of modern development" associated with the Anthropocene's onset (Demos 2017, 16–17).

Ultimately, Demos exposes Anthropocene visualization's role in perpetuating colonial values associated with capitalism, which he suggests is primarily responsible for the advent of this problematic new geologic epoch.

Another strain of Anthropocene visual analysis concerns itself explicitly with climate change. This research primarily assesses the effects of visual imagery on climate change communication (DiFrancesco and Young 2011; Yusoff and Gabrys 2011; O'Neill 2013; O'Neill and Smith 2014). This approach examines visual imagery across a wide variety of media sources. Such scholarship advances understanding of climate change imagery's content and framing. Another style of climate change visual analysis considers ruins. Herrmann's (2019) critique of the "fallen house" photograph in Shishmaref, Alaska, integrates wider discussions of climate change visual communication with aesthetic conceptualizations.

A third category of Anthropocene visual analysis centered on environmental reform relies on the closely related concept of the sublime. For instance, Dunaway (2018) uses "ecological sublime" in reference to Subhankar Banerjee's photograph *Snow Geese 1*, whereas Demos (2017) references "apocalyptic sublime" when describing a photograph of the Deepwater Horizon oil platform engulfed in flames. The most prevalent use of the sublime within Anthropocene visual analysis, however, is the "toxic sublime" (Peeples 2011; Gatlin 2015; Ray 2016). Used to make sense of a new tradition of landscape photography (particularly work by Edward Burtynsky) and associated with industrial sites of pollution, the toxic sublime is most clearly defined as "the tensions that arise from recognizing the toxicity of a place, object, or situation, while simultaneously appreciating its mystery, magnificence and ability to inspire awe" (Peeples 2011, 375). Importantly, this understanding of Anthropocene visual imagery draws from a deep aesthetic tradition.

The concept of the sublime has roots in the first century CE, when an unknown Greek author dubbed Longinus wrote of "height" in literature (Malm 2000). Late Renaissance scholars debated Longinian sublime within the contexts of creative literary practice (Malm 2000). By the eighteenth century, European philosophers shifted the sublime's meanings to "heightened" aesthetic experiences derived from encountering objects in the world. For continental thinkers such as Kant, sublime experience emphasized the limitlessness of human thought and action. In England, Edmund Burke redefined Longinian sublime to describe experiences that force reckoning with human limitedness. Burkean sublime incites both terror and delight and galvanizes individuals to action (Ryan 2001). Both continental and English sublime increasingly coalesced around shared motifs characterizing objects and places thought to cultivate a sublime state of mind (Malm 2000).

These objects—plummeting waterfalls, rugged mountains, vast seascapes—became sources of the natural sublime and were identified as "those rare places on earth where one had more chance than elsewhere to glimpse the face of God" (Cronon 1995, 73). This iteration of the sublime came to the United States during the nineteenth century. Sublime and wilderness were artistically connected through Northeastern intellectuals. Later, artists and literati working within the U.S. West transformed these connections. For example, the writings of John Muir and paintings by Albert Bierstadt and Thomas Moran created a distinct tradition of the natural sublime associated with national park wilderness landscapes (Allen 1992). Instead of depicting astonishing horror associated with European sublime, these artists presented national park wilderness landscapes where the sublime was both "tamed" and "domesticated" (Cronon 1995, 75). The preservation of national park wilderness landscapes is irrevocably tied to this romanticized version of the natural sublime (Nash 2001).

Sublime experiences also expanded in other directions. During the nineteenth century, technological advancements extended the sublime beyond nature and into human-built environments. Although sharing a similar intellectual lineage, the technological sublime developed into a separate aesthetic tradition as industrial advancements such as steamboats and railroads transformed relationships between humans and nature (Nye 1995). Current scholarship considers the technological sublime a key source of the toxic sublime. Specifically, Peeples (2011) visualizes the toxic sublime as "counterpart and required 'other' to the technological sublime," because "[i]t shares with the technological sublime a marvel at human accomplishments," but "acts to counter that marvel with alarm for the immensity of destruction one witnesses" (380).

We propose that *National Parks 2050* forces scholars to confront stronger relationships between natural and toxic sublimes. Although the technological sublime has a derivative presence in her work,

Rothstein's depictions of toxic national park landscapes visually play off conventions (e.g., color, scale, and abstractness) associated with natural sublime. Furthermore, Rothstein's work represents a toxic wilderness infused with Burkean sublime's heightened aesthetic experiences. Far from the romantic natural sublime of Muir or Bierstadt, Rothstein re-instills terror, horror, and despair. In the following sections, we identify specific visual strategies that Rothstein employs to complicate existing understandings of the toxic sublime as a ruined landscape of industrial extraction and pollution.

Case Study: *National Parks 2050*

National Parks 2050 is a collection inspired by Works Progress Administration (WPA)-style depictions of national park landscapes. This early-twentieth-century aesthetic proliferated during the 1930s, when the New Deal's WPA Federal Art Project facilitated the creation of approximately 35,000 unique poster images promoting many aspects of U.S. culture (DeNoon 1987). Several artists painted poster imagery promoting visitation to national park landscapes. This occurred in two separate collections: (1) See America and (2) National Park Service (Pillen 2008). Specifically, the WPA produced fourteen posters in its National Park Service collection (Bennett 2016). Primarily designed by WPA artist Chester Don Powell and screen printer Dale Miller, these posters largely disappeared until the 1970s. At that time, Doug Leen—a seasonal ranger at Grand Teton National Park—salvaged a poster from a burn pile and then later tracked down a set of black-and-white negatives. After this discovery, Leen formed a company in 1993—Ranger Doug's Enterprises (hereafter Ranger Doug's)—that created reproductions of the fourteen original National Park Service collection poster art images and two original reproductions from the See America collection. The company also created WPA-style designs for many other national parks not included in the original WPA artwork (Ranger Doug's Enterprises 2019).

Rothstein based *National Parks 2050* on seven of Ranger Doug's productions. Two of Ranger Doug's poster art images—Yellowstone and Great Smokey National Parks—are reproductions of WPA originals, and five poster art images—Crater Lake, Mount McKinley (hereafter Denali), Everglades, Redwoods, and Saguaro—are Ranger Doug's originals created in the WPA style. Although Rothstein was initially unaware that these images were not WPA originals, Ranger Doug's expanded collection allowed her to choose from a more diverse selection of national park landscapes (H. Rothstein, personal communication, 10 April 2020).

Furthermore, as both an environmental activist and independent artist, Rothstein created *National Parks 2050* with the goal of increasing public awareness and action toward climate change. Using a variety of digital tools, Rothstein began work on the collection shortly after climate change skeptic Donald Trump was elected president of the United States in November 2016. She completed the collection in time for Earth Day 2017. Primarily disseminated through online platforms and media coverage, her Web site's *National Parks 2050* page received several hundred thousand clicks in the first few weeks after its release. Although it is difficult to quantify public knowledge of the collection, it is clear that the collection's online presence and subsequent media coverage mean that it is engaged with by thousands. In our communications, Rothstein estimated that two out of ten random strangers might know something about it or recognize it. In addition to her personal Web site, Rothstein noted that *National Parks 2050* garnered widespread media attention. Given its wide distribution and media coverage, *National Parks 2050* undoubtedly has shaped popular geographical understandings of national parks (H. Rothstein, personal communication, 10 April 2020).

In the following sections, we describe selected examples from Rothstein's work and establish how they represent images of a toxic sublime. We build on and complicate previous conceptualizations of the toxic sublime, including Peeples's (2011) identification of tensions at work in producing the toxic sublime and Gatlin's (2015) discussion of a toxic sublime aesthetic. *National Parks 2050* presents a toxic sublime that emerges from the natural sublime (rather than the technological sublime) and demonstrates that the toxic sublime in the Anthropocene relies on two underexplored themes within literature examining the sublime: (1) dying and disappearance and (2) scale and the toxic.

Dying and Disappearance

Davies (2018) suggests that conceptualizing linkages between death and toxicity is challenging because of toxicity's invisibility and shifting temporalities that range from the quick to the gradual. Existing scholarship does not directly treat visual representations of dying or disappearance related to toxicity. For example, Peeples (2011) suggests that absence—particularly of victims—within the toxic sublime provides space for viewers to place themselves within the image as casualties. *National Parks 2050* depicts death and dying directly, as in the image of a starving grizzly in Yellowstone National Park. The collection also captures disappearances, or what Peeples (2011) called "monstrous absences" (384). To highlight how the collection demonstrates monstrous absence, we pair its Everglades National Park with Ranger Doug's more traditional WPA-era inspired counterpart (Figure 2).

WPA-era imagery typically develops place images through portraying vibrant and healthy landscapes. Ranger Doug's Everglades National Park, for example, uses charismatic subtropical flora and fauna to present a delightful, more beautiful vision of the romantic sublime. Egrets, great blue herons, and ibises hunt in a rich wetland, and large alligators sun themselves beneath a large and healthy tree adorned by orchids. Mangroves provide a verdant subtropical background beneath a bright sky (Figure 2A). In contrast, Rothstein's revisionist artwork paints a different picture that draws on the toxic sublime's trope of monstrous absence. The birds and alligators have vanished, and the tree has been reduced to a blackened stump. Beyond, the lines between dark skies and blackened mangrove forests no longer matter. Although the water retains its blue and green coloring from Ranger Doug's WPA-style art, Rothstein's Everglades landscape is desolate and uninhabitable, a subtle nod to the invisibility of toxicity (Figure 2B).

Keeping with the traditional WPA-style approach, *National Parks 2050* also makes use of promotional text. In contrast to photographic representations of the toxic sublime, which are devoid of text, Rothstein's text (often modified and twisted from Ranger Doug's posters) makes toxicity, dying, and disappearance more legible. In her Everglades image, Rothstein exhorts viewers to visit dying mangroves and dead freshwater pines. Paradoxes occur: How does one visit extinct tropical orchids or lost alligator habitat? In these instances, text presents impossible realities of the toxic sublime and makes demands on Rothstein's viewers to come to grip with themes of horrifying absence and loss.

A)

B)

Figure 2. Everglades National Park. (A) Ranger Doug's Everglades National Park, Doug Leen and Brian Maebius (artists), 2011. Copyright Douglas Verner Leen. All rights reserved. Used with permission. (B) Everglades National Park 2050, Hannah Rothstein (artist), 2017. Copyright Hannah Rothstein, 2017. All rights reserved. Used with permission.

Furthermore, the contrast between Rothstein's toxic image of Yellowstone National Park and its nontoxic counterpart highlights dying as a prominent theme (Figure 3). In Ranger Doug's restoration of the original WPA poster, Old Faithful erupts into an evening sky streaked with the orange light of a Rocky Mountain sunset (Figure 3B). Although the color palette from Rothstein's image is more funereal and darkened than Ranger Doug's recoloring (Figure 3C), it stands in even higher contrast to the salubrious greens and blues of the original image (Figure 3A). In Rothstein's Yellowstone, Old Faithful sputters weakly. A starving grizzly, ribs and its formerly powerful frame now standing out against its ragged fur, investigates the former geothermal icon. The text hanging above the scene tells us that Yellowstone National Park is where dying trout and starving grizzlies struggle to eke out their lives.

Multiple dimensions of the Anthropocene's toxic sublime are articulated through Rothstein's presentation of features within the image. The dying grizzly and its emaciated frame are a palette swap of the trope of starving polar bears, often viewed as the "poster boys" of climate change in mass media (Herrmann 2019, 861). Rothstein's image presents the slow, inevitable dying of a charismatic creature, infusing the scene with a sense of horror and despair. The image invokes other dimensions of the Anthropocene. Old Faithful is clearly an example of the dying geysers named in the poster's text, but it is also symbolically representative of wider disruptions characterizing the Anthropocene. No longer is it faithfully following long-familiar patterns of eruption and rest. Instead, it is implied to have reached a new point of no return, and one that is clearly at odds with previous understandings of human–nature relations.

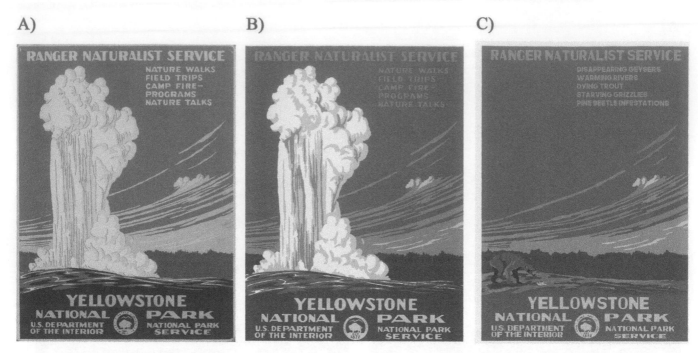

Figure 3. Yellowstone National Park. (A) WPA's original poster art of Yellowstone National Park, C. Don Powell (artist), ca. 1938. Retrieved from the Library of Congress online database. (B) Ranger Doug's derivative of the WPA original, C. Don Powell (artist), ca. 1938. Copyright Douglas Verner Leen as derivative art, 1995. All rights reserved. Used with permission. (C) Yellowstone National Park 2050, Hannah Rothstein (artist), 2017. Copyright Hannah Rothstein, 2017. All rights reserved. Used with permission. WPA = Works Progress Administration.

Scale and the Toxic

Seeing through scale is a key strength of geographical thinking (Hanson 2004). Previous toxic sublime scholarship fails to completely capture the utility of geographic scale as a framing device. The full force of the Anthropocene cannot be recognized without reference to the interlocking and immense scales at which it operates. These geographic scales range from site-specific to global and also include a temporal element. Rather than focusing on the vastness of geologic time described in other versions of the sublime (Mackay 1998), however, the Anthropocene's toxic sublime relies on the astonishing rates at which irrevocable processes (e.g., extinction) occur.

One way the toxic sublime operates is through notions of magnitude. For Peeples (2011), photographers convey magnitude through deft manipulation of subject matter, selective cropping of images, and immense presentations in oversized formats. Peeples argues that photographer Edward Burtynsky conveys magnitude by presenting strategically cropped images of enormous industrial sites (principally strip-mined mountains, mining shafts, and tailings ponds) on oversized printing paper. Rothstein draws on these same strategies to present magnitude. Her choice of subject matter (the wilderness landscapes of national parks) and other presentation strategies (e.g., text), though, brings an element of scale to toxic sublime that is conspicuously absent from simply attributing it to magnitude alone.

Perhaps no other image of Rothstein's collection conveys magnitude quite like Denali National Park and Preserve (Figure 4A). Earth tones dominate. Denali fills the frame, its snowless slopes a swathe of dark tans and grays. Its hulking presence appears as a massif rather than a clearly defined peak. Earth—bulky, solid, heavy as coal—frames the image, surrounding a lake that mirrors the ochre sky above. In the foreground sits a skull, tying the image into the theme of dying and disappearance in the toxic sublime. Across the top of the image, text invites viewers to visit melted permafrost, snowless peaks, and vanished tundra. This image of Denali aligns with photographic strategies of conveying the toxic sublime through selective framing.

Rothstein's artwork also develops the toxic sublime by subtly connecting it to an interlocking set of geographic scales. She develops presentation strategies to introduce a vision of the toxic that captures not only site scale dynamics but also those operating

A)

B)

Figure 4. Scale and the toxic in *National Parks 2050*. (A) Mount McKinley (Denali) National Park 2050, Hannah Rothstein (artist), 2017. Copyright Hannah Rothstein, 2017. All rights reserved. Used with permission. (B) Saguaro National Park 2050, Hannah Rothstein (artist), 2017. Copyright Hannah Rothstein, 2017. All rights reserved. Used with permission.

at the scale of landscape, region, and globe. In the Denali image, an iconic portion of the park is cropped, and the viewer is left to assume that the peak extends beyond the frame in either direction. Text helps convey geographic scales as well. It invites viewers to landscape features found throughout the park and encourages them to visit snowless peaks (without specific names or otherwise defining traits) and engage with other landscape features (e.g., tundra and permafrost) that transcend park boundaries. Through these interscalar features, viewers' understanding of the park's environmental degradation is expanded to include recognition of human impacts on wider Arctic and sub-Arctic geographies.

Similarly, Rothstein's Saguaro National Park conveys a toxic sublime at several geographic scales (Figure 4B). Two riders traverse a spoiled desert landscape beset by invasive bufflegrass. Nameless, this landscape might be found anywhere within the park's boundaries. Text, meanwhile, tells us that drought has struck the U.S. Southwest. Its associated species loss presumably ignores the park's artificial boundaries. Hanging above the desiccated landscape is a grim sky. Its sickly brown and yellow hues are a visual reminder that atmospheric conditions are neither static nor local phenomena. Rather, they are vulnerable to human activities at many geographic scales.

A sense of sublime is intimately bound into notions of time (Mackay 1998; White 2012). Within the current literature on the toxic sublime, however, the relationship between toxic sublime and its temporal dimensions remains largely unexamined. The romantic natural sublime traditionally relies on geologic time (White 2012). McPhee (1980) famously called this "deep time" in *Basin and Range*. Deep time is the planetary metronome through which Earth system processes operate. In this planetary timescale, the heightened experience of the romantic sublime is in part produced out of humans' inability to directly experience the full rhythm of changes to Earth systems and processes. Our lives are too short. Coming to grips with the concept of deep time imparts a sense of astonishment and marvel. In contrast, *National Parks 2050* imparts astonishment and horror at the massively accelerated pace of the Anthropocene. These Earth system processes—whether climatic changes, or ocean acidification, or even geological composition—are now occurring at hyperaccelerated rates in timescales we actively live and experience (Steffen et al. 2011; Menga and Davies 2019). The problem—and the horror—lies in the realization that the changes we have put into effect are now beyond our individual control. Yet we will see those changes come.

Conclusion

Extending analysis of the toxic sublime to depictions of national park wilderness landscapes enables a deeper understanding of Anthropocene visualizations and furthers theorization of the toxic sublime itself. Whereas Peeples (2011) defines the toxic sublime in relation to the technological sublime and through aesthetic tensions between toxic and awe-inspiring magnificence, we argue for positioning the toxic sublime primarily as a mechanism for re-instilling the Burkean sense of heightened awareness—through horror and despair—into the romantic natural sublime. The Anthropocene provides an opportunity to rethink twentieth-century consequences associated with a romantic natural sublime as pleasant and benign. Ironically, Rothstein's *National Parks 2050*—through its manipulation of dying and disappearance as well as scale and the toxic—creates a vision of future national park wilderness landscapes more closely associated with a premodern wilderness sense of fear, foreboding, and desolation than contemporary notions of beauty.

This article takes an initial step toward extending toxic sublime scholarship beyond a narrow focus on industrial landscape photography and the technological sublime. Further opportunities remain to expand understandings of the toxic sublime. For instance, interested scholars should seek to include other representational mediums in their analyses as well as landscapes beyond industrial and national park wildernesses. Moreover, clearer understandings of emotional responses to toxic sublime imagery are needed. Although certain climate change communication literature evaluates the utility of visual imagery for stimulating salience and efficacy (O'Neill and Nicholson-Cole 2009; O'Neill et al. 2013), we are not aware of any comparable studies specific to the toxic sublime. With respect to Rothstein's use of poster art to re-instill horror and despair into national park wilderness landscapes, further evaluation is needed to determine whether or not this is an effective strategy for stimulating social and political responses to threats from climate change.

It is also critically important to remember that visual images are agents actively shaping environmental perceptions (Dunaway 2018). *National Parks 2050* paints a provocative picture of potential future consequences for national park wilderness landscapes if climate change is not adequately addressed. Its singular focus, however, does privilege environmental problems at the expense of other challenges facing national parks. From sexual harassment transgressions within the National Park Service to the construction and reification of national park spaces for a predominately white, middle-class population, Rothstein's work does not render visible other significant social issues facing national parks. Ultimately, however, the stunning images of *National Parks 2050* do make perceptible the invisibility of climate change, do challenge the myth of national parks as an unimpaired resource for the enjoyment of future generations, and do complicate contemporary romantic understandings of wilderness through depicting a possible future characterized by the toxic sublime.

Acknowledgments

We extend our gratitude to editors David Butler and Jennifer Cassidento as well as the two anonymous reviewers for their feedback and support. We also thank Frank Bergmann for his copyediting prowess. Generous support from Doug Leen, Angie Bunker, and Ranger Doug's Enterprises allowed for the use of two images. We owe our greatest debt to artist Hannah Rothstein, who generously granted us permission to use her artwork and shared important details involved with the creation and production of *National Parks 2050*.

ORCID

Nicolas T. Bergmann ⓘ http://orcid.org/0000-0002-0850-0021

Robert M. Briwa ⓘ http://orcid.org/0000-0002-5915-4114

References

Allen, J. 1992. Horizons of the sublime: The invention of the romantic west. *Journal of Historical Geography* 18 (1):27–40. doi: 10.1016/0305-7488(92)90274-D.

Balayannis, A. 2019. Routine exposures: Reimaging the visual politics of hazardous sites. *GeoHumanities* 5 (2):572–90. doi: 10.1080/2373566X.2019.1624189.

Bennett, J. 2016. The forgotten history of those iconic national park posters. *Popular Mechanics*, August 25. Accessed November 23, 2019. https://www.popularmechanics.com/adventure/outdoors/a22536/national-parks-posters/.

Briwa, R., and N. Bergmann. 2020. Picturing the national parks through postcards at the National Park Service Centennial. *FOCUS on Geography* 63. doi: 10.21690/foge/2020.63.1f.

Butler, D. R. 2014. *Fire lookouts of Glacier National Park*. Charleston, SC: Arcadia.

Byrne, J., and J. Wolch. 2009. Nature, race, and parks: Past research and future directions for geographic research. *Progress in Human Geography* 33 (6):743–65. doi: 10.1177/0309132509103156.

Castree, N. 2014. The Anthropocene and geography 1: The back story. *Geography Compass* 8 (7):436–49. doi: 10.1111/gec3.12141.

Cosgrove, D. 1994. Contested global visions: *One-world, whole-Earth*, and the *Apollo* space photographs. *Annals of the Association of American Geographers* 84 (2):270–94. doi: 10.1111/j.1467-8306.1994.tb01738.x.

Cronon, W. 1995. The trouble with wilderness; or, getting back to the wrong nature. In *Uncommon ground: Rethinking the human place in nature*, ed. W. Cronon, 69–90. New York: Norton.

Davies, T. 2018. Toxic space and time: Slow violence, necropolitics, and petrochemical pollution. *Annals of the American Association of Geographers* 108 (6):1537–53. doi: 10.1080/24694452.2018.1470924.

Davis, J. 2019. Black faces, black spaces: Rethinking African American underrepresentation in wildland spaces and outdoor recreation. *Environment and Planning E: Nature and Space* 2 (1):89–109. doi: 10.1177/2514848618817480.

Demos, T. 2017. *Against the Anthropocene: Visual culture and the environment today*. Cambridge, MA: The MIT Press.

DeNoon, C. 1987. *Posters of the WPA*. Los Angeles: Wheatley.

DiFrancesco, D., and N. Young. 2011. Seeing climate change: The visual construction of global warming in Canadian national print media. *Cultural Geographies* 18 (4):517–36. doi: 10.1177/1474474010382072.

Dilsaver, L. 2009. Research perspectives on national parks. *Geographical Review* 99 (2):268–78. doi: 10.1111/j.1931-0846.2009.tb00430.x.

Dilsaver, L. 2014. Preserving lands for future generations: The U.S. experience. In *North American odyssey: Historical geographies for the twenty-first century*, ed. C. Colten and G. Buckley, 177–93. Lanham, MD: Rowman & Littlefield.

Dunaway, F. 2005. *Natural visions: The power of images in American environmental reform*. Chicago: The University of Chicago Press.

Dunaway, F. 2018. Reconsidering the sublime: Images and imaginative geographies in American environmental history. In *The American environment revisited: Environmental historical geographies of the United States*, ed. G. Buckley and Y. Youngs, 277–94. Lanham, MD: Rowman & Littlefield.

Gatlin, J. 2015. Toxic sublimity and the crisis of human perception: Rethinking aesthetic, documentary, and political appeals in contemporary wasteland photography. *Interdisciplinary Studies in Literature and Environment* 22 (4):717–41. doi: 10.1093/isle/isv032.

Hanson, S. 2004. Who are "we?" An important question for geography's future. *Annals of the Association of American Geographers* 94 (4):715–22.

Herrmann, V. 2019. Rural ruins in America's climate change story: Photojournalism, perception, and agency in Shishmaref, Alaska. *Annals of the American Association of Geographers* 109 (3):857–74. doi: 10.1080/24694452.2018.1525272.

Jacobson, B. 2019. *The Shadow of Progress* and the cultural markers of the Anthropocene. *Environmental History* 24 (1):158–72. doi: 10.1093/envhis/emy123.

Kane, C. 2018. The toxic sublime: Landscape photography and data visualization. *Theory, Culture & Society* 35 (3):121–47. doi: 10.1177/0263276417745671.

Lorimer, J. 2017. The Anthropo-scene: A guide for the perplexed. *Social Studies of Science* 47 (1):117–42. doi: 10.1177/0306312716671039.

Mackay, M. 1998. Singularity and the sublime in Australian landscape representation. *Literature & Aesthetics* 8:113–27.

Malm, M. 2000. On the technique of the sublime. *Comparative Literature* 52 (1):1–10. doi: 10.2307/1771516.

McPhee, J. 1980. *Basin and range*. New York: Farrar, Strauss, and Giroux.

Menga, F., and D. Davies. 2019. Apocalypse yesterday: Posthumanism and comics in the Anthropocene. *ENE: Nature and Space* 1–25. doi: 10.1177/2514848619883468.

Mirzoeff, N. 2014. Visualizing the Anthropocene. *Public Culture* 26 (2):213–32. doi: 10.1215/08992363-2392039.

Moore, J., ed. 2016. *Anthropocene or Capitalocene? Nature, history, and the crisis of capitalism*. Oakland, CA: PM Press.

Nash, R. 2001. *Wilderness and the American mind*. 4th ed. New Haven, CT: Yale University Press. doi: 10.1086/ahr/73.5.1612.

Nye, D. 1995. *American technological sublime*. Cambridge, MA: MIT Press. doi: 10.1086/ahr/101.2.550.

O'Neill, S. 2013. Image matters: Climate change imagery in U.S., UK, and Australian newspapers. *Geoforum* 49:10–19. doi: 10.1016/j.geoforum.2013.04.030.

O'Neill, S., M. Boykoff, S. Niemeyer, and S. Day. 2013. On the use of imagery for climate change engagement. *Global Environmental Change* 23 (2):413–21. doi: 10.1016/j.gloenvcha.2012.11.006.

O'Neill, S., and S. Nicholson-Cole. 2009. "Fear won't do it": Promoting positive engagement with climate change through visual and iconic representations. *Science Communication* 30 (3):355–79. doi: 10.1177/1075547008329201.

O'Neill, S. J., and N. Smith. 2014. Climate change and visual imagery. *Wiley Interdisciplinary Reviews: Climate Change* 5 (1):73–87. doi: 10.1002/wcc.249.

Peeples, J. 2011. Toxic sublime: Imaging contaminated landscapes. *Environmental Communication* 5 (4):373–92. doi: 10.1080/17524032.2011.616516.

Pillen, C. 2008. See America: WPA posters and the mapping of a New Deal democracy. *The Journal of American Culture* 31 (1):49–65. doi: 10.1111/j.1542-734X.2008.00663.x.

Ranger Doug's Enterprises. 2019. Home page. Accessed November 23, 2019. https://www.rangerdoug.com.

Ray, S. 2016. Environmental justice, vital materiality, and the toxic sublime in Edward Burtynsky's *Manufactured Landscapes*. *GeoHumanities* 2 (1):203–19. doi: 10.1080/2373566X.2016.1167615.

Rose, G. 2016. *Visual methodologies: An introduction to researching with visual materials*. 4th ed. London: Sage.

Runte, A. 2010. *National parks: The American experience*. 4th ed. Lanham, MD: Taylor Trade.

Ryan, V. 2001. The physiological sublime: Burke's critique of reason. *Journal of the History of Ideas* 62 (2):265–79. doi: 10.2307/3654358.

Schwartz, J., and J. Ryan, eds. 2003. *Picturing place: Photography and the geographical imagination*. London: I. B. Tauris.

Steffen, W., J. Grinevald, P. Crutzen, and J. McNeill. 2011. The Anthropocene: Conceptual and historical perspectives. *Philosophical Transactions: Series A. Mathematical, Physical, and Engineering Sciences* 369 (1938):842–67. doi: 10.1098/rsta.2010.0327.

Weber, J., and S. Sultana. 2013. Why do so few minority people visit national parks? Visitation and the accessibility of "America's best idea." *Annals of the Association of American Geographers* 103 (3):437–64. doi: 10.1080/00045608.2012.689240.

White, P. 2012. Darwin, Concepción, and the geological sublime. *Science in Context* 25 (1):49–71. doi: 10.1017/S0269889711000299.

Wilson, R. 2020. *America's public lands: From Yellowstone to Smokey Bear and beyond*. 2nd ed. Lanham, MD: Rowman & Littlefield.

Wyckoff, W., and L. Dilsaver. 1997. Promotional imagery of Glacier National Park. *Geographical Review* 87 (1):1–26. doi: 10.2307/215655.

Youngs, Y. 2012. Editing nature in Grand Canyon National Park postcards. *Geographical Review* 102 (4):486–509. doi: 10.1111/j.1931-0846.2012.00171.x.

Youngs, Y., 2018. Wild, unpredictable, and dangerous: A historical geography of hazards and risks in U.S. national parks. In *The American environment revisited: Environmental historical geographies of the United States*, ed. G. Buckley and Y. Youngs, 59–80. Lanham, MD: Rowman & Littlefield.

Yusoff, K., and J. Gabrys. 2011. Climate change and the imagination. *Wiley Interdisciplinary Reviews: Climate Change* 2 (4):516–34. doi: 10.1002/wcc.117.

Ziegler, S., and D. Kaplan. 2019. Forum on the Anthropocene. *Geographical Review* 109 (2):249–51. doi: 10.1111/gere.12336.

Climate Necropolitics: Ecological Civilization and the Distributive Geographies of Extractive Violence in the Anthropocene

Meredith J. DeBoom

The declaration of the Anthropocene reflects the magnitude of human-caused planetary violence, but it also risks disguising the inequitable geographies of responsibility and sacrifice that underlie its designation. Similarly, many existing strategies for climate change mitigation, including the development of low-carbon energy, are critical to reducing carbon emissions and yet simultaneously risk deepening extractive violence against marginalized communities. If the uneven distribution of historical and contemporary climate violence is not recognized and redressed, climate change solutions may increase the burdens borne by the very people, places, and environments expected to experience some of the worst effects of climate change itself. To aid in identifying and analyzing the distributive geographies of geo-power capable of facilitating this perverse outcome, this article develops a theoretical framework –climate necropolitics –for revealing the multiscalar processes, practices, discourses, and logics through which Anthropocenic imaginaries can be used to render extractive violence legitimate in the name of climate change response. Drawing on field work using multiple methods, I illustrate the applied value of climate necropolitics through a case study of the Chinese Communist Party's Ecological Civilization. The analysis reveals how the utopian "green" vision of Ecological Civilization, as promoted by both Chinese and Namibian state actors, has been used to legitimate intensified extractive violence against minority communities living near uranium mines in Namibia. I conclude by discussing how geographers can use multiscalar frameworks like climate necropolitics to develop integrated analyses of the uneven distribution of both social and environmental violence in the Anthropocene. *Key Words:* Anthropocene, climate change, extraction, just transition, necropolitics.

Epochal declarations like that of the Anthropocene have transformative potential. In acknowledging humans as a climatic, biotic, and geologic force (Crutzen and Stoermer 2000), the Anthropocene recognizes the magnitude of human-caused environmental violence—manifesting most notably as climate change—that has become so banal as to be rendered nearly invisible. Understanding and mitigating this violence will require geophysical and environmental research as well as technical and practical innovation across a wide range of fields and professions. Such approaches alone, however, are not sufficient to rewrite the entrenched geographies of violence that have impelled the declaration of this new epoch. Beyond calling attention to the violence enacted by humans upon the Earth, the Anthropocene's declaration prompts the interrogation of the social relations –including racial capitalism (Robinson [2000] 1983; Gilmore 2007), statist geopolitics (Grove 2019), and the human-environment binary (Swyngedouw 2015) –through which humans enact violence not only against the planet, but also against one another.

History suggests that the realization of lasting transformation from formal declarations of epochal change is far from guaranteed. The formal end of colonialism, for example, signaled both the emergence of a new era and the continuation of violence by other means and new (as well as old) actors. Like the human-driven violence that has caused fundamental changes to the Earth's ecosystems and atmosphere, human-embodied climate violence is unlikely to be resolved by the mere declaration of the Anthropocene. To the contrary, by lumping all of humanity together as a "singular undifferentiated force" (Yusoff 2013, 782), the Anthropocene risks enabling collective amnesia about the uneven historical responsibilities for climate change (Chakrabarty 2012) and obscuring its uneven implications. Its planetary declaration can also promote the false assumption that "all men are potentially homines sacri" (Agamben [1995] 1998, 84), ignoring the reality that not all of humanity is likely to be branded as equally sacrificial in the name of righting past and present planetary wrongs.

The Anthropocene's possibilities and pitfalls as a scalar project have received significant attention from geographers (McKittrick 2013; Castree 2014; Braun et al. 2015; Derickson and MacKinnon 2015; H. Davis and Todd 2017; Pulido 2018; J. Davis et al. 2019; Yusoff 2019; Dalby 2020). There has been less attention to how more specific imaginaries associated with the Anthropocene's declaration might affect existing distributions of violence associated with resource extraction. From the mining of rare earth elements for solar panels and electric vehicles (Klinger 2017) to the enclosure of land for biofuel plantations (Fairhead, Leach, and Scoones 2013), climate change mitigation strategies often rely on existing geographies of extractive violence and marginalization (Sultana 2014). Many of those most affected by climate change are also among the most affected by intensified demands for the raw materials of "green" energy. Far from facilitating an end to socioenvironmental violence, "anthropocenic imaginaries"—a term I use here, in an extension of the term's introduction by Reszitnyk (2015) and development by Mostafanezhad and Norum (2019), to refer not only to visions of environmental apocalypse but also to visions of environmental salvation—could be used to justify such violence as long as it facilitates reduced carbon emissions.

This article introduces *climate necropolitics* as a theoretical framework for analyzing the implications of Anthropocenic imaginaries for the distribution of climate change–related violence and for identifying the multiscalar processes, practices, discourses, and logics through which such imaginaries can be used to render violence legitimate in the name of climate change response. Climate necropolitics integrates Mbembe's (2003, 2019) concept of necropower, the capacity to "make die and let live," with geographic scholarship on the spatiality of violence (Pratt 2005; Gilmore 2007; Yates 2011; Tyner 2019) and the scalar (geo)politics of the Anthropocene (O'Lear and Dalby 2015; Mann and Wainwright 2018). Although I have focused this article on climate change mitigation broadly and nuclear energy in particular, I anticipate that climate necropolitics could also be applied to analyze the distributive geographies and legitimation strategies of other forms of climate violence, including those associated with

"climate security" (O'Lear and Dalby 2015), "green" militarization (Bigger and Neimark 2017), adaptation (Thomas and Warner 2019), and vulnerability (Ribot 2014). I illustrate the applied value of climate necropolitics through a case study of Chinese state-based investments in uranium mining in Namibia. This analysis reveals both the sacrificial geographies that underlie the Chinese Communist Party's (CCP) utopian Anthropocenic imaginary of Ecological Civilization and the logics and discourses through which such sacrificial violence is rendered legitimate in the name of climate change mitigation. Before turning to theory and the case study, I begin with a discussion of the broader, intertwined geographies of climate change and extractive violence in Africa.

Climate Change and Extractive Violence in Africa

Nowhere is the possibility that responses to climate change might reinforce rather than redress the violent social relations underlying the declaration of the Anthropocene clearer than in Africa. Although there is no singular "African Anthropocene" (Hecht 2018), it is possible to speak to patterns. A historical sacrifice zone par excellence, the continent is a site of simultaneous climate crisis and solution. Africans contributed just 4 percent of global carbon dioxide emissions in 2018, a figure that declines to roughly 1.5 percent after excluding South Africa and the six countries of North Africa (Global Carbon Project 2018). Despite the miniscule contributions of the vast majority of Africans to climate change, Africans are expected to experience some of its worst effects. Fears that "climate wars" will engulf the continent often stray into environmental determinism, but the high levels of inequality and poverty, rapid population growth, political characteristics, and reliance on food imports, among other factors, that characterize many African countries are certainly likely to increase the socioecological strains posed by climate change in the coming years (Raleigh, Linke, and O'Loughlin 2014; Abrahams and Carr 2017).

The climate strains faced by Africans are not limited to vulnerability, adaptation, and conflict. Many of the largest sources of raw materials essential to current climate change mitigation strategies are also located in Africa. The Democratic Republic of the Congo represents roughly 60 percent of global production of cobalt, a key component in electric vehicle batteries (Frankel 2016). Globally significant sources of additional battery components, including coltan, graphite, and lithium, are located in the Democratic Republic of the Congo, Ivory Coast, Mali, Mozambique, Namibia, Niger, and Zimbabwe. Six countries—Botswana, Gabon, Namibia, Niger, South Africa, and Tanzania—together constitute roughly 20 percent of global uranium resources (World Nuclear Association [WNA] 2019a). Even Malawi, a relatively mineral-poor country by African standards, is expected to become a major producer of neodymium and praseodymium, two rare earth elements used in wind turbines and electric vehicles (Scheyder and Shabalala 2019). Beyond minerals, carbon sequestration and biofuels plantations exist across the continent, most notably in Ghana, Kenya, Sierra Leone, Tanzania, Uganda, Zambia, and Zimbabwe (Fairhead, Leach, and Scoones 2013). Almost all of these projects—sometimes characterized as "green grabbing" (Fairhead, Leach, and Scoones 2013) or "carbon colonialism" (Lyons and Westoby 2014)—capitalize on the relatively low cost of land, mineral, and water rights in African countries to reduce carbon emissions and ensure a continued supply of cheap but "green" energy beyond the continent.

There is both continuity and change in these distributive geographies. Extractive projects that undermine the sustainability of African political economies for the benefit of outside actors continue a long history of wealthier people and places outsourcing the negative externalities of accumulation and consumption to poorer people and places. Yet whereas "dirty" extractive endeavors have been rendered legitimate in the name of colonialism, nationalism, racial capitalism, or energy security, among other justifications, "green" extractive projects are often justified in the name of universal climate salvation, including for the very populations most likely to bear their costs. Indeed, instead of the clichéd "starving African child" used as motivation for Western children to finish their dinners, there is now the "climate change–besieged African child" used as motivation to buy a new electric car—a car with a battery, as documented by O'Driscoll (2017), containing ten to twenty pounds of cobalt perhaps mined by several such African children. Making sense of this perverse outcome, in which many of the greatest climate sacrifices are borne by the very populations on whose behalf they are claimed to be made, requires attending to the uneven geographies of violence that are too easily hidden beneath Anthropocenic imaginaries and the multiscalar logics and processes through which such violence is rendered legitimate.

Necropolitics for the Anthropocene: Death-Worlds in the Name of Planetary Life

Identifying and explaining how and why violence against whom is rendered legitimate by whom is a topic of significant interest in political theory (Fanon 1963; Arendt 1968; Agamben [1995] 1998) as well as in geography (Pratt 2005; Gilmore 2007; Yates 2011; McKittrick 2013; Tyner 2019). Among geographers, Mbembe's (2003, 2019) necropolitics has emerged as a particularly useful approach to studying the spatiality and justification of violence (Wright 2011; Cavanagh and Himmelfarb 2015; Davies 2018; Alexis-Martin 2019; Margulies 2019). Necropolitics draws its name from the concept of necropower, defined by Mbembe (2003) as "the power and the capacity to dictate who may live and who must die" (11). Like biopower ("the power to 'make' live and 'let' die"; Foucault [1976] 2003), the target of necropower is the population rather than the individual. Necropower is distinguished from biopower, however, by the relationship between the targeted population and the sovereign. Whereas sovereigns justify biopower in the name of protecting life for the population on which biopower is imposed, necropower is enacted against "them" in the name of protecting life for "us." Biopower, in other words, is a tool to govern the sovereign's "subjects." Necropower, by contrast, is a tool to control and render disposable "savage" (Mbembe 2019, 92) populations that might otherwise endanger (or be perceived to endanger) the sovereign's subjects or authority.

The suitability of necropolitics for analyzing the distribution and legitimation of violence is enhanced by Mbembe's attention to where necropolitical violence is enacted as well as against whom. His work here builds on that of Fanon (1963), who in *Wretched of the Earth* described the colonizer's perspective on colonized neighborhoods as follows: "[t]hey are born *there*, it matters little where or how; they die *there*, it matters not where, nor how" (38, italics added). Similarly, for Mbembe (2003), colonized territories are "the zone where the violence of the state of exception is deemed to operate in the service of 'civilization'" (24). Contemporary iterations of these spaces—"death-worlds" in Mbembe's (2019, 92) terminology—are a spatial strategy through which sovereigns render violence legitimate without outright colonization, because such necropower is exercised not against subjects "here", but rather against "savage" populations (the "living dead") "there". In addition to immobilization ("the camp"), sovereigns can exercise necropower through the spatial strategy of "scattering" (Mbembe 2019, 86), as exemplified by the violence endured by refugees and displaced persons (Davies, Isakjee, and Dhesi 2017). In contrast to classical sovereign violence, which entails immediate death for the individual, the disposability experienced by those relegated to the status of the living dead rarely entails a quick demise. Rather than immediate death for individuals at the hands of the sovereign, death-worlds subject entire populations to extended conditions of death-in-life.

Climate violence might be diffuse, but it is not randomly distributed. Whether through adaptation (K. J. Grove 2010; Thomas and Warner 2019), conflict (Raleigh, Linke, and O'Loughlin 2014), or mitigation (Curley 2018; Mulvaney 2019; Riofrancos 2019), climate change often manifests as violence against populations that have already been violently marginalized (Sultana 2014). Geographers have illustrated the value of necropolitics for analyzing the uneven geographies of violence associated with colonial dispossession by conservation (Cavanagh and Himmelfarb 2015), chemical exposure (Davies 2018), nuclear imperialism (Alexis-Martin 2019), and human–wildlife relations (Margulies 2019), among other environmental topics. In these cases, state actors exercised or allowed the exercise of necropower within the state's territorial boundaries in the name of protecting the state's chosen subject population. The planetary declaration of the Anthropocene, however, challenges conceptualizations of environmental sovereignty limited to territorial states (O'Lear and Dalby 2015). How might this scalar shift affect the legitimation of necropolitical violence and the geographies of its distribution?

Mann and Wainwright (2018) propose that climate change could provoke the emergence of planetary sovereignty: a state of exception "proclaimed in the name of preserving life on Earth" (15). The promise of planetary salvation is a powerful justification, one that a sovereign could seemingly use to legitimate an extensive range of action, including violence. Indeed, although Mann and Wainwright (2018) do not engage explicitly with Mbembe's necropolitics, they anticipate that planetary sovereignty would entail the determination of "what measures are necessary and what and who must be sacrificed in the interests of life on Earth" (15). By explaining how climate change could be used to justify the rescaling of sovereignty, Mann and Wainwright have identified a pathway through which climate change could be used to legitimate the borderless exercise of climate necropower, rendering some populations sacrificial so that others—and the planet—might live. My argument builds on this pathway by developing a framework, climate necropolitics, for identifying and analyzing the processes, practices, discourses, and logics through which Anthropocenic imaginaries can be used to render extractive violence legitimate within and beyond the borders of the territorialized state in the name of planetary salvation. In the next section, I apply climate necropolitics to the case study of Ecological Civilization to reveal how Chinese and Namibian officials have leveraged this imaginary to justify intensified necropolitical violence and, in so doing, have magnified the unevenly distributed violence of climate change itself.

Placing Climate Necropolitics: The Extractive Violence of Ecological Civilization in Namibia

Methods

The case study that follows draws on over two cumulative years of research on uranium mining in Namibia between 2011 and 2019, with a focus on Chinese state investments. Although I conducted research across the country, most of the empirical data presented here were collected in Namibia's uranium mining region of Erongo or in Windhoek, Namibia's capital city. I used a multimethod approach consisting of participant observation, interviews, focus groups, and textual analysis.[1] My participant observation entailed data collection at more than sixty events, including government and community meetings, public forums, industry conferences, and protests. I also made four visits to uranium mines, which allowed me to observe mining operations, albeit under the close supervision of mine representatives. My interviews involved 102 Namibians, including residents of mining-proximate communities, government officials (local, regional, and national), mine employees, and mining industry representatives. These interviews were semistructured with the exception of interviews with government officials, who often required a list of questions in advance. I also conducted seventeen focus groups, which included similar participants to the interviews. To enhance participant comfort in the cultural context of Namibia, I organized each focus group on the basis of previous alignment, such as members of a community organization or miners employed at the same pay grade. Finally, I analyzed more than 800 documents related to mining policies and licenses, environmental regulations and conditions, and minority communities in Namibia. These documents included reports by government, mining industry stakeholders, and civil society groups; speeches by political leaders; and media coverage of mining, the environment, China, and the specific communities where I conducted research.

Ecological Civilization and Chinese State Investments in Uranium Mining

Despite its domestic deployment of necropower against minority populations (Alexis-Martin 2019), the CCP's historical defense of state-defined sovereignty makes it seemingly an unlikely candidate for extraterritorial climate necropolitics. As I have argued elsewhere though (DeBoom 2020a), the CCP has become more assertive in exercising extraterritorial sovereignty in recent years, including in the environmental realm. This shift is reflected in the rising prominence of Ecological Civilization (Yeh 2009; Hansen, Li, and Svarverud 2018; Pow 2018), a CCP-promoted Anthropocenic imaginary that envisions a global transition, led by China, from "Western industrial civilization" to "socialist ecological civilization" (Geall and Ely 2018). Like the Anthropocene, Ecological Civilization recognizes that the well-being of humanity depends on the well-being of the planet. President Xi (2017) characterized it as a "global endeavor" that "will benefit generations to come" through the implementation of a "new model of modernization with humans developing in harmony with nature" (47). The aim of Ecological Civilization, in other words, is to rectify

the environmental wrongs recognized by the declaration of the Anthropocene.

Fittingly, the same energy source that fueled the 1945 Trinity test in New Mexico—one proposed starting date for the Anthropocene—is also a foundational fuel for Ecological Civilization: nuclear energy. Despite connecting its first nuclear reactor to the grid only in 1991—a year in which the United States had 112 operating reactors—China is the world's fastest growing generator of nuclear energy (International Energy Agency 2018). It is expected to surpass the United States as the largest nuclear energy producer by 2040 (International Energy Agency 2018). The CCP's current nuclear energy plans suggest that its annual uranium consumption by 2050 might rival the world's total uranium consumption in 2015. This is not an immediate impediment: China has the world's eighth-largest uranium reserves (Zhang 2015). Yet instead of primarily pursuing intensified domestic extraction, the CCP is sourcing most of its uranium from abroad. China's uranium imports have grown faster than those of any other country over the past decade (WNA 2020). Rather than relying solely on international market purchases to source this uranium, China's two nuclear state-owned enterprises (SOEs)—China General Nuclear Power Corporation (CGN) and China National Nuclear Corporation (CNNC)—have purchased ownership stakes in foreign uranium mines since 2006, the first year in which the country's domestic demand outpaced supply (WNA 2020). The overseas uranium holdings of Chinese SOEs accounted for roughly 80 percent of global foreign investments in uranium between 2005 and 2015 and are estimated to be three times the size of China's domestic proven reserves (WNA 2020).

As shown in Table 1, most of the foreign uranium production controlled by Chinese SOEs is located in Namibia. These investments have catalyzed significant changes in Namibia's uranium mining sector over the past fifteen years (DeBoom 2020b). Between 1976 and 2006, Namibia had only one operational uranium mine: Rössing, then owned by Rio Tinto. In 2006, the global uranium rush catalyzed the development of Namibia's first new uranium mine in thirty years (Conde and Kallis 2012). By early 2012, however, the Fukushima nuclear disaster had precipitated a crash in global uranium prices that has persisted into early 2020. All of

Table 1. Chinese SOE ownership in uranium mines located overseas

Country	Mine	Chinese SOE	2018 uranium production (tonnes U)	Chinese SOE ownership stake (%)
Namibia	Husab	CGN	3,159	90.0
Namibia	Rössing	CNNC	2,102	68.6
Kazakhstan	Semizbai	CGN	560	49.0
Namibia	Langer Heinrich	CNNC	394	25.9
Niger	Azelik	CNNC	Inactive	37.2
Uzbekistan	Boztau	CGN	In development	50.0
Kazakhstan	Zhalpak	CNNC	In development	49.0
Niger	Abokorum	CNNC	In development	25.0
Canada	Patterson Lake	CGN	In development	20.0

Note: SOE = state-owned enterprise; CGN = China General Nuclear Power Corporation; CNNC = China National Nuclear Corporation. Data source: WNA (2019b, 2020).

Namibia's active and under-development uranium mines have subsequently either been mothballed or sold (in part or in whole) to Chinese SOEs, which have leveraged low global prices to enhance China's uranium supply security. In 2012, CGN purchased a 90 percent stake in Husab, the world's second largest operating uranium mine. CGN's intra-state competitor CNNC followed suit in 2014, when it purchased a 25 percent stake in Langer Heinrich. Most recently, in 2019, CNNC purchased a 90 percent ownership stake in Rössing, the world's longest running open pit uranium mine. This last purchase prompted significant controversy in Namibia, because it means that Chinese SOEs now have at least a 25 percent ownership in each of Namibia's active uranium mines. Together, these three mines account for 12 percent of global uranium production and more than 90 percent of production from foreign mines in which Chinese SOEs have ownership stakes (WNA 2020).

Radioactive and Socioenvironmental Violence in Namibia

For most residents of the rural, agricultural communities located near Chinese SOE-owned uranium mines in Namibia's Namib desert, the lived experience of Ecological Civilization is a far cry from the CCP's green utopian vision. Namibia is the world's second-least densely populated country, and few places in Namibia are more sparsely populated than rural Erongo. Despite having two of Namibia's four largest cities, Erongo's population density is 6.1 people per square mile (Erongo Regional Council 2011). With the exception of the occasional tourist bound for an exclusive campsite or ecoreserve, the villages of rural Erongo lie distant from major roadways and

attract few visitors. My interviews with Namibians living elsewhere, including in urban Erongo, suggested that many Namibians are unaware that these communities even exist. Their invisibility reflects not only their physical isolation but also their demographic composition and socioeconomic status.[2] Most residents of rural, agricultural communities rely on subsistence livelihoods, including subsistence farming, goat herding, and artisanal mining, all of which are relatively low-status livelihoods in the local and national contexts. Most residents also identify as members of the Nama and Damara minority groups, which respectively comprise roughly 5 and 7 percent of Namibia's population.[3]

It is fitting that one of the earliest uses of the term *sacrifice zone* was to describe the radioactive landscapes created through nuclear weapons testing (Kirsch 2005; Pitkanen and Farish 2018). From the South African state's pursuit of nuclear weapons to the CCP's nuclear energy strategy, rural communities near uranium mines in Namibia have been rendered sacrificial in the name of outside actors' geopolitical and political ambitions. Unlike most mine employees, who live in the provided housing of company towns or distant from the mines in Erongo's coastal cities, residents of rural, minority communities largely live in unsealed homes. Their outdoor livelihoods involve daily, direct interactions with soil, exposing them to the pervasive dusty winds of the Namib desert. The open-pit design of each of Namibia's uranium mines (Figure 1A) aggravates this situation. Given the dry, windswept character of the desert environment and the heavy equipment used in mining—and despite significant water usage, as detailed later—dust control at these mines is nearly impossible (Figure 1B). In addition to transporting dust from the open pit to downwind

Figure 1. The Rössing uranium mine in Namibia. (A) A view of the open pit from the mine's southern edge. (B) Dust from the movements of mining equipment. There was no active blasting, which would have created far more extensive dust, at the time of this photograph. (C) Mine tailings deposits large enough to be traversed by several roads. (D) A pipeline transporting water to the mine. Photos by author.

communities, the Namib's intense winds can dislodge radioactive particles from the expansive deposits of mine tailings (unused material remaining after uranium extraction) stored around the edges of Namibia's mines (Figure 1C). These tailings, which retain most of the radioactivity of uranium itself (National Research Council 2011) are particularly profuse in Namibia due to the low grades of its granite-housed uranium deposits; many more tons of host ore body must be detonated and removed to extract the same volume of uranium when compared with higher grade mines such as those in Canada, Kazakhstan, and even China (DeBoom 2020b).

Yet due to the logics of what Hecht (2009) has termed "nuclearity," or the "apparently immutable ontology [that] has long distinguished nuclear things from non-nuclear things" (897), many government authorities and industry leaders treat the radioactive risks faced by residents of uranium mining–proximate

communities as if wholly separate from the risks associated with nuclear energy. This distinction is reflected in the scientific and regulatory radiation exposure threshold of 100 millisieverts (mSv) per year, which distinguishes "man-made" nuclear weapons and energy-based exposures from "natural" uranium-based exposures (Hecht 2012). Despite this distinction, the likelihood of negative health outcomes, including cancer, rises in association with increasing cumulative radiation exposure (National Research Council 2011). These health outcomes can take decades—or, in the case of hereditary disorders, generations—to emerge (Kreuzer et al. 2008). The "slow violence" (Nixon 2013, 2) of long-term, low-level radiation exposure is rendered even more invisible by the compounding risk factors faced by residents of local minority communities, including their inherent exposure to environmental dust, inadequate access to preventative care, and low socioeconomic status. These characteristics make it difficult to definitively attribute negative health outcomes to uranium mining (DeBoom 2020b).

The violence of uranium mining extends beyond radioactivity to the intertwined environmental and sociocultural violence of water scarcity. Because the Namib receives only five to ten inches of annual rainfall, rural communities rely on aquifers and ephemeral rivers—filtered through the potentially contaminated sand of the Namib desert—for their water supply. The combination of intensified uranium mining and climate change–attributed drought, which has affected southwestern Africa since 2013, has depleted those water sources. One herder reflected on his community's future with frustration during a focus group in late 2015. The mines build desalination plants and pipelines (Figure 1D) when they need water, he explained, but his community cannot afford to buy water, let alone pipe water in. "Water belongs to our culture," he explained. "What happens to us when the water here is no more?" The man's village signed an emergency contract with NamWater, the Namibian government's water utility, for a water "loan" roughly one year later. The community has not yet been able to repay this loan and, as of my last contact with the community, there was little hope that it would be able to do so. Other communities owe NamWater upwards of $6,500, a nearly insurmountable debt in the local context. NamWater has threatened to close the taps of "debtor" villages it deems unable to pay, escalating intragroup tensions in communities already under tremendous strain.

The conditions faced by these rural Erongo communities exemplify what Mbembe (2003) characterized as "death-in-life"; it is being "kept alive but in a *state of injury*" (21). Residents face an agonizing choice. They can continue to reside in the deathworld, or they can "scatter," abandoning their livelihoods, communities, and identities in the process. A young man who had relocated to look for work in a nearby city told me that, as a farmer, he "could not see a future there [in his home community]. ... We wonder where the water disappears? Husab [uranium mine] takes from the Swakop River. It never reaches my community." His anger was echoed in interviews and focus groups with minority Namibians across the region. Although most of the relocated individuals I interviewed maintained at least sporadic contact with their home communities in the Namib, many

indicated that they felt a fundamental part of their identity had been lost.

Legitimating Violence: Climate Necropower

Although surely a sign of geopolitical change, the embodied violence of uranium mining for minority communities in Namibia casts doubt on whether Ecological Civilization departs from the extractive violence that has facilitated the Anthropocene. Chinese leaders have carefully contrasted their Anthropocenic imaginary with colonization by linking Ecological Civilization to the CCP's geopolitical emphasis on "south-south solidarity." "China is a socialist country," then-Vice Minister of Environmental Protection Pan Yue argued in 2006, "and cannot engage in environmental colonialism, nor act as a hegemony, so it must move towards a new type of civilization" (Zhou 2006). Yet the CCP's promotion of Ecological Civilization as China's "global duty and mission" (Pan Yue, quoted in Zhou 2006) is reminiscent of the "civilizing" missions used to justify the necropolitical violence of colonization. Like those colonial projects, Ecological Civilization is far from self-sacrificing. Its carbon benefits might be globally significant, but its most concentrated benefit is likely to be the enhancement of the CCP's geopolitical reputation. Its costs, meanwhile, appear likely to borne disproportionately by marginalized communities for whom the violence of uranium mining in the name of climate change mitigation is rivaled only by the violence of climate change itself.

The necropolitical benefits of intensified uranium mining in Namibia are not limited to China. Indeed, it was life rather than death, or even exploitation, that featured most prominently in my interviews with Namibian officials as well as in my analysis of political leaders' speeches and government press releases. Mining is the largest contributor to the Namibian government's annual revenue, and Chinese investments have increased that contribution. One mine alone—Husab—is expected to generate between $170 million and $200 million in annual government revenues at full production, an amount roughly equivalent to 5 percent of the Namibian government's annual revenues prior to its opening. Beyond revenues, Namibian officials argue that intensified uranium mining will benefit minority communities. One official described mining as a "lifeline" for a region he characterized as "jobless."

Bemoaning the fact that I had spent time in such "hamlets," he asked "Is that [what you saw] a life?" He shook his head, answering his own question before I could. Namibian President Hage Geingob expressed similar sentiments in a 2015 speech, when he described the Husab mine as follows: "The mine was opened in a desolate area characterized by barren hills and mountains amongst which a modern highway has been built, leading to *life*" (italics added). Chinese diplomats and mine executives expressed similar sentiments. At an event commemorating the Husab uranium mine's first yellowcake in 2016, for example, a Chinese SOE representative stated, "We can now proudly declare that the Husab mine is in production, bringing new vigor and vitality to the ancient Namib desert."

The violence associated with uranium extraction in Namibia predates Chinese investments. The profitability of Namibia's first uranium mine—Rössing, then owned by Rio Tinto—was facilitated by South Africa's imposition of apartheid in Namibia prior to its 1990 independence (Hecht 2012). Today, however, this violence is rendered legitimate not in the explicit name of racial capitalism, as espoused by the apartheid South African state, but rather in the name of Ecological Civilization and its promise of planetary salvation. Several Namibian officials even adopted the language of Ecological Civilization in my interviews, expressing hope that high-profile Chinese investments will attract other "eco-minded" investors to Namibia's "green" uranium mines. Government officials have subsequently taken this rhetoric on the road, using the slogan of "green" uranium to solicit interest among other would-be nuclear investors at events like the Abu Dhabi Sustainability Week.

The statements just described suggest that the minority communities of rural Erongo are not the subjects of Ecological Civilization but rather its "savages," the "living dead" (Mbembe 2019, 92) against whom the exercise of necropower can be rendered legitimate "in the service of 'civilization'" (Mbembe 2003, 24). For representatives of the Chinese and Namibian states, intensified uranium mining is not life-taking; it is life-giving. Namibian officials in particular often described minority communities to me as if they were "part of nature … 'natural' human beings who lacked the specifically human character" (Arendt 1968, 192). Reflecting the definition of "slow violence" developed by

Nixon (2013), the radioactive violence embodied by these communities "occurs gradually and out of sight … dispersed across time and space" (2) from the lofty Anthropocenic imaginary of Ecological Civilization. Local Namibians' understandings of Ecological Civilization reflect not the transcendence of extractive violence but rather its entrenchment through "green" energy manifesting as violence against those already bearing the violence of climate change itself.

Conclusion

Will the declaration of the Anthropocene catalyze a radical revision of the violent geographies that have facilitated climate change? Or will it usher in a "status quo utopia"—a "green" future that is merely a "thinly disguised version of the present" (Günel 2019, 13)? Epochal moments can sow transformative possibilities, but reaping their potential requires more than declarations. The violence of colonialism did not "magically disappear after the ceremony of trooping the national colors" (Fanon 1963, 60). Likewise, the necropolitical violence that preceded the Anthropocene's declaration has not magically disappeared upon its introduction. Like the iterations of necropower that have come before, the violence of climate necropower is accumulative. Existing strategies of social debridement—of protecting "us" from "them" while extracting climate solutions from "them" as has been done historically—might reduce carbon emissions, but they are also likely to reinforce rather than rectify existing geographies of violence. Left to fester too long, such sacrificial necrosis risks inciting an endless cycle of inversion—a planet of ever-expanding, diffuse death-worlds created in the name of planetary life.

The distributive geographies of climate change and its associated Anthropocenic imaginaries—including which strategies for mitigation are pursued, where, and based on which priorities decided by whom—are likely to set the foundational conditions for the transformative potential or lack thereof of the Anthropocene's declaration. Understanding these geographies necessitates a multiscalar approach that does not lose sight of the uneven distribution of Anthropocenic violence in the pursuit of existential, technical, and geophysical understanding of the Anthropocene itself. Geographers can contribute to this task by disaggregating the Anthropocene and asking "what is being secured and for whom" through climate change response strategies and, of course, where (O'Lear and Dalby 2015, 207). Trained in approaches that transcend both scale and traditional disciplinary boundaries between the social and the physical or the environmental and the cultural, geographers are well-suited to holding "*the planet* and *a place on the planet* on the same analytic plane" (Hecht 2018, 112, italics in original). Such strategies can reconcile the Anthropocene as both a much-needed recognition of human-induced violence on a planetary scale and a scalar project that risks obscuring not only "who pays the price for humanity's planetary footprints" (Hecht 2018, 135) but also who pays the price for mitigating those footprints.

Toward that end, this article introduced climate necropolitics as a theoretical framework for analyzing the multiscalar processes, practices, discourses, and logics through which Anthropocenic imaginaries like Ecological Civilization can be used to render intensified extractive violence legitimate in the name of climate change response. I illustrated the applied value of this approach through a case study of investments by Chinese SOEs in uranium mining in Namibia. Applying climate necropolitics to this case study revealed the logics through which socioenvironmental violence, as embodied by minority communities in Namibia, is reinforced and rendered legitimate by both Chinese and Namibian state actors through the Anthropocenic imaginary of Ecological Civilization. Although my analysis focused on climate change mitigation—namely, nuclear energy—I anticipate that climate necropolitics could also be applied to analyze the distributive geographies and legitimation strategies associated with other forms of climate violence, including in the realms of climate security (O'Lear and Dalby 2015) and "green" militarization (Bigger and Neimark 2017) as well as adaptation (Thomas and Warner 2019) and vulnerability (Ribot 2014).

The argument presented here should not be interpreted as a call to abandon either the Anthropocene or pursuits of climate change mitigation. Indeed, my argument is to the contrary. The framework of climate necropolitics is a call not to abandon mitigation but rather to expand our definition of mitigation to include the mitigation of violent social relations. It is a call not to abandon the Anthropocene but rather to investigate and seek to rectify its violent manifestations at scales beyond the planetary. As a framework for understanding how and why current strategies for climate change mitigation risk legitimating sacrificial violence against marginalized communities in the name of preserving planetary life, climate necropolitics echoes Ruddick's (2015) call for an Anthropocene that "renders neither people nor the planet as disposable" (1126). I hope it will prompt my fellow geographers—whether self-identified as physical or human, cultural or environmental, or, like most, some combination thereof—to attend not only to the practical questions of what must be done to ensure planetary life but also to the ethical questions of what should be done and by whom.

Acknowledgments

I am grateful to the individuals and communities in Namibia who shared their experiences of and insights about uranium mining, climate change, violence, and the state with me. I also extend my thanks to Natalie Koch, the members of the Critical Ecologies Lab at the University of South Carolina, and the participants of panels at the 2019 and 2020 American Association of Geographers annual meetings as well as the 2020 Dimensions of Political Ecology conference for their suggestions on previous iterations of this article. Finally, thank you to two anonymous referees for their constructive suggestions and to David Butler for his thoughtful editorial insights. Any errors or omissions are mine alone.

Funding

This research was funded by a Doctoral Dissertation Research Improvement Grant (1536313), a Graduate Research Fellowship from the National Science Foundation, and an American Dissertation Fellowship from the American Association of University Women.

Notes

1. Details on the demographic and geographic composition of focus group and interview participants are available in Appendix 1 and Appendix 2 of DeBoom (2018), which is publicly available via the University of Colorado's open access research repository (CU Scholar).

2. This section focuses on nonmining communities in rural Erongo. It is important to distinguish these communities from mining company towns like Arandis, because the two types of communities have distinct cultural, socioeconomic, and demographic characteristics. For an analysis of the risks faced by mine employees and residents of company towns in Erongo, see DeBoom (2020b).

3. Namibia's current president, Hage Geingob, belongs to the Damara group. He is, however, an exception in the upper echelon of Namibian national politics, which are dominated by the SWAPO political party and its base of support in the Owambo majority group.

ORCID

Meredith J. DeBoom ⓘ http://orcid.org/0000-0001-8748-5634

References

Abrahams, D., and E. R. Carr. 2017. Understanding the connections between climate change and conflict: Contributions from geography and political ecology. *Current Climate Change Reports* 3 (4):233–42.

Agamben, G. [1995] 1998. *Homo sacer: Sovereign power and bare life*. Palo Alto, CA: Stanford University Press.

Alexis-Martin, B. 2019. The nuclear imperialism-necropolitics nexus: Contextualizing Chinese–Uyghur oppression in our nuclear age. *Eurasian Geography and Economics* 60 (2):152–76.

Arendt, H. 1968. *The origins of totalitarianism*. Orlando, FL: Harcourt.

Bigger, P., and B. D. Neimark. 2017. Weaponizing nature: The geopolitical ecology of the U.S. Navy's biofuel program. *Political Geography* 60:13–22.

Braun, B., M. Coleman, M. Thomas, and K. Yusoff. 2015. Grounding the Anthropocene: Sites, subjects, struggles in the Bakken oil fields. Antipode Foundation, November 3. Accessed December 1, 2019. https://antipodefoundation.org/2015/11/03/grounding-the-anthropocene.

Castree, N. 2014. The Anthropocene and geography I: The back story. *Geography Compass* 8 (7):436–449.

Cavanagh, C. J., and D. Himmelfarb. 2015. "Much in blood and money": Necropolitical ecology on the margins of the Uganda protectorate. *Antipode* 47 (1):55–73. doi: 10.1111/anti.12093.

Chakrabarty, D. 2012. Postcolonial studies and the challenge of climate change. *New Literary History* 43 (1):1–18.

Conde, M., and G. Kallis. 2012. The global uranium rush and its Africa frontier: Effects, reactions and social movements in Namibia. *Global Environmental Change* 22 (3):596–610.

Crutzen, P., and F. Stoermer. 2000. Have we entered the Anthropocene? *International Geosphere-Biosphere Program Newsletter* 41:17–18.

Curley, A. 2018. A failed green future: Navajo green jobs and energy "transition" in the Navajo Nation. *Geoforum* 88:57–65.

Dalby, S. 2020. *Anthropocene geopolitics: Globalization, security, sustainability*. Ottawa: University of Ottawa Press.

Davies, T. 2018. Toxic space and time: Slow violence, necropolitics, and petrochemical pollution. *Annals of the American Association of Geographers* 108 (6):1537–53.

Davies, T., A. Isakjee, and S. Dhesi. 2017. Violent inaction: The necropolitical experience of refugees in Europe. *Antipode* 49 (5):1263–84.

Davis, H., and Z. Todd. 2017. On the importance of a date, or decolonizing the Anthropocene. *ACME: An International Journal for Critical Geographies* 16 (4):761–80.

Davis, J., A. A. Moulton, L. Van Sant, and B. Williams. 2019. Anthropocene, Capitalocene, … Plantationocene?: A manifesto for ecological justice in an age of global crises. *Geography Compass* 13 (5):e12438. doi: 10.1111/gec3.12438.

DeBoom, M. J. 2018. Developmental fusion: Chinese investment, resource nationalism, and the distributive politics of uranium mining in Namibia. PhD diss., University of Colorado-Boulder. https://scholar.colorado.edu/concern/graduate_thesis_or_dissertations/hd76s013x.

DeBoom, M. J. 2020a. Sovereignty and climate necropolitics: The tragedy of the state system goes "green." In *Handbook on the changing geographies of the state: New spaces of geopolitics*, ed. S. Moisio, A. Jonas, N. Koch, C. Lizotte, and J. Luukkonen, 276–286. Northampton, UK: Edward Elgar.

DeBoom, M. J. 2020b. Toward a more sustainable energy transition: Lessons from Chinese investments in Namibian uranium. *Environment: Science and Policy for Sustainable Development* 62 (1):4–14.

Derickson, K. D., and D. MacKinnon. 2015. Toward an interim politics of resourcefulness for the Anthropocene. *Annals of the Association of American Geographers* 105 (2):304–12.

Erongo Regional Council. 2011. Demographics. Accessed July 12, 2020. http://www.erc.com.na/erongo-region/demographics/.

Fairhead, J., M. Leach, and I. Scoones, eds. 2013. *Green grabbing: A new appropriation of nature?* London and New York: Routledge.

Fanon, F. 1963. *The wretched of the Earth*. New York: Grove.

Frankel, T. D. 2016. The cobalt pipeline. *Washington Post*, September 30. https://www.washingtonpost.com/graphics/business/batteries/congo-cobalt-mining-for-lithium-ion-battery/.

Foucault, M. [1976] 2003. *Society must be defended: Lectures at the Collège de France, 1975–1976*. London: Allen Lane.

Geall, S., and A. Ely. 2018. Narratives and pathways towards an Ecological Civilization in contemporary China. *The China Quarterly* 236:1175–96.

Gilmore, R. W. 2007. *Golden gulag: Prisons, surplus and crisis in globalizing California*. Oakland: University of California Press.

Global Carbon Project. 2018. Supplemental data of global carbon budget 2018. https://www.icos-cp.eu/global-carbon-budget-2018.

Grove, J. V. 2019. *Savage ecology: War and geopolitics at the end of the world*. Durham, NC: Duke University Press.

Grove, K. J. 2010. Insuring "our common future"? Dangerous climate change and the biopolitics of environmental security. *Geopolitics* 15 (3):536–63.

Günel, G. 2019. *Spaceship in the desert: Energy, climate change, and urban design in Abu Dhabi*. Durham, NC: Duke University Press.

Hansen, M. E., H. Li, and R. Svarverud. 2018. "Ecological Civilization": Interpreting the Chinese past, projecting the global future. *Global Environmental Change* 53:195–203.

Hecht, G. 2009. Africa and the nuclear world: Labor, occupational health, and the transnational production of uranium. *Comparative Studies in Society and History* 51 (4):896–926.

Hecht, G. 2012. *Being nuclear: Africans and the global uranium trade*. Cambridge, MA: MIT Press.

Hecht, G. 2018. Interscalar vehicles for an African Anthropocene. *Cultural Anthropology* 33 (1):109–41.

International Energy Agency. 2018. *World energy outlook 2018*. https://www.iea.org/reports/world-energy-outlook-2018.

Kirsch, S. 2005. *Proving grounds: Project Plowshare and the unrealized dream of nuclear earthmoving*. Newark, NJ: Rutgers University Press.

Klinger, J. M. 2017. *Rare earth frontiers: From terrestrial subsoils to lunar landscapes*. Ithaca, NY: Cornell University Press.

Kreuzer, M., L. Walsh, M. Schnelzer, A. Tschense, and B. Grosche. 2008. Radon and risk of extrapulmonary cancers: Results of the German uranium miners' cohort study, 1960–2003. *British Journal of Cancer* 99 (11):1946–53. doi: 10.1038/sj.bjc.6604776.

Lyons, K., and P. Westoby. 2014. Carbon colonialism and the new land grab: Plantation forestry in Uganda and its livelihood impacts. *Journal of Rural Studies* 36:13–21.

Mann, G., and J. Wainwright. 2018. *Climate Leviathan: A political theory of our planetary future*. London: Verso.

Margulies, J. D. 2019. Making the "man-eater": Tiger conservation as necropolitics. *Political Geography* 69:150–61.

Mbembe, A. 2003. Necropolitics. *Public Culture* 15 (1):11–40.

Mbembe, A. 2019. *Necropolitics*. Durham, NC: Duke University Press.

McKittrick, K. 2013. Plantation futures. *Small Axe: A Caribbean Journal of Criticism* 17 (3):1–15.

Mostafanezhad, M., and R. Norum. 2019. The Anthropocenic imaginary: Political ecologies of tourism in a geological epoch. *Journal of Sustainable Tourism* 27 (4):421–35.

Mulvaney, D. 2019. *Solar power: Innovation, sustainability, and environmental justice*. Oakland: University of California Press.

National Research Council. 2011. *Scientific, technical, environmental, human health and safety, and regulatory aspects of uranium mining*. Washington, DC: National Academy of Sciences.

Nixon, R. 2013. *Slow violence and the environmentalism of the poor*. Cambridge, MA: Harvard University Press.

O'Driscoll, D. 2017. Overview of child labour in the artisanal and small-scale mining sector. https://assets.publishing.service.gov.uk/media/5a5f34feed915d7dfb57d02f/209-213-Child-labour-in-mining.pdf.

O'Lear, S., and S. Dalby, eds. 2015. *Reframing climate change: Constructing ecological geopolitics*. London and New York: Routledge.

Pitkanen, L., and M. Farish. 2018. Nuclear landscapes. *Progress in Human Geography* 42 (6):862–80. doi: 10.1177/0309132517725808.

Pow, C. P. 2018. Building a harmonious society through greening: Ecological Civilization and aesthetic governmentality in China. *Annals of the American Association of Geographers* 108 (3):864–83.

Pratt, G. 2005. Abandoned women and spaces of the exception. *Antipode* 37 (5):1052–78.

Pulido, L. 2018. Racism and the Anthropocene. In *Future remains: A cabinet of curiosities for the Anthropocene*, ed. G. Mitman, M. Armiero, and R. S. Emmett, 116–28. Chicago: University of Chicago.

Raleigh, C., A. Linke, and J. O'Loughlin. 2014. Extreme temperatures and violence. *Nature Climate Change* 4 (2):76–77.

Reszitnyk, A. 2015. Uncovering the Anthropocenic imaginary: The metabolization of disaster in contemporary American culture. PhD dissertation, McMaster University.

Ribot, J. 2014. Cause and response: Vulnerability and climate in the Anthropocene. *The Journal of Peasant Studies* 41 (5):667–705.

Riofrancos, T. 2019. What green costs. *Logic*, December 9. https://logicmag.io/nature/what-green-costs/.

Robinson, C. 1983 [2000]. *Black Marxism: The making of the black radical tradition*. Chapel Hill: University of North Carolina Press.

Ruddick, S. 2015. Situating the Anthropocene: Planetary urbanization and the anthropological machine. *Urban Geography* 36 (8):1113–30.

Scheyder, E., and Z. Shabalala. 2019. Pentagon eyes rare earth supplies in Africa. https://www.reuters.com/article/us-usa-rareearths-pentagon-exclusive/exclusive-pentagon-eyes-rare-earth-supplies-in-africa-in-push-away-from-china-idUSKCN1T62S4.

Sultana, F. 2014. Gendering climate change: Geographical insights. *The Professional Geographer* 66 (3):372–81.

Swyngedouw, E. 2015. Depoliticized environments and the Anthropocene. In *The international handbook of political ecology*, ed. R. L. Bryant, 131–46. Northampton, UK: Edward Elgar.

Thomas, K. A., and B. P. Warner. 2019. Weaponizing vulnerability to climate change. *Global Environmental Change* 57:101928.

Tyner, J. A. 2019. *Dead labor: Toward a political economy of premature death.* Minneapolis: University of Minnesota Press.

World Nuclear Association. 2019a. Supply of uranium. Accessed February 20, 2020. https://www.world-nuclear.org/information-library/nuclear-fuel-cycle/uranium-resources/supply-of-uranium.aspx.

World Nuclear Association. 2019b. Uranium mining. Accessed February 20, 2020. https://www.world-nuclear.org/information-library/nuclear-fuel-cycle/uranium-resources/supply-of-uranium.aspx.

World Nuclear Association. 2020. China's nuclear fuel cycle. Accessed February 26, 2020. https://www.world-nuclear.org/information-library/country-profiles/countries-a-f/china-nuclear-fuel-cycle.aspx.

Wright, M. W. 2011. Necropolitics, narcopolitics, and femicide: Gendered violence on the Mexico–U.S. border. *Signs* 36 (3):707–31. doi: 10.1086/657496.

Xi, J. 2017. Secure a decisive victory in building a moderately prosperous society. http://www.xinhuanet.com/english/download/Xi_Jinping's_report_at_19th_CPC_National_Congress.pdf.

Yates, M. 2011. The human-as-waste, the labor theory of value and disposability in contemporary capitalism. *Antipode* 43 (5):1679–95.

Yeh, E. 2009. Greening western China: A critical view. *Geoforum* 40 (5):884–94.

Yusoff, K. 2013. Geologic life: Prehistory, climate, futures in the Anthropocene. *Environment and Planning D: Society and Space* 31 (5):779–95.

Yusoff, K. 2019. *A billion black Anthropocenes or none.* Minneapolis: University of Minnesota Press.

Zhang, H. 2015. Uranium supplies: A hitch to China's nuclear energy plans? Or not? *Bulletin of the Atomic Scientists* 71 (3):58–66.

Zhou, J. 2006. The rich consume and the poor suffer the pollution. *ChinaDialogue,* October 27.

Cultures and Concepts of Ice: Listening for Other Narratives in the Anthropocene

Harlan Morehouse and Marisa Cigliano

The Anthropocene is marked not only by significant environmental changes massively distributed in space and time but also by a substantial proliferation of scientific data. From Intergovernmental Panel on Climate Change reports to growing extinction lists, there is neither a shortage of environmental crises nor data to serve as official evidence of crises. As crucial as these data are, however, questions remain as to how science data-driven approaches police the boundaries of what counts as evidence and risk marginalizing other ways of encountering, knowing, and narrating environmental change. In this article, we address how certain narratives are not being told, or heard, amid European-American climate discourses. Drawing on nature–society studies, political ecology, and environmental philosophy, this article focuses on how prevailing discussions around glacier recession ignore the cultural and conceptual consequences of glacier loss. As glaciers become increasingly iconized in the Anthropocene, they become more detached from cultural and conceptual contexts. Such detachment overlooks how the fate of glaciers is not only a matter of quantifiable loss but is also implicated in everyday encounters, generational experiences, and stories spun at the nexus of ice and culture(s). *Key Words: Anthropocene, glacier recession, Indigenous perspectives, more-than-human sentience, storytelling.*

In Iceland, on 18 August 2019, a funeral was held for a glacier. The glacier—*Okjökul*, or commonly referred to as Ok—officially lost glacier status in 2014 after melting to 0.386 square miles in area from its original size of 5.8 square miles (Engel 2019). By August 2019 all that remained were a few small ice patches on a brown, rock-strewn landscape. Affixed to a prominent boulder is a plaque inscribed with the following words from Icelandic writer Andri Snaer Magnason (Agence France-Presse 2019):

A letter to the future:

Ok is the first Icelandic glacier to lose its status as a glacier. In the next 200 years, all our glaciers are expected to follow the same path. This monument is to acknowledge that we know what is happening and know what needs to be done. Only you know if we did it.

August 2019

415ppm CO_2

Magnason's words confirm what is well known: Glacier recession is tied to CO_2 emissions and Ok's circumstances are not unique to that glacier alone. Prevailing scientific reporting provides a staggering view of quantitative loss of glaciers around the world (Intergovernmental Panel on Climate Change 2018). These once heavy and sluggish forms have become lighter and faster, shedding mass as they go. Their decline shows up as newly exposed terrains, in creeping waters lapping at the edges of coastal cities, and as a persistent blue splotch southeast of Greenland affixed to surface temperature maps increasingly dominated by red and orange. There is no shortage of data or forecast models to offer a sobering sense of the implications that glacier recession has for habitat destruction, coastal erosion, population displacement, and death. Given the consensus across the scientific community, it might appear that the science is settled and there is little debate regarding the general state of present and future environmental conditions.

Still, despite the appearance of general agreement, we maintain that there are various and unique points of departure that open unto a range of possible futures. Ok's funeral offers one such point. Indeed, what is unique about Ok was that it was the first glacier to be given a funeral (at least to our knowledge). In one sense, the funeral seems appropriate for bringing to light the solemnity of Ok's loss. Ok is gone and this deserves our attention. Yet, in another sense the funeral is a peculiar kind of event that raises the question of what is lost when a glacier disappears. Typically, funerals evoke a different experience of loss—more qualitative than quantitative, to put it coarsely. A funeral notes the death of something or someone. It typically carries a spiritual connotation and inspires reflection on the passage from one mode of existence to another. Thus, Ok's funeral seems both appropriate for marking measurable loss yet also strange for evoking a different sense of loss not commonly extended to ostensibly inanimate objects like glaciers. As we argue, this event's strangeness is nevertheless fitting for an era of deep uncertainty. It provides a departure point for differently coping with the staggering loss of glaciers. Moreover, Ok's funeral poses an important question: If the ceremony was held to mark Ok's death, does this mean that Ok once lived?

In this article we take seriously the idea that Ok lived. In what follows we work through some conceptual challenges involved in recognizing more-than-human sentience in their various animate and inanimate forms. Our aim is to draw attention to the importance of stories that are attuned to the world but that do not commonly circulate in prevailing discourses of environmental change. We argue that gathering insights, perspectives, and stories flung far outside these prevailing narratives is a strategy for enacting better futures.

New Animacies and the Living Dead

Although the science might be settled, it is not the case that *science* is settled. The production, practice, and application of scientific knowledge remain a matter of debate. It is not our aim to dismiss the importance of scientific research on glacier recession or the climate emergency. Although it provides crucial observations of environmental change, scientific knowledge nevertheless carries colonial pasts that routinely crop up in present contexts, no matter how seemingly progressive the concern (see Haraway 1988; Seitler 2008; Alaimo 2010). Indeed, the production and pursuit of scientific knowledge can remain averse to cultural contexts, which are often considered of secondary importance to the object of inquiry. Glacier recession, for example, is commonly rendered a matter of melting ice, recorded by rate over time through calculated standards. The cultural experience of glacier recession, however, might be considered immaterial or irrelevant. Our concerns lie precisely here and revolve around what counts as evidence in scientific knowledge. Put differently, our issue is how prevailing scientific discourses police the boundaries of what qualifies as evidence and, in doing so, bracket off other ways of knowing and experiencing worlds undergoing rapid and catastrophic change. If whole worlds are disappearing, their loss cannot be explained through quantitative metrics alone. Loss is also experienced by those whose lives are entangled with disappearing others. At stake, then, is much more than mere numbers. At stake is the unraveling of connections that make a world.

Here, we are reminded of Davis and Todd's (2017) challenge to dominant Anthropocene origin theories. They argued, "Evidence [for Anthropocene origins] does not, generally, entail the fleshy story of *kohkoms* (the word for grandmother in Cree) and the fish they fried up over hot stoves in prairie kitchens to feed their large families. ... But these fleshy philosophies and fleshy bodies are precisely the stakes of the Anthropocene" (767). Davis and Todd taught us that when "fleshy bodies" disappear so, too, do practices and stories that inhered in the presence of these disappearing others. This rich sense of connectivity evokes Alaimo's (2016) concept of *transcorporeality*, wherein "humans are interconnected not only with one another but also with the material interchanges between body, substance, and place" (77). The possibilities of living, in other words, emerge through relations among many forms of life situated within broader environmental and cultural contexts. To isolate forms of life disregards how these contexts not only bring lives into being but also hold them together. It follows that the loss of one species ripples through various others and cuts knots of relations that bind together living worlds (see Yusoff 2013; Heise 2016).

At present these cuts are distressingly widespread (Ceballos, Ehrlich, and Dirzo 2017). Such distress is tangible in the writing of Tsing (2015), who argued, "Without collaborations, we all die" (28). van Dooren (2014) voiced similar concern. Responding to the lack of public mourning over extinction, van Dooren (2014) wrote: "At the core of the answer ... is our inability to really get ... the multiple connections and dependencies between ourselves and these disappearing others: a failure to appreciate all the ways in which we are at stake in one another, all the ways in which we share a world"

(140). Such views find common ground with others (e.g., Bennett 2010; Collard, Dempsey, and Sundberg 2015; Alaimo 2016; Haraway 2016; Singh 2017; Weston 2017; Greyson 2019) who challenge a European-American cosmogony that sets nature apart from society, articulating instead a relational onto-epistemology (Barad 2007) that recognizes the role that more-than-human natures play in productions, understandings, and experiences of the world.

These approaches challenge long-standing nature–society dualisms by emphasizing the constitutive power of more-than-human natures. There remains a general, but not strict, vitalist tendency, however, that risks inscribing divisions between living and dead matter, privileging the former over the latter. It is easier, but not uncomplicated, to understand distress over severed connections with the living. But what about severed connections with that which is often perceived as dead? Admittedly, seeking companionship in flesh rather than stone is understandable. Fostering relations across species is conceptually easier with dogs, sharks, crows, or mollusks than it is with mountains, fossils, rivers, or fog. Thus, to recognize something like a glacier as a living entity is a challenge not only within the boundaries of European-American cosmogony but also in counter approaches that attempt to break free of those boundaries yet still carry vitalist traces that demarcate the living and the dead. As Weston (2017) put it, "It is not easy to convey to people heavily invested in Euro-American conceptions of dead matter what it means to live in a world where trees ruminate, baskets talk, ancestral spirits inhabit palisade fortifications, elk decide whether to offer themselves to the hunter" (85).

We question whether this vitalist orientation delimits how relations can be articulated and enacted in a more inclusive way that blends animacy and inanimacy, liveliness and lifelessness, vibrancy and dormancy. It is not that the aforementioned discourses are misguided but rather that they do not go far enough. That is, in looking to what is recognizably living they risk overlooking what is considered dead. Following Macfarlane's (2019b) appeal to "extend being and sentience respectfully and flexibly beyond the usual bearers of such qualities" (112), we argue that rethinking divisions between living and dead matter is a powerful strategy, especially amid environmental crisis.

Rethinking this division helps to better approximate the multiple senses of loss invoked in mourning for disappearing glaciers. If a glacier can die it must have once lived. And, if it can live, might it also think, desire, or express anger? If so, are these expressions legible to humans? At first glance, such questions seem novel outgrowths of emerging debates around nonhuman sentience. There is, however, rich cultural precedence for recognizing glacier sentience (see Cruikshank 1981; Aporta 2011; Berkes 2012). We turn to Cruikshank (2001), who amplified Indigenous perspectives on glaciers. According to Cruikshank (2001), Athapaskan and Tlingit oral traditions "attribute to glaciers characteristics rather different from those discovered through science" (378). Glaciers, Cruikshank (2001) continued, "are characterized by sentience: they listen, pay attention, and respond to human behavior" (378). In *Do Glaciers Listen?*, Cruikshank (2005) relayed how "women … portrayed glaciers as conscious and responsive to humans. Glaciers, they insisted, are willful, sometimes capricious, easily excited by human intemperance but equally placated by quick-witted human responses. Glaciers engage all the senses" (8). Glaciers, thus, do not appear as dead matter. Their advances and retreats are not reducible to measurement alone but are also an effect of their personalities.

These perspectives on glacier liveliness are, perhaps, expected to emerge where ice and culture intersect. Carefully traversing shifting land requires respect for the power land holds. It is a small step to consider this power as an expression of the will, sentience, and intelligence of the land itself. Yet, despite multigenerational persistence, these stories of ice do not count as evidence within European-American frameworks. Historically, European-American colonizers viewed Arctic landscapes as vast stretches of "inanimate nature subject to empirical investigation and measurement. They were heartened by the possibility that their Enlightenment categories and scientific instruments might help them to pry nature from culture" (Cruikshank 2005, 10). Cruikshank argued that early industrialists "were not about to stand for objects that act willfully, for how could manufacturing proceed under such circumstances? The new sciences came to see alchemy as too animistic and willful objects as too troublesome, and they opted instead for mechanistic models" (Cruikshank 2005, 143; see also Povinelli 1993, 1995). Similarly, for Weston (2017), "The econometric propensity to turn the world into a collection of dead yet measurable and manipulated 'resources' is itself symptomatic [of colonialism]" (194). Not only did colonialism exert violence over indigenous peoples and their lands, it also exerted violence over the stories people told of their relations to land. Colonialism attempted to render their stories mere mythologies and quaint curiosities (which, of note, nevertheless carried the threat of interfering with extraction and accumulation).

As mentioned, legacies of cultural and narrative suppression are carried into the present and reflected in discourses that reject non-Western traditions and knowledge as evidence. This practice continues despite the clear value that such perspectives retain for understanding rapid and catastrophic environmental change. Addressing this tension, Laidler (2006) argued, "Inuit knowledge is based on extensive, repeated observation and experience that is further verified, shared, and improved in a collective context. This implies both rigor and confidence in local understandings of complex systems" (411; see also Riedlinger and Berkes 2001; Leduc 2007; Weatherhead, Gearheard, and Barry 2010; Cameron, Mearns, and McGrath 2015). Recent reporting on climate change has attempted to incorporate indigenous perspectives in view of the fact that indigenous populations are on the front lines of catastrophic environmental change (see Inuit Circumpolar Conference 1998; Kelman 2017; Intergovernmental Panel on Climate Change 2018). In some respects, this is a step forward in recognizing the generational knowledge held by inhabitants of landscapes undergoing catastrophic change. Yet, as Nadasdy (1999) argued, attempts to integrate traditional ecological knowledge (TEK) with scientific resource management meet several challenges. Nadasdy (1999) argued, "TEK must be expressed in forms that are compatible with the already existing institutions and processes of scientific resource management. ... [This] approach ... ignores the cultural processes in which different 'ways of knowing' are embedded and treats traditional knowledge ... as simply another type of information or source of data" (5). Compartmentalization marginalizes Indigenous lived realities. "A whole array of stories, values, social relations and practices," argued Nadasdy (1999), "must be 'distilled out' of TEK before it can be incorporated into the institutional framework of scientific resource management" (7). TEK is thus stripped of context and made to fit the contours of scientific convention rather than modifying it. This is less an equitable power-sharing arrangement than one of domination.

Cruikshank (2005) raised similar concerns arguing against the "premise that different cultural perspectives are bridgeable by concepts in English language ... and within scientific discourse (256). Despite attempts to translate across difference, like in Nunavut's glossary of climate-related terms (Cameron, Mearns, and McGrath 2015), Indigenous terms are not easily translated into English "without losing the breadth of their specific cultural meaning" (Leduc 2007, 243). These challenges recall Kimmerer's (2013) discussions around "the grammar of animacy" (48). For Kimmerer (2013), "The language scientists speak, however precise, is based on a profound error in grammar, an omission, a grave loss in translation from the native languages of these shores" (49). English, Kimmerer (2013) noted, is a noun-based language "somehow appropriate to a culture obsessed with things" (53). With only 30 percent of its vocabulary being verbs, English is arguably ill equipped to address the processual quality of change over time, preferring instead to organize the world into discrete categories, thereby refusing the complex relations that cut across longstanding divisions between nature and society, human and more-than-human, and living and dead. For Kimmerer's Potawatomi language, "A bay is a noun only if water is *dead*. When *bay* is a noun, it is defined by humans, trapped between its shores and contained by the word. But the verb *wiikwegamaa*—to *be* a bay—releases the water from bondage and lets it live" (Kimmerer 2013, 55).

Depictions of environmental change are thus filtered through inequitable modes of representation. One implication of the tensions among Indigenous and European-American grammars is that the former's stories do not get told. Or, rather, their circulation is delimited via diminished relevance and subsequent inadmissibility as "evidence" within sanctioned discourses. This poses a problem, because entire ways of relating to the world are hidden from view at a moment of great need. Granted, they persist and thrive against considerable odds but most often where they are already known or where one knows where to look, which is not always obvious.

On Telling, and Listening for, Other Stories

There is something about scientific discourse that, although instrumental, fails to generate the empathy or public outcry that van Dooren (2014) felt is appropriate to this time. When it inspires outcry, it often seems of the technocratic variety, replete with default appeals to governing bodies to steer humanity through this crisis. Outrage for the severed relations amid environmental emergency seems relegated to the all-too-emotional margins, unfit for sanctioned and civil discourses. If we are struck silent in the face of all this, such silence partly emerges from the inadequacy of language to properly frame environmental crisis. This, as discussed, however, is less a matter of circumstances transcending language and more so an issue around dominant grammars and their own shortcomings. Indeed, silence is also an effect of disappearing worlds. When worlds disappear, so do the stories told about those worlds.

To refuse the conditions of this silencing, however, is to insist on the continuation of livable worlds. This is precisely why stories matter (see Cronon 1992; Cameron 2012; Haraway 2016). Stories are capable of shaping outrage and inspiring empathy for disappearing others. They can unlock imaginations and strategies for bringing about better futures. As van Dooren (2014) put it, "Telling stories has consequences: one of which is that we will inevitably be drawn into new connections and, with them, new accountabilities and obligations" (10).

Thus, there is an urgent need to tell stories that can enroll people to enact the kind of radical and systemic change worryingly absent at present. Put differently, if we are going to have to fight for this world, we better learn to fall in love with it first. Falling in love with it requires that we understand how we are part of it. Given the seemingly novel sociopolitical and environmental circumstances of the Anthropocene, there is a need for stories that decenter the human as the sole determinant of meaning and thoughtfully consider the grammar they marshal to make sense of the world. These are the stories of Skywoman's Fall, trickster coyotes, or sentient glaciers—stories that acknowledge and unfold with the world by maintaining attunement to the world. We also need stories of life thriving in circumstances that are far from ideal. These stories can tend to the margins where the fleshy bodies of more-than-humans congregate—for example, minnows (Todd 2016), mushrooms (Tsing 2015), and crows (van Dooren 2019). There is a need for stories that speak the terror of disappearing lands and underscore the silencing of other-than-human voices. Such stories are not new, of course. Stories of disappearance and dispossession are known, although often ignored.

That this great range of stories persists raises the question of who gets to tell them. Far be it from us, two European-American scholars situated on unceded Abenaki territory, to grant permission. It strikes us, though, that those in the best position to tell these stories are those who have been doing so all along. One task for settlers who directly benefit from colonialism is to listen and learn from the stories that precede settler colonialism yet that nevertheless endure (see Nungak and Arima 1988; Moses and Goldie 1998). As Davis and Todd (2017) remarked, "Industrialized capitalism might make us forget our entwined relations and dependency on this body of the Earth, but we are surrounded by rich traditions and many people that have not forgotten this vital lesson. If we are to adapt with any grace to what is coming, those with power … would do well to begin to listen to those voices" (776). Maybe after time listening will feel like remembering. If we remember, then we might better enact other futures (but let us not wait too long).

Telling stories is important. Listening is equally so. Listening well involves humility and the willingness to cede territory (literally and figuratively) that many in positions of power resist. "Power," Rose (2015) noted, "lies in the ability to not hear what is being said, not to experience the consequences of one's actions, but rather to go one's own self-centric and insulated way" (128). Thus, listening well requires decentering ourselves as fixed and referential subjects and shifting our attention to the relations within which we are entangled. It is a strategy that asks us to listen to human and nonhuman storytellers. None of this is easy. Yet, we draw insight from Tsing's (2015) "arts of noticing," which work to collapse distances between humans and more-than-humans through recognition of complexly mixing bodies. We are inspired by Kanngieser and Beuret (2017), who advocated for active listening "designed to draw the more-than-human background into the foreground of thought" (370). We learn from Rose's (2013) "tellers" scattered about the world, "those who provide information: they give news of what is happening in the world" (103). Rose offered the following examples:

- When the march flies bite, the crocodiles are laying their eggs.

- When the cicadas sing, the figs are ripe and the turtles are fat.
- When the fireflies come, the conkerberries are ripe. (Rose 2013, 103)

The world is continually telling stories. Lands are disclosing old histories while revealing new tellers. What is the Arctic telling us? In the Arctic, rising temperatures bring mosquitoes. Their incessant feasting drives caribou to colder snowbanks where mosquitoes are fewer, but so are available nutrients (Culler, Ayres, and Virginia 2015). Aggressive mosquitoes and withering caribou herds tell this story. The Arctic is a pollution sink (Vinogradova 2000). By wind and water current, industrial contaminants end up in Arctic environments. Some are trapped in glaciers and permafrost. As temperatures rise, ice melts and pollutants leech out, threatening communities and wildlife (Liboiron 2013). The contaminants tell a story. In August 2019, Donald Trump expressed interest in purchasing Greenland, claiming that it would provide economic and geopolitical gains (Baker & Haberman 2019). Power is attracted to what lies below the ice: valuable minerals. Disappearing ice reveals opportunities for extraction (Johnson 2010). New mining operations, newly open sea routes, and geopolitical maneuvering all tell a story. Glaciers announce retreats with cracks, groans, rumbles, and rushing waters. As they shed mass, records of past atmospheres dissolve. As glaciers retreat inland, their absence is marked by silence. Silence, too, tells a story. We would do well to listen to what the world is telling us and what it is no longer able to.

A Ghost Story

We began with a funeral, so it is fitting to end with ghosts. What makes a good ghost story? We pose this question not in reference to the kind of ghost story that causes one—with rapid heartbeat and clenched fists—to jump in fright at "the reveal." Such stories have their place, for sure. Rather, we have in mind ghost stories that suggest something is not altogether right. That something about a given present is broken in a way to suggest reality is not as rationally ordered as assumed. We call on Fisher's (2016) discussion of "the eerie." For Fisher, "the eerie" is "constituted by a failure of absence or by a failure of presence. [It] occurs either when there is something present when there should be nothing, or

there is nothing present when there should be something" (61). For us, dead glaciers are eerie. Their absence haunts our present and marks a silence that is beginning to resonate around the world. This haunting is most obvious in spaces where glaciers ought to be but are not. (Ok's absence haunts the very landscape it has revealed.) Yet, the ghosts of other glaciers haunt other and sometimes less obvious places. Rising seas are haunted by the dissolution of these masses once living, thriving, and pulsing at their latitudes. Although the effects of glacier recession are unevenly distributed, in time these ghosts of glaciers will haunt everywhere to indicate their catastrophic absence.

We close with a curious account: In Kalaallit Nunaat (Greenland), the Inuit carved portable maps out of driftwood for navigating coastal waters. The pieces are shaped to represent coastlines, are compact, are buoyant, and can be read in the dark (Decolonial Atlas 2019). The driftwood coastlines no longer match their territorial counterparts. In *Underland*, Macfarlane (2019b) recounted how "small craft hugging the Greenland coastline will sometimes find their GPS [Global Positioning System] navigation devices screaming in alarm, warning of collision" (344). Yet, there is no ice to be found in close proximity. Macfarlane (2019b) continued: "The coordinates of the former extent of glaciers have been inputted into the mapping, but the retreat rate has been so fast that they are sailing into and *through* the digital phantom left behind by the ice" (344). Both driftwood cartography and GPS maps are rendered artifacts in this quickly changing world, which is to say that the ghosts of disappearing others haunt things far and wide. The severity of future hauntings, however, has yet to be determined, provided that we alter course and summon the courage and will to bring about better futures. Once more, Magnason's (Agence France-Presse 2019) haunting words come to mind:

> This monument is to acknowledge that we know what is happening and know what needs to be done. Only you know if we did it.

Acknowledgments

We thank the two anonymous reviewers of the article and Dr. David R. Butler for their perceptive, constructive comments, and suggestions. All errors and omissions are ours alone.

References

Agence France-Presse. 2019. Iceland holds funeral for first glacier lost to climate change. *The Guardian*, August 18.

Alaimo, S. 2010. *Bodily natures: Science, environment, and the material self*. Bloomington: Indiana University Press.

Alaimo, S. 2016. *Exposed: Environmental politics and pleasures in posthuman times*. Minneapolis: University of Minnesota Press.

Aporta, C. 2011. Shifting perspectives on shifting ice: Documenting and representing Inuit use of the sea ice. *The Canadian Geographer / Le Géographe Canadien* 55 (1):6–19. doi: 10.1111/j.1541-0064.2010.00340.x.

Baker, P., and M. Haberman. 2019. Trump's interest in buying Greenland seemed like a joke. Then it got ugly. *The New York Times*, August 21.

Barad, K. 2007. *Meeting the universe halfway: Quantum physics and the entanglement of matter and meaning*. Durham, NC: Duke University Press.

Bennett, J. 2010. *Vibrant matter: A political ecology of things*. Durham, NC: Duke University Press.

Berkes, F. 2012. *Sacred ecology*. London and New York: Routledge.

Cameron, E. 2012. New geographies of story and storytelling. *Progress in Human Geography* 36 (5):573–92. doi: 10.1177/0309132511435000.

Cameron, E., R. Mearns, and J. T. McGrath. 2015. Translating climate change: Adaptation, resilience, and climate politics in Nunavut, Canada. *Annals of the Association of American Geographers* 105 (2):274–83. doi: 10.1080/00045608.2014.973006.

Ceballos, G., P. R. Ehrlich, and R. Dirzo. 2017. Biological annihilation via the ongoing sixth mass extinction signaled by vertebrate population losses and declines. *Proceedings of the National Academy of Sciences* 114 (30):E6089–E6096. doi: 10.1073/pnas.1704949114.

Collard, R.-C., J. Dempsey, and J. Sundberg. 2015. A manifesto for abundant futures. *Annals of the Association of American Geographers* 105 (2):322–30. doi: 10.1080/00045608.2014.973007.

Cronon, W. 1992. A place for stories: Nature, history, and narrative. *The Journal of American History* 78 (4):1347–76. doi: 10.2307/2079346.

Cruikshank, J. 1981. Legend and landscape: Convergence of oral and scientific traditions in the Yukon Territory. *Arctic Anthropology* 18 (2):67–93. doi: 10. 17863/CAM.12841.

Cruikshank, J. 2001. Glaciers and climate change: Perspectives from oral tradition. *Arctic* 54 (4):377–93. doi: 10.14430/arctic795.

Cruikshank, J. 2005. *Do glaciers listen? Local knowledge, colonial encounters, and social imagination*. Vancouver, BC, Canada: UBC Press.

Culler, L. E., M. P. Ayres, and R. A. Virginia. 2015. In a warmer Arctic, mosquitoes avoid increased mortality from predators by growing faster. *Proceedings of the Royal Society B: Biological Sciences* 282 (1815):20151549:1–8. doi: 10.1098/rspb.2015.1549.

Davis, H., and Z. Todd. 2017. On the importance of a date, or, decolonizing the Anthropocene. *ACME: An International Journal for Critical Geographies* 16 (4):761–80.

Decolonial Atlas. 2019. Inuit cartography 2016. Accessed November 25, 2016. https://decolonialatlas.wordpress.com/2016/04/12/inuit-cartography/.

Engel, C. 2019. Scientists unveil memorial to Iceland's "first" dead glacier lost to climate change. *Time*, July 22.

Fisher, M. 2016. *The weird and the eerie*. London: Repeater Books.

Greyson, L. 2019. *Vital reenchantments: Biophilia, Gaia, cosmos, and the affectively ecological*. Brooklyn, NY: Punctum Books.

Haraway, D. 1988. Situated knowledges: The science question in feminism and the privilege of partial perspective. *Feminist Studies* 14 (3):575–99. doi: 10.2307/3178066.

Haraway, D. 2016. *Staying with the trouble: Making kin in the Chthulucene*. Durham, NC: Duke University Press.

Heise, U. K. 2016. *Imagining extinction: The cultural meanings of endangered species*. Chicago: The University of Chicago Press.

Intergovernmental Panel on Climate Change [IPCC]. 2018. Summary for policymakers: Global warming of 1.5° C. In *An IPCC special report on the impacts of global warming of 1.5°C*, ed. V. Masson-Delmotte, P.Zhai, and H. O. Pörtner. Geneva, Switzerland: IPCC: 3–24.

Inuit Circumpolar Conference. 1998. Inuit Circumpolar Conference Charter 1998. Accessed December 2, 2019. https://www.inuitcircumpolar.com/icc-international/icc-charter/

Johnson, L. 2010. The fearful symmetry of Arctic climate change: Accumulation by degradation. *Environment and Planning D: Society and Space* 28 (5):828–47. doi: 10.1068/d9308.

Kanngieser, A., and N. Beuret. 2017. Refusing the world: Silence, commoning, and the Anthropocene. *South Atlantic Quarterly* 116 (2):363–80. doi: 10.1215/00382876-3829456.

Kelman, I., ed. 2017. *Arcticness: Power and voice from the North*. London: UCL Press.

Kimmerer, R. W. 2013. *Braiding sweetgrass: Indigenous wisdom, scientific knowledge and the teachings of plants*. Minneapolis: Milkweed Editions.

Laidler, G. J. 2006. Inuit and scientific perspectives on the relationship between sea ice and climate change: The ideal complement? *Climatic Change* 78 (2–4):407–44. doi: 10.1007/s10584-006-9064-z.

Leduc, T. B. 2007. Sila dialogues on climate change: Inuit wisdom for a cross-cultural interdisciplinarity. *Climatic Change* 85 (3–4):237–50. doi: 10.1007/s10584-006-9187-2.

Liboiron, M. 2013. Plasticizers: A twenty-first-century miasma. In *Accumulation: The material politics of plastic*, ed. J. Gabrys, G. Hawkins, and M. Michael, 134–49. London and New York: Routledge.

Macfarlane, R. 2019a. Should this tree have the same rights as you? *The Guardian*, November 2.

Macfarlane, R. 2019b. *Underland: A deep time journey*. New York: Norton.

Moses, D. D., and T. Goldie, eds. 1998. *An anthology of Canadian native literature in English*. Toronto: Oxford University Press.

Nadasdy, P. 1999. The politics of TEK: Power and the "integration" of knowledge. *Arctic Anthropology* 36 (1/2):1–18.

Nungak, Z., and E. Arima. 1988. *Inuit stories*. Hull, QC, Canada: Canadian Museum of Civilization.

Povinelli, E. A. 1993. *Labor's lot: The power, history, and culture of Aboriginal action*. Chicago: University of Chicago Press.

Povinelli, E. A. 1995. Do rocks listen? The cultural politics of apprehending Australian Aboriginal labor. *American Anthropologist* 97 (3):505–18. doi: 10.1525/aa.1995.97.3.02a00090.

Riedlinger, D., and F. Berkes. 2001. Contributions of traditional knowledge to understanding climate change in the Canadian Arctic. *Polar Record* 37 (203):315–28. doi: 10.1017/S0032247400017058.

Rose, D. B. 2013. Val Plumwood's philosophical animism: Attentive interactions in the sentient world. *Environmental Humanities* 3 (1):93–109. doi: 10.1215/22011919-3611248.

Rose, D. B. 2015. Dialogue. In *Manifesto for living in the Anthropocene*, ed. K. Gibson, D. B. Rose, and R. Fincher, 127–31. Brooklyn, NY: Punctum Books.

Seitler, D. 2008. *Atavistic tendencies: The culture of science in American modernity*. Minneapolis: University of Minnesota Press.

Singh, J. 2017. *Unthinking mastery: Dehumanism and decolonial entanglements*. Durham, NC: Duke University Press.

Todd, Z. 2016. An indigenous feminist's take on the ontological turn: "Ontology" is just another word for colonialism. *Journal of Historical Sociology* 29 (1):4–22. doi: 10.1111/johs.12124.

Tsing, A. L. 2015. *The mushroom at the end of the world: On the possibility of life in capitalist ruins*. Princeton, NJ: Princeton University Press.

van Dooren, T. 2014. *Flight ways: Life and loss at the edge of extinction*. New York: Columbia University Press.

van Dooren, T. 2019. *The wake of crows: Living and dying in shared worlds*. New York: Columbia University Press.

Vinogradova, A. A. 2000. Anthropogenic pollutants in the Russian Arctic atmosphere: Sources and sinks in spring and summer. *Atmospheric Environment* 34 (29–30):5151–60. doi: 10.1016/S1352-2310(00)00352-6.

Weatherhead, E., S. Gearheard, and R. G. Barry. 2010. Changes in weather persistence: Insight from Inuit knowledge. *Global Environmental Change* 20 (3):523–28. doi: 10.1016/j.gloenvcha.2010.02.002.

Weston, K. 2017. *Animate planet: Making visceral sense of living in a high-tech ecologically damaged world*. Durham, NC: Duke University Press.

Yusoff, K. 2013. Insensible worlds: Postrelational ethics, indeterminacy and the (k)nots of relating. *Environment and Planning D: Society and Space* 31 (2):208–26. doi: 10.1068/d17411.

Ruins of the Anthropocene: The Aesthetics of Arctic Climate Change

Mia M. Bennett

In the Anthropocene, ruin appreciation is shifting its focus from crumbling architecture to the deteriorating planet. Whereas Romantic and modern ruin gazing privileged nature's reconquest of the built environment, now, the carbon-intensive infrastructures of global capitalism are turning nature itself to ruin. By critiquing popular representations of the melting Arctic—a visual trope within Anthropocene aesthetics involving images of shrinking icebergs, melting glaciers, and drowning polar bears—this article explicates how both conceptions of ruins and actual, material processes of ruination are shifting away from manmade infrastructure toward the natural environment. I argue that ruins in the Anthropocene are distinct in that natural ruins, especially icy ones, will not persist on the landscape, particularly as environmental degradation accelerates and is upscaled to encompass entire regions like the Arctic, if not the whole planet. By applying Romantic aesthetic principles, I critique the two dominant categories of representations of the current geological epoch: the picturesque and the sublime. As with Romantic and modern ruin iconography, depictions of Anthropocene ruins harness these elements to induce feelings of awe, melancholy, and resignation. These reactions might now be more problematic, however, because helplessness and passive voyeurism could inhibit action on climate change. I thus conclude that refocusing the Anthropocene gaze on the third aesthetic principle—the beautiful, which emphasizes the tangible and comprehensible—might be more conducive to transforming aesthetics into action and fostering an effective rather than affective ethics of planetary care and stewardship. *Key Words: aesthetics, Anthropocene, Arctic, climate change, ruins.*

In 2007, a substantial chunk of the Arctic's sea ice went missing. By summer's end, when the icy blanket draped over the top of the planet shrinks to its annual minimum, its extent had dropped 24 percent below the previous record (Comiso et al. 2008). The ice had disappeared, with most of it melting into the warming, acidifying ocean. The decaying bergy bits and growlers still remaining form some of the most recognizable visuals of the ruins of the Anthropocene, the geological epoch in which humans have arguably become the largest single force affecting the planet (Crutzen 2006).

The scientific jury is still out on whether the Anthropocene constitutes a geological break from the Holocene. Nonetheless, there is growing consensus within the social sciences and humanities that it marks an "aesthetic event" (Davis and Turpin 2015, 11), one dominated by rapid ruination. The epoch "speaks to a specific ruining: the ruining of modernity" (Beuret and Brown 2017, 332). The Anthropocene's power to challenge beliefs in irreversible progress and mankind's ability to control nature has compelled significant artistic and literary exploration (Möllers 2013). Photographer Edward Burtynsky, in books like *Anthropocene* (Burtynsky et al. 2018), captures what he called "the ruins of our society" (Zehle 2008, 111). Literary scholar Jason McGrath (2014), too, explored the "logic of late Anthropocene ruins." For these artists and scholars, ruins in the Anthropocene are typically infrastructural rather than natural. As moments such as the calving of the Greenland ice sheet in the climate change documentary *Chasing Ice* (Orlowski 2012) depict, however, the natural environment is crumbling now, too. Such depictions of a "fallen" Arctic extend appreciation for ruins born out of both Romantic and modern aesthetics. Yet these sublimely apocalyptic visualizations also suggest something distinctly environmental about ruin imaginaries in the Anthropocene.

Recognition of nature's ongoing ruination bears implications for how people both react to climate change and think about place. Carruth and Marzec (2014) suggested that visuals shape "ways of seeing and also perceiving twenty-first-century ecological realities" (207). Yet visuals and, more broadly, aesthetics have the power to mold those very realities, too. They can also redistribute political agency in ways that can help or hinder marginalized communities (Engelmann and McCormack 2018). Journalistic accounts of the Arctic's sinking houses and relocated villages perpetuate the victimization and "othering" of gazed-upon communities typical of both ancient and post(modern) ruin consumption (Herrmann 2017, 2019), while narratives portraying the region's 500,000 Indigenous people as vulnerable undermine their autonomy (Martello 2008; Haalboom and Natcher 2012). The aesthetics of climate change are therefore instrumental to shaping human attitudes and actions and the futures of places themselves.

Fostering a more empowering Anthropocene aesthetics first requires defining aesthetics. For Plato, the founder of philosophical aesthetics, art constitutes an essential expression of the human spirit and represents an attempt at ordering the world in pursuit of beauty (Hofstadter and Kuhns 1964). The Neoplatonist philosopher Plotinus accorded art an even more central position in his thinking, believing that it could "reveal the form of an object more clearly than ordinary experience" (Wilson 2006, 21). Scrutinizing aesthetic experience, or the attitudes, emotions, and perceptions wrought by art, is thus critical to making sense of the real world. Philosophical aesthetics bears parallels with affectual geography, which views emotions, rather than being exclusive to individuals, as existing between them and their "perceptual environments" (Pile 2010, 13). Affect is malleable and can be instrumentalized to reconfigure actual environments, too, especially because "affective imagery"—sensorial depictions to which positive or negative feelings become attached over time through repeated exposure—can affect attitudes toward climate change policies (Leiserowitz 2006).

To reconcile the Anthropocene as both aesthetic event and geological epoch, the ideas of twentieth-century U.S. philosopher John Dewey, who was deeply concerned with aesthetic experience and everyday aesthetics (Haskins 1992), might prove insightful. Dewey (1925), recognizing that science itself is an art, asserted that art could ultimately and ideally become "the complete culmination of nature" (358). Just as the Anthropocene might have its origins in aesthetics, with the discovery of combustion—the very process now indirectly melting Arctic sea ice via the burning of fossil fuels—possibly emerging from artistic experiments in firing ceramics (Clark 2015), solutions for the epoch might be born of them as well.

This brief article has three goals. First, I aim to shift theories of ruins in the Anthropocene away from the built environment toward the natural environment, demonstrating how ruination has expanded and accelerated. Although this heuristic assumes an overly dualistic categorization of the built and natural environments, it is helpful for understanding the changes occurring to both aesthetic representations and material processes of Anthropocene ruination and for addressing the epoch's consequences for conceptions of place. Second, I critique the predominance of the "picturesque" and the "sublime"—two fundamental aesthetic categories rooted in the Romantic era—within representations of Arctic climate change. In over a decade of researching the Arctic through both critical discourse analysis and ethnography, I have witnessed changes to how popular narratives represent the region. These observations have enabled me to select several vignettes that exemplify the aesthetics of Arctic climate change. Third, I contend that if climate change visualizations emphasize the third aesthetic category of the "beautiful," which privileges comprehensibility and familiarity, over the picturesque or sublime, they may more readily transform aesthetics into action.

The Upscaling and Acceleration of Ruination in the Anthropocene

Contemporary ruin aesthetics can be traced to Romanticism in eighteenth-century Europe, which was gripped by a "cult of ruins" (Huyssen 2006, 7). Sites like vine-strangled Gothic churches and disintegrating Viking farms—"half building, half nature" (Stead 2003, 52)—exuded the sublime, which Burke ([1757] 1990, 67) called "delightful terror." In the twentieth and twenty-first centuries, a new iteration of ruin appreciation gained cultural traction as recognition of the failures of capitalism and political interventions arose (Martin 2014). Interest in "modern ruins" manifests in voyeuristic phenomena such as "dark tourism," in which individuals spectate northeast English industrial wreckage and radioactive Soviet landscapes (Lennon and Foley 2000). As with Romantic ruin-gazing, which was limited to those in a position of imperial or colonial privilege, modern ruin-gazing is confined to those possessing "scopic mastery" (Hell 2009, 292): individuals who can ephemerally enjoy the sight of a ruin rather than those who still reside within it (M. M. Bennett 2020).

In Romantic and modern ruin aesthetics, the built environment is lost to nature. From this vantage point, the decaying natural environment might not be perceived as a ruin. The aesthetics of Anthropocene ruins, however—devastation, individual powerlessness, and a reversal of progress—overlap with those of Romantic and modern ruins. Yet whereas they privilege the archaeological, architectural, and the slow return to the geological, Anthropocene ruins cut to (and through) the chase. Images of melting ice caps and burning rainforests immediately bring to light the "social as composed through the geologic" (Yusoff 2013, 780). The spectated ruin emblematizes the fall of infrastructure and nature, which have now converged (Chester, Markolf, and Allenby 2019).

Ruination in the Anthropocene has two distinct features: a spatial shift and upscaling from the built to the natural environment and a temporal acceleration. First, ruins are no longer just buildings but rather whole regions, ecosystems, and potentially even the entire "ruined Earth of the Anthropocene" (Beuret and Brown 2017, 346). A comparison of two *New York Times* nonfiction bestsellers from 2007 and 2019 reveals this shift and upscaling. In 2007, Weisman's *The World Without Us* portrayed a planet from which humans had suddenly disappeared. In contrast, in 2019, Wallace-Wells's *The Uninhabitable Earth: Life after Warming* painted a lyrically grim picture of a planet rendered imminently unlivable due to extreme temperatures, sea-level rise, and food shortages. The crux of the matter has become not what will happen to all of the things humans have built on top of the planet but rather how humans will cope with an unrecognizable Earth whose ecosystem we have eviscerated.

Second, with the positive feedback loops of climate change becoming clearer, ruination is accelerating. As Chakrabarty (2009) contended, "The geologic now of the Anthropocene has become entangled with the now of human history" (212). Large-scale human activity has hastened the normally creeping pace of geologic time so much that the connotation of "moving glacially" might speed up. At this century's outset, scientists predicted that the Arctic would lose its summer sea ice by century's end (Gregory et al. 2002). By 2020, using the latest generation of climate models, scientists estimated that due to warming already set in motion, summer sea ice will almost certainly disappear before 2050

even if we were to immediately halt greenhouse gas emissions (SIMIP Community 2020). Ice-free summers might happen much sooner than scientists estimated just twenty years ago, portending a certain finitude for a place that was frozen throughout all of human history and long before it, too.

There is something deeply disorienting about losing an entire place—or perhaps all of Earth as we know it—not only as it has been socially constructed but as it has naturally emerged, too. In both their Romantic and modern conceptions, it was assumed that ruins would endure on the landscape in perpetuity. In the Anthropocene, the landscape itself is in peril, and the Arctic might be one of the first ruins with an end date. Once summer sea ice disappears, this would mark the first time in 2.6 million years that the Arctic lacks ice cover (Knies et al. 2014). The sublime, as theorized by Kant, suggests that the Earth is indifferent to human survival (Grove and Chandler 2017). Meillassoux (2010) reworked this idea in his exploration of "ancestrality": The Earth existed before us and it will continue to exist after us. Although the Arctic, at least in its glaciated iteration, predated humanity, it might not survive after or even during the species' existence. The cryosphere could experience more than just ruination: It might disappear altogether, with only lines on a map like the Arctic Circle to remind us where the region once lay as the icy features that defined how the region was inhabited and imagined melt away.

Arctic Ruins: Picturesque Funerals and Sublime Resurrections

As conceived by the Romantics, visual aesthetics comprise three categories: the picturesque, the sublime, and the beautiful (Trott 1999). Whereas the picturesque is a style or affect that captures and curates sights worth seeing, the sublime inheres in spectated objects. In Kant's conceptualization, things that are sublime may possess either mathematical sublimity, with nature's infinite extent transcending beyond the realm of human comprehension, or dynamic sublimity, representing nature's uncontrollable force (Budd 1998). Climate change aesthetics reproduce both characteristics, emphasizing nature's magnitude and might in ways that could contradict efforts to promote feelings of personal empowerment or responsibility. In contrast, the beautiful appears controllable and contained, inspiring devotion and affection. Moving away from picturesque and sublime depictions of ruination and summoning the beautiful in their place might produce affective encounters that strengthen an ethics of stewardship and care and spark possibilities for action in the Anthropocene (Roelvink and Zolkos 2011; Beacham 2018).

The Problem of the Picturesque

The contemporary prominence of visual culture has led the picturesque to accrue value relative to the sublime and the beautiful. In contemporary parlance, *aesthetic* has become an adjective meaning "pleasing to look at," as in, "That's so aesthetic." Instagramming adolescents are not the only ones seeking visual pleasure: Anthropocene-era tourists also scout out places with exotic, photogenic elements (Smith 2018)—perhaps those that future generations might not be able to capture. The Arctic is a travel destination possessing such traits, exemplified by the explosion of tourism in Iceland. One unexpected factor contributing to the country's sudden popularity was pop singer Justin Bieber's music video for a song fittingly called "I'll Show You" (2015), watched nearly half a billion times on YouTube (Huddart and Stott 2016). In the video, the scenic film locations of which attracted visitors in droves to previously untrammeled Icelandic canyons and cliffs, the singer gazes down from mossy promontories in shots reminiscent of German Romantic artist Caspar David Friedrich's painting *Wanderer above the Sea of Fog* (1818). Stark Icelandic vistas, however, are more common than icebergs and polar bears. These icons of Arctic climate change represent even more highly sought souvenirs of "last chance tourism" (Dawson et al. 2011), which also constitutes "ruin tourism" in the Anthropocene. Amidst calving ice sheets and thawing glaciers, as consumption of spectated sites shifts from ruined buildings to the eviscerating Earth, Urry's (1992) "tourist gaze" merges with what might be called the "Anthropocene gaze."

The proliferation of picturesque iconography parallels the externalization, mediatization, and commodification of popular sites within "processes of social remembering" (Edensor 2005, 830). In certain external narratives, the Arctic is not only depicted as on the cusp of disappearing: It has already died. In 2016, environmental nongovernmental organization Greenpeace built a floating white platform on which Italian pianist Ludovico Einaudi played a song he composed titled "Elegy for the Arctic." The platform drifted in front of a picturesque tidewater glacier in Svalbard, playing precisely to the epoch's aesthetics. Acts of remembrance like performing an elegy—a lament for the dead—reify the idea that the Arctic is already lost. As an *Economist* headline lamented, "The Arctic as it is known today is almost certainly gone" (2017). Funereal portrayals make it seem as if nothing can be done to change the future, embodying what Boym (2001) identified as reflective nostalgia, which "dwells in *algia* (aching), in longing and loss, the imperfect process of remembrance" (141).[1]

The aesthetic category of the picturesque in the Anthropocene evokes longing and mourning but rarely anger. In fact, depictions that are not picturesque provoke the most heated reactions. In 2019, a video of a polar bear—a climate change icon (Ziser and Sze 2007)—with graffiti spray-painted onto its fur in Chukotka, Russia, went viral, "spark[ing] outrage," according to a *Guardian* headline (Roth 2019). The video disrupted the aesthetics of the Anthropocene that show nature in ruins without identifying a culprit. The emblazoned polar bear makes painfully evident that at least one individual committed this act of environmental vandalism. The ruined ursine violates visual expectations of environmental destruction in the Anthropocene, for there is nothing picturesque or sublime about the bear.

The Daunting Sublime

In forcing reconsideration of our place in the world, the Anthropocene represents "a uniquely sublime moment in the history of the species" (Williston 2016, 171). Previously, the sublime inhered in unfettered nature, with the Arctic an exemplar of the aesthetic category. In the nineteenth century, as polar exploration—an imperial endeavor formative to nation-building in places like Britain and Norway—took off, the "Arctic sublime" suffused novels such as Mary Shelley's *Frankenstein* (in which it was enhanced by the "Gothic sublime") and landscape painting (Morgan 2016). Even empirical descriptions of Arctic sea ice from the time exude sublimity. Sailing near Alaska in 1778, Captain James Cook (1784) observed, "We were, at this time, close to the edge of the ice, which was as compact as a wall; and seemed to be ten or twelve feet high at least."

Figure 1. Sled dogs dash through meltwater floating atop a frozen inlet in northwest Greenland. Photo: Steffen Olsen (2019), reproduced by permission.

Shrinking to a brittle shell, the ice is eliciting a return to yet another sublime: that of the nineteenth-century ocean. Its vast and uncharted depths were a leitmotif for artists of the time, with its "deathly attraction" satisfying Romantic sensibilities reacting against Enlightenment rationality (Levine 1985, 389). Anthropocene aesthetics reawaken such sentiments. Maps accompanying a *National Geographic* article titled "What the World Would Look Like If All the Ice Melted" (2013), depict a waterlogged Earth in which an average sea-level rise of 65.8 m has inundated the Mississippi Delta, the Amazon Basin, and eastern China. The monumentality of these anticipated ruins, constituting entire regions rather than individual buildings or even cities, at first arouse anxiety. Still, the accompanying article underscores, an ice-free planet is still at least 5,000 years away. In this temporal distancing, the reader, generations away from the global deluge, experiences the sublime and might feel uninspired to act on climate change.

One photograph taken in northwest Greenland simultaneously captures the Arctic, oceanic, and Anthropocene sublime. Shot in June 2019 by Danish climatologist Steffen Olsen, the image depicts a team of sled dogs dashing across cerulean meltwater floating atop a frozen inlet with the musher absent from view (Figure 1). With its striking colors and biblical imagery of sled dogs appearing to walk on water, the image went viral. The photograph depicts an ice-free Arctic already arrived, inducing awe and helplessness while upholding a sense of safety. The striding sled dogs appear able to reach the shore; perhaps humans—or, more specifically, the scientists and their team of local hunters—will still triumph over the rising seas with their perseverance, knowledge, and technology. Yet the supernatural (or rather supranatural) scene might also engender thinking that only *deus ex machina* solutions such as geoengineering fixes mobilizing the "technological sublime" (Nye 1994) offer an escape from a diluvial future. By overpowering the viewer, sublime aesthetics diminish the potential for individual action. To that end, instead, the aesthetic category of the beautiful might be more conducive.

On Beauty and Transforming Aesthetics into Action

Indian novelist Amitav Ghosh has asked why no great novel about climate change has been written. He argued that if literature neglects to engage with climate change, its "failures will have to be counted as an aspect of the broader imaginative and cultural failure that lies at the heart of the climate crisis" (Ghosh 2016). Yet gradually, popular culture is reckoning with climate change, expressing it in visual

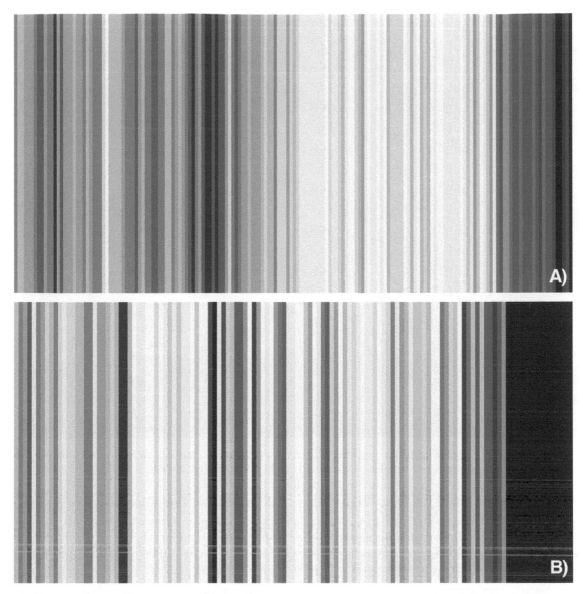

Figure 2. (A) Warming Stripes for the entire planet depicting annual global temperatures (1850–2019). (B) Warming Stripes for the Arctic Ocean depicting annual regional temperatures (1893–2019). Figures by Professor Ed Hawkins (University of Reading) and licensed under CC BY 4.0 (https://showyourstripes.info).

and acoustic (if still not literary) ways accentuating the beautiful. To encourage individual human agency to affect change within the Anthropocene, this aesthetic category might be more productive than the picturesque or sublime.

A popular climate data visualization called "Warming Stripes" or "Climate Stripes" reveals beauty's affective power. For one week in September 2019, as the northern hemisphere's hottest summer yet on record was finally winding down and as international climate strikes took to the streets, media outlets worldwide featured the simple, clear graphic on their covers and front pages. Designed by Ed Hawkins, a climate scientist

at the University of Reading, the stripes represent the average global temperature in each year between 1850 and 2019 as a single line (Figure 2A). There are no labels or scales: The bluer hues corresponding to earlier years simply ominously redden as time marches toward the present and the Earth warms. The graphic lacks the picturesque qualities that lead viewers to consume landscapes and the sublime aspects that overwhelm. The stripes, which can be customized online for different regions (see https://showyourstripes.info) such as the Arctic (Figure 2B), provide a promising and scalable alternative to the conventional Anthropocene gaze myopically trained on a ruined Earth.

Figure 3. At the 2019 Reading Festival, headline act The 1975 performed their new single featuring a recording of Swedish teenage climate activist Greta Thunberg urging action on climate change. Photo: Mia M. Bennett (2019).

One month prior to the 2019 Climate Week, the Warming Stripes entered the pop pantheon at the Reading Festival, Britain's longest running music festival, which takes place each summer in the city where Hawkins is based. As British rock band Enter Shikari's set began, the climate stripes were projected on stage in front of 100,000 spectators. The band's climate consciousness was not anomalous: Two days prior, festival headliner The 1975, another English band, played their new single featuring a five-minute recording of Swedish teenage climate change activist Greta Thunberg intoning over a tinkling piano and synthesizers, "It is time to rebel," as placid stock images of nature flickered on the Jumbotron (Figure 3). On current trends, it seems that the first great climate change song or album will be written before the first great climate change novel. In 2020, Canadian indie artist Grimes released her fifth album, *Miss Anthropocene*—wordplay befitting the epoch (Figure 4). Discussing the album's genesis, she reflected, "I want to make climate change fun. People don't care about it because we're being guilted. I see the polar bear and want to

kill myself. No one wants to look at it, you know? I want to make a reason to look at it. I want to make it beautiful" (Bradley 2019).

If something is beautiful without being tantalizingly picturesque or dauntingly sublime, people, driven by affection rather than terror, might feel more compelled to act. Macfarlane (2016) found that in the darker subgenres of Anthropocene writing, "categories such as the picturesque or even the beautiful congeal into kitsch." Yet the beautiful need not be maudlin, as the Warming Stripes illustrate. Rather, it is depictions of picturesque and sublime ruins that summon the sentimental by spotlighting victimized Indigenous Peoples, forlorn flora and fauna, and melted-down, burnt-out landscapes. Striving to instead foster a beautiful or "charming Anthropocene" (Buck 2015) resonates with trying to find "some kind of joy" (Morton 2018, 161) amidst the epoch's existential dread. Echoing calls for a vital materialism recognizing the liveliness and agency of matter (J. Bennett 2010; Lorimer 2012), a vital aesthetics might be required, too. Rediscovering and representing the beauty that remains on Earth

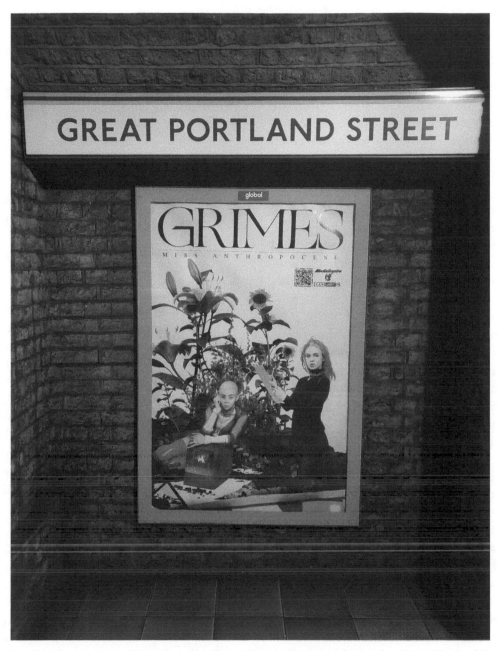

Figure 4. A poster for Canadian musician Grimes's studio album, *Miss Anthropocene*, displayed at Great Portland Street tube station in the London Underground. Photo: Mia M. Bennett (2020).

rather than fixating on ecological, geological, and glaciological ruins might encourage an ethics of care and stewardship for what we still have.

Conclusion: From Affective to Effective Aesthetics

A paradigm shift in ruin aesthetics and material ruination is occurring in the Anthropocene. First, ruins are shifting from the built to the natural environment as the carbon-intensive infrastructures of global capitalism undermine nature's foundations. Whereas infrastructural ruins once lingered on the landscape, many ruins of the Anthropocene like the Earth's icy edges will not last, at least in certain seasons. Ruination is also accelerating and scaling up to encompass the planet. Entire regions risk losing their ecological underpinnings, challenging the basis for how they are both imagined and inhabited as places.

Second, by emphasizing the picturesque and sublime, Anthropocene ruin aesthetics exacerbate the problems of Romantic and modern ruin-gazing. Whereas ruin appreciation in Manchester or Chernobyl risks voyeurism, consuming imagery of elegiac icebergs and nearly-swimming sled dogs might hasten climate change if such visualizations condone passivity rather than action. Refocusing narratives on the beautiful and promoting a comprehensible rather than extraordinary aesthetics could more effectively inspire action on climate change. Surprisingly, pop music is one place where the beautiful is making itself heard, if not seen. In the absence of optics, it might be easier for the beautiful to transcend the sublime and the inherently visual picturesque. Elevating the beautiful could also find inspiration within Antiquity's early explorations of philosophical aesthetics, which associated the quality with order, harmony, and balance (Diessner 2019)—all attributes of a planet in sync.

The representations of the Anthropocene described in this article reflect a worldview emblematic of Western media and pop culture while marginalizing many others. Even the epoch's name, "in its reassertion of universality … implicitly aligns itself with the colonial era" (Davis and Todd 2017, 763). Nevertheless, critiquing Anthropocene aesthetics—an "unintended supplement to imperial aesthetics" (Mirzoeff 2014, 220)—can help reveal how to influence those most invested with the power to alter planetary conditions. This is imperative because although "not every human is responsible for bringing on the Anthropocene, every human is destined to live in it" (Hamilton 2017, 77). To more fully represent the wide range of Anthropocene aesthetics, scholars should equally address how the people living within the landscapes, icescapes, and seascapes represented as ruins interpret such visualizations. Future research should also heed the alternative and subaltern stories they tell about their homes.

The emergence of subfields such as the geohumanities and environmental humanities points to the growing convergence of geographical, environmental, and artistic concerns in the Anthropocene. Interventions at this nexus might offer a chance to build a more radical Anthropocene aesthetics, for art in its truest sense constitutes not just a "radical political act" offering alternative sociopolitical configurations (Tuan 1989, 239). It can be a radical environmental act, too, which serves to realize the commensurability between art and nature for which

Dewey hoped. Diversifying Anthropocene aesthetics might produce livelier, more sensitive depictions of a climatically altered planet whose most vulnerable places are often still inhabited. Fostering an effective as opposed to affective aesthetics in the Anthropocene might also increase the possibility that these places will remain so.

Epilogue: After the Ice

If all the ice melted—the Arctic's summer ice cap perhaps by 2050, and the entire cryosphere many more millennia from now—no ruins would remain. The ice would simply transform into water or vapor. The Arctic, then, might not only be a region with an end date. One day, there could be nothing visible to mourn. In remembering loved ones who have vanished without leaving corporeal remains, humans substitute touchstones like flags or coffins (Sturken 2004). What will stand in, though, for vanished ice?

A recent funeral in Iceland may provide some indication. In August 2019, dignitaries, artists, and residents bid farewell to one of the first glaciological victims of climate change: Ok(jökull), an 800-year-old glacier described in the sagas. During the ceremony, a small bronze plaque was unveiled. Engraved on it was a message in Icelandic and English entitled, "A letter to the future":

> Ok is the first Icelandic glacier to lose its status as a glacier.
>
> In the next 200 years all our glaciers are expected to follow the same path.
>
> This monument is to acknowledge that we know what is happening and what needs to be done.
>
> Only you know if we did it.
>
> Ágúst 2019
>
> 415ppm CO_2

Note

1. The second form she distinguished, restorative nostalgia, emphasized "*nostos* (returning home) and proposes to rebuild the lost home and patch up the memory gaps" (Boym 2001, 41). In the context of climate change, restorative nostalgia manifests in efforts such as the pursuit of geoengineering fixes to restore the ice cap.

References

The Arctic as it is known today is almost certainly gone. 2017. *The Economist*, April 29. Accessed May 31, 2020. https://www.economist.com/leaders/2017/04/29/the-arctic-as-it-is-known-today-is-almost-certainly-gone.

Beacham, J. 2018. Organising food differently: Towards a more-than-human ethics of care for the Anthropocene. *Organization* 25 (4):533–49. doi: 10.1177/1350508418777893.

Bennett, J. 2010. *Vibrant matter: A political ecology of things.* Durham, NC: Duke University Press.

Bennett, M. M. 2020. The making of post-post-Soviet ruins: Infrastructure development and disintegration in contemporary Russia. *International Journal of Urban and Regional Research*, 1–16. doi: 10.1111/1468-2427.12908.

Beuret, N., and G. Brown. 2017. The walking dead: The Anthropocene as a ruined Earth. *Science as Culture* 26 (3):330–54. doi: 10.1080/09505431.2016.1257600.

Boym, S. 2001. *The future of nostalgia.* New York: Basic Books.

Bradley, R. 2019. The life and death of Grimes. *The Wall Street Journal Magazine.* Accessed November 9, 2020. https://www.wsj.com/articles/the-life-and-death-of-grimes-11553084548

Buck, H. J. 2015. On the possibilities of a charming Anthropocene. *Annals of the Association of American Geographers* 105 (2):369–77. doi: 10.1080/00045608.2014.973005.

Budd, M. 1998. Delight in the natural world: Kant on the aesthetic appreciation of nature. Part III: The sublime in nature. *The British Journal of Aesthetics* 38 (3):233–50. doi: 10.1093/bjaesthetics/38.3.233.

Burke, E. [1757] 1990. *A philosophical enquiry into the origin of our ideas of the sublime and the beautiful.* Oxford, UK: Oxford University Press.

Burtynsky, E., J. Baichwal, N. De Pencier, S. Boettger, C. Waters, J. A. Zalasiewicz, M. Atwood, Art Gallery of Toronto, National Gallery of Canada, and Fondazione MAST. 2018. *Anthropocene.* Göttingen, Germany: Steidl.

Carruth, A., and R. P. Marzec. 2014. Environmental visualization in the Anthropocene: Technologies, aesthetics, ethics. *Public Culture* 26 (2):205–11. doi: 10.1215/08992363-2392030.

Chakrabarty, D. 2009. The climate of history: Four theses. *Critical Inquiry* 35 (2):197–222. doi: 10.1086/596640.

Chester, M. V., S. Markolf, and B. Allenby. 2019. Infrastructure and the environment in the Anthropocene. *Journal of Industrial Ecology* 23 (5):1006–15. doi: 10.1111/jiec.12848.

Clark, N. 2015. Fiery arts: Pyrotechnology and the political aesthetics of the Anthropocene. *GeoHumanities* 1 (2):266–84. doi: 10.1080/2373566X.2015.1100968.

Comiso, J. C., C. L. Parkinson, R. Gersten, and L. Stock. 2008. Accelerated decline in the Arctic sea ice cover. *Geophysical Research Letters* 35 (1):1–6. doi: 10.1029/2007GL031972.

Cook, J. 1784. *A voyage to the Pacific Ocean. Undertaken, by the command of His Majesty, for making discoveries in the Northern Hemisphere, to determine the position and extent of the west side of North America; its distance from Asia; and the practicability of a northern passage to Europe.* London: Rare Books Division.

Crutzen, P. J. 2006. The Anthropocene. In *Earth system science in the Anthropocene,* ed. E. Ehlers and T. Krafft, 13–18. Berlin: Springer.

Davis, H. M., and Z. Todd. 2017. On the importance of a date, or decolonizing the Anthropocene. *ACME* 16 (4):761–80.

Davis, H. M., and E. Turpin. 2015. Art & death: Lives between the Fifth Assessment & the Sixth Extinction. In *Art in the Anthropocene: Encounters among aesthetics, politics, environments and epistemologies,* ed. H. M. Davis and E. Turpin, 3–30. London: Open Humanities Press.

Dawson, J., M. J. Johnston, E. J. Stewart, C. J. Lemieux, R. H. Lemelin, P. T. Maher, and B. S. R. Grimwood. 2011. Ethical considerations of last chance tourism. *Journal of Ecotourism* 10 (3):250–65. doi: 10.1080/14724049.2011.617449.

Dewey, J. 1925. *Experience and nature.* New York: Dover.

Diessner, R. 2019. *Understanding the beauty appreciation trait: Empirical research on seeking.* Cham, Switzerland: Palgrave Macmillan.

Edensor, T. 2005. The ghosts of industrial ruins: Ordering and disordering memory in excessive space. *Environment and Planning D: Society and Space* 23 (6):829–49. doi: 10.1068/d58j.

Engelmann, S., and D. McCormack. 2018. Elemental aesthetics: On artistic experiments with solar energy. *Annals of the American Association of Geographers* 108 (1):241–59. doi: 10.1080/24694452.2017.1353901.

Ghosh, A. 2016. Where is the fiction about climate change? *The Guardian,* October 28. Accessed December 4, 2019. https://www.theguardian.com/books/2016/oct/28/amitav-ghosh-where-is-the-fiction-about-climate-change-.

Gregory, J. M., P. A. Stott, D. J. Cresswell, N. A. Rayner, C. Gordon, and D. M. H. Sexton. 2002. Recent and future changes in Arctic sea ice simulated by the HadCM3 AOGCM. *Geophysical Research Letters* 29 (24):28-1–28-4. doi: 10.1029/2001GL014575.

Grove, K., and D. Chandler. 2017. Introduction: Resilience and the Anthropocene: The stakes of "renaturalising" politics. *Resilience* 5 (2):79–91. doi: 10.1080/21693293.2016.1241476.

Haalboom, B., and D. C. Natcher. 2012. The power and peril of "vulnerability": Approaching community labels with caution in climate change. *Arctic* 65 (3):319–27. doi: 10.14430/arctic4219.

Hamilton, C. 2017. *Defiant Earth.* Cambridge, UK: Polity.

Haskins, C. 1992. Dewey's "Art as Experience": The tension between aesthetics and heticism. *Transactions of the Charles S. Peirce Society* 28 (2):217–59.

Hell, J. 2009. Katechon: Carl Schmitt's imperial theology and the ruins of the future. *Germanic Review* 84 (4):283–326. doi: 10.1080/00168890903291443.

Herrmann, V. 2017. America's first climate change refugees: Victimization, distancing, and disempowerment in journalistic storytelling. *Energy Research & Social Science* 31 (July):205–14. doi: 10.1016/j.erss.2017.05.033.

Herrmann, V. 2019. Rural ruins in America's climate change story: Photojournalism, perception, and agency in Shishmaref, Alaska. *Annals of the American Association of Geographers* 109 (3):857–74. doi: 10.1080/24694452.2018.1525272.

Hofstadter, A., and R. Kuhns. 1964. *Philosophies of art and beauty: Selected readings in aesthetics from Plato.* Chicago: University of Chicago Press.

Huddart, D., and T. Stott. 2016. *Adventure tourism: Environmental impacts and management.* Cham, Switzerland: Palgrave Macmillan.

Huyssen, A. 2006. Nostalgia for ruins. *Grey Room* 23:6–21. doi: 10.1162/grey.2006.1.23.6.

Knies, J., P. Cabedo-Sanz, S. T. Belt, S. Baranwal, S. Fietz, and A. Rosell-Melé. 2014. The emergence of modern sea ice cover in the Arctic Ocean. *Nature Communications* 5 (1):7. doi: 10.1038/ncomms6608.

Leiserowitz, A. 2006. Climate change risk perception and policy preferences: The role of affect, imagery, and values. *Climatic Change* 77 (1–2):45–72. doi: 10.1007/s10584-006-9059-9.

Lennon, J. J., and M. Foley. 2000. *Dark tourism.* London: Continuum.

Levine, S. Z. 1985. Seascapes of the sublime: Vernet, Monet, and the oceanic feeling. *New Literary History* 16 (2):377–400. doi: 10.2307/468752.

Lorimer, J. 2012. Multinatural geographies for the Anthropocene. *Progress in Human Geography* 36 (5):593–612. doi: 10.1177/0309132511435352.

Macfarlane, R. 2016. Generation Anthropocene: How humans have altered the planet forever. *The Guardian*, April 1. Accessed December 5, 2019. https://www.theguardian.com/books/2016/apr/01/generation-anthropocene-altered-planet-for-ever.

Martello, M. L. 2008. Arctic Indigenous peoples as representations and representatives of climate change. *Social Studies of Science* 38 (3):351–76. doi: 10.1177/0306312707083665.

Martin, D. 2014. Introduction: Towards a political understanding of new ruins. *International Journal of Urban and Regional Research* 38 (3):1037–46. doi: 10.1111/1468-2427.12116.

McGrath, J. 2014. Apocalypse, or, the logic of late Anthropocene ruins. *Cross-Currents: East Asian History and Culture Review* 10:113–19.

Meillassoux, Q. 2010. *After finitude: An essay on the necessity of contingency.* London: Bloomsbury.

Mirzoeff, N. 2014. Visualizing the Anthropocene. *Public Culture* 26 (2):213–32. doi: 10.1215/08992363-2392039.

Möllers, N. 2013. Cur(at)ing the planet—How to exhibit the Anthropocene and why. *RCC Perspectives* 3:57–66.

Morgan, B. 2016. After the Arctic sublime. *New Literary History* 47 (1):1–26. doi: 10.1353/nlh.2016.0000.

Morton, T. 2018. *Dark ecology: For a logic of future coexistence.* New York: Columbia University Press.

Nye, D. E. 1994. *Technological sublime.* Cambridge, MA: The MIT Press.

Orlowski, J. (dir.) 2012. *Chasing ice.* DVD. New York: Submarine Deluxe.

Pile, S. 2010. Emotions and affect in recent human geography. *Transactions of the Institute of British Geographers* 35 (1):5–20. doi: 10.1111/j.1475-5661.2009.00368.x.

Roelvink, G., and M. Zolkos. 2011. Climate change as experience of affect. *Angelaki: Journal of the Theoretical Humanities* 16 (4):43–57.

Roth, A. 2019. Sight of polar bear daubed with graffiti sparks outrage. *The Guardian*, December 3. Accessed December 6, 2019. https://www.theguardian.com/world/2019/dec/03/sight-of-polar-bear-daubed-with-graffiti-sparks-outrage.

SIMIP Community. 2020. Arctic sea ice in CMIP6. *Geophysical Research Letters* 47: e2019GL086749. doi: 10.1029/2019GL086749.

Smith, S. P. 2018. Instagram abroad: Performance, consumption and colonial narrative in tourism. *Postcolonial Studies* 21 (2):172–91. doi: 10.1080/13688790.2018.1461173.

Stead, N. 2003. The value of ruins: Allegories of destruction in Benjamin and Speer. *Form/Work: An Interdisciplinary Journal of the Built Environment* 6 (6):51–64.

Sturken, M. 2004. The aesthetics of absence: Rebuilding Ground Zero. *American Ethnologist* 31 (3):311–25. doi: 10.1525/ae.2004.31.3.311.

Trott, N. 1999. The picturesque, the beautiful and the sublime. In *A companion to Romanticism*, ed. D. Wu, 79–98. Malden, MA: Blackwell.

Tuan, Y.-F. 1989. Surface phenomena and aesthetic experience. *Annals of the Association of American Geographers* 79 (2):233–41. doi: 10.1111/j.1467-8306.1989.tb00260.x.

Urry, J. 1992. The tourist gaze and the environment. *Theory, Culture & Society* 9:1–26.

Wallace-Wells, D. 2019. *The uninhabitable Earth: Life after warming.* New York: Penguin Random House.

Weisman, A. 2007. *The world without us.* New York: Thomas Dunne Books/St. Martin's.

What the world would look like if all the ice melted. 2013. *National Geographic* September. Accessed November 5, 2020. https://www.nationalgeographic.com/magazine/2013/09/rising-seas-ice-melt-new-shoreline-maps/

Williston, B. 2016. The sublime Anthropocene. *Environmental Philosophy* 13 (2):155–74.

Wilson, N. 2006. *Encyclopedia of Ancient Greece.* New York and London: Taylor & Francis.

Yusoff, K. 2013. Geologic life: Prehistory, climate, futures in the Anthropocene. *Environment and Planning D: Society and Space* 31 (5):779–95. doi: 10.1068/d11512.

Zehle, S. 2008. Dispatches from the depletion zone: Edward Burtynsky and the documentary sublime. *Media International Australia* 127 (1):109–15. doi: 10.1177/1329878X0812700114.

Ziser, M., and J. Sze. 2007. Climate change, environmental aesthetics, and global environmental justice cultural studies. *Discourse* 29 (2):384–410.

The New (Ab)Normal: Outliers, Everyday Exceptionality, and the Politics of Data Management in the Anthropocene

Katherine R. Clifford and William R. Travis

The Anthropocene affects how we manage the environment in many ways, perhaps most importantly by undermining how past conditions act as baselines for future expectations. In a period when historical analogues become less meaningful, we need to forge new practices and methods of environmental monitoring and management, including how to categorize, manage, and analyze the deluge of environmental data. In particular, we need practices to detect emerging hazards, changing baselines, and amplified risk. Some current data practices, however, especially the designation and dismissal of outliers, might mislead efforts to better adapt to new environmental conditions. In this article we ask these questions: What are the politics of determining what counts as "abnormal" and is worthy of exclusion in an era of the ever-changing "normal"? What do data exclusions, often in the form of outliers, do to our ability to understand and regulate in the Anthropocene? We identify a recursive process of distortion at play where constructing categories of abnormal–normal allows for the exclusion of "outliers" from data sets, which ultimately produces a false rarity and hides environmental changes. To illustrate this, we draw on a handful of examples in regulatory science and management, including the Exceptional Event Rule of the Clean Air Act, beach erosion models for nourishment projects, and the undetected ozone hole. We conclude with a call for attention to the construction of "normal" and "abnormal" events, systems, data, and natures in the Anthropocene. *Key Words: climate adaptation, data exclusions, environmental change, extreme events, rarity.*

One expected consequence of continuing global change is that the behavior of environmental systems will transform, particularly in ways that increase extreme events (Walsh et al. 2014; Hayhoe et al. 2018). Extremes—those rare, high-magnitude events that stretch out into the tail of a distribution—destroy property and infrastructure, strain government functioning and budgets, and in many cases lead to injury or death. To make decisions about how to protect populations—how high to build a levee, designate evacuation routes, or restrict zoning for new development—regulators rely on historical data to calculate probabilities and risk and to compare costs and benefits. Although using historic ranges or baselines is a well-established practice (Ruhl and Salzman 2011), it is losing efficacy as historic variability becomes a less reliable guide to future variability (Hirsch 2020).

The Anthropocene is a recognition that we have entered a phase in which human–environment interactions pervade most major earth systems (Crutzen and Steffen 2003; Zalasiewicz et al. 2010; S. L. Lewis and Maslin 2015). In many ways, this is a step by physical geographers closer to the work and critiques of human geographers to recognize the powerful role that social processes play in shaping the physical environment. The process of designating, and even naming, the Anthropocene has drawn thoughtful critique (Buck 2015; Haraway 2015, 2016; Moore 2016); here, rather than delve into its driving forces and debates over definition and onset, we focus on how society might engage with the impacts of the Anthropocene on environmental management.

Even in a past when distributions of environmental variables were seen as stable, determining what data counted as "abnormal" or "outlier" was difficult and consequential work, and the challenge only increases in an era of environmental change. As the Anthropocene brings novel conditions, it is worth asking how scientific practices, like the exclusion of outliers meant to clean up and normalize data by removing "bad" or "atypical" observations, might fundamentally alter our understanding and management of changing earth systems. In an era where normal is a moving target and baselines will shift, determining what is abnormal, atypical, and bad data will be difficult. It is an overstatement to claim as some have that "outliers are now the norm" (Coleman 2019), but how do we evolve our thinking about both outliers and normals, as well as our corresponding data management practices, to effectively apprehend the Anthropocene?

In the Anthropocene we need practices that acknowledge changing baselines to detect emerging hazards and better adapt to new environmental conditions. What are the stakes of determining what counts as "abnormal" and is worthy of inclusion or exclusion in environmental analyses during an era of change, of ever-shifting "normal"? How do data outliers affect our ability to understand and manage the environment in the Anthropocene?

We argue for attention to the construction of normal and abnormal events, data, and natures in the Anthropocene, an era in which environmental change makes seemingly simple divisions between normal and abnormal more complex. More specifically, scholars need to look at how constructions of abnormality influence data exclusions and understandings of rarity in a way that focuses on the relational and mutually dependent processes of (mis)understanding a changing environment. This type of inquiry requires combining approaches often not in conversation: quantitative risk assessment and critical theory. In this article we, as geographers steeped in two different subfields, attempt to build a bridge between hazards and risk analysis and science and technology studies (STS) and other critical geographies of the environment. Our aim is that this article speaks to both sets of scholars and makes ideas from one subfield accessible to the other.

New Normals, Shifting Baselines, and Outliers

Most systems for managing the environment, from stormwater infrastructure to forestry to agriculture, are predicated on the notion of an expected or "normal" range of conditions, usually arrayed in a statistical distribution of values huddled around the mean (in normal distributions) or the median (in skewed distributions; Figure 1). Distributions might be empirically derived (a histogram of observed values) or theoretical (a distribution with a shape that fits theoretical understanding of a system's behavior

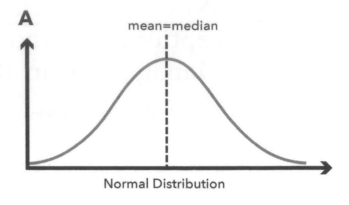

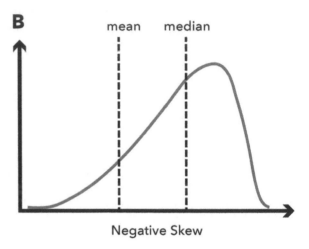

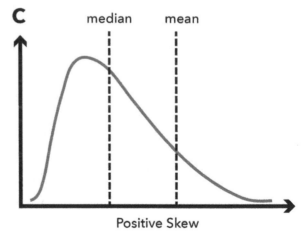

Figure 1. A range of statistical distributions: (A) normal, (B) negative skew, and (C) positive skew.

and can be used to estimate values not yet observed). In both cases the past is prologue, and although environmental scientists recognize that systems evolve over time, many analytical and management approaches still rely on the future behaving like the past. The death knell of stationarity (Milly et al. 2008) was perhaps sounded too early. This is especially true, we argue, in approaches that crave the longest historical records to flesh out the statistical distributions on which risk assessments for extremes and intervention plans are based.

Alternatively, a key intellectual pursuit of the Anthropocene should be to understand how environmental systems are changing, to better envision how different they will look in the future. An army of climate and earth systems modelers has been engaged in precisely this complicated and important work, building new worlds within their models to offer glimpses of what is to come (Edwards 2001). One of their challenges is not to allow parameters based on past data to overly constrain the future conditions their models can predict, keeping in mind that a transformed future suffers from the "no analogue" problem (J. W. Williams and Jackson 2007): Models trained on the past struggle with novel futures.

The sense that the future will differ markedly from the past is signaled by emerging conditions, like greenhouse gas concentrations in the atmosphere, not observed for millennia, but also by the quotidian experience of extreme weather and climate. Such palpable changes are especially difficult to foretell; the distributions of climate variables will shift in uncertain ways and could usher in many different "normals." Even for temperature change, about which earth scientists are most confident (Collins et al. 2013; Kirtman et al. 2013), warming could manifest in many possible future distributions, with longer or shorter tails (Figure 2), perhaps a shift to higher temperatures while maintaining the same relationships between extreme and average events (Figure 2A), or to a different normal in which the average persists but the range expands, offering greater variability and more extremes (Figure 2B). Perhaps both average (shifting the mean) and variance (altering the shape of the distribution and relationship between average and extreme events) could change (Figure 2C).

Such simple illustrations of statistical distributions for current and future climate reveal a lot but also hide much. First, the distributions are typically graphed as normal curves, neglecting the fact that many climate variables exhibit skewed distributions. Climate projections point both to more extreme high temperatures (Collins et al. 2013) and a skew toward more intense precipitation events (Kirtman et al. 2013). Projections also point to increasing contrasts between wet and dry spells (Kirtman et al. 2013), yielding the awkward notion that global warming brings both more floods and droughts. Expectations of bigger changes in extremes than in means have long been part of the climate literature (Wigley 2009) but are difficult to translate into environmental management protocols.

These technical twists are echoed in the popular alliterative phrase "new normal," which S. C. Lewis, King, and Perkins-Kirkpatrick (2017) noted is "widely used in mainstream media reports to succinctly categorize observed extreme weather and climate events as both unusual and influenced, in some regard, by anthropogenic climate change" (1139). Alliteration is not always veracity, however, and the contradictory notion of abnormal conditions coming to be considered normal works in some ways and not in others. S. C. Lewis, King, and Perkins-Kirkpatrick (2017), who sought to define "new normal" in a scientific manner, admitted that it is "used ambiguously without precise definition in both scientific literature and public commentary on climate change" (1140). Moreover, they noted that a "system under the influence of anthropogenic warming is nonstationary and exhibits a nonconstant mean," and thus "in a true statistical sense 'new' and 'normal' are essentially oxymoronic" (S. C. Lewis, King, and Perkins-Kirkpatrick 2017, 1141).

To make some sense of the new normal in the Anthropocene—indeed, to make sense of the Anthropocene—we must attend to the matter of the baseline on which any expectation is founded. Because the new normal is so often (and in contradiction to its prima facie meaning) tied to extremes, we must also evaluate how exceptional events, outliers, and atypicality are defined. We must also recognize that, in environmental management regimes, baselines and outliers are both epistemic and legal (Hirsch 2020) and steeped in the coproduction of science and regulatory law (Jasanoff 2004). Establishing a baseline is not a straightforward nor necessarily objective act (Ureta, Lekan, and von Hardenberg 2020), and the same is true for establishing the range of normal, which is needed to evaluate the abnormal. Just as the production of baselines is often framed as objective yet is value-laden and frequently the site of legal battles (Hirsch 2020), so, too, is distinguishing normal from abnormal.

One of the valuable contributions from STS is the recognition that data management is always political (Hacking 1990; Porter 1996; Desrosières 2002; Bowker 2008; Gitelman 2013; Pine and Liboiron 2015; Dillon et al. 2019), and politics are

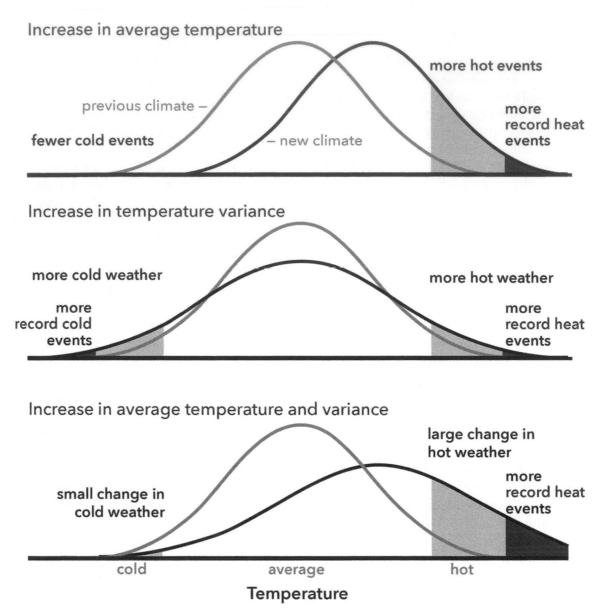

Figure 2. Different temperature distribution shifts. The three graphs show how changes in average, variance, or both alter the amount of extremes.

heightened in the Anthropocene. In this article, then, "politics" is not just referring to partisan politics, power grabs, or intentional actions. We use politics in a much broader, more encompassing way to include both those overt actions as well as more covert and even unintentional politics. Drawing on the work of STS scholars who explore the politics of science, we understand politics to mean a process that is not neutral or inevitable and instead one that is situated, subjective, and full of small but important choices that can be seen as technical but are often shaped by value-laden assumptions and judgments of scientists.[1]

Of course, definitions of politics vary across and among different intellectual communities; even within geography, subfields interpret politics to mean many things. We find it useful to distinguish between the uses of politics, specifically between Big-P Politics and small-p politics (King and Tadaki 2018). Big-P Politics refers to politicized science, with most of the politics touching science after its produced to determine how it is used or whether people believe it. This type of Big-P Politics is intentional and explicit and usually harnessed to support the end goals of certain actors. Small-p politics, however, refers to choices scientists make while conducting science, including choices of theories, data collection methods, and statistical analyses. King and Tadaki (2018) defined small-p politics of science "as the ways in which scientists make (intentional or unintentional) value-laden choices within the scientific realm that produce distinct consequences (social meanings, inequalities, power relations) for real people and environments" (72). Assumptions and judgments by scientists about how systems work shape how they set out to study and track them. Each choice has trade-offs and cannot be looked at as neutral regardless of intention; these small-p political choices embedded in science make some things known and others not. Pine and Liboiron (2015) insisted that "data—and attendant processes of measurement, database production, stabilization, curation, maintenance and use—reproduce power dynamics, knowledge systems, and culturally-based assumptions" (1). In this way, "the scientific method is *inherently* political" (King and Tadaki 2018, 68) and data are never "raw" (Ribes and Jackson 2013).

Establishing what is normal also establishes what is an outlier. When these two categories are unsettled by shifting and morphing system-level changes, they often are redefined, and redefining one category (e.g., normal or abnormal) redefines the other. Both definitions are dependent, relational, and mutually constructive.

Outliers can take many forms, are called different names, and are justified by different logics. Data are considered "outliers" when they are markedly different from the rest of the sample, sometimes based on statistical metrics like standard deviation cutoffs. They might then be thrown out. The assumption here is that outlying data are so different from typical data that they might be influenced by other variables or reflect different underlying causes or patterns. For example, if a dam breaches and results in flooding, it might be excluded from the hydrologic data set for calculating flood frequency because it reflected an "unnatural" driver or influence. Outliers could also be categorized as "errors," assuming they were a product of faulty measurement. For example, an instrument can produce errors due to location, monitor drift over time, or malfunction. In some cases, data are excluded based on less technical criteria, sometimes simply because the data do not seem to fit or do not appear normal; sometimes data are excluded because they are deemed rare or exceptional.

Examining the production of outliers also requires a stand on what counts as normal. All of these categories rely on normal for their own defining criteria. Although categories of "natural" have long been critiqued and shown to rely on a false divide (R. Williams 1977; Cronon 1996; Mansfield et al. 2015; Cantor 2016; Davis 2016), normal is equally a subjective category based on similar false dualisms. Normal requires constructing a boundary between the normal and abnormal as though there are clear lines rather than arbitrary elements of normativity. Canguilhem (2008) argued that normal is a category that "has no absolute meaning" (132–33), outside of narrow statistical definition. Yet when applied to data analysis, the category of normal often is framed as a technical assessment, especially when statistical demarcations (e.g., standard deviation) are applied. How to divide up data into normal and outlier or error, however, is a judgment that often has political outcomes in terms of, for example, what in the environment gets monitored, what conditions are considered acceptable or not, and when some management intervention is triggered.

Outlying: Making and Remaking Fictional Rarity through Data Management

The politics of data management (i.e., the stakes of data decisions) are heightened in the Anthropocene. If we do not examine our understanding of abnormal as environmental systems transform, we might exclude data telling us about change, resulting in distorted understandings of the evolving environment. Here we argue that it is important to attend to a chain of distortion that could affect our ability to track change, with consequences that reverberate through science and management spheres, and that will become more consequential in the Anthropocene.

In the process of distortion, three key elements work together in a recursive way (Figure 3). First comes the boundary work of categorizing normal and, importantly, the abnormal. Second, outliers may be excluded, based on their abnormal categorization. Third, this exclusion distorts depictions of the environment and renders a "false rarity" as the extremes disappear from the data. That false rarity can then actually work to help bolster new events to be excluded from the data set as abnormal as their

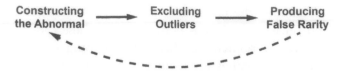

Figure 3. The process of distortion. Categorizing certain data as abnormal then allows those data to be excluded as outliers, which in turn produces a false rarity. This process can be recursive as the false rarity feeds back into how we categorize data.

previous peers have been excluded and rendered invisible. Thus, this process is always at work, making and remaking normal and abnormal environments. We suspect that the speed, scope, and level of distortion will expand in the Anthropocene. False rarity has staying power as it is stabilized and maintained, even as it increasingly differs from the true state of the system. Many scholars have documented how false understandings and representations take on a life of their own—for example, Simon (2010) in the case of 100th meridian and Sayre (2017) for rangeland management—and false representations of rarity could harden through this process, go unquestioned, and find their way into important legal and management decisions.

Inaccurate understandings of variability are not necessarily a product of environmental change; even without environmental change, distorted representations create messy, problematic, and risky outcomes. For example, Stakhiv (2011) argued that a standard statistical distribution used in the design of water management projects in the United States downplays hydrologic extremes, both floods and drought. He concluded that, although safety buffers have prevented failures, "underdesign" will become more of a problem in a changing climate. Similarly, false rarity hides the realities of variability and the potential for increasing extremes, allowing managers to approve activities and land uses that can lead to dangerous outcomes. Even in an environment that is not changing and where normals are not new, excluding data as outliers makes it easier and easier to exclude future data because, in comparison to a revised data set, new extremes appear even more abnormal.

Here we highlight each of the three elements of the process just described and examine how they could work together to distort and hide certain system behavior.

1. Constructing the Abnormal

When written into laws and regulations, categories for abnormal data become even more powerful and consequential because they ultimately shape actions and interventions in environmental systems (Cantor 2016). Categories of abnormal data shape how regulatory frameworks understand and manage environmental systems. Extreme events—the observations likely to be considered outliers due to their infrequency—are the most valuable data for risk assessment and preparedness. As we move toward expecting greater levels of change, outliers—particularly those enshrined in laws and regulation—will only amplify the complications of the already difficult tasks of environmental monitoring and management.

We can see the work done by categorizing certain environmental events—and their data—as abnormal in current air quality regulations. The Exceptional Event Rule (EER), which is part of the Clean Air Act, allows for "exceptional" air quality events to be excluded from state regulatory data sets nationwide. High pollution days can be nullified by removing those events, and this similarly erases regulatory violations (i.e., the fines, increased governance, and costs of not meeting standards). An "exceptional" event is defined as either conditions that are unlikely to reoccur again (aligning well with common understandings of exceptional) or an event that is natural (regardless of its frequency), like a dust storm. Of course, natural is a highly debated category (R. Williams 1977; Cronon 1996; Mansfield et al. 2015; Cantor 2016; Davis 2016) because, especially in setting regulations, its definition often involves drawing arbitrary boundaries between nature and society rather than recognizing how most elements are hybrid. This challenge increases in the Anthropocene as it becomes even harder to disentangle the two (Harden 2012; Castree 2014; Purdy 2015; Mansfield and Doyle 2017). Further, both types of events—the truly extreme and the common, everyday, but bothersome "natural" air quality events—are important data for apprehending how a system behaves.

Once events are categorized as exceptional, they can be removed from regulatory data sets; this has the potential to hide important environmental risk as well as the true conditions of poor air quality. For example, Maricopa County, Arizona (where Phoenix is located), categorized more than twenty days with dust as "exceptional" in 2011 alone and explicitly stated that they would use the rule to avoid nonattainment status (Maricopa Association of Governments n.d.). Dust storms are caused by environmental factors like increasing aridity but also a slew of land use practices surrounding the desert city. Through the use of this rule, the county is attempting to be in attainment (i.e., meet air quality standards) for the first time in more than twenty years.

Exclusions based on the EER also can hide changes in air quality, which are particularly important to track because the Anthropocene is likely to exhibit increased dust (Romm 2011) and smoke (McKenzie et al. 2014) from elevating temperatures and aridity. Yet data for those increases will be hidden, contradicting the obvious clues in plain sight. It is through the creation of exceptional events, constructing this category of abnormal or outlier, that the EER underwrites this exclusion and changes the story coming from the data. Ultimately the EER and its exclusions of exceptional events are working against the goals of the Clean Air Act, to protect human health and public welfare, because they allow high levels of pollution to go unregulated and hidden from analysis (Clifford 2020).

2. Excluding Outliers

Excluding extreme events due to their infrequency, or outliers due to their abnormality, can erase the variability of a system, artificially constructing a distribution with little variance and small standard deviations that renders events or changes invisible. In a sense, detectable (and maybe dangerous) extreme events disappear, at least in the data. Many and repeated alterations to the data transform how we understand a system in more than just discursive ways because distributions are heavily relied on for studies of risk, regulations of environmental hazards, and other critical management decisions.

Blinders constructed by data exclusions have already affected our tracking of the Anthropocene. For years, scientists missed the growing "ozone hole" over Antarctica despite technological advancements in monitoring atmospheric processes and increasing efforts by the National Aeronautics and Space Administration to measure and understand earth systems (von Hobe 2007). Ironically, it was not technological advancements that made the ozone hole detectable but instead reverting back to old, "obsolete" methods (Farman, Gardiner, and Shanklin 1985). Nor was it the complexity of the hole that left it invisible but instead data management practices, specifically exclusions.

The data exclusions that hid the ozone hole were not the act of an individual who reviewed the data and threw them out but instead were built into the data management algorithms that acted as an intermediary between the instruments collecting data and the models and scientific assessments using the data. The algorithms were set to exclude data that fell outside of what was expected in the system (what was envisioned as possible). In this case, outliers took the role of "errors." Errors are different than other types of outliers: Instead of theoretically referring to an accurate depiction of a rare event that is deemed beyond or outlying typical system functioning, they signal a malfunction of the instrument or incorrect information. Yet the algorithms' categorization of errors highlights the very subjective and dangerous politics of thinking about normal. We often think of normal as historic (i.e., what happened in the past) or present (i.e., how systems behave now), but this forecloses the possibility of how the system might shift and evolve.

Ironically, the ozone hole was eventually detected by basic instruments that did not have the automated exclusions. At first, the scientists observing the hole (Farman, Gardiner, and Shanklin 1985) faced some skepticism because their data were collected by less advanced instruments and it was only once the National Aeronautics and Space Administration scientists reran their models with the "error" data that the ozone hole became legible (Stolarski et al. 1986). It should have been clearly detectable for some time but was hidden through data exclusions that allowed an inaccurate representation of the atmosphere to persist.

3. Producing False Rarity and Erasing Variability

Bias against including plausible extreme events in the planning of projects that could fail given those events could also make the Anthropocene more dangerous than we expect. The statistics of natural system behavior essentially ensure that a place that experiences a significant range of intensity of natural events will also experience events several times larger than, say, those at one or two standard deviations from the mean. The costly implications of including extremes in project planning might encourage their neglect or exclusion and also yield a false sense of their rarity. When included in plans, extremes usually support larger and more costly hazard protection (e.g., higher sea walls, more robust seismic building protections, etc.). Studies of the 2011 Fukushima nuclear power plant disaster found that the target wave height used for tsunami protection at the plant was inadequate; even when a historic tsunami comparable to the 2011 event was recognized by seismologists, it was neglected in the risk assessment when plant infrastructure was upgraded in 2002 (Earthquake Engineering Research Institute 2011; The Fukushima Nuclear Accident Independent Investigation Commission 2012; National Research Council 2014; Wheatley, Sovacool, and Sornette 2017).

A tendency to discount the likelihood of large events in a project lifetime is another form of data truncation, one that creates false rarity. Critics of coastal engineering on U.S. shores argue that large storm events are more common than models and plans allow for. Pilkey and Pilkey-Jarvis (2007; see also Pilkey, Young, and Cooper 2013) contended that the concept of beach erosion taking place through slow processes was constructed by treating coastal storm events as unusual or abnormal, when in fact they are common. This is especially important in beach nourishment projects, the multibillion-dollar efforts to rebuild beaches, especially on the U.S. East Coast, after storms and long-term erosion have worn them (and the residents and tourists that use them) away. Static numbers reified into models for calculating key variables, like sand transport rates, can have the effect of making variability disappear in projections of beach erosion (Young et al. 1995). Pilkey and Pilkey-Jarvis (2007) argued that

> when artificial beaches are lost more rapidly than predicted by the models, the most common excuse is that the storm that caused the beach loss was unusual and unexpected. Certainly the unusual storm can occur, but the label "unusual" is used so frequently with lost artificial beaches as to imply that the last few decades have been truly extraordinary in their storminess. (135)

No beach lasts forever in an era of rising sea level, and coastal storms able to significantly whittle away beach width are not rare. Beach nourishment can slow this loss but not always as well as advertised. To improve beach project plans, the U.S. Army Corps of Engineers and others invested significant modeling efforts to include storms in erosion models (National Research Council 1995; Thieler et al. 2000). Estimating how likely those storms are over some project duration remains a big challenge

(Toimil et al. 2020), however, and critics claim that beach prediction models produce numbers that might overstate beach stability, tip benefit–cost analyses in favor of nourishment, and hide trends in beach erosion caused by storms and rising sea level.

Outlying in the Anthropocene

The stakes of false rarity increase in the Anthropocene. Removing "outliers" or other abnormal data critically shapes our understanding of the system not only because we miss events but because historical distributions, and changing distributions over time, are what we use to understand whether a system is changing. In other words, removing data about current extreme events cripples our ability to detect change and understand future environments. Moreover, as Pine and Liboiron (2015) reminded us, "the interplay of inclusion and exclusion *makes* things" (3). The normal makes the outlier and vice versa. Distorted pictures of environmental systems rely on data exclusions; those alternative data would rebuke notions of rarity. The consequences of false understandings about dynamic environmental systems is that they often lead to management strategies that do not work or work against the stated goals (Sayre 2017). Data exclusion is an Achilles' heel in the Anthropocene.

All three of the examples presented exemplify the consequences of how constructing outliers can produce a skewed understanding of environmental behavior, particularly inaccurate claims of low variability. The stakes increase in the Anthropocene because they are not just associated with missing current system behavior but also with delaying recognition of larger changes and hindering adaptation responses.

Treating dust storms in the U.S. Southwest as "exceptional" or outliers, despite their frequency and historical analogues, makes it harder to see signals of emerging threats. Little imagination is required to think about how increasing air pollution events like dust storms or wildfire smoke could be disastrous; many have called the Dust Bowl the nation's greatest "natural" disaster. Yet, we know that the Dust Bowl was driven at least in part by land use and allowed to amplify by officials (and farmers) ignoring signs of increased soil erosion (Worster 2004). Discarding "exceptional" events from regulatory analysis could halt proactive action that might be able to slow changes or facilitate adaptation.

The undetected ozone hole might be one of the most powerful analogues we have for thinking about outliers in the Anthropocene. We missed a significant change—one that we had data about—and that slowed our response. We run the same risk of missing important and likely high-consequence environmental changes in the Anthropocene if we exclude data that depart from a historical normal. In a time where we are increasingly automating and outsourcing decisions to algorithms and artificial intelligence, the ozone hole offers a cautionary note on how automating exclusions might make emerging risks less visible.

In the case of beach nourishment, treating severe storms as outliers in models can handicap analysis and undermine interventions, particularly if models overstate the effectiveness of beach nourishment. Sea level rise and warming oceans that could lead to worsened storms (Bindoff et al. 2013) mean that the gap between environmental realities and model outputs will only widen and likely lead to greater "surprises" as beach nourishments endure a dwindling fraction of projected life spans. The Anthropocene nullifies beaches but not beach simulation models.

Conclusion: Warning Signs

The chances of successful climate adaptation will improve if we can avoid false depictions of rarity and monitor the changing environment more accurately. Thus, we need to return to previously asked questions of normality: How does thinking about the "new normal" influence our understanding of outliers and other abnormal data? Conversely, how does excluding outliers affect our ability to apprehend the evolving Anthropocene?

An era of accelerating environmental change raises the stakes of outliers and false rarity. How distributions will evolve in the Anthropocene remains uncertain, so our monitoring must be open to surprise, to the abnormal. Early clues of how a system is changing provide critical information for management strategies to respond and protect communities. Combined, false representations of rarity and increasing variability have the potential for greater distortion of system behavior and more likely surprises and unanticipated events, ultimately undermining climate adaptation and environmental laws.

We invite others to examine these questions in other environmental systems. As scholars engaged in studying this new epoch and tracking transformation, we need to wrestle with how our own (and others') notions of normal and abnormal affect inquiries into environmental change. Such inquiries are integral to knowing (or not knowing) environmental change, because change itself is again a departure from some defined normal state or baseline. Future work should further examine how outliers are produced and excluded, as well as the consequences of false rarity in risk management, legal paradigms, and climate adaptation efforts. How are data boundaries between normal and abnormal being negotiated? Has false rarity been produced through debatable exclusions and problematic characterizations? How is this process—and categories of normal–abnormal, data exclusions, and false rarity—reproduced and reified in legal and policy spheres? These questions can provide important insight into understanding and mitigating risk in the Anthropocene.

Funding

Parts of this work was supported by the Western Water Assessment, funded by the U.S. National Oceanic and Atmospheric Administration under Climate Program Office grant #NA10OAR4310214WWA.

Acknowledgments

Thanks to Laura Nash for helpful comments on draft versions and to Ami Nacu-Schmidt and Erica Clifford for graphics. We also appreciate the careful reading and comments by anonymous reviewers, which improved the article.

Note

1. By referring to the situatedness of science, we are not critiquing science or its findings but illuminating the decisions, assumptions, and trade-offs in the process. Scientists and regulators who crunch numbers and analyze systems make a slew of decisions in the collection and analysis of data; it is impossible to make science without such decisions, but it is also important to note that these decisions do imbue scientific data and analysis, our understanding of a system, and decisions about how to intervene in that system. STS scholars and

geographers building on this thinking have long documented how the decisions scientists make during the production of science are influenced by neoliberal forces (Lave 2012), available instruments and technologies (Clarke and Fujimura 1992; Frickel and Vincent 2007), disciplinary forces (Clarke and Fujimura 1992; Murphy 2006; Kleinman and Suryanarayanan 2013), problem orientation (Frickel et al. 2010), time frame of analysis (Sedell 2019), and use of categories (Bowker and Star 2000; Duvall, Butt, and Neely 2018).

References

Bindoff, N. L., P. A. Stott, K. M. AchutaRao, M. R. Allen, N. Gillett, D. Gutzler, K. Hansingo, et al. 2013. Detection and attribution of climate change: From global to regional. In *Climate Change 2013: The physical science basis. Contribution of Working Group I to the Fifth Assessment Report of the Intergovernmental Panel on Climate Change*, ed. T. F. Stocker, D. Qin, G.-K. Plattner, M. Tignor, S. K. Allen, J. Boschung, A. Nauels, Y. Xia, V. Bex, and P. M. Midgley. Cambridge, UK: Cambridge University Press.

Bowker, G. C. 2008. *Memory practices in the sciences*. Cambridge, MA: The MIT Press.

Bowker, G. C., and S. L. Star. 2000. *Sorting things out: Classification and its consequences*. Cambridge, MA: MIT Press.

Buck, H. J. 2015. On the possibilities of a charming Anthropocene. *Annals of the Association of American Geographers* 105 (2):369–77. doi:10.1080/00045608. 2014.973005.

Canguilhem, G. 2008. *Knowledge of life*, trans. S. Gerouolanos and D. Ginsburg. New York: Fordham University Press.

Cantor, A. 2016. The public trust doctrine and critical legal geographies of water in California. *Geoforum* 72:49–57. doi:10.1016/j.geoforum.2016.01.007.

Castree, N. 2014. The Anthropocene and geography I: The back story. *Geography Compass* 8 (7):436–49. doi:10.1111/gec3.12141.

Clarke, A. E., and J. H. Fujimura, eds. 1992. *The right tools for the job: At work in twentieth-century life sciences*. Princeton, NJ: Princeton University Press.

Clifford, K. R. 2020. Problematic Exclusions: Analysis of the Clean Air Act's Exceptional Event Rule Revisions. *Society & Natural Resources*. doi: 10.1080/ 08941920.2020.1780358.

Coleman, H. 2019. Climate is getting more extreme in every possible way. *Massive Science* (blog), January 8. Accessed April 13, 2020. https://massivesci.com/ articles/climate-change-wildfires-hurricanes/.

Collins, M., R. Knutti, J. Arblaster, J.-L. Dufresne, T. Fichefet, P. Friedlingstein, X. Gao, et al. 2013. Long-term climate change: Projections, commitments and irreversibility. In *Climate Change 2013: The physical science basis. Contribution of Working Group I to the Fifth Assessment Report of the Intergovernmental Panel on Climate Change*, ed. T. F. Stocker, D. Qin, G.-K. Plattner, M. Tignor, S. K. Allen, J. Boschung, A. Nauels, Y. Xia, V. Bex, and P. M. Midgley. Cambridge, UK: Cambridge University Press.

Cronon, W. 1996. The trouble with wilderness: Or, getting back to the wrong nature. *Environmental History* 1 (1):7–28. doi:10.2307/3985059.

Crutzen, P. J., and W. Steffen. 2003. How long have we been in the Anthropocene era? *Climatic Change* 61 (3):251–57. doi:10.1023/B:CLIM.0000004708.74871.62.

Davis, D. K. 2016. *The arid lands: History, power, knowledge*. Cambridge, MA: MIT Press.

Desrosières, A. 2002. *The politics of large numbers: A history of statistical reasoning*. Cambridge, MA: Harvard University Press.

Dillon, L., R. Lave, B. Mansfield, S. Wylie, N. Shapiro, A. S. Chan, and M. Murphy. 2019. Situating data in a Trumpian era: The environmental data and governance initiative. *Annals of the American Association of Geographers* 109 (2):545–55. doi:10.1080/24694452. 2018.1511410.

Duvall, C. S., B. Butt, and A. Neely. 2018. The trouble with savanna and other environmental categories, especially in Africa. In *The Palgrave handbook of critical physical geography*, eds. Lave, R., Biermann, C., & Lane, S. N. 107–27. Cham, Switzerland: Palgrave Macmillan.

Earthquake Engineering Research Institute. 2011. The Japan Tohoku Tsunami of March 11, 2011. EERI Special Earthquake Report, Oakland, CA: Earthquake Engineering Research Institute.

Edwards, P. N. 2001. Representing the global atmosphere: Computer models, data, and knowledge about climate change. In *Changing the atmosphere: Expert knowledge and environmental governance*, ed. C. A. Miller and P. N. Edwards, 31–33. Cambridge, MA: MIT Press.

Farman, J. C., B. G. Gardiner, and J. D. Shanklin. 1985. Large losses of total ozone in Antarctica reveal seasonal ClOx/NOx interaction. *Nature* 315 (6016):207–10. doi:10.1038/315207a0.

Frickel, S., S. Gibbon, J. Howard, J. Kempner, G. Ottinger, and D. J. Hess. 2010. Undone science: Charting social movement and civil society challenges to research agenda setting. *Science, Technology, & Human Values* 35 (4):444–73. doi:10.1177/ 0162243909345836.

Frickel, S., and M. B. Vincent. 2007. Hurricane Katrina, contamination, and the unintended organization of ignorance. *Technology in Society* 29 (2):181–88. doi:10.1016/j.techsoc.2007.01.007.

The Fukushima Nuclear Accident Independent Investigation Commission. 2012. The national diet of Japan. https://www.nirs.org/wp-content/uploads/ fukushima/naiic_report.pdf.

Gitelman, L. 2013. *Raw data is an oxymoron*. Cambridge, MA: MIT Press.

Hacking, I. 1990. *The taming of chance*. Cambridge, UK: Cambridge University Press.

Haraway, D. 2015. Anthropocene, capitalocene, plantationocene, chthulucene: Making kin. *Environmental Humanities* 6 (1):159–65. doi:10.1215/22011919-3615934.

Haraway, D. 2016. *Staying with the trouble: Making kin in the Chthulucene*. Durham, NC: Duke University Press.

Harden, C. P. 2012. Framing and reframing questions of human–environment interactions. *Annals of the Association of American Geographers* 102 (4):737–47. doi:10.1080/00045608.2012.678035.

Hayhoe, K., D. J. Wuebbles, D. R. Easterling, D. W. Fahey, S. Doherty, J. Kossin, W. Sweet, R. Vose, and M. Wehner. 2018. Our changing climate. In *Impacts, risks, and adaptation in the United States: Fourth National Climate Assessment*, ed. D. R. Reidmiller, C. W. Avery, D. R. Easterling, K. E. Kunkel, K. L. M. Lewis, T. K. Maycock, and B. C. Stewart, vol. II, 72–144. Washington, DC: U.S. Global Change Research Program. doi:10.7930/NCA4.2018.CH2.

Hirsch, S. L. 2020. Anticipatory practices: Shifting baselines and environmental imaginaries of ecological restoration in the Columbia River Basin. *Environment and Planning E: Nature and Space* 3 (1):40–57. doi:10.1177/2514848619857523.

Jasanoff, S., ed. 2004. *States of knowledge: The co-production of science and the social order*. London and New York: Routledge.

King, L., and M. Tadaki. 2018. A framework for understanding the politics of science (Core Tenet #2). In *The Palgrave handbook of critical physical geography*, 67–88, ed. Lave, R., Biermann, C., & Lane, S. N. Cham, Switzerland: Palgrave Macmillan.

Kirtman, B., S. B. Power, J. A. Adedoyin, G. J. Boer, R. Bojariu, I. Camilloni, F. J. Doblas-Reyes, et al. 2013. Near-term climate change: Projections and predictability. In *Climate Change 2013: The physical science basis. Contribution of Working Group I to the Fifth Assessment Report of the Intergovernmental Panel on Climate Change*, 953–1028, ed. T. F. Stocker, D. Qin, G. K. Plattner, M. Tignor, S. K. Allen, J. Boschung, A. Nauels, Y. Xia, V. Bex, and P. M. Midgley. Cambridge, UK: Cambridge University Press.

Kleinman, D. L., and S. Suryanarayanan. 2013. Dying bees and the social production of ignorance. *Science, Technology, & Human Values* 38 (4):492–517. doi:10.1177/0162243912442575.

Lave, R. 2012. Neoliberalism and the production of environmental knowledge. *Environment and Society* 3 (1):19–38. doi:10.3167/ares.2012.030103.

Lewis, S. C., A. D. King, and S. E. Perkins-Kirkpatrick. 2017. Defining a new normal for extremes in a warming world. *Bulletin of the American Meteorological Society* 98 (6):1139–52. doi:10.1175/BAMS-D-16-0183.1.

Lewis, S. L., and M. A. Maslin. 2015. Defining the anthropocene. *Nature* 519 (7542):171–80. doi:10.1038/nature14258.

Mansfield, B., C. Biermann, K. McSweeney, J. Law, C. Gallemore, L. Horner, and D. K. Munroe. 2015. Environmental politics after nature: Conflicting socioecological futures. *Annals of the Association of American Geographers* 105 (2):284–93. doi:10.1080/00045608.2014.973802.

Mansfield, B., and M. Doyle. 2017. Nature: A conversation in three parts. *Annals of the American Association of Geographers* 107 (1):22–27. doi:10.1080/24694452.2016.1230418.

Maricopa Association of Governments. n.d. PM-10 monitoring data. http://azmag.gov/Programs/Maps-and-Data/Air-Quality.

McKenzie, D., U. Shankar, R. E. Keane, E. N. Stavros, W. E. Heilman, D. G. Fox, and A. C. Riebau. 2014. Smoke consequences of new wildfire regimes driven by climate change. *Earth's Future* 2 (2):35–59. doi:10.1002/2013EF000180.

Milly, P. C. D., J. Betancourt, M. Falkenmark, R. M. Hirsch, Z. W. Kundzewicz, D. P. Lettenmaier, and R. J. Stouffer. 2008. Stationarity is dead: Whither water management? *Science* 319 (5863):573–74. doi:10.1126/science.1151915.

Moore, J. W., ed. 2016. *Anthropocene or Capitalocene? Nature, history, and the crisis of capitalism*. Oakland, CA: PM Press.

Murphy, M. 2006. *Sick building syndrome and the problem of uncertainty: Environmental politics, technoscience, and women workers*. Durham, NC: Duke University Press.

National Research Council. 1995. *Beach Nourishment and Protection*. Washington, DC: The National Academies Press. doi: 10.17226/4984.

National Research Council. 2014. *Lessons learned from the Fukushima nuclear accident for improving safety and security of U.S. nuclear plants*. Washington, DC: National Academies Press.

Pilkey, O. H., and L. Pilkey-Jarvis. 2007. *Useless arithmetic: Why environmental scientists can't predict the future*. New York: Columbia University Press.

Pilkey, O. H., R. Young, and A. Cooper. 2013. Quantitative modeling of coastal processes: A boom or a bust for society? *Special Papers of the Geological Society of America* 502 (7):135–44. doi:10.1130/2013.2502(07).

Pine, K. H., and M. Liboiron. 2015. The politics of measurement and action. In *Proceedings of the 33rd Annual ACM Conference on Human Factors in Computing Systems*, 1–10. ACM. doi:10.1145/2702123.2702298.

Porter, T. M. 1996. *Trust in numbers: The pursuit of objectivity in science and public life*. Princeton, NJ: Princeton University Press.

Purdy, J. 2015. *After nature: A politics for the Anthropocene*. Cambridge, MA: Harvard University Press.

Ribes, D., and S. J. Jackson. 2013. Data bite man: The work of sustaining a long-term study. In *"Raw data" is an oxymoron*, 147–66, ed. R. Kitchin. Cambridge, MA: MIT Press.

Romm, J. 2011. Desertification: The next dust bowl. *Nature* 478 (7370):450–51. doi:10.1038/478450a.

Ruhl, J. B., and J. E. Salzman. 2011. Gaming the past: The theory and practice of historic baselines in the administrative state. *Vanderbilt Law Review* 64 (1):1–57. 10.2139/ssrn.1553484.

Sayre, N. F. 2017. *The politics of scale: A history of rangeland science*. Chicago, IL: University of Chicago Press.

Sedell, J. K. 2019. No fly zone? Spatializing regimes of perceptibility, uncertainty, and the ontological fight over quarantine pests in California. *Geoforum*. doi:10.1016/j.geoforum.2019.04.008.

Simon, G. L. 2010. The 100th meridian, ecological boundaries, and the problem of reification. *Society & Natural Resources* 24 (1):95–101. doi:10.1080/08941920903284374.

Stakhiv, E. Z. 2011. Pragmatic approaches for water management under climate change uncertainty. *JAWRA: Journal of the American Water Resources Association* 47 (6):1183–96. doi:10.1111/j.1752-1688.2011.00589.x.

Stolarski, R. S., A. J. Krueger, M. R. Schoeberl, R. D. McPeters, P. A. Newman, and J. C. Alpert. 1986. Nimbus 7 satellite measurements of the springtime Antarctic ozone decrease. *Nature* 322 (6082):808–11. doi:10.1038/322808a0.

Thieler, E. R., O. H. Pilkey, R. S. Young, D. M. Bush, and F. Chai. 2000. The use of mathematical models to predict beach behavior for U.S. coastal engineering: A critical review. *Journal of Coastal Research* 16 (1):48–70.

Toimil, A., I. J. Losada, R. J. Nicholls, R. A. Dalrymple, and M. J. F. Stive. 2020. Addressing the challenges of climate change risks and adaptation in coastal areas: A review. *Coastal Engineering* 156:103611. doi:10.1016/j.coastaleng.2019.103611.

Ureta, S., T. Lekan, and W. G. von Hardenberg. 2020. Baselining nature: An introduction. *Environment and Planning E: Nature and Space* 3 (1):3–19. doi:10.1177/2514848619898092.

von Hobe, M. 2007. Atmospheric science. Revisiting ozone depletion. *Science* 318 (5858):1878–79. doi:10.1126/science.1151597.

Walsh, J., D. Wuebbles, K. Hayhoe, J. Kossin, K. Kunkel, G. Stephens, P. Thorne, et al. 2014. Our changing climate. In *Climate change impacts in the United States: The Third National Climate Assessment*, ed. J. M. Melillo, T. C. Richmond, and G. W. Yohe, 19–67. Washington, DC: U.S. Global Change Research Program. doi:10.7930/J0KW5CXT.

Wheatley, S., B. Sovacool, and D. Sornette. 2017. Of disasters and dragon kings: A statistical analysis of nuclear power incidents and accidents. *Risk Analysis: An Official Publication of the Society for Risk Analysis* 37 (1):99–115. doi:10.1111/risa.12587.

Wigley, T. M. L. 2009. The effect of changing climate on the frequency of absolute extreme events. *Climatic Change* 97 (1–2):67–76. doi:10.1007/s10584-009-9654-7.

Williams, J. W., and S. T. Jackson. 2007. Novel climates, no-analog communities, and ecological surprises. *Frontiers in Ecology and the Environment* 5 (9):475–82. doi:10.1890/070037.

Williams, R. 1977. *Keywords: A vocabulary of culture and society*. London and New York: Routledge.

Worster, D. 2004. *Dust bowl: The southern plains in the 1930s*. New York: Oxford University Press. doi:10.1086/ahr/85.3.732.

Young, R. S., O. H. Pilkey, D. M. Bush, and E. R. Thieler. 1995. A discussion of the generalized model for simulating shoreline change (GENESIS). *Journal of Coastal Research* 11 (3):875–86.

Zalasiewicz, J., M. Williams, W. Steffen, and P. Crutzen. 2010. The new world of the Anthropocene. *Environmental Science & Technology* 44 (7):2228–31. doi:10.1021/es903118j.

Part 6

The Anthropocene and Geographic Education

What Does That Have to Do with Geology? The Anthropocene in School Geographies around the World

Péter Bagoly-Simó

Based on the growing body of research and vivid scientific discourse, the Anthropocene is slowly making its way into the curricula of geography programs across the world. Unlike academic geography, the school subject seems to be more reluctant when it comes to the Anthropocene's implicit implementation. Using content analysis, this article explores how geography curricula or compound subjects containing geography for lower secondary education in fifty countries represented the Anthropocene. The results showed that most curricula detached the Anthropocene from geological time and focused, in a disconnected manner, on three of its descriptors, namely, population growth, industrialization, and globalization. Greenhouse gases played a subordinate role. Also, most curricula operated at the national or global scale, leaving little room to navigate processes on several scales. The results also revealed few differences between curricula prescribing geography as an independent subject or as part of compound subjects, such as social studies or social sciences. *Key Words: Anthropocene, comparative study, curriculum, education for sustainable development, geographical knowledge.*

In his seminal paper "Geology of Mankind," Crutzen (2002, 23), reinforcing the initial ideas of Crutzen and Stoermer (2019), attributed the concept "'Anthropocene' to the present, in many ways human-dominated, geological epoch, supplementing the Holocene—the warm period of the past 10–12 millennia." Taking on these thoughts, the International Commission on Stratigraphy's Anthropocene Working Group voted in favor of treating the Anthropocene as a formal chrono-stratigraphic unit based on stratigraphic signals originating from as early as the mid-twentieth century (Anthropocene Working Group 2019).

Despite the modest chances for its addition to the Geological Time Scale soon, in a synoptic triptych, Castree (2014a, 2014b, 2014c) examined the potential of the buzzword Anthropocene to mature into a keyword of great societal importance. Looking beyond the conceptual framework offered by Earth system science and the planetary boundaries concept, Castree (2014c) sketched the role of geography, a broad discipline dedicated to human-not-human-relationships, in exploring "the full spectrum of problem definitions and suggested responses reflective of human disagreements about the right way to live on Earth" (474).

Indeed, finding responses to the right way(s) to live on Earth has been at the heart of geography both as an academic discipline and as a school subject. As the International Charter on Geographical Education (IGU-CGE 2016) states, in schools, geography provides the knowledge to "live sustainably in this world [and to] understand human relationships and their responsibilities to both the natural environment and to others" (5). Nevertheless, how far are we down this road? What are the curricular prescriptions of the school subject of geography around the world concerning the Anthropocene? This article aims to explore how fifty geography curricula for lower secondary education discuss the Anthropocene. Following a brief discussion of curricular research in geography education, the article describes the methods and sample, proceeds to present the results and their discussion, and closes by formulating some conclusions.

Geography Curricula and (Education for) Sustainable Development

For geography as a school subject, "the right way to live on Earth" (Castree 2014c, 474) has traditionally been connected to environmental education and, later on, education for sustainable development (ESD; Bagoly-Simó 2014). Signed in Lucerne, the International Declaration on Geographic Education for Sustainable Development (Haubrich, Reinfried, and Schleicher 2007) reinforced the role geography plays in formal education in shaping a more sustainable future.

Following the Lucerne Declaration's ratification, Lidstone and Stoltman (2007) identified the need for more comparative research on how geography curricula addressed ESD. A decade later, Chang and Kidman (2018) concluded that "we are still awaiting this research," although, "what we do have are a loose collection of anecdotes and examples on practices across Higher Educational Institutes and from K-12 contexts" (281). In general terms, this diagnosis might be accurate. As early as 2014, however, Bagoly-Simó (2014) published in the same journal a comparative curricular analysis of Bavarian, Mexican, and Romanian geography curricula and found a broad and deep implementation of both the concept of sustainable development (SD) and ESD-relevant topics. In a larger project, the author compared all subjects included in the curriculum of the three countries, measuring the topic-based implementation of ESD into lower secondary education (Bagoly-Simó 2013, 2014). The evidence showed that geography, closely followed by biology and technology, exhibited the broadest and deepest implementation of ESD topics in all three countries (Bagoly-Simó 2013, 2014).

Geography's contribution to ESD rests on both its conceptual affinity with SD and the intimate ties between geographical knowledge and ESD topics (Bagoly-Simó 2014). In most countries, geography reflects on human–environment interactions from multiple perspectives, such as ecologic, economic, social, cultural, and political ones. Also, the school subject emphasizes scale as one of its central concepts that enables the exploration of intragenerational or global equity from the local to the global. In addition to its conceptual affinity with the three poles of SD and the matter of generational equity (Bagoly-Simó 2014), geography connects ESD to its subject-specific knowledge through ESD topics (e.g., hunger, water, climate change, demographic development, consumption, waste, biodiversity) rather than implementing ESD as an add-on.

Large-scale international comparative work on geography curricula, particularly concerning ESD,

and, implicitly, the issue of how to negotiate the right ways to live on the planet, have remained an open issue. This article aims to explore how fifty geography curricula for lower secondary education discuss the Anthropocene as one specific facet of the subject's contribution to a much broader ESD.

Method and Sample

Content analysis served to analyze the representation of the Anthropocene based on nine predefined categories (see Table 1) derived from the definitions of the Anthropocene by Crutzen (2002), Crutzen and Stoermer (2019), and the Anthropocene Working Group (2019). The first category targeted all segments entailing the Anthropocene's explicit mentioning and aimed at uncovering its direct implementation in the individual curricula. The remaining eight categories covered various facets of the Anthropocene concept and enabled it to explore its implicit implementation in the curricular documents. The second category (geological time) addresses together with the third category (golden spike) the claim to formalize the Anthropocene as a unit within the Geological Time Scale via a global boundary stratotype section and point, or golden spike. Both categories originate from Crutzen's (2002) definition and also constitute the core of the Anthropocene Working Group's (2019) argument to formally end the Meghalayan Age of the Holocene and replace it with the human-dominated Anthropocene. Categories 4 through 6 (population growth, industrialization, and globalization) stand for the geological proxy signals accumulated within recent geological strata that describe the unprecedented impact humans had on the planet. All three processes continue to shape the strata and, consequently, record the impact of societies on the Earth, the reason why the Anthropocene Working Group (2019) selected them as proxies for the age of Great Acceleration. Artificial radionuclides, the seventh category, represented the Anthropocene's primary marker at the global scale due to the thermonuclear bomb tests of the 1950s. As the Anthropocene Working Group (2019) argued, the thermonuclear tests carried out with artificial radionuclides left the sharpest and globally most synchronous effect in the geological strata marking the beginning of the Anthropocene. The last two categories, namely carbon dioxide and greenhouse gases, are other

Table 1. Definition and source of the predefined empirical categories

Category no.	Category name	Source	Definition
1	Anthropocene	Crutzen (2002, 23)	The present, in many ways human-dominated, geological epoch, supplementing the Holocene—the warm period of the past 10–12 millennia.
2	Geological time	Anthropocene Working Group (2019)	The place of the Anthropocene in the Geological Time Scale at series/epoch level (its base/beginning would terminate the Holocene Series/Epoch as well as the Meghalayan Stage/Age).
3	Golden spike	Anthropocene Working Group (2019)	The place of the Anthropocene in the Geological Time Scale defined by the standard means for a unit of the Geological Time Scale, colloquially known as a golden spike.
4	Population growth	Anthropocene Working Group (2019)	Geological proxy signals preserved within recently accumulated strata resulting from the Great Acceleration of population growth placing the beginning of the Anthropocene in the mid-twentieth century.
5	Industrialization	Anthropocene Working Group (2019)	Geological proxy signals preserved within recently accumulated strata resulting from the Great Acceleration of industrialization placing the beginning of the Anthropocene in the mid-twentieth century.
6	Globalization	Anthropocene Working Group (2019)	Geological proxy signals preserved within recently accumulated strata resulting from the Great Acceleration of globalization placing the beginning of the Anthropocene in the mid-twentieth century.
7	Artificial radionuclides	Anthropocene Working Group (2019)	The sharpest and most globally synchronous signal forming a primary marker is made by the artificial radionuclides spread worldwide by the thermonuclear bomb tests from the early 1950s.
8	Carbon dioxide	Crutzen (2002); Crutzen and Stoermer (2019)	Atmospheric and stratigraphic proxy signal of the Anthropocene resulting from the increasing carbon dioxide emission over the course of the twentieth century and ongoing.
9	Greenhouse gases	Crutzen (2002); Crutzen and Stoermer (2019)	Atmospheric and stratigraphic proxy signal of the Anthropocene resulting from the increasing greenhouse gas (other than carbon dioxide) emissions over the course of the twentieth century and ongoing.

atmospheric and stratigraphic proxy signals of the Anthropocene (Crutzen 2002; Crutzen and Stoermer 2019).

Four main reasons support the conceptualization of the Anthropocene, along with the nine previously described categories. First, Crutzen's (2002), as well as Crutzen and Stoermer's (2019) definition, despite various (re)conceptualizations resulting from the growing interest across a multitude of disciplines, continues to represent the most influential conceptual framework for the Anthropocene. Second, the arguments put forward by the Anthropocene Working Group (2019) both synthesize the magnitude of the human impact on the planet and claim, based on their impact, the formal inclusion of the Anthropocene in the Geological Time Scale. Third, the emphasis on human–environment interaction in general and human impact on the planet, in particular, lies at the heart of geography as a school subject (IGU-CGE 2016) and stands for its contribution to ESD (Haubrich, Reinfried, and Schleicher 2007). Fourth, the nine categories operationalize the explicit and implicit implementation of the Anthropocene concept into geography curricula.

The first analytical step consisted of software-assisted (MAXQDA) lexical retrieval of all segments entailing the predefined categories. Thereby, segment retrieval rested on both truncated concepts and their synonyms, in each of the eight languages represented in the sample. During the second step, the qualitative analysis of each retrieved segment led to semantic disambiguation. Finally, document mapping served to analyze each segment's position in the curricular document concerning those coded in the nine categories, revealing both their clusters and dispersal representations. During this step, the analytical process focused on both the number and relative position of segments belonging to the nine categories to diagnose the Anthropocene's specific conceptualization in the curricular document. For example, balanced conceptualizations rested on several proxies (categories 4–9) of similar importance located in relative proximity within the curricular document. In contrast, unbalanced conceptualizations reflected a unilateral emphasis on one or only a few proxies scattered across the various elements of the curricular architecture. In addition, the analysis of document maps enabled the distinction of curricula that featured (some of) the categories as part of their mandatory or elective content. Document maps also allowed us to analyze the relative position of segments representing the individual categories within the curricular architecture and disclosed information on the Anthropocene's implementation as part of, for example, skills, competencies, content, suggested learning activities, or assessment.

The sample consisted of fifty-two national or state lower secondary geography curricula representing fifty countries. Curricula Worldwide, the international curriculum database of the Georg Eckert Institute for International Textbook Research (https://curricula-workstation.edumeres.net/en/curricula/), served to identify geography curricula for lower secondary education available online or in the institute's library. In addition to geography as an independent school subject, the database considered all compound subjects that geography was part of (e.g., social studies, social science). In the last step, the sample constitution considered all curricula accessible to the author in the following languages: English, French, German, Hungarian, Italian, Portuguese, Romanian, and Spanish. Also, a final selective step targeted the broadest global distribution. In the case of countries with federal educational systems, curricula selection followed national particularities. The curricula of the provinces of Ontario (largest in terms of [English-speaking] population) and Québec (largest community of French Canadians) represented Canada. North Rhine–Westphalia (largest population) and Berlin/Brandenburg (joint curriculum of a former socialist region and one of the three city-states of the country) represented Germany, following the usual sampling of federal educational systems. The final sample contained twenty-one geography curricula and twenty-nine curricula of compound subjects with geographical elements (see Appendix).

Results

Semantic disambiguation identified 558 of the 3,327 initially retrieved segments that contain five of the nine predefined categories. The distribution of the segments (Table 2) showed an emphasis on population growth, followed by globalization and industrialization. Although the curricula of the Canadian provinces entailed five and the Liechtenstein curriculum four categories, 16 percent of the fifty curricula contained three, one third two, one quarter one, and one fifth none of the nine categories.

Except for the Canadian curricula, document mapping showed a dispersed distribution of the five identified categories constituting a thematic island within the curricular documents.

The concept of the Anthropocene was missing from the analyzed curricula. The eight categories derived from its definitions, however, showed different patterns, as described next.

Geological Time and Stratigraphy

A total of two of the fifty analyzed curricula mentioned geological time. In Moldova, the curriculum required students to apply their knowledge of geological time to determine the age of geomorphological structures. In contrast, the Tanzanian curriculum prescribed the simplified geological time as a mandatory content element of the chapter dedicated to Earth's structure and its crust's petrographic composition. Information on the golden spike was missing from all curricula.

Table 2. Distribution of segments across the sample

Country	Anthropocene	Geological time	Golden spike	Population growth	Industrialization	Globalization	Artificial radionuclides	Carbon dioxide	Greenhouse gases
Angola	0	0	0	4	0	0	0	0	0
Argentina	0	0	0	3	1	1	0	0	0
Austria	0	0	0	2	0	2	0	0	0
Belize	0	0	0	0	0	3	0	0	0
Bermuda	0	0	0	0	0	3	0	0	0
Bolivia	0	0	0	0	0	9	0	0	0
Brazil	0	0	0	0	7	37	0	0	2
Cape Verde	0	0	0	11	0	0	0	0	0
Cameroon	0	0	0	0	0	1	0	0	0
Canada	0	0	0	13	1.5	6.5	0	1	2
Chile	0	0	0	0	1	0	0	0	0
Colombia	0	0	0	1	0	5	1	0	0
Congo	0	0	0	13	0	0	0	0	0
Costa Rica	0	0	0	3	4	0	0	0	0
France	0	0	0	2	2	0	0	0	0
Germany	0	0	0	5.5	0	10.5	0	1	5
Ghana	0	0	0	2	0	0	0	0	1
Guyana	0	0	0	2	0	0	0	0	0
Honduras	0	0	0	10	0	13	0	0	0
Hungary	0	0	0	2	0	8	0	0	0
Ireland	0	0	0	8	1	0	0	0	1
Italy	0	0	0	0	0	0	0	0	0
Ivory Coast	0	0	0	0	0	0	0	0	0
Jamaica	0	0	0	23	0	0	0	0	0
Liberia	0	0	0	0	0	0	0	0	0
Liechtenstein	0	0	0	6	3	3	0	0	1
Mauritius	0	0	0	0	0	2	0	0	1
Mexico	0	0	0	8	0	2	0	0	0
Mozambique	0	1	0	1	0	1	0	0	0
Moldova	0	0	0	10	0	0	0	0	0
Namibia	0	0	0	18	0	1	0	0	0
New Zealand	0	0	0	0	0	0	0	0	0
Niger	0	0	0	11	3	0	0	0	0
Panama	0	0	0	11	0	20	0	0	0
Paraguay	0	0	0	0	0	0	0	0	0
Portugal	0	0	0	9	0	0	0	0	0
Romania	0	0	0	4	0	0	0	0	0
Rwanda	0	0	0	17	25	0	0	0	0
Samoa	0	0	0	2	0	0	0	0	0
Singapore	0	0	0	15	0	0	0	1	1

South Africa	0	0	4	0	1	10	0	0	0
South Sudan	0	0	0	0	0	0	0	0	0
Spain	0	0	0	11	8	11	0	0	0
Sweden	0	0	0	0	0	0	0	0	0
Tanzania	0	0	0	0	0	4	0	1	0
Thailand	0	0	0	0	0	0	0	0	0
Trinicad and Tobago	0	0	0	18	6	0	0	0	0
Uganda	0	0	0	17	14	44	0	0	0
Uruguay	0	0	0	0	0	0	0	0	0
Zimbabwe	0	0	0	0	0	3	0	0	0
Total	14	3	5	171	77.5	285.5	0	2	0

Population Growth

The thirty-two curricula discussing demographic dynamics embedded the phenomenon into specific conceptual networks, often left its historical development unconsidered, and addressed the topic at various scales. Whereas three curricula only mentioned the concept (Colombia, Costa Rica, Mozambique, and Sāmoa), the curricula of five countries (Canada, Cape Verde, Honduras, Jamaica, and Uganda) dedicated extensive attention to population growth. The remaining twenty-three curricula contained a balanced discussion of the topic.

With the majority prescribing population growth as a mandatory concept, only a few curricula (e.g., Moldova and Tanzania) entailed specific demographic content, such as fertility and mortality. Also, population growth appeared nested in fourteen conceptual networks across the sample (Figure 1). Regarding the Anthropocene, particularly relevant curricular content reflected on the interrelation between population growth and the environment (especially addressing the greenhouse effect); the interplay between population dynamics, resources, and economic activities (including human development); as well as the individual responsibility (reproductive behavior). Population growth was part of such content in eight curricula.

The majority of the thirty-two curricula explored population growth in the present. Only four case studies emphasized its historical development, and only the French curriculum considered the global population at the expense of the national demographic development.

Most curricula addressed population growth at several scales (Table 3). Nevertheless, the global and national scales prevailed. Most case studies, however, introduced demographic development at various scales in different thematic units, with only a few curricula prescribing comparative perspectives across various scales.

Industrialization

The second process leaving stratigraphic evidence is the ongoing process of industrialization. The concept shaped the content of fifteen curricula as part of specific conceptual matrices presented at various scales, mainly in combination with historical development and current status. The Rwandan curriculum extensively discussed the concept, but its

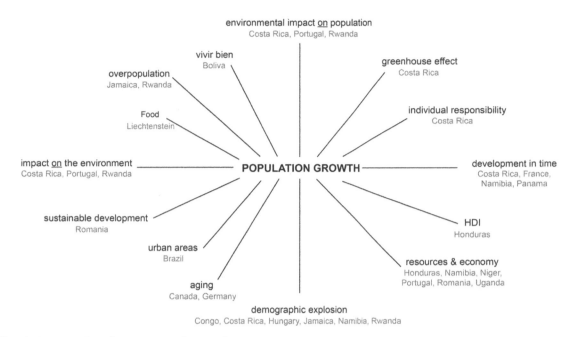

Figure 1. Population growth and its conceptual network.

Argentinian, Irish, and South African counterparts briefly mentioned it. The remaining curricula offered a balanced presentation of the phenomenon.

Nine main conceptual networks integrated the concept of industrialization (Figure 2). In addition, the temporal perspective tracing back the stratigraphic and atmospheric impact of industrialization to its early days during the Industrial Revolution is essential. Nine curricula explicitly addressed the history of industrialization, and six countries (Chile, Costa Rica, France, Niger, Rwanda, and Spain) opted to discuss their industrial past within the framework of global industrialization. As a result, the prevailing scale tied to the process of industrialization was global, followed by the national. In addition, the curricula often prescribed comparative perspectives based on scale.

Globalization

The third component describing the Great Acceleration is globalization. The content of twenty curricula included the concept of globalization focusing on the present, along with four main conceptual dimensions, and predominantly following a comparative view. Except for five curricula (Argentina, Canada, Cameroon, Mozambique, and Namibia), all case studies exhibited a balanced introduction of the concept.

All twenty curricula introduced globalization by discussing at least one of its four main dimensions: economic, political, cultural, and financial (Figure 3). In addition, the Austrian curriculum connected the topic to the overarching objective of global learning (learning for the one world), and several Latin American curricula were very critical of globalization or even adopted an antiglobalization stance.

The temporal emphasis of all curricula was in the present. Nevertheless, they also offered an explanatory background of the process, revealing its development in time. Concerning scale, all curricula focused on the national territory and its place in the globalized world. As a result, scale appeared as a constant contrast between the national and the global. In addition, some curricula complemented this bipolar scalarity by a subnational scale (Canada and Uganda) and a supranational regional scale (Latin America in Honduras and Panama and developing countries in Honduras).

Artificial Radionuclides

Except for the Colombian and South African curriculum, the topic of artificial radionuclides was missing from the sample. The Colombian curriculum focused on nuclear waste and the challenges connected to its disposal and management. In contrast, the South African curriculum discussed the Nuclear

Table 3. Population growth according to scale

	Scale		
	National	Continental	Global
Angola	✓	✓	✓
Argentina	✓		✓
Austria	✓		
Belize			
Bermuda			
Bolivia			
Brazil			
Cape Verde	✓		
Cameroon			
Canada	✓		✓
Chile			
Colombia			
Congo	✓		✓
Costa Rica	✓	✓	
France			✓
Germany	✓		✓
Ghana			
Guyana			
Honduras	✓	✓	
Hungary			
Ireland			✓
Italy			
Ivory Coast			
Jamaica	✓		✓
Liberia			
Liechtenstein	✓		✓
Mauritius			
Mexico	✓		✓
Mozambique			
Moldova	✓		✓
Namibia	✓		✓
New Zealand			
Niger		✓	✓
Panama	✓	✓	
Paraguay			
Portugal			✓
Romania			✓
Rwanda	✓		✓
Samoa			
Singapore	✓	✓	
South Africa			✓
South Sudan			
Spain	✓	✓	
Sweden			
Tanzania			
Thailand			
Trinidad and Tobago			
Uganda	✓	✓	✓
Uruguay			
Zimbabwe			

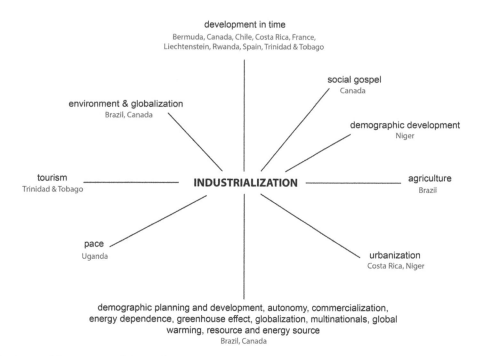

Figure 2. Industrialization and its conceptual network.

Figure 3. Globalization and its conceptual network.

Age and the Cold War as part of its chapters dedicated to history.

Carbon Dioxide and Other Greenhouse Gases

Three curricula prescribed CO_2 as mandatory content. Whereas the Singaporean curriculum referred to carbon dioxide in discussing photosynthesis in the tropical rainforest, its Ghanaian counterpart listed CO_2 as one of the greenhouse gases produced during farming. Similarly, the Ontario curriculum linked CO_2 emissions to human activities resulting from an interplay of available resources, population density, and prevailing economic activities.

Four curricula addressed other greenhouse gases when introducing the carbon offset (Canada), global warming (Guyana and Mauritius), and climate change (Ghana). It was exclusively the Ghanaian curriculum that listed CO_2, CH_4, water vapor, and nitrous oxides as exemplary greenhouse gases. In addition, five curricula (Brazil, Canada, Ireland, Liechtenstein, and Singapore) prescribed the greenhouse effect as mandatory content. Direct references to greenhouse gases, however, were missing from these curricula.

Discussion and Outlook

The results of this study lead to five main characteristics of the ways fifty curricula introduced the Anthropocene. First, all curricula implicitly addressed the Anthropocene. Given the relevance of the concept to geography as an academic discipline (Castree 2014a, 2014b, 2014c) on the one hand and the centrality of human–environment interactions to geography as a school subject (IGU-CGE 2016) as well as its crucial role for ESD (Haubrich, Reinfried, and Schleicher 2007; Bagoly-Simó 2013) on the other hand, the missing explicit mentioning of the Anthropocene in the curricular documents reflects the mismatch between the progress of the academic discipline and the conceptual update in geography as a school subject. Educating for a more sustainable future, however, requires, especially in the field of human–environmental interaction, students to apply the best available expert knowledge when designing alternatives to nonsustainable processes and structures they encounter.

Second, the representation of the Anthropocene remains, in the majority of the analyzed curricula, fragmented and limited to a few of its features. The results showed an equal share (one fifth) of curricula entailing at least three or none of the nine categories defining the Anthropocene. With population growth, globalization, and industrialization accounting for two thirds of all segments, the analyzed curricula address traditional topics of geography as a school subject (Bagoly-Simó 2013). Surprising is the modest share (11 percent) of the segments dedicated to greenhouse gases and CO_2 (Crutzen and Stoermer 2019; Crutzen 2002), given the substantial contribution of the subject to discussing climate change (Bagoly-Simó 2013; Chang and Kidman 2018). Despite referencing at least three of the categories defining the Anthropocene, the curricula rarely linked these categories—a vital prerequisite to understanding the

Anthropocene. Consequently, students seem to be exposed to disparate knowledge on the essential processes that induced and have sustained the Anthropocene. Future curricular reform should explore ways to better link these processes, including principles of ESD (Haubrich, Reinfried, and Schleicher 2007; IGU-CGE 2016).

Third, geological time is a neglected topic in all but two geography curricula (Moldova and Tanzania). In addition, the analyzed curricula predominantly focused on the present, which might originate from the traditional and artificial division between history (time) and geography (space) underlying curriculum development. Understanding the magnitude and speed of human impact on the planet, however, requires its contextualization in (geological) time. Therefore, based on the results, a reevaluation of the key concept time (Clifford et al. 2009) in geography curricula might facilitate students' access to the chronology of change inherent to population dynamics, industrialization, globalization, development, system, environment, and other processes.

Fourth, addressing the Anthropocene requires both a flexible and comparative approach to scale. The results of this article indicate that most curricula prefer the national or the global scale, with the fewest examples comparing, in a systematic manner, structures and processes at least at two scales (e.g., Mexico and the world or Panama in the Americas and the world). Of course, negotiating a more flexible use of the key concept scale in the teaching and learning of geography reaches beyond its immediate necessity for a better understanding of the Anthropocene and for ESD implementation (Bagoly-Simó 2013, 2014).

Fifth, this study's results uncovered only a few differences between geography as an independent subject or as part of a compound school subject. Interestingly, it was compound subjects that featured the highest number of categories describing the Anthropocene, dominated the group of curricula void of any such categories, and discussed artificial radionuclides. Despite the very different curricular cultures, geography in compound subjects often seems to connect with history and political science by stressing its human geographical side to the detriment of physical geography. The independent subject of geography has a limited contribution to the discussion of greenhouse gases or geological time and strata. Based on the data presented in this article, when geography was to assume its role as the subject exploring "human-not-human relationships," as

Castree (2014c, 474) emphasized, the not-human needs to be reinforced. Of course, this translates into a strengthening of physical geography to enable the exploration of new connections between geographical subdisciplines that "transcend the all-too-familiar (and thankfully unrealistic) aspiration for a holistic approach" (Castree 2014c, 474) inherent to integrative dilutions of the discipline. Linking back to the debates on a new physical geography, Bagoly-Simó and Uhlenwinkel (2016) described a range of challenges that physical geography faces in school curricula. Understanding the Anthropocene, however, requires both solid physical geographical knowledge and clear links between the human and the nonhuman—a pressing matter for both geography and compound subject curricula.

This article offers a first insight into the ways in which geography curricula for lower secondary education addressed the Anthropocene. In addition, it contributes, based on a large sample, to comparative curricular research in geography education (Chang and Kidman 2018). Nevertheless, both its conceptualization and operationalization bear limitations. On the conceptual side, alternative definitions of the Anthropocene might uncover a different contribution to teaching this central topic. Also, opting for a more geoscientific framework instead of ESD would lead to other findings that would be relevant for geography education in other ways. On the methodological side, the content analysis revealed a rich conceptual network that integrated the concepts of population growth, industrialization, and globalization. Future qualitative work based on an in-depth analysis (e.g., Bagoly-Simó 2014) could identify additional conceptual links supporting both ESD and the Anthropocene concept. Such studies could also be more sensitive to the approach (thematic, regional, or combined) the individual curricula followed as well as to their architecture covering aspects, such as educational aims, key concepts, standards, content, and presentation (in a brief schematic or extensive textual format). In doing so, they would contribute to much-needed comparative curricular studies (Lidstone and Stoltman 2007; Bagoly-Simó 2017; Chang and Kidman 2018), particularly regarding geography's contribution to and role in compound subjects. Finally, the sampling process would require an exploration of curricula published in other languages. The work presented in this article could not consider India, China, Japan, and a range of countries with curricula published in various Slavic, African, Asian, and Australasian languages.

Summing up, this study's results offered a first insight into the way geography curricula of fifty selected countries discussed the Anthropocene. Based on the results, several recommendations were formulated that could create a minimal framework to discuss this timely concept. Future work addressing the limitations of this study could offer in-depth information and specific recommendations matching the formal requirements of national or regional curricular traditions, approaches, and political objectives tied to the curriculum and the expert geographical knowledge to which future generations should receive access.

Funding

The author acknowledges support from the Open Access Publication Fund of Humboldt-Universität zu Berlin.

References

Anthropocene Working Group. 2019. Results of bidding vote by AWG. Accessed June 10, 2020. http://quaternary.stratigraphy.org/working-groups/anthropocene/.

Bagoly-Simó, P. 2013. Tracing sustainability: An international comparison of ESD implementation into lower secondary education. *Journal of Education for Sustainable Development* 7 (1):95–108. doi: 10.1177/0973408213495610.

Bagoly-Simó, P. 2014. Tracing sustainability: Concepts of sustainable development and education for sustainable development in lower secondary geography curricula of international selection. *International Research in Geographical and Environmental Education* 23 (2):126–41. doi: 10.1080/10382046.2014.908525.

Bagoly-Simó, P. 2017. Exploring comparative curricular research in geography education. *Documents D'Anàlisi Geogràfica* 63 (3):561–73. doi: 10.5565/rev/dag.493.

Bagoly-Simó, P., and A. Uhlenwinkel. 2016. Why physical geography? An analysis of justifications in teacher magazines in Germany. *Nordidactica: Journal of Humanities and Social Science Education* 2016 (1):23–27.

Castree, N. 2014a. The Anthropocene and geography I: The back story. *Geography Compass* 8 (7):436–49. doi: 10.1111/gec3.12141.

Castree, N. 2014b. Geography and the Anthropocene II: Current contributions. *Geography Compass* 8 (7):450–63. doi: 10.1111/gec3.12140.

Castree, N. 2014c. The Anthropocene and geography III: Future directions. *Geography Compass* 8 (7):464–76. doi: 10.1111/gec3.12139.

Chang, C.-H., and G. Kidman. 2018. Editorial: The future of education for sustainable development—Where next after a decade of discourse? *International Research* in *Geographical and Environmental Education* 27 (4):281–82. doi: 10.1080/10382046.2018.1511040.

Clifford, N., S. Holloway, S. Rice, and G. Valentine. 2009. *Key concepts in geography*. London: Sage.

Crutzen, P. 2002. Geology of mankind. *Nature* 415 (6867):23. doi: 10.1038/415023a.

Crutzen, P., and E. F. Stoermer. 2019. The Anthropocene. *Global Change Newsletter* 41:17–18.

Haubrich, H., S. Reinfried, and Y. Schleicher. 2007. Lucerne Declaration on geographical education for sustainable development. In *Geographical views on education for sustainable development*, ed. S. Reinfried, Y. Schleicher, and A. Rempfler, 243–50. Weingarten, Germany: HGD.

International Geographical Union. Commission on Geographical Education (IGU-CGE). 2016. 2016 International charter in geographical education. Accessed June 10, 2020. https://www.igu-cge.org/wp-content/uploads/2019/03/IGU_2016_eng_ver25Feb2019.pdf.

Lidstone, J., and J. Stoltman. 2007. Editorial: Sustainable environments or sustainable cultures. Research priorities. *International Research in Geographical and Environmental Education* 16 (1):1–4. doi: 10.2167/irg16.1.0.

Appendix

Angola: Ministério da Educação. 2019. *Programas de geografia. 7ª, 8ª e 9ᵉ classes*. Luanda, Angola: Editora Moderna.

Argentina: Ministerio de Educación. 2011. *Núcleos de aprendizajes prioritarios: Ciencias sociales. Ciclo básico. Educación secundaria. 1° y 2°/2° y 3°años*. Buenos Aires: Ministerio de Educacíon.

Austria: Bundesministerium für Bildung. 2017. Lehrplan für die neuen Mittelschulen. *Bundesgesetzblatt II Nr. 111/2017*. Vienna: Bundesministerium für Bildung.

Belize: Ministry of Education, Culture, Youth & Sports. 2005. *Resource guide for teaching social studies in primary schools*. Belize City, Belize: Ministry of Education, Culture, Youth & Sports.

Bermuda: Ministry of Education. 2008. *Essential curriculum 2008: Social studies*. Hamilton, Bermuda: Ministry of Education, Curriculum & Instructional Leadership Office.

Bolivia: Ministerio de Educación. 2012. *Currículo base del sistema educativo plurinacional*. La Paz, Bolivia: Ministerio de Educación, Estado Plurinacional de Bolivia.

Brazil: Ministério da Educação e do Desporto. Secretaria de Educação Fundamental. 1998. *Parâmetros curriculares nacionais: Geografia*. Brasília, Brazil: Ministério da Educação e do Desporto, Secretaria de Educação Fundamental.

Cameroon: Ministry of Basic Education. 2018. *Cameroon primary school curriculum. English subsystem. Level II: Class 3 & Class 4*. Yaoundé, Camaroon: Ministry of Basic Education.

Canada (Ontario): Ministry of Education. 2018. *The Ontario curriculum Grade 9 and 10. Canadian and world studies. Geography. History. Civics (Politics)*. Toronto, ON, Canada: Queen's Printer for Ontario.

Canada (Québec): Québec Education Program. 2010. *Geography*. Québec, QC, Canada: Québec Education Program.

Canada (Québec): Québec Education Program. 2010. *Progression of learning in secondary school. Geography. Cycle One*. Québec, QC, Canada: Québec Education Program.

Canada (Québec): Québec Education Program Frameworks for the Evaluation of Learning. 2011. *Framework for the evaluation of learning. Geography. Secondary school. Cycle One*. Québec, QC, Canada: Québec Education Program.

Cape Verde: Ministério da Educação. Direção Nacional de Educação. 2018. *Programa da Disciplina da História e Geografia de Cabo Verde (HGCV) 5° 6° anos. 2° Ciclo do Ensino Básico Obrigatório*. Praia, Cape Verde: Ministério da Educação. Direção Nacional de Educação.

Chile: Ministerio de Educación. Unidad de Currículum y Evaluación. 2020. *Priorización curricular Covid-19. Historia, geografía y ciencias sociales y educación ciudadana. 1° básico a 4° medio*. Santiago, Chile: ME, República de Chile.

Colombia: Ministerio de Educación Nacional. 2015. *Lineamientos curriculares. Ciencias sociales*. Bogotá, Colombia: Ministerio de Educación Nacional.

Costa Rica: Ministerio de Educación Pública. 2003. *Programas de estudio estudios sociales. Tercer ciclo educación general básica y educación diversificada*. San José, Costa Rica: Ministerio de Educacíon Pública.

DR Congo: Ministère de L'Enseignement Primaire, Secondaire et Professionnel, Direction des Programmes Scolaires et Matériel Didactique. 2005. *Programme national de géographie*. Kinshasa, DR

Congo: Ministère de L'Enseignement Primaire, Secondaire et Professionnel, Direction des Programmes Scolaires et Matériel Didactique.

France: Ministère de l'Éducation Nationale, de l'Enseignement Supérieur et de la Recherche. 2015. *Programmes pour les cycles 2, 3, 4.* Paris: Ministère de l'Enseignement Primaire, Secondaire et Professionnel, Direction des Programmes Scolaires et Matériel Didactique.

Germany (Berlin/Brandenburg): Senatsverwaltung für Bildung, Jugend und Familie and Ministerium für Bildung, Jugend und Sport Brandenburg. 2015. *Rahmenlehrplan für die Jahrgangsstufen 1–10. Teil C. Geografie. Jahrgangsstufen 7–10.* Berlin/Potsdam: Senatsverwaltung für Bildung, Jugend und Familie and Ministerium für Bildung, Jugend und Sport Brandenburg.

Germany (North Rhine-Westphalia): Ministerium für Schule und Weiterbildung des Landes Nordrhein-Westfalen. 2007. *Kernlehrplan für das Gymnasium–Sekundarstufe I (G8) in Nordrhein-Westfalen.* Düsseldorf, Germany: Ministerium für Schule und Weiterbildung.

Ghana: Ghana Education Service. National Council for Curriculum & Assessment of Ministry of Education. 2019. *Our world and our people: Curriculum for primary schools (Basic 4–6).* Accra, Ghana: National Council for Curriculum & Assessment of Ministry of Education.

Guyana: Ministry of Education. 2009. *Secondary school: Transitional curriculum guide. Social studies. Grades 6–7.* Georgetown, Guyana: Ministry of Education.

Honduras: Secretaría de Educación. Subsecretaría Técnico Pedagogía. Dirección General de Currículo. 2009. *Diseño curricular nacional para la educación básica. Tercer ciclo.* Tegucigalpa, Honduras: Secretaría de Educación. Subsecretaría Técnico Pedagogía. Dirección General de Currículo.

Hungary: Magyarország Kormánya. 2012. *Nemzeti alaptanterv. 110/2012. (VI. 4.) kormányrendelet. Földrajz.* Budapest, Hungary: Magyarország Kormánya.

Ireland: National Council for Curriculum and Assessment. 2017. *Junior cycle geography.* Dublin, Ireland: National Council for Curriculum and Assessment.

Italy: Ministero dell'istruzione, dell'università e della ricerca. 2012. Indicazioni nazionali per il curricolo della scuola dell'infanzia e del primo ciclo d'istruzione. *Annali della Pubblica Istruzione* 2012:1–88.

Ivory Coast: Ministère de l'Éducation Nationale et de l'Enseignement Technique. 2012. *Domaine de l'univers social. Programmes educatifs et guides d'exécution. Histoire et géographie. 6eme/5eme.* Yamoussoukro, Ivory Coast: Ministère de l'Éducation Nationale et de l'Enseignement Technique.

Ivory Coast: Ministère de l'Éducation Nationale et de l'Enseignement Technique. 2012. *Domaine de l'univers social. Programmes educatifs et guides d'exécution. Histoire et géographie. 4eme/3eme.* Yamoussoukro, Ivory Coast: Ministère de l'Éducation Nationale et de l'Enseignement Technique.

Jamaica: Ministry of Education and Culture. 1998. *Curriculum guide. Grade 7–9. Career education, language, arts, mathematics, science, social studies.* Kingston, Jamaica: Ministry of Education and Culture.

Liberia: Ministry of Education. 2011. *National curriculum for Grades 7 to 9. Social studies.* Monrovia, Liberia: Ministry of Education.

Liechtenstein: Schulamt des Fürstentums Liechtenstein. 2018. *Liechtensteiner Lehrplan. Natur, Mensch, Gesellschaft. Kompetenzaufbau 1. und 2. Zyklus.* Vaduz, Liechtenstein: Schulamt des Fürstentums Liechtenstein.

Mauritius: Mauritius Institute of Education & Ministry of Education & Human Resources. 2011. *Syllabus forms I, II & III.* Port Louis, Mauritius: Mauritius Institute of Education & Ministry of Education & Human Resources.

Mexico: Secretaría de Educación Pública. 2011. *Programas de estudio 2011: Guía para el maestro. Educación básica. Secundaria. Geografía de México y del mundo.* Mexico City, Mexico: Secretaría de Educación Pública.

Moldova: Ministerul Educației al Republicii Moldova. 2010. *Geografia. Curriculum pentru învățământul gimnazial (clasele V–IX).* Chișinău, Moldova: Ministerul Educației al Republicii Moldova.

Mozambique: Ministério da Educação e Desenvolvimento Humano, Instituto Nacional do Desenvolvimento da Educação. 2015. *Programas do ensino primário. Língua Portuguesa, matemática, ciências naturais, ciências sociais e educação física 2° Ciclo (3^a, 4^a e 5e Classes).* Maputo, Mozambique: Ministério da Educação e Desenvolvimento Humano, Instituto Nacional do Desenvolvimento da Educação.

Namibia: Ministry of Education, National Institute of Educational Development. 2010. *Junior secondary phase: Geography syllabus. Grades 8–10.* Okahandja, Namibia: Ministry of Education, National Institute of Educational Development.

New Zealand: Ministry of Education, Te Tāhuhu o te Mātauranga. 2017. *Social sciences.* Wellington, New Zealand: Ministry of Education.

Niger: Ministère des Enseignements Secondaire et Supérieur de la Recherche et de la Technologie. Direction des Enseignements des Cycles de Base II et Moyen. Inspection Pédagogique Nationale. 2009. *Programmes officiels de l'enseignement des cycles base II et moyen. Histoire et géographie.* Niamey, Niger: Ministère des Enseignements Secondaire et Supérieur de la Recherche et de la Technologie. Direction des Enseignements des Cycles de Base II et Moyen. Inspection Pédagogique Nationale.

Panama: Ministerio de Educación, Dirección Nacional de Currículo y Tecnología Educativa. 2014. *Educación básica general: Programa de geografía. 7°, 8° y 9°.* Panana City, Panama: Ministerio de Educación, Dirección Nacional de Currículo y Tecnología Educativa.

Paraguay: Ministerio de Educación y Cultura. 2011. *Programa de estudio educación básica bilingüe para personas jóvenes y adultas 3° ciclo.* Asunción, Paraguay: Ministerio de Educación y Cultura.

Portugal: Ministério da Educação, Departamento de Educação Básica. 2018. *Geografia. Orientações curriculares. 3° ciclo.* Lisbon, Portugal: Ministério da Educação, Departamento de Educação Básica.

Romania: Ministerul Educaţiei Naţionale. 2017. *Programa şcolară pentru disciplina Geografie. Clasele a V-a – a VIII-a.* Bucharest, Romania: Ministerul Educaţiei Naţionale.

Rwanda: Ministry of Education, National Curriculum Development Centre. 2008. *Ordinary level geography curriculum for Rwanda.* Kigali, Rwanda: Ministry of Education, National Curriculum Development Centre.

Samoa: Ministry of Education, Sports and Culture, Curriculum Development Unit. 2004. *Social studies. Year 9–11. Sāmoa secondary school curriculum.* Apia, Samoa: Ministry of Education, Sports and Culture, Curriculum Development Unit.

Singapore: Curriculum Planning and Development Division. 2014. *Lower secondary geography: Teaching syllabuses. Express course. Normal (academic) course.* Singapore: Curriculum Planning and Development Division.

South Africa: Department of Basic Education. 2011. *Curriculum and assessment policy statement: Grades 7–9. Social sciences.* Pretoria, South Africa: Department of Basic Education.

South Sudan: Ministry of Education, Science and Technology. 2017. *Subject overviews: South Sudan.* Juba, South Sudan: Ministry of Education, Science and Technology.

Spain: Ministerio de la Presidencia, Relaciones con las Cortes e Igualdad. 2015. Real Decreto 1105/2014, de 26 de diciembre, por el que se establece el currículo básico de la educación secundaria obligatoria y del bachillerato. *Boletín Oficial del Estado (BOE)* 3:169–546.

Sweden: Skolverket. 2018. *Curriculum for the compulsory school, preschool class and school-age education: Revised 2018.* Stockholm: Skolverket.

Tanzania: Ministry of Education and Vocational Training. Tanzania Institute of Education. 2013. *Curriculum for ordinary level secondary education in Tanzania.* Dar es Salaam, Tanzania: Ministry of Education and Vocational Training.

Thailand: Ministry of Education. 2008. *The basic education core curriculum.* Bangkok, Thailand: Ministry of Education.

Trinidad and Tobago: Ministry of Education, Curriculum Planning and Development Division. 2008. *Secondary school curriculum: Forms 1–3. Social studies.* Port of Spain, Trinidad and Tobago: Ministry of Education, Curriculum Planning and Development Division.

Uganda: National Curriculum Development Centre. 2013. *Lower secondary curriculum, assessment and examination reform programme: Social studies learning area syllabus.* Kampala, Uganda: National Curriculum Development Centre.

Uruguay: Consejo de Educación Secundaria. 2006. *Programas 1er año. Geografía. Primer año de ciclo básico. Reformulación 2006.* Montevideo, Uruguay: Consejo de Educación Secundaria.

Uruguay: Consejo de Educación Secundaria. 2006. *Programas 2° año. Geografía. Segundo año de ciclo básico. Reformulación 2006.* Montevideo, Uruguay: Consejo de Educación Secundaria.

Uruguay: Consejo de Educación Secundaria. 2006. *Programas 3er año. Geografía. Tercer año de ciclo básico. Reformulación 2006.* Montevideo, Uruguay: Consejo de Educación Secundaria.

Zimbabwe: Ministry of Education, Sport and Culture, Curriculum Development Unit. 2007. *Geography syllabus. Zimbabwe junior certificate.* Harare, Zimbabwe: Ministry of Education, Sport and Culture, Curriculum Development Unit.

Geographic Education in the Anthropocene: Cultivating Citizens at the Neoliberal University

Lindsay Naylor and Dana Veron

In the Anthropocene, the charge to address climate change has been taken up by youth. From the landmark climate lawsuit filed in 2015 by twenty-one young people to secure the legal right to a safe climate to the thousands of climate marches and school strikes that took place in 2018 and 2019, young people are making their voices heard. Many undergraduate students enter universities passionate about solving problems related to the changing climate; however, they arrive at a site of conflicting values about education. As the university increasingly treats students as customers, the ability to educate and foster a new generation of climate change–informed and action-oriented citizens is challenged. In this article, we ask how we cultivate citizens in the geography classroom in the Anthropocene. Drawing on our experience cocreating and teaching a climate change and food security class for sophomores, we examine this question to understand the place of geographic education in the neoliberal university in the Anthropocene. Here, we detail an example of coteaching and using a problem-based learning task based on pressing climate change issues as a way of modeling and practicing problem solving that reinforces students' identification as agents who can act in a climate-changing world. We find that teaching from both a climate science and social or cultural perspective is an important component of students considering where they fit in the Anthropocene. In this article, we offer a model for geography educators for integrative educational experiences, translating student passion for solving problems into momentum toward change. *Key Words: Anthropocene, climate change, geographic education, knowledge production, neoliberal university.*

In the epoch of the Anthropocene, there is tension in higher education whereby neoliberal universities are increasingly treating their students as consumers, creating an atmosphere of commodification where students exchange tuition for employability. Yet simultaneously, many students enter universities with a deep desire to learn how to solve the problems of a climate-changing world and do not see themselves as the customer (cf. Bratman et al. 2016). In 2019, ahead of the United Nations Climate Change Summit, youth from more than 185 countries participated in the school strike for climate popularized by Swedish sixteen-year-old Greta Thunberg (Laville and Watts 2019).[1] It is estimated that up to 6 million people participated in the strikes on 20 September; the Friday school strikes continue (even moving to social media during the COVID-19 pandemic), engaging thousands more. This generation of young people is leading the charge to address climate change. There are other examples, from the *Juliana v. United States* case, where twenty-one youth plaintiffs are suing the federal government to act on climate change, to organized youth climate movements around the world such as the Seed Movement, a group of aboriginal youth in Australia who are actively campaigning against mining and fracking in their communities.[2] The actions of this generation center on inciting systemic change, not reliance on technological fixes.

In 2017, the *New York Times* reported on a climate change poll directed at adults (age twenty-five or older) in the United States, which showed that although the vast majority think climate change is happening, they do not think it will affect them (Popovich, Schwartz, and Schlossberg 2017).[3] The respondents see climate change and its effects but do not see themselves as part of the story. In contrast, younger students see themselves as an integral part of the climate change story and desire to be part of systemic adaptation and mitigation strategies. Simultaneously, the era of the Anthropocene changes the character of engagement for young people, as it sets up an anxiety-inducing, existential crisis. Moreover, in the neoliberal university in the United States, students find a mismatch between desires to solve the problems precipitated by capitalist-driven climate change and the push to participate in the very same market processes responsible for environmental degradation. Indeed, Petersen and Barnes (2019) noted that the neoliberal tendencies of university programming actually make it more difficult to advance teaching that is hopeful and inspires change. Complicating this paradoxical educational atmosphere is the decision making that educators must undertake in developing pedagogy that is both attentive to detail and provides practical knowledge. Much of the focus of today's neoliberal education is on providing skills that translate to employability. Frequently, however, students are not trained to put the larger puzzle together. In effect, the neoliberal university presents a roadblock to the next generation of problem solvers. In this article, we argue that to develop effective teaching in the Anthropocene that explores these tensions we must engage students not as customers but as problem-solving citizens of the planet. We aim to demonstrate that new approaches are not only necessary in the Anthropocene epoch; they are also part of an existing toolkit in geography departments that can be used now. This article contributes to the larger body of geographic work on the Anthropocene more broadly and draws our attention to teaching; we offer the example of team teaching to promote geographic education in the Anthropocene (see also O'Brien et al. 2013). Specifically, we note that the diverse training of geographers makes possible deeper and broader classroom experiences when addressing complex topics such as the Anthropocene and climate change.

Drawing on theoretical frameworks developed in geographic writing on the Anthropocene demands that we conduct our academic activities differently. We begin by discussing the paradox of the neoliberal university in the Anthropocene. Following this foundation, we delve into our pedagogical approach to teaching geography in the Anthropocene, situating it in broader discussions of effective interdisciplinary education. We then turn to the case study, where we discuss, through the lens of a climate change and food security course taught in 2016, our coteaching and problem-based learning approach in geographic education. Finally, we offer some suggestions for future directions in higher education, where we reframe the university as a site of fostering citizenship, not meeting consumer demand.

Neoliberal Education in the Anthropocene

Students are passionate about solving problems to "make the world a better place." Many see climate change as the defining problem of their generation. The training to integrate knowledges and use their education toward solving those problems is lacking for many students, however. A neoliberal university setting presents students with a path toward employability rather than a set of ideas that can be deployed to solve the problems that they are passionate about. If we consider this conundrum through the lens of the Anthropocene, it demands that we reconsider knowledge production and our pedagogical approach.

The Human Era

Crutzen and Stoermer launched the debate about a human era in 2000, which emphasized the role of humankind in planetary change (Crutzen and Stoermer 2000). Scholarly narratives about the Anthropocene proliferated in the early 2010s following debate in the broader scientific community about a human-changed world and a possible transition from the Holocene. In geography, this transition was engaged and is now the subject of dozens of papers, which primarily focus on the impact for the discipline (see Castree 2014c); recently others have used it as a context for understanding long-standing geographic questions (see Ziegler and Kaplan 2019). Ultimately, raising questions about and officially declaring the epoch of the human has led scholars in a variety of fields to address what this means for our planetary futures. There are deeper implications for what it means to write and teach about, and in the context of, the Anthropocene. The existential crisis and stress on students suggests that more attention must be paid to how we discuss the human and nonhuman world and whether that binary serves us (cf. Pawson 2015; Fagan 2019), as well as the sociopolitical trajectories of the impacts of the new epoch (cf. Ernstson and Swyngedouw 2018; Loftus 2019).

Due to the wide-ranging engagement with the Anthropocene, the production of knowledge is a site where human geographers in particular are raising questions about how, where, and by whom ontological and epistemological understandings of a human era are understood and adopted. As Gibson-Graham (2014) noted, to think about the loss of earth others, and the threat to human life through the lens of the Anthropocene raises existential questions. Others urge a consideration of "the utility of the knowledge we are creating and the reasons *why* and for *whom* this knowledge is produced" (Leichenko and Mahecha 2015, 330, italics in original). Indeed, the long subjugation of multiple knowledges and the need to break down the geopolitics of knowledge in the face of extreme inequities is paramount in this moment (see Naylor et al. 2018). Cook, Rickards, and Rutherfurd (2015) argued that the Anthropocene illuminates knowledge politics, noting, "as an object of knowledge, the Anthropocene is complex, ambiguous, and contested" (238–39). Moreover, they argued that because it is an inherently interdisciplinary concept, there is a fluidity and mobility that makes it uniquely positioned to be taken up as a knowledge-building project that cuts across science, technology, engineering, and math; social sciences; and the humanities. As scholars, we are urged to participate in inclusive, dynamic, and plural knowledge building so that we can cocreate a planetary future that is for all inhabitants.

Yet, as Knitter et al. (2019) argued, we must do more than simply attach the word Anthropocene to our work. A deeper engagement requires not only critical approaches to the Anthropocene from those geographers conducting research and teaching on or about the Anthropocene but also attention to power structures in human, natural, and more-than-human systems. When operating in an educational system that is increasingly focused on exchanging diplomas for tuition dollars, academics face an uphill battle in putting this type of scholarship and pedagogy into practice. Although much work has been done on what the Anthropocene means to geographers as researchers, here we draw attention to what it means for teaching.

There Is No Alternative in Higher Education

A key facet of the Anthropocene is not just facing the impact of the production of goods for a neoliberal market but the neoliberal production of knowledge. As Connell (2013) noted, "Neoliberalism has a definite view of education, understanding it as human capital formation" (104). It is a system meant to cultivate consumers and laborers, rather than citizens (see also Pawson 2015). In the political–economic context of "there is no alternative" stemming from Thatcher's and Reagan's neoliberal push, public institutions of higher education became prime instruments of neoliberalism (Slaughter and Rhoades 2000). The market became the engine of the educational system with universities rebranded as sites that trained

students for jobs. Simultaneously, students became data points to be used as metrics that are measured and reported. The market logics of efficiency, productivity, and quotas now define the pathway to success (see Mountz et al. 2015).[4] As within all neoliberal institutions, the only things that count are those things that can be counted.

Neoliberal logics are bound to open up new spaces for capital. Connell (2013) discussed this marketization of higher education at length, arguing that education in the neoliberal system is rationed, thereby instituting restrictions on who has access, effectively reframing higher education as a commodity rather than a "citizen right" (102). The pay-to-play model of higher education, focused on the output of workers, reifies the structures that generate privilege and oppression. As Slaughter and Rhoades (2000) noted, students are now seen as revenue producers for the university—the upper administration refers to students as "customers", however, what this system does is turn students into "products" for the market (74).

With higher education entrenched in neoliberal settings, how are we to take seriously our roles as academics in the face of global environmental degradation? It is clear, as Dalby (2016) articulated, that we cannot use the same neoliberal logics that are part of the problem to create a solution to the challenges identified in the Anthropocene. Yet, "the point about the Anthropocene is that *it is the next time, not the end time*, and hence focusing on making the future, rather than responding to danger, has to be the pedagogic priority" (Dalby, cited in Johnson et al. 2014, 444, italics added). We fully recognize that our one experimental course does not break out of the constraints of the neoliberal university setting; in part, it directly feeds into them by maintaining institutional feedback processes, uneven labor regimes, and merit review. Simultaneously, we continue to exist as potential agents of change, whereby we can use our agency to push back on these structures. With this challenge in mind, we turn to a discussion of geographic education in the context of teaching climate change in the human era.

Climate Change Education in the Anthropocene

The Intergovernmental Panel on Climate Change (IPCC) provides consensus reports on Earth's changing climate; impacts and risks across natural, political, and economic sectors; and adaptation and mitigation possibilities (Weart 2019). These reports generated by scholars and government representatives and published over the last three decades state with increasing clarity and urgency that Earth is warming, and human activities are largely responsible. Although the international scientific community had discussed anthropogenic warming since the 1960s, the publication of the IPCC's (2007) Fourth Assessment report, which stated that global warming is "unequivocal" and due to human activities (>90 percent; Weart 2019) both called for action by governments to limit carbon emissions and had a significant impact on attitudes about climate change in the United States (Borick and Rabe 2010). The global scale of the problem and the mismatch between the countries causing climate change and those experiencing impacts in the late 2000s, however, led to failures in developing an international treaty to address this issue (Monastersky and Sousanis 2015). In 2006 and 2007, in the United States, the American College & University Presidents' Climate Commitment (ACUPCC) was launched to increase research and teaching in higher education on stabilizing Earth's climate. This high-visibility effort was developed in part to educate students to envision and prepare for a sustainable future (Second Nature 2007; Dyer and Dyer 2017).

Embedding climate change into curricula at institutions of higher education to facilitate development of "future-proof" students is challenging for several reasons, however (Fahey 2012). Information about the physical climate system is commonly presented in courses offered in natural science curricula, but the preparation is often piecemeal (Veron et al. 2016) and largely focuses on the fundamental processes as opposed to the felt impact of climate change or mitigation and adaptation strategies (Dyer and Dyer 2017). The impact of climate change and the assessment of risk is taught in social science curricula to a lesser degree (Cortese 2012), but without an overarching institutional strategy these parallel educational efforts are isolated and do not address the learning experience holistically (Fahey 2012). A further challenge is that the inherent complexity of the climate system requires knowledge from a variety of subjects as well as an understanding of probability and predictions (Hestness et al. 2011). In addition, information about climate change is contradictory outside academia influencing the sense of urgency

and public acceptance of the role that humans play (Hestness et al. 2011; Palm, Lewis, and Feng 2017).

Recent studies of universities in the United States and abroad detail the haphazard exposure university students have to climate change foundational principles.[5] Fahey (2012) determined that courses developed and taught in isolation from each other do not lead to deepened understanding of climate change. Instead, they found that an institutionally supported, coherent effort, providing content across courses, is needed to foster a comprehensive understanding of the climate system and the ability to propose solutions. Many higher education institutions are currently employing a more scattershot approach, though. For example, Veron et al. (2016) surveyed 119 faculty at four U.S. institutions about course alignment with the fundamental climate literacy principles (U.S. Global Change Research Program 2009) and determined that whereas undergraduate biology courses frequently include basic climate information, chemistry courses often exclude core climate principles. Very few courses examined the difference between weather and climate, a concept crucial to understanding climate change. Dyer and Dyer (2017) similarly determined that ACUPCC member institutions developed curriculum that primarily emphasized "environmental considerations" as opposed to "social justice, health, and economic dimensions" (115). Earth's climate system is complex, with multiple components that influence and interact with each other and include human systems and power dynamics.

Understanding climate science requires training in physics, chemistry, biology, geology, meteorology, and oceanography. Simultaneously, students must understand the uneven distribution of resources and how differently situated people experience this unevenness in distinctive ways. Teaching climate science is challenging because of this complexity and multidisciplinarity (Ekborg and Areskoug 2006). Yet, Veron et al. (2016) showed that most faculty teaching courses containing substantial climate content were not formally trained in teaching about climate (>80 percent). For Pharo et al. (2012), the greatest improvements in student understanding of climate change were demonstrated when coordinated across disciplines, but this effort was only sustainable with institutional support, including a paid coordinator. In the neoliberal university, professional development of faculty and employment of a coordinator to embed climate change into curricula are investments that will not directly generate profit and so are unlikely to occur.

The preformed opinions that students have on climate change, strongly influenced by political ideology, present another challenge in climate change education.[6] What this means for higher education is that we can only more thoroughly educate those who are already concerned about the changing climate in the Anthropocene and cocreate knowledge about how to apply what they have learned to produce solutions and develop strategies for adaptation and mitigation. Here we argue that geography is an ideal place to capture needed interdisciplinary approaches that might transcend some of the institutional barriers of the neoliberal university.[7]

The Global Climate Encounter

In 2014, the University of Delaware (2019b) began modifying the general education requirements that all undergraduate students at the university must satisfy. Beyond revising general education goals to reflect current education research and changes in workplace skills, the university supported piloting core courses required for all students that emphasized engaged citizenship and ethical reasoning (University of Delaware 2019a). As a team of geographers, one a climate scientist and the other a political geographer, we proposed a course that focused on feeding 9 billion in the context of a changing climate. With a climate scientist, trained in meteorology and oceanography, providing expertise and training on the climate system and the environmental impacts of climate change, and a political geographer, trained in the dynamics of food justice, structural oppression, and geopolitics, providing expertise on climate change social impacts, risks, ethics, and engagement, our proposed core course met the university's goal, engaging students with a grand challenge.[8]

The course, titled the Global Climate Encounter, was developed as a blended lecture and problem-based learning (PBL) course.[9] Although we purposely selected PBL topics related to food scarcity in a climate-changed world due to our professional areas of expertise, the course design was flexible so that any combination of geography faculty members could take on the overall topic of the environmental and social impacts, as well as risks and responses to climate change. As noted in the Introduction, this

Figure 1. Design of the course in three climate system components, with formative assessments (black) built in at the end of each module and summative assessments (gray) scattered throughout the course after the first two full weeks of class. PBL = problem-based learning.

team teaching environment is intended to emphasize the multifaceted character of climate change in the Anthropocene—we did not aim to cross-train students but to give a broader scope to allow them to develop problem-solving tools that were informed by both. Lectures were limited to twenty minutes; we emphasized learning through discussions of assigned readings and in-class activities, such as carbon cycling and carbon footprint calculation. The course explored issues surrounding food and climate change through readings from a variety of perspectives, including a memoir on the loss and grief experienced as Earth loses glacial ice (Jackson 2015) and an environmental journalist's catalog of the environmental catastrophes caused by global climate change (Kolbert 2015). We often hosted experts as guest speakers, including the authors of the two books, to directly interact with and respond to students. Both of us attended all classes, facilitating discussion through commentary and clarification from our two perspectives and challenging students to incorporate prior knowledge and explore foreign and uncomfortable ideas. The multidisciplinary perspectives of our guest speakers, combined with input from us and our students, led to a transdisciplinary learning environment where we could engage with the various complex issues that are an integral part of climate change studies.

PBL was introduced in the sixth week of the fifteen-week course and developed over the rest of the semester (see Figure 1). We selected PBL as the primary pedagogical method for student engagement because it incorporated open-ended questions, critical thinking and creative skills, and the transfer of knowledge that is necessary for dealing with complex topics like climate change (Fahey 2012; Pawson 2015). We divided the students into four groups with the goal of making connections between the environmental, political, social, and economic, with each group creating a concept map and presentation that could communicate the multifaceted impacts of climate change. Each group had a reader related to

Table 1. Course learning objectives and assessments

	Course objective	Assessment
1	Demonstrate knowledge of fundamental concepts in climate change science	Lead discussion on assigned readings (F), exams on climate system components (S)
2	Analyze, interpret, and evaluate quantitative and qualitative data in the context of global climate change	Exams on climate system components (S), develop reading summaries and prompts for discussion (S), problem-based learning concept map (F/S)
3	Evaluate human impacts and responses to climate change	Personal food diary (F), problem-based learning concept maps, write-up, and presentation (S)

Note: F = formative; S = summative.

their topic that provided a foundation for research for their concept maps. Group composition leveraged students' self-identified strengths and incorporated disciplinary breadth. Students were majors in environmental science, environmental studies, business and economics, women's studies, and agricultural and natural resources.

In Pharo et al.'s (2012) project to develop and maintain an interdisciplinary approach to teaching climate change, they chose to evaluate the nascent project by looking at the learning opportunities created instead of assessing learning outcomes, using student written reflections for data, following Eisenman et al. (2003). We provided a similar opportunity for our students to reflect on the impact of PBL on their learning through semester-long journaling, group development of concept maps, and writings submitted both during and at the end of the semester. Student products were graded both for accuracy and for demonstrated understanding of the inherent complexity of the human and natural systems in Earth's climate. Similar to Pharo et al.'s (2012) findings, we were able to model interdisciplinary collaboration and learning for the students as we interacted while teaching. There was never a time during the semester where students only saw one of us; we both attended and participated in all classes, and we held joint office hours. The effectiveness of the core classes was not institutionally evaluated, but we evaluated the students through a suite of formative and summative assessments (see Table 1).

Through class discussions and required work, we aided students in developing several essential discipline-specific proficiencies as specified by the U.S. National Geographic Society, including collaboration, communication, critical thinking, ethics, information literacy, and quantitative literacy (Levia and Quiring 2008). The course objectives and assessments were

distributed throughout the semester (Figure 1). Learning objectives 1 and 2 (Table 1), which emphasize acquisition and evaluation of climate information, were attained through student-led discussions and in-class exams on material in lectures and in the readings. The students were each assigned two readings for which they were the discussion leader, which required guiding the class on the readings and developing a written summary. We used the creation of concept maps to show the interrelation between meat production and climate change to further develop the ability to assess and synthesize climate change knowledge (objective 2). Food diaries were required twice in the semester, which allowed initial exploration of how food choices have climate impacts (objective 3). The PBL activity helped students evaluate the climate impacts of continuing to produce meat at an industrial scale.

The concept map activity allowed students to explore the interrelation between the social and natural parts of the climate system (Figure 2). They were developed collaboratively and were refined throughout the six weeks of the PBL activity as students' knowledge deepened. On the concept maps, these changes were particularly evident as original assumptions about the relative importance of certain issues, like soil degradation, were challenged and often reassigned new locations on the concept map to reflect a changed priority.

In terms of summative assessments, there were three exams that tested knowledge of the components of the climate system. In general, students performed well, with a mean of 83 percent. There was significant improvement in both correctness of answer and demonstrated understanding of climate complexity over the course of the semester in these exams, as well as in the journal entries and concept maps. Summative assessments of the final deliverables from

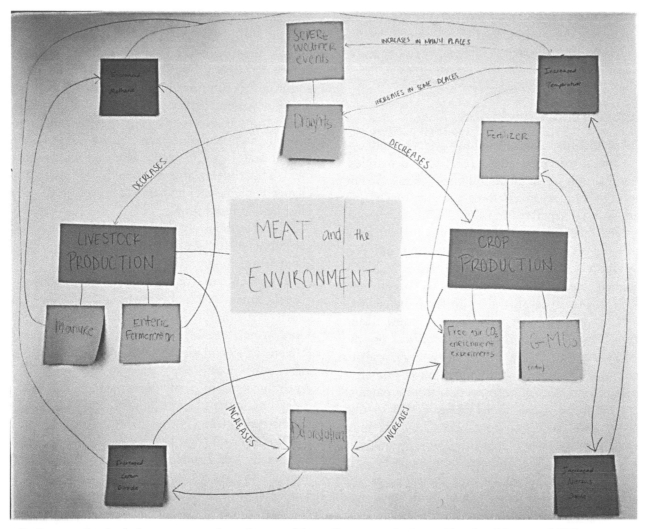

Figure 2. An example of a draft concept map drawn by one of the student groups for the problem-based learning (PBL) activity.

the PBL and presentations suggest effective synthesis of the climate knowledge gained in the class.

In addition to course-designed assessments, we were provided institutional student evaluations; we fully acknowledge that these evaluations are a key component of the "counting" that takes place in the neoliberal university. We share these comments not as an unbiased data set but to give insight into student responses to this form of teaching. These reports reflected the value that students placed on the interdisciplinary, coteaching model; a small sample is provided here:

I do think more people should be required to take courses like this that are colead [sic] by professors that show different perspectives on the same issue. Career-wise, I identified with one of the instructors a lot more than the other and have mostly only taken courses by similar professors, but having two of them really made me think about the problems in a more holistic way.

I am a scientist by training and want to be a scientist throughout the rest of my life, and this course has really made me question assumptions I had about science. We had one class in particular where we talked about subjectivity and objectivity and which one science was based on. At the beginning of the class I was dead set on science being objective. ... That class, though ... we had a discussion [that] was probably the most impactful single class, period, I have had in my seven semesters here at UD. It really made me think about science differently and I have referenced it in conversations I have elsewhere, including in another course.

The instructor had points which expanded upon the ideas of the other instructors. Allowed us to formulate new ways of considering problems as well as solutions.

The interdisciplinary character of the class changed the way students thought about their roles and the impact of climate change. Further, it should be noted that the class was taught in geography, from a geographic perspective, to an audience of nongeography majors.

Discussion and Conclusion

One of the first course readings we assigned was written by a freelance journalist, who demanded: "Stop screaming at us about climate change." The author urged educators to forego an approach that focused on the negative impacts of climate change, such as sea level rise and drought, and instead focus on inspiring stories, such as tree planting and personal carbon footprint reductions (Neimark 2016). Although it is easy to see limitations to this author's approach, it also speaks to a desire for moving away from "catastrophe narratives" to "stewardship narratives" when addressing the politics of the Anthropocene (see Dalby 2016). Many argue that to facilitate this change we must move away from fear and grief toward deeper introspection from a place of hope (see Gibson, Rose, and Fincher 2015) so that we can rewrite the Earth together. Thus, we must challenge ourselves when engaging students, many of whom can already see and feel the impacts of a warming world, to move beyond focusing on the problems to coconstructing ideas and narratives about addressing climate change in the era of deep human interference.

Data from the Yale Program on Climate Change Communication and the George Mason Center for Climate Change Communication show that attention to climate change in the United States is returning to the level observed after the release of the IPCC's Fourth Assessment Report (Yale Program on Climate Change Communication 2018; Ballew et al. 2019; George Mason Center for Climate Change Communication 2019).[10] It is in this social context that we are seeing youth speaking passionately for action to counter the impacts of climate change. Despite the positive lessons learned in our pilot course, the university experiment with the new core courses was ultimately unsuccessful due to the inherent rigidity and slow pace at which higher education responds to change (Dyer and Dyer 2017). Although the course was well aligned with institutional priorities, there was a lack of support at the administrative level to integrate the new courses into current curricular structures, a common barrier for multidisciplinary

material (Pharo et al. 2012), which obviates students voluntarily enrolling in pilot courses. This issue further illuminates the paradox of the neoliberal university—how to create "future-proof" students within a structure that has built-in resistance to cross-disciplinary approaches. The students who enrolled in this course self-selected to participate in a pilot course that purposely examined the climate change environmental and social impacts associated with food scarcity. They were already convinced that climate change is happening and that it is a critical issue that must be addressed now.

As the articles in this special issue demonstrate, geographers are ideally placed to address questions about human–environment interactions in the Anthropocene in their research and teaching (see also Lorimer 2012; Lorimer and Driessen 2014; Castree 2014a, 2014b, 2014c, 2015; Dalby 2016; Johnson et al. 2014; Pawson 2015; Ziegler and Kaplan 2019). Human and physical geographers alike are taking up the key queries of the Anthropocene in their work, which suggests that geography is the place to unravel the core challenges in this epoch (see Castree 2014a, 2014b, 2014c; Ellis 2017; Ziegler and Kaplan 2019). As geographers, we are already well trained in how to make connections and put the pieces of the puzzle together; in some cases, the work of recognizing multiple epistemological approaches is underway. As Pawson (2015) noted, a geographic education is ideal for creating the conditions for addressing the anxieties of the Anthropocene in a holistic way. The geographical approach employed in our pilot course used several strategies that have been successful in delivering meaningful, stimulating education on interdisciplinary topics, namely, coteaching across disciplines, employing active learning approaches such as PBL, and cocreating a narrative about the actions we can take in a climate-changing world. We suggest that those who can provide a geographic education take on inter- and multidisciplinary coteaching approaches to work with students to cocreate knowledges, address existential anxieties, and find new and creative ways to teach, learn, and act.

The number of students who want to be active in addressing climate change is on the rise. They are entering universities ready to learn the skills that society desperately needs to adapt to and mitigate climate change impacts. The deep, integrative, interdisciplinary education needed to develop such a skill set, however, is challenging to foster and maintain in the neoliberal university, where education is progressively compartmentalized and focused on deliverables that can be counted. As we move further into the Anthropocene, geography is uniquely poised to provide the model by which scholars across disciplines can come together with these motivated students to break down the confines of the neoliberal university and envision the solutions necessary to survive and thrive in the Anthropocene.

Acknowledgments

We are grateful to the writing group in the Department of Geography & Spatial Sciences and Del Levia for their initial review of this article. We are also grateful to the students who participated in the core class. As we were revising this article, COVID-19 disrupted life as we have known it—we want to acknowledge all of the volunteer academic labor that is ongoing. It is essential and undervalued. We value you and your labor and we thank you. We also value the labor of the editors and the reviewers, who not only helped us make this article better but who worked with us during a time of enormous upheaval in labor demands. We also thank our families for support during time spent working and living under shelter-in-place orders in our state.

Notes

1. It should be noted that indigenous groups, particularly indigenous youth are—past and present—informing on, protesting, and putting their lives on the line to bring about structural change related to environmental degradation. The recent and well-publicized protests over the Keystone XL and Dakota Access pipelines are not the only examples, nor are they isolated incidents of contestation. That the efforts of largely white students are receiving the most media attention points to long-standing racial bias, racism, and structural oppression.
2. At the time of writing, a petition had been filed for rehearing *en banc* with the Ninth Circuit Court of Appeals. This petition asks the Court of Appeals to convene a new panel of eleven circuit court judges to review a sharply divided opinion by the Ninth Circuit Court about whether their case can go to trial. Several attempts have been made by the 45th presidential administration to have the case dismissed.
3. For this news article, county and district-level opinion data are estimates based on survey responses from more than 18,000 U.S. adults (age twenty-five and older) collected between 2008 and 2016.
4. This phenomenon reinforces unequal academic subjectivities for faculty that value institutional productivity in some sectors (grant winning, publication, upper administrative roles) and undervalue it in others (academic advising, teaching, mentoring). On the impact on women's time specifically, see Mountz et al. (2015).
5. Although we focus more specifically on studies in the United States, it should be noted that there is excellent work from both U.S. and international scholars on this topic; see, for example, Beck, Sinatra, and Lombardi (2014), Burandt and Barth (2010), Correia et al. (2010), Davison et al. (2014), Hess and Collins (2018), Howlett, Ferreira, and Blomfield (2016), O'Brien et al. (2013), Pharo et al. (2012), and Sipos, Battisti, and Grimm (2008), and references therein.
6. See Huxster, Uribe-Zarain, and Kempton (2015), Kahan (2015), Palm, Lewis, and Feng (2017), Popovich, Schwartz, and Schlossberg (2017), and Shealy et al. (2019).
7. We fully recognize that this work is already well underway in geography, as traditions in political ecology and the emergence of subdisciplines such as critical physical geography (Lave et al. 2014) demonstrate.
8. This approach was important to our curriculum because food scarcity is one of the issues raised by Dyer and Dyer (2017) as an area where the ACUPCC fell short.
9. It is important to note that we are not experts in PBL and we used it as one of a suite of activities in our experimental course. For in-depth discussions of PBL, see Dolmans et al. (2016), Levia and Quiring (2008), Lyon and Teutschbein (2011), Raath and Golightly (2017), and Weiss (2017).
10. The number of people who think global warming is due to human activities has increased by 8 percent since 2011, and perception that scientists have reached a consensus on the occurrence of global warming has jumped to 53 percent since 2010.

References

Ballew, M. T., A. Leiserowitz, C. Roser-Renouf, S. A. Rosenthal, J. E. Kotcher, J. R. Marlon, E. Lyon, M. H. Goldberg, and E. W. Maibach. 2019. Climate change in the American mind: Data, tools, and trends. *Environment: Science and Policy for Sustainable Development* 61 (3):4–18. doi: 10.1080/00139157.2019.1589300.

Beck, A., G. M. Sinatra, and D. Lombardi. 2013. Leveraging higher-education instructors in the climate literacy effort: Factors related to university faculty's propensity to teach climate change. *The International Journal of Climate Change: Impacts and Responses* 4 (4):1–17. doi: 10.18848/1835-7156/CGP/v04i04/37181.

Borick, C. P., and B. G. Rabe. 2010. A reason to believe: Examining the factors that determine individual views on global warming. *Social Science Quarterly* 91 (3):777–800. doi: 10.1111/j.1540-6237.2010.00719.x.

Bratman, E., K. Brunette, D. C. Shelly, and S. Nicholson. 2016. Justice is the goal: Divestment as climate change resistance. *Journal of Environmental Studies and Sciences* 6 (4):677–90. doi: 10.1007/s13412-016-0377-6.

Burandt, S., and M. Barth. 2010. Learning settings to face climate change. *Journal of Cleaner Production* 18 (7):659–65. doi: 10.1016/j.jclepro.2009.09.010.

Castree, N. 2014a. The Anthropocene and geography I: The back story. *Geography Compass* 8 (7):436–49. doi: 10.1111/gec3.12141.

Castree, N. 2014b. The Anthropocene and geography III: Future directions. *Geography Compass* 8 (7):464–76. doi: 10.1111/Gec3.12139.

Castree, N. 2014c. Geography and the Anthropocene II: Current contributions. *Geography Compass* 8 (7):450–63. doi: 10.1111/gec3.12140.

Castree, N. 2015. Coproducing global change research and geography: The means and ends of engagement. *Dialogues in Human Geography* 5 (3):343–48. doi: 10.1177/2043820615613265.

Connell, R. 2013. The neoliberal cascade and education: An essay on the market agenda and its consequences. *Critical Studies in Education* 54 (2):99–112. doi: 10.1080/17508487.2013.776990.

Cook, B. R., L. A. Rickards, and I. Rutherfurd. 2015. Geographies of the Anthropocene. *Geographical Research* 53 (3):231–43.

Correia, P. R. M., B. Xavier do Valle, M. Dazzani, and M. E. Infante-Malachias. 2010. The importance of scientific literacy in fostering education for sustainability: Theoretical considerations and preliminary findings from a Brazilian experience. *Journal of Cleaner Production* 18 (7):678–85. doi: 10.1016/j.jclepro.2009.09.011.

Cortese, A. D. 2012. Promises made and promises lost: A candid assessment of higher education leadership and the sustainability agenda. In *The Sustainable University: Green Goals and New Challenges for Higher Education Leaders*, ed. J. Martin and J. E. Samels, 17–31. Baltimore, MD: Johns Hopkins University Press.

Crutzen, P. J., and E. F. Stoermer. 2000. Global change newsletter. *The Anthropocene* 41:17–18.

Dalby, S. 2016. Framing the Anthropocene: The good, the bad and the ugly. *The Anthropocene Review* 3 (1):33–51. doi: 10.1177/2053019615618681.

Davison, A., P. Brown, E. Pharo, K. Warr, H. McGregor, S. Terkes, D. Boyd, and P. Abuodha. 2014. Distributed leadership. *International Journal of Sustainability in Higher Education* 15 (1):98–110. doi: 10.1108/IJSHE-10-2012-0091.

Dolmans, D. H., S. M. M. Loyens, H. Marcq, and D. Gijbels. 2016. Deep and surface learning in problem-based learning: A review of the literature. *Advances in Health Sciences Education: Theory and Practice* 21 (5):1087–112. doi: 10.1007/s10459-015-9645-6.

Dyer, G., and M. Dyer. 2017. Strategic leadership for sustainability by higher education: The American college & university presidents' climate commitment. *Journal of Cleaner Production* 140:111–16. doi: 10.1016/j.jclepro.2015.08.077.

Eisenman, L., D. Hill, R. Bailey, and C. Dickison. 2003. The beauty of teacher collaboration to integrate curricula: Professional development and student learning opportunities. *Journal of Vocational Education Research* 28 (1):85–104. doi: 10.5328/JVER28.1.85.

Ekborg, M., and M. Areskoug. 2006. How student teachers' understanding of the greenhouse effect develops during a teacher education programme. *Nordic Studies in Science Education* 2 (3):17–29.

Ellis, E. C. 2017. Physical geography in the Anthropocene. *Progress in Physical Geography: Earth and Environment* 41 (5):525–32. doi: 10.1177/0309133317736424.

Ernstson, H., and E. Swyngedouw, eds. 2018. *Urban political ecology in the Anthropo-Obscene*. London and New York: Routledge.

Fagan, M. 2019. On the dangers of an Anthropocene epoch: Geological time, political time and post-human politics. *Political Geography* 70:55–63. doi: 10.1016/j.polgeo.2019.01.008.

Fahey, S. J. 2012. Curriculum change and climate change: Inside outside pressures in higher education. *Journal of Curriculum Studies* 44 (5):703–22.

George Mason Center for Climate Change Communication (Mason 4C). 2019. Center for Climate Change Communication. Accessed November 29, 2019. https://www.climatechangecommunication.org/.

Gibson, K., D. B. Rose, and R. Fincher, eds. 2015. *Manifesto for living in the Anthropocene*. Goleta, CA: Punctum Books.

Gibson-Graham, J. K. 2011. A feminist project of belonging for the Anthropocene. *Gender, Place & Culture* 18 (1):1–21. https://doi.org/ doi: 10.1080/0966369X.2011.535295..

Gibson-Graham, J. K. 2014. Being the revolution, or, how to live in a "more-than-capitalist" world threatened with extinction. *Rethinking Marxism* 26 (1):76–94. doi: 10.1080/08935696.2014.857847.

Hess, D. J., and B. M. Collins. 2018. Climate change and higher education: Assessing factors that affect curriculum requirements. *Journal of Cleaner Production* 170:1451–58. doi: 10.1016/j.jclepro.2017.09.215.

Hestness, E., J. Randy McGinnis, K. Riedinger, and G. Marbach-Ad. 2011. A study of teacher candidates' experiences investigating global climate change within an elementary science methods course. *Journal of Science Teacher Education* 22 (4):351–69.

Howlett, C., J.-A. Ferreira, and J. Blomfield. 2016. Teaching sustainable development in higher education. *International Journal of Sustainability in Higher Education* 17 (3):305–21. doi: 10.1108/IJSHE-07-2014-0102.

Huxster, J. K., X. Uribe-Zarain, and W. Kempton. 2015. Undergraduate understanding of climate change: The influences of college major and environmental group membership on survey knowledge scores. *The Journal of Environmental Education* 46 (3):149–65. doi: 10.1080/00958964.2015.1021661.

Intergovernmental Panel on Climate Change. 2007. *Climate Change 2007: The Physical Science Basis. Contribution of Working Group I to the Fourth*

Assessment Report of the Intergovernmental Panel on Climate Change, ed. S. Solomon, D. Qin, M. Manning, Z. Chen, M. Marquis, K. B. Averyt, M. Tignor and H. L. Miller. Cambridge, UK, and New York: Cambridge University Press.

Jackson, M. 2015. *While glaciers slept: Being human in a time of climate change*. Brattleboro, VT: Green Writers Press.

Johnson, E., H. Morehouse, S. Dalby, J. Lehman, S. Nelson, R. Rowan, S. Wakefield, and K. Yusoff. 2014. After the Anthropocene: Politics and geographic inquiry for a new epoch. *Progress in Human Geography* 38 (3):439–56. doi: 10.1177/0309132513517065.

Knitter, D., K. Augustin, E. Biniyaz, W. Hamer, M. Kuhwald, M. Schwanebeck, and R. Duttmann. 2019. Geography and the Anthropocene: Critical approaches needed. *Progress in Physical Geography: Earth and Environment* 43 (3):451–61. doi: 10.1177/0309133319829395.

Kolbert, E. 2015. *Field notes from a catastrophe: Man, nature, and climate change*. Rev. ed. New York: Bloomsbury.

Lave, R., M. W. Wilson, E. S. Barron, C. Biermann, M. A. Carey, C. S. Duvall, L. Johnson, K. M. Lane, N. McClintock, D. Munroe, et al. 2014. Intervention: Critical physical geography. *The Canadian Geographer/Le Géographe Canadien* 58 (1):1–10. doi: 10.1111/cag.12061.

Laville, S., and J. Watts. 2019. Across the globe, millions join biggest climate protest ever. *The Guardian*, September 20. Accessed September 28, 2019. https://www.theguardian.com/environment/2019/sep/21/across-the-globe-millions-join-biggest-climate-protest-ever.

Leichenko, R., and A. Mahecha. 2015. Celebrating geography's place in an inclusive and collaborative Anthropo(s)Cene. *Dialogues in Human Geography* 5 (3):327–32. doi: 10.1177/2043820615613252.

Levia, D. F., and S. M. Quiring. 2008. Assessment of student learning in a hybrid PBL capstone seminar. *Journal of Geography in Higher Education* 32 (2):217–31. doi: 10.1080/03098260701514041.

Loftus, A. 2019. Political ecology III: Who are "the people"? *Progress in Human Geography*. doi: 10.1177/0309132519884632.

Lorimer, J. 2012. Multinatural geographies for the Anthropocene. *Progress in Human Geography* 36 (5):593–612.

Lorimer, J., and C. Driessen. 2014. Wild experiments at the Oostvaardersplassen: Rethinking environmentalism in the Anthropocene. *Transactions of the Institute of British Geographers* 39 (2):169–81.

Lyon, S. W., and C. Teutschbein. 2011. Problem-based learning and assessment in hydrology courses: Can non-traditional assessment better reflect intended learning outcomes? *Journal of Natural Resources and Life Sciences Education* 40 (1):199–205. doi: 10.4195/jnrlse.2011.0016g.

Monastersky, R., and N. Sousanis. 2015. The fragile framework. *Nature News* 527 (7579):427–35. doi: 10.1038/527427a.

Mountz, A., A. Bonds, B. Mansfield, J. Loyd, J. Hyndman, M. Walton-Roberts, R. Basu, R. Whitson, R. Hawkins, T. Hamilton, et al. 2015. For slow scholarship: A feminist politics of resistance through collective action in the neoliberal university. *ACME: An International Journal for Critical Geographies* 14 (4):1235–59.

Naylor, L., M. Daigle, S. Zaragocin, M. M. Ramírez, and M. Gilmartin. 2018. Interventions: Bringing the decolonial to political geography. *Political Geography* 66:199–209. doi: 10.1016/j.polgeo.2017.11.002.

Neimark, J. 2016. Stop screaming at us about climate change–and start inspiring us to take action. *Quartz*. Accessed May 16, 2016. http://qz.com/685269/stop-screaming-at-us-about-climate-change-and-start-inspiring-us-to-take-action/

O'Brien, K., J. Reams, A. Caspari, A. Dugmore, M. Faghihimani, I. Fazey, H. Hackmann, D. Manuel-Navarrete, J. Marks, R. Miller, et al. 2013. You say you want a revolution? Transforming education and capacity building in response to global change. *Environmental Science & Policy* 28:48–59. doi: 10.1016/j.envsci.2012.11.011.

Palm, R., G. B. Lewis, and B. Feng. 2017. What causes people to change their opinion about climate change? *Annals of the American Association of Geographers* 107 (4):883–96.

Pawson, E. 2015. What sort of geographical education for the Anthropocene? *Geographical Research* 53 (3):306–12. doi: 10.1111/1745-5871.12122.

Petersen, B., and J. R. Barnes. 2019. From hopelessness to transformation in geography classrooms. *Journal of Geography* 1–12. doi: 10.1080/00221341.2019.1566395.

Pharo, E. J., A. Davison, K. Warr, M. Nursey-Bray, K. Beswick, E. Wapstra, and C. Jones. 2012. Can teacher collaboration overcome barriers to interdisciplinary learning in a disciplinary university? A case study using climate change. *Teaching in Higher Education* 17 (5):497–507. doi: 10.1080/13562517.2012.658560.

Popovich, N., J. Schwartz, and T. Schlossberg. 2017. How Americans think about climate change, in six maps. *The New York Times*, March 21. Accessed September 22, 2017. https://www.nytimes.com/interactive/2017/03/21/climate/how-americans-think-about-climate-change-in-six-maps.html.

Raath, S., and A. Golightly. 2017. Geography education students' experiences with a problem-based learning fieldwork activity. *Journal of Geography* 116 (5):217–25. doi: 10.1080/00221341.2016.1264059.

Second Nature. 2007. The presidents' climate leadership commitments. Accessed September 29, 2019. https://secondnature.org/signatory-handbook/the-commitments/.

Shealy, T., L. Klotz, A. Godwin, Z. Hazari, G. Potvin, N. Barclay, and J. Cribbs. 2019. High school experiences and climate change beliefs of first year college students in the United States. *Environmental Education Research* 25 (6):925–35.

Sipos, Y., B. Battisti, and K. Grimm. 2008. Achieving transformative sustainability learning: Engaging head. *International Journal of Sustainability in Higher Education* 9 (1):68–86. doi: 10.1108/14676370810842193.

Slaughter, S., and G. Rhoades. 2000. The Neo-Liberal University. *New Labor Forum* (6):73–79.

University of Delaware. 2019a. Core curriculum proposal. Accessed October 7, 2019. https://sites.udel.edu/gened/sample-page/general-education-reform/core-curriculum-proposal/.

University of Delaware. 2019b. General education reform. Accessed October 7, 2019. https://sites.udel.edu/gened/sample-page/general-education-reform/.

U.S. Global Change Research Program. 2009. Climate literacy: The essential principles of climate. Washington, DC: Author. Accessed January 16, 2015. http://www.globalchange.gov.

Veron, D. E., G. Marbach-Ad, J. Wolfson, and G. Ozbay. 2016. Assessing climate literacy content in higher education science courses: Distribution, challenges, and needs. *Journal of College Science Teaching* 45 (6):43–49. doi: 10.2505/4/jcst16_045_06_43.

Weart, S. 2019. Climate change: Discovery of global warming. Accessed October 7, 2019. https://history.aip.org/history/exhibits/climate/.

Weiss, G. 2017. Problem-oriented learning in geography education: Construction of motivating problems. *Journal of Geography* 116 (5):206–16. doi: 10.1080/00221341.2016.1272622.

Yale Program on Climate Change. 2018. Climate change in the American mind: National survey data on public opinion (2008–2017). 10.17605/OSF.IO/W36GN.

Ziegler, S. S., and D. H. Kaplan. 2019. Forum on the Anthropocene. *Geographical Review* 109 (2):249–51. doi: 10.1111/gere.12336.

Index

Note: Page numbers in **bold** refer to tables, page numbers in *italics* refer to figures and page numbers followed by "n" refer to end notes.

9 781032 076690